Self-Organized Criticality in Astrophysics

The Statistics of Nonlinear Processes in the Universe

Markus Aschwanden

Self-Organized Criticality in Astrophysics

The Statistics of Nonlinear Processes in the Universe

Springer

Published in association with
Praxis Publishing
Chichester, UK

PRAXIS

Professor Dr Markus Aschwanden
Lockheed Martin Advanced Technology Center
Solar and Astrophysics Laboratory
Palo Alto
California
USA

Cover images courtesy of NASA (detailed image credits are given under Figures 1.4a, 1.12b, 1.16, and 8.22 in Appendix on Image Credit: Public Websites).

SPRINGER–PRAXIS BOOKS IN ASTRONOMY AND PLANETARY SCIENCES
SUBJECT *ADVISORY EDITORS*: Philippe Blondel, C.Geol., F.G.S., Ph.D., M.Sc., F.I.O.A., Senior Scientist, Department of Physics, University of Bath, UK; John Mason, M.B.E., B.Sc., M.Sc., Ph.D.

ISBN 978-3-642-15000-5 e-ISBN 978-3-642-15001-2
DOI 10.1007/978-3-642-15001-2
Springer Heidelberg Dordrecht London New York

Cover design: Jim Wilkie
Project copy editor: Mike Shardlow
Author-generated LaTex, processed by EDV-Beratung Herweg, Germany

Printed on acid-free paper

Springer is part of Springer Science + Business Media (www.springer.com)

To Carol J. Kersten

Contents

Preface

How did this book come about? In 1985, Brian Dennis published a review on solar flares and presented a stunning figure that showed a perfect powerlaw distribution in the occurrence of solar flares that extended over almost 4 orders of magnitude, with a slope of -1.8, for which no explanation could be found. Just two years later in 1987, Per Bak, the father of *self-organized criticality (SOC)*, published his landmark paper on the interpretation of the ubiquitous powerlaw distributions, observed also in sandpile avalanches and earthquakes (the so-called Gutenberg–Richter law), by relating the scale-free behavior to the $1/f$-flicker noise. A few years later, Per Bak gave a colloquium at the NASA Goddard Space Flight Center (GSFC), where he met Brian Dennis and heard about solar flare statistics; but he admitted in his book *How Nature Works* that he did not really understand how solar flares work. Intuitively, there was the notion that the intricate details of the underlying physical processes could not provide the answer to the fundamental understanding of the observed powerlaws. In 1991, the two students Ed Lu and Russ Hamilton at Stanford University wrote the first paper where self-organized criticality was applied to solar flare statistics, which was interpreted and modeled with a cellular automaton model. This approach offered an explanation of the observed powerlaws in terms of statistics of next-neighbor interactions of complex dissipative systems in a critical state. This universal aspect fascinated me more and more and I gave a number of colloquia on self-organized criticality applied to solar flares at the ETH Zurich, NASA GSFC, and the University of Maryland in 1991–1993. Since powerlaw distributions were also observed for stellar flares, pulsar glitches, lunar craters, and asteroid sizes, I speculated that these may all be dissipative systems with self-organized criticality. During one of the seminars at the University of Maryland I remember that Lucy McFadden, an expert in solar system small bodies, commented that this was the most fascinating model she had ever heard of and asked whether it applied also to the powerlaw distributions of asteroids and Saturn rings. I did not know the answer at this time but an answer is given in this book. A textbook that explains the fundamental aspects of self-organized criticality in terms of the statistics of nonlinear events has never been written in astrophysics, which motivated me to undertake such an endeavor. One of the major aims of this book is to convey a deeper understanding of the statistics of nonlinear processes that is common to solar flares, sandpile avalanches, and earthquakes, although the underlying physics is completely different.

This textbook is intended to be an introduction to the relatively new subject of self-organized criticality (SOC), suitable for students and post-docs, as well as for researchers who want to know all the relevant literature references. The main applications are astrophysical phenomena, although we include also a few other phenomena from geophysics or social sciences that provided important basic models, later applied to astrophysical phenomena. In Chapter 1 we give an introductory broad overview of SOC phenomena observed in the entire universe, wherever publications with SOC interpretations were found in the scientific literature. The theoretical modeling of SOC phenomena can be pursued in 3 different approaches: by numerical (mostly cellular automaton) simulations (Chapter 2), by analytical modeling of statistical distributions (Chapter 3), or by physical modeling (Chapter 9). The temporal aspects of SOC statistics includes random statistics (Chapter 4), waiting-time statistics (Chapter 5), and event-detection methods (Chapter 6). Using these basic prerequisites, we can then model and understand the occurrence frequency distributions of SOC events, which reveal the ubiquitous powerlaws that are the hallmark of SOC (Chapter 7). The spatial aspects of SOC events entail the geometry of fractal structures (Chapter 8). Finally, we arrive at a general physics-free definition of SOC phenomena (Section 9.1). Individual physical processes for astrophysical SOC phenomena are summarized in Table 9.1 and discussed case by case in the remainder of Chapter 9, qualitatively for astrophysical observations, and somewhat more quantitatively for solar physics applications. Alternatives to SOC processes are discussed in Chapter 10, which may also exhibit powerlaw distributions but can be discriminated from pure SOC processes using the criteria of our physics-free SOC definition (Table 10.1).

Do we understand SOC completely now? Although we hope to have established a deeper understanding of SOC phenomena in this book, there are still a lot of open questions that can only be answered by large statistics of observations and by more detailed modeling. For instance, how does the statistics of next-neighbor interactions result in the exponential growth characteristics of SOC avalanches? What determines the powerlaw slopes? How much is the powerlaw slope determined by mathematical statistics, and how much by physical scaling laws? The relatively new scientific discipline of self-organized criticality is a very interdisciplinary field and we hope that this book stimulates a cross-fertilization in the data analysis and development of methods among the disciplines of astrophysics, geophysics, biophysics, and social sciences.

The author is most indebted to invaluable discussions with, comments from, and reviewing by colleagues and friends, who are listed in alphabetical order: Eric Buchlin, Anne Cristina Cadavid, Sandra Chapman, Paul Charbonneau, Norma Crosby, Pablo Dmitruk, Manuel Güdel, Henrik Jeldtoft Jensen, Debbie Leddon, Yuri Litvinenko, William Liu, Nadege Meunier, Laura Morales, Jeff Scargle, Virginia Trimble, Astrid Veronig, Nicolas Watkins, and Mike Wheatland. The author wishes to acknowledge the efficient and most helpful support provided by Springer/Praxis, especially by the publishers Clive Horwood (Praxis) and Ramon Khanna (Springer), who encouraged and supported the publication of this book. Extensive usage of scientific literature was enabled by the NASA Astrophysics Data System (ADS), operated by the Smithsonian Astrophysical Observatory (SAO), as well as by numerous *Wikipedia* and *Google* searches. Special thanks go also to my family, to my children Pascal Dominique and Alexander Julian, and particularly to my wife, Carol J. Kersten, for their enthusiastic support of this project.

Palo Alto, California, July 2010 *Markus J. Aschwanden*

1. Self-Organized Criticality Phenomena

How can the universe start with a few types of elementary particles at the big bang, and end up with life, history, economics, and literature? The question is screaming out to be answered but it is seldom even asked. Why did the big bang not form a simple gas of particles, or condense into one big crystal ?

Per Bak (1996), *How Nature Works*

In this introductory chapter we want to get a flavor of physical processes that are governed by self-organized criticality, starting from small experiments in our laboratories, proceeding to nature phenomena on our planet, all the way to the remotest astrophysical realms of our universe. We will discover that most complex systems with many interacting components display some nonlinear behavior that is governed by self-organized criticality. The remainder of the book will focus on numerical, analytical, and physical modeling of self-organized criticality in astrophysical systems.

1.1 The Concept of Self-Organized Criticality

This book is all about nonlinear systems in nature. What do we mean by a "system"? A *system* is a set of interacting (or interdependent) components or entities that are combined into an integrated whole, such as a car (a mechanical system), a coupled pendulum (a physical system), tectonic plates (a geological system), a hurricane (a weather system), the stock market (an economic system), or an accretion disk in a binary star system (an astrophysical system).

Our whole universe is governed by nonlinear systems or nonlinear dynamics. In principle, a system can exhibit linear or nonlinear behavior, depending on how the output is causally related to the input of the system. The property of linearity is well-defined in mathematics: a linear function (or equation) obeys the properties of additivity, $f(x+y) = f(x) + f(y)$, and homogeneity, $f(a \times x) = a \times f(x)$. A car, for instance, is a linear mechanical system by design: the directional change of the wheels is proportional

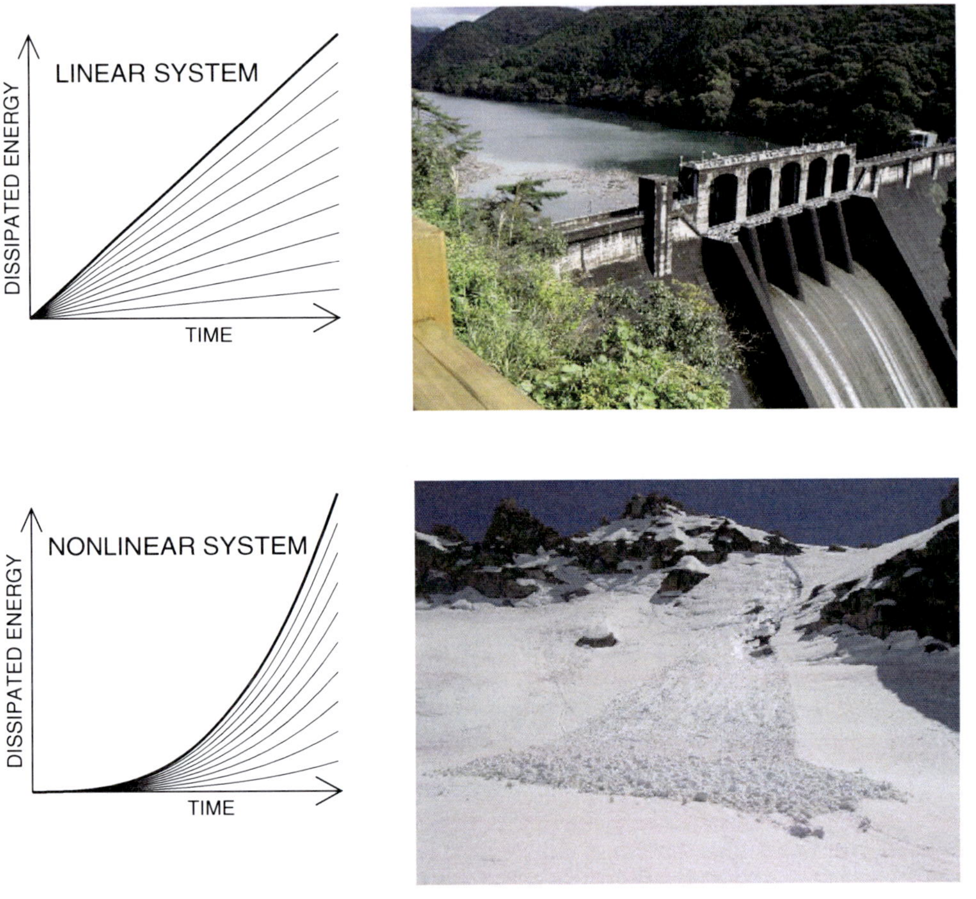

Fig. 1.1 The output or dissipated energy in a linear system grows linearly with time, for a constant input rate (top left), while the output is highly unpredictable and not correlated with the input rate in a nonlinear dissipative system (bottom left). A practical example of a linear system is a hydroelectric plant, where the produced electric energy is proportional to the water input, as depicted by the water-storage dam at Yaotsu, Gifu, Japan (top right). A classical example of a dissipative nonlinear system is a snow avalanche, as shown in the large wet-snow avalanche at Deadman Canyon in the Sierra Nevada range (bottom right).

to the angle you turn on the steering wheel. Also a hydropower plant can be described as a linear system, since the produced electric energy is proportional to the water input that flows out of a dam into the pipeline of the power plant (Fig. 1.1, top). The electric output of multiple power plants requires a proportional amount of water supply, so the properties of additivity and homogeneity typical for linear systems is fulfilled. Of course, the linear behavior occurs only in a limited parameter range.

Many systems, however, consist of a large number of entities that interact in a complex way and exhibit nonlinear behavior, which are called *nonlinear dissipative systems*, such as coupled pendulums, avalanches (Fig. 1.1, bottom), earthquakes, hurricanes, the stock market, or star-forming molecular clouds. Coupled pendulums pull and kick each other with rapidly changing amplitudes and phases, so that the resulting motion becomes chaotic (for a discussion of chaotic systems see Section 10.7). Tectonic plates have complicated interactions with cracking, shearing, or sliding motions, so that earthquakes happen at very irregular time intervals. A hurricane originates far out in the ocean as a result of apparently insignificant fluctuations of thermal gradients, air pressure, and circular motions, which evolve into powerful monsters that amass more and more angular momentum until they cause a catastrophe at landfall. Edward Lorenz, a pioneer of chaos theory, coined the term *"butterfly effect"* (inspired by a 1952 science fiction story by Ray Bradbury), which refers to the idea that a butterfly's wing flap might be sufficient to cause a change in the atmosphere that ultimately could result in a tornado. The stock market can behave quite regularly over many days, but sporadic glitches that lead to a Wall Street crash can happen unpredictably. Molecular clouds in our galaxy condense as a result of angular momentum and gravity (or triggered by a supernova shock), until a gravitational collapse sets in and leads to star formation. So, there are many nonlinear dissipative systems in our universe, which exhibit inherent nonlinear behavior just as a result of complex interactions that occasionally lead to instabilities with subsequent catastrophes. However, although we humans pay most attention to the largest events, the catastrophes, the myriads of smaller events share the same statistical properties, which can be described with the concept of *self-organized criticality.*

What special condition is needed to enable self-organized criticality? Since there exists no *perpetuum mobile*, a system that works without external energy input ad infinitum, we obviously need a source of energy. This input of external energy often occurs randomly in nature, causing local disturbances of a system, to which a linear system will respond with a proportional change in output, while a nonlinear dissipative system will respond sporadically with a little or largely amplified output. A classical paradigm for self-organized criticality is a sandpile. If we continuously drip sand grains onto the same place, a conical sandpile will grow in a stable manner, with a steepening surface shape, until a critical slope (with an angle of $\approx$ 34–37 degrees, depending on the consistency, granularity, and humidity of the sand) is reached, after which the state of self-organized criticality sets in (Fig. 1.2, right panels). The continuously trickled sand will produce large or small avalanches of random sizes that have no relation to the input rate of sand (at least for low rates). This is the critical state that is needed to observe SOC phenomena. As long as the critical slope is not reached, the sand will be in a stable equilibrium, such as in the flat sand beach in northern California shown in Fig. 1.2 (left). So, we need two things, a continuous energy input source and a nonlinear dissipative system. The energy dissipation of sandpile avalanches corresponds to the kinetic energy and change in gravitational potential. The nonlinearity results from the highly complex interactions of colliding sand grains, which act on each other by collisions and friction. In contrast, water flowing through a pipeline (Fig. 1.1 top), has a much more linear characteristic, at least for laminar flows, because of the inherent physical properties of homogeneity and viscosity in fluids. Nonlinearity occurs in almost all systems with many components. The physics is only simple for one- or two-component systems, say under the influence of one gravitational or electric force (classical two-body descriptions), while the physical and mathematical treatment becomes immensely complex for n-body problems (for $n \geq 3$). In fluids, turbulence can already occur in a single fluid, and many instabilities can occur in two-fluid systems.

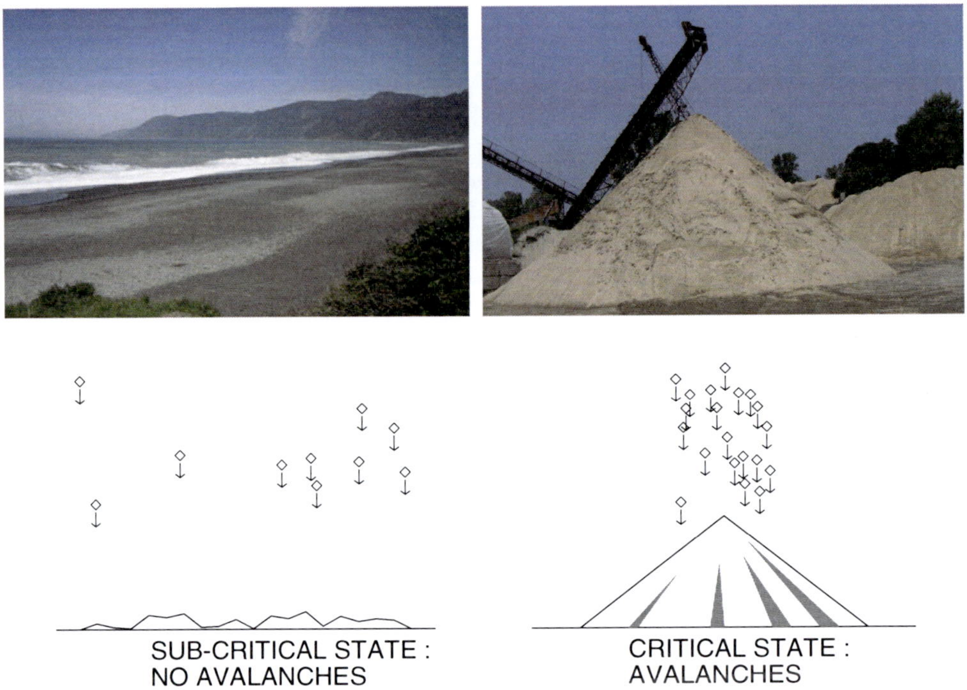

Fig. 1.2 A static equilibrium produces no avalanche events (bottom left panel), such as the flat sand beach in northern California (top left panel), while randomly dripping sand onto a sandpile produces a state of self-organized criticality where avalanches occur (bottom right panel), such as with the conveyer belt of the Indian River Enterprises (top right panel).

The concept of self-organized criticality has been pioneered by Per Bak and was first published in the seminal paper by Bak, Tang, and Wiesenfeld (1987), which has been cited already over 2000 times. Their brief abstract succinctly summarizes the quintessence of SOC: *We show that certain extended dissipative dynamical systems naturally evolve into a critical state, with no characteristic time or length scales. The temporal "fingerprint" of the self-organized critical state is the presence of 1/f noise; its spatial signature is the emergence of scale-invariant (fractal) structure.* The authors demonstrate the principle of SOC with a simple sandpile automaton model, which produces avalanches of arbitrary sizes that can be statistically sampled and exhibit a size distribution close to a powerlaw function. The powerlaw shape of size distributions became the hallmark and principal diagnostic of SOC phenomena (for which we will give a mathematical explanation in Chapter 3). However, as we will see later, powerlaw distributions are a necessary, but not a sufficient condition for SOC processes (Chapter 10; Sornette (2004), chapter 14 therein). Popular accounts of SOC phenomena can be found in the lucidly written book *How nature works* by Bak (1996), and in the article "Self-Organized Criticality" by Bak and Chen (1991) in *Scientific American*. Mathematical treatments of SOC phenomena can be found in the textbooks of Hendrik Jeldtoft Jensen (1998) and Didier Sornette (2004).

Before we proceed to practical examples of SOC phenomena, we should also clarify the difference between the terms *self-organization* and *self-organized criticality*. *Self-organization* refers to a broad range of pattern formation processes in both physical and biological systems. Pattern formation occurs through interactions internal to the system, without intervention of external directing influences (Camazine et al. 2001), such as zebra stripes, the bee's honeycomb structure, lichen growth, the hexagonal Bénard convection cells in boiling liquids, the granulation of the solar photosphere, in dusty space plasmas, or spheromaks. So, although the principle of *self-organization* is also concerned with complex interactions of neighboring components in a nonlinear system, it focuses on the resulting spatial (fractal) patterns, while the principle of *self-organized criticality* is concerned with the dynamical aspects. The dynamic behavior produces spatio-temporal events, whose statistical distributions of energy, temporal, and spatial scales can be sampled and quantitatively modeled.

1.2 SOC Laboratory Experiments

The principle of *self-organized criticality* was introduced by Bak, Tang, and Wiesenfeld (1987) as a theoretical concept, but the authors illustrated it also with the following practical example in the introduction of Bak et al. (1988): *To illustrate the basic idea of self-organized criticality in a transport system, consider a simple "pile of sand." Suppose we start from scratch and build the pile by randomly adding sand, a grain at a time. The pile will grow, and the slope will increase. Eventually, the slope will reach a critical value (called the "angle of repose"); if more sand is added it will slide off. Alternatively, if we start from a situation where the pile is too steep, the pile will collapse until it reaches the critical state, such that it is barely stable with respect to further perturbations.* At the end of their article, they suggest: *Finally, we invite the reader to perform the following home experiment. To demonstrate self-organized criticality, one needs a shoebox and a cup or two of sand – sugar or salt will do in a pinch. Wet the sand with a small amount of water, mix, and gather the sand into the steepest possible pile in one corner of the box. The angle of repose (i.e., the threshold slope) is larger for wet sand. So as the water evaporates, one observes a sequence of slides – some very small, others quite large – occurring at random places on the pile. (The evaporation process can be sped up by placing the box on a warm surface, or under direct sunlight.)*

Since experiments are always the toughest judge of new theories, several researchers started to test the SOC theory with sandpiles in their laboratory. The first experiment was performed by Jaeger et al. (1989) at the University of Chicago, who filled a cylindrical drum with grains, and rotated the drum slowly, like a concrete blender. Rotating the drum produces a one-sided sandpile with a critical slope inside the drum. The slow rotation steepens the slope and indeed created avalanches of all sizes. However, the authors did not find a powerlaw distribution of avalanche sizes. While the small-size and intermediate-size avalanches produced a powerlaw distribution, large-scale avalanches occurred in periodic time intervals when the slope became too steep. Bak (1996, p.68) blamed the failure to demonstrate SOC behavior in this experiment on the inertial effects of the rotation-induced periodic large-scale avalanches.

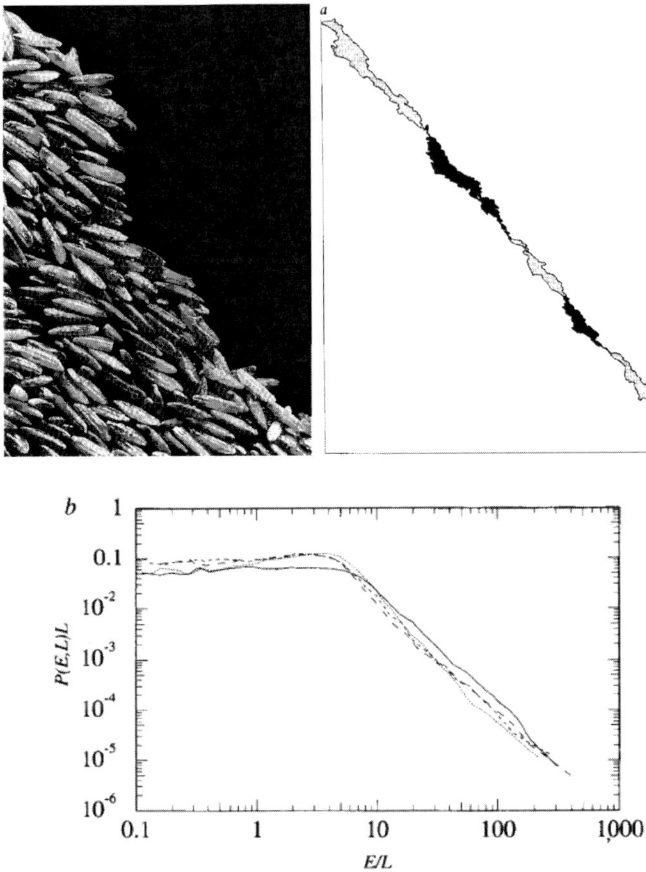

Fig. 1.3 The Norwegian rice pile experiment: a close-up photograph of a pile of rice, confined between two glass plates (top left); an electronically processed record of an avalanche, where loss and gains of rice mass is indicated with gray and black color (top right); and frequency distribution of the probability density $P(E,L)$ of events as a function of the energy dissipation E, normalized by the size L of the system. The powerlaw slope is $\alpha \approx 2.04$ (Frette et al. 1996).

A more accurate experiment closer to the suggestion of Bak et al. (1989) was carried out by Held et al. (1990) at IBM. They built a sandpile on top of a circular plate of an electronic precision scale. Sand grains were dropped slowly on top of the sandpile and the size of the resulting avalanche, if any, was recorded by the weight scale after each added sand grain. Again, the authors measured a powerlaw distribution for sufficiently large avalanche sizes only, but not for small to intermediate avalanche sizes. Bak (1996, p.68) comments that the experiment records only those avalanches that fall off the plate, but ignores the avalanches that stop along the sandpile, because they do not introduce a change on the weight scale, and thus the distribution is incomplete.

An experimental setup with a inclined Plexiglas box, similar to the rotating drum of the Chicago group, was used by Bretz et al. (1992), but the avalanches of dry, noncohesive granular material were recorded with a video camera. Again, the authors found a powerlaw distribution for small avalanches only, and Bak (1996, p.69) thought that the system they used was too small and the experiment was interrupted.

The ultimate and most careful experiment was performed by Frette et al. (1996) at the University of Oslo, who used rice piles instead of sand. Two-dimensional rice piles confined between two glass plates were monitored with video cameras to record the motion and size of individual rice grain avalanches, while additional rice grains were slowly added at the top of the pile. The experiment was performed for different plate distances, different system sizes (from a few cm to a few m), and for different rice types. The experimenters found that the occurrence distribution of energy (measured by the gravitational potential of the height difference between the beginning and end of an avalanche) dissipated in the rice avalanches obeyed a powerlaw function over up to 1.5 orders of magnitude, as expected for SOC behavior. However, the researchers found that the SOC behavior worked better for one type of rice (for grains with a large aspect ratio) than for others (for less elongated grains), and thus concluded that SOC is not universal, but depends on the detailed mechanism of energy dissipation. Note that this experiment records a complete sample of avalanches that stop on the pile or reach the end of the pile. A snapshot of the rice pile is shown in Fig. 1.3 (top left), the recorded area of a rice avalanche is displayed in Fig. 1.3 (top right)), and the frequency distribution of avalanche energies E, normalized by the system size L, is shown in Fig. 1.3 (bottom), which exhibits a powerlaw slope of $\alpha \approx 2.04$.

On a larger scale, if you combine a number of sandpiles, you end up with an entire landscape. Somfai et al. (1994a,b) at the Eötvös University in Budapest (Hungary) built their own mini-landscape in a laboratory and mimicked the landscape formation in nature by subjecting it to artificial erosion. A ridge-like landscape made of silica and potsoil was sprayed with water sprinklers, which produced mudslides of various sizes. They recorded the distribution of spatial sizes of mudslides and found a powerlaw with an exponent of $\alpha = 0.78 \pm 0.05$, which can be interpreted as evidence for SOC behavior. The resulting landscape was found to exhibit a fractal dimension, similar to that found in Norwegian fjords.

SOC behavior was also found in a number of laboratory experiments and in physical sciences, such as in plasma physics and material physics. A selection of phenomena that exhibit SOC behavior or to which SOC models have been applied is listed in Table 1.1 (adapted from Turcotte (1999)).

1.3 SOC in Human Activities

The unpredictability of outcomes in nonlinear dissipative systems is an inherent property of randomness, which is also called a Poisson process (see Chapter 4). The randomness occurring in our daily human life is one of the basic experiences that everybody accepts as a natural fact. A few examples of such Poisson processes in human life are: the number of cars that pass through a certain point on a road (sufficiently distant from traffic lights) during a given period of time; the number of spelling mistakes one makes while typing a single page; the number of phone calls at a call center per minute; the number of times a web server is accessed per minute; the number of light bulbs that burn out in a certain

Table 1.1 SOC behavior in material physics and laboratory plasma physics.

Phenomenon	References
sandpiles	Bak et al. (1989)
	Jaeger et al. (1989)
	Held et al (1990)
	Bretz et al. (1992)
rice piles	Frette et al. (1996)
silica and potsoil landscape	Somfai et al. (1994a,b)
fracture of fibrous materials	Bernardes and Moreira (1995)
microfracturing	Petri et al. (1994)
friction	Ciliberto and Laroche (1994)
random directed polymers	Jogi and Sornette (1998)
ceramics (Andrade creeps)	Cottrell (1996)
autocatalytic surface reactions	Drossel and Schwabl (1995)
annealed disorders	Caldarelli et al. (1996)
foam rheology	Okuzono and Kawasaki (1995)
	Kawasaki and Okuzono (1996)
dislocation networks	Marchesoni and Patriarca (1994)
lattice models of oscillators	Corral et al. (1995)
	Mousseau (1996)
pinned flux lattices	Pla and Nori (1991)
interface dynamics	Sneppen and Jensen (1993)
magnetic domain patterns	Babcock and Westervelt (1990)
	Che and Suhl (1990)
DC glow discharge plasma	Nurujjaman and Sekar-Iyenbgar (2007)
Barkhausen effect	Cote and Meisel (1991)
	O'Brien and Weissman (1994)
vortices in superconductors	Field et al. (1995)
	Zieve et al. (1996)
	Olson et al. (1997)
	Bassler and Paczuski (1998)
	Prozorov and Giller (1999)
plasma confinement	Carreras et al. (1996)
	Medvedev et al. (1996)
transport in tokamak plasmas	Kishimoto et al. (1996)
turbulence in tokamak plasmas	Rhodes et al. (1999)

amount of time; the number of roadkill (animals killed) found per unit length of road; the inventivity of inventors over their career; or the number of publications written by a scientist. Going back in time, the randonmness was even noted in the number of soldiers killed by horse-kicks each year in each corps in the Prussian cavalry, an example that was made famous in the book *The Law of Small Numbers* by Ladislaus Josephovich Bortkiewicz (1898).

Exploring the randomness in nature we will find Poisson processes everywhere: such as in the number of stars in a given volume of space, in the number of pine trees per unit area of mixed forest, in the number of viruses that can infect a cell in cell culture, in the number of hematopoietic stem cells in a sample of unfractionated bone marrow cells, in the distribution of visual receptor cells in the retina of the human eye, in the number of mutations in a given stretch of DNA after a certain amount of radiation, in the number of unstable nuclei that decay within a given period of time in a piece of radioactive substance, or in the number of particles that scatter off a target in a nuclear or high energy physics experiment.

Since we can easily gather statistics on event sizes that occur in such random processes, we can test whether the statistical distributions match the mathematical powerlaw distributions that are expected in the state of self-organized criticality (Chapter 3). Such occurrence distributions, where the number N of events is statistically sampled versus the size S of the event, preferentially on a log-log scale, i.e., log(N) vs. log(S), so that the powerlaw function appears as a straight line, have been sampled for city sizes, word counts in English literature (Zipf's law; Saichev et al. 2009), cotton prices, stock-market, lottery wins, or traffic jams, as they are eloquently described in Bak's book, *How Nature Works*. We will look at some of these examples in more detail in the following.

Let us consider the sizes of settlements, villages, towns, and cities on a continent. There are obviously a lot of small towns, while there are only a few large cities, such as New York or Los Angeles. If we look at a geographical map of North America (Fig. 1.4, left panel), there are millions of settlements and villages with small communities, which aggregated apparently randomly at almost any livable place, near rivers (to have water support), near roads (to have access to traffic), or near coasts (to take advantage of ocean transportation). So, people gathered in small communities for some economic or survival benefit. In some places, economic growth was more favorable than in others, which attracted more people who left a poor countryside and moved to small towns where lifestyle was more promising. Improvement of lifestyle and economic growth caused more urban sprawl so that large cities grew, up to the limit of the *carrying capacity*. Thus the size of a community is the result of complex human interactions between many members, which can be considered as a *nonlinear dissipative system*. So, we expect a powerlaw distribution of city sizes, if urban growth is in the state of *self-organized criticality*. First statistics of city sizes was already

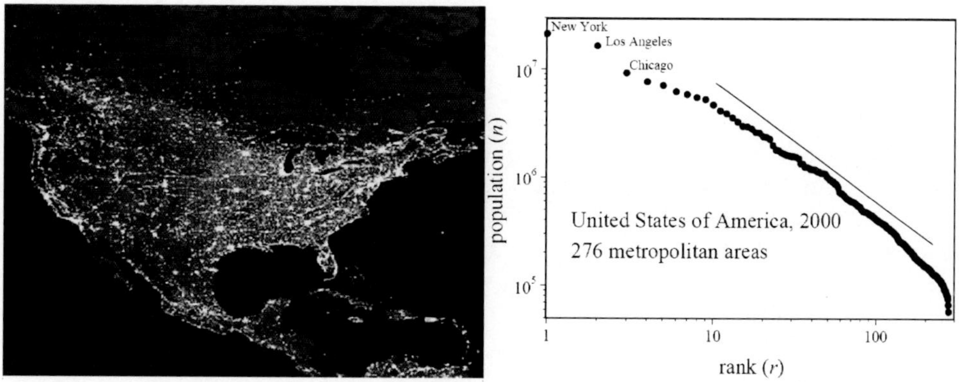

Fig. 1.4 *Left:* Satellite picture of North America at night, taken from the orbiting International Space Station (courtesy of NASA). The sizes of the light dots scale with the sizes of cities. *Right:* Zipf rank plot for 276 metropolitan areas in the United States, after results of the census in 2000. Source: http://factfinder.census.gov (Zanette 2007).

gathered by George Kingsley Zipf in 1920, when he produced a log-log histogram of the number of inhabitants per city versus the rank number of cities (which is essentially the order in a cumulative distribution), published in the book *Human Behavior and the Principle of Least Effort* (Zipf 1949). There were only a few cities larger than 8 million, 10 cities larger than 1 million, and 100 larger than 200,000. So, he found a powerlaw distribution with a slope of about $\log(N_2/N_1)/\log(S_2/S_1) \approx \log(100/10)/\log(2 \times 10^5/10^6) \approx -1.4$ for the cumulative occurrence distribution. A modern version of Zipf's plot is shown in Fig. 1.4 (right panel), based on a census in 2000 on 276 metropolitan areas in the United States
indexcensus (Zanette 2007). An up-to-date tutorial review on multiplicative processes in urban growth that lead to Zipf's law of city sizes is given in the same paper by Zanette (2007). Zipf's law applies also to the distribution of family names, or the number of individuals that speak the same language, because they are subject to the same multiplicative growth or inheritance process that is common to all biological systems.

Zipf (1949) also investigated the complexity of a language. In particular, he counted how often each word is used in a text of English literature, such as in James Joyce's *Ulysses* or in a collection of American newspapers. The most frequently used words in English texts are, in order of frequency, *"the"*, *"of"*, *"and"*, *"in"*, *"to"*, *"a"*, *"is"*, *"that"*, *"it"*, *"as"*, *"this"*, *"by"*, *"for"*, *"be"*, *"not"*, etc. If one plots the number of these words versus their rank, similarly to the Zipf plots of city ranks, one finds invariably a powerlaw distribution with a slope of ≈ -1. Computer programs that can be downloaded from the web (e.g., http://www.hermetic.ch/index.php) allow the user to produce such a Zipf plot for any arbitrary text. What is the reason for this powerlaw or SOC behavior? The use of every word is the result of complex thinking processes in our brain that involve associations of word concepts with perceived objects and logical connections that are expressed in verbal sentences. Associations or connections have a multiplicative functionality, and thus the word frequency or usage is proportional to the number of (meaningful) connections. Multiplicative behavior is an inherent characteristics of nonlinear systems, and thus enables SOC behavior. So, building up a complex language with rules of semantics, we end up with a word frequency that depends on the number of possible (i.e., meaningful) combinations, which is somehow multiplicative, based on the generality and applicability of a word. Some words that describe very rare applications, e.g., abacinate (to blind by putting a hot copper basin near someone's eyes), abcedarian (a person who teaches the alphabet), abderian (given to incessant or idiotic laughter), etc., as you can find in a grandiloquent dictionary (http://www.islandnet.com/˜egbird/dict/dict.htm), are obviously at the bottom of a Zipf rank plot, because the people who have heard of these words and even use them is the smaller the more specific the word content is. So, ultimately, the word usage is the result of an avalanche-like chain-reaction in communication, similar to the nonlinear-growth interactions in sand avalanches. Other applications of Zipf's Law to economics, especially the births and deaths of firms, can be found in the textbook of Saichev, Malevergne, and Sornette (2009).

SOC behavior was also found in economy (financial market, lottery wins, random drawings), which is governed by random input and nonlinear system dynamics. Mandelbrot (1963) collected data of monthly cotton prices over several years and plotted a log-log histogram of the monthly cotton price fluctuations, finding a Lévy distribution with a pow-

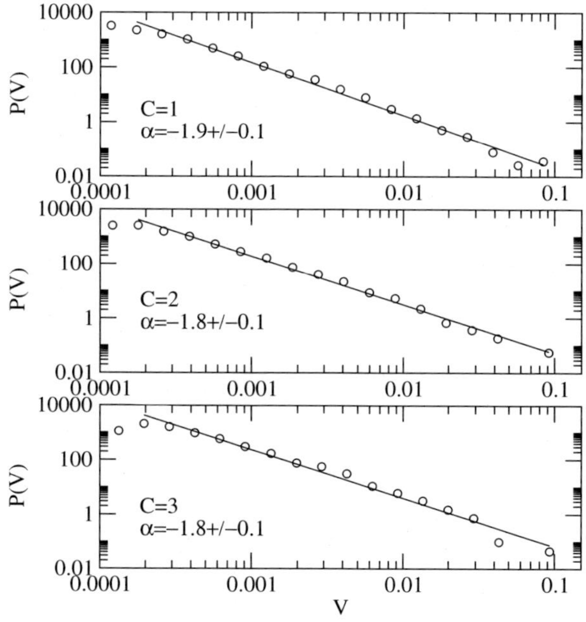

Fig. 1.5 Probability distribution function of "returns" (avalanche sizes) for the Dow Jones index daily closures from 1939/2/2 to 2004/4/13, computed for different cutoff parameters (in the selection of time intervals of high activity) using a wavelet method with three different settings (C = 1,2,3) (Bartolozzi et al. 2005).

erlaw tail with a slope of ≈ -1.0. More recently, fluctuations of the stock market have been investigated in great detail, which often show marginal variations of the daily Dow-Jones index, but occasionally can escalate into catastrophic events, such as the Wall Street crash in October 1929, the October 1987 market crash, or the big world-wide economic crisis in 2008 (at the time of writing). Stock market crashes, which can be triggered by speculation or political events, represent large avalanches in a SOC system (Scheinkman and Woodford 1994; Mantegna and Stanley 1997). The multiplicative chain reaction inherent to SOC systems has been modeled in terms of interacting producers and vendors by Bak et al. (1993, 1997). The avalanche-like evolution of stock market crashes is believed to be preceded by precursors with log-periodic fluctuations (Feigenbaum and Freund 1996; Sornette et al. 1996; Sornette and Johansen 1997). Recent studies (Bartolozzi et al. 2005) of the Nasdaq100, the (Standard and Poor's) S&P ASX50, and the Dow Jones index revealed powerlaws and SOC behavior for the logarithmic returns of these indices (avalanche sizes, see Fig. 1.5), the high activity periods (avalanche durations), and the quiet "laminar" times (waiting times). A recent progress report on financial physics reviews the application of SOC to economics, the Cont-Bourchaud percolation model, multiple-strategy agent-based models of financial markets, the minority game (i.e., the El Farol problem), and log-periodic precursors to financial crashes (Feigenbaum 2003).

A traffic jam is another driven dynamic system with random input that exhibits SOC behavior. Cars enter a highway at random times. If the traffic rate is low, such as on a

Sunday morning, the system is subcritical because there is plenty of space between subsequent cars so that they do not bother each other. During a rush hour, however, everybody is slowed down by the cars in front, and the spacings between subsequent cars is irregular due to different driving speeds, braking manoeuvres, car passings, delays in human reaction, or interfering weather conditions, etc. The distribution of car spacings will exhibit SOC behavior for a busy traffic situation with maximum throughput. If there is too high a traffic rate, the traffic slows down to a bumper-to-bumper situation. In fact, SOC is the most efficient state of traffic, because too low a rate is a waste in terms of under-utilized streets, while too high a rate leads to a permanent jam (Bak 1996). Emergent traffic jams were simulated by Nagel and Paczuski (1995), who found a powerlaw distribution of $N(t) \propto t^{-3/2}$ for the lifetimes of jams and $1/f$-noise in the power spectrum. Further models for traffic jams were studied by Nagel and Herrmann (1993), and Nagatani (1995a,b,c,d).

Other human activities, where a large number of individuals are involved and where random factors govern, are wars. The size or intensity of wars was quantified with the number of battle deaths and the statistical distribution of the number of wars versus the number of casualties was found to be a powerlaw distribution (Richardson 1941, 1960; Levy 1983; Roberts and Turcotte 1998; Turcotte 1999). But why are wars examples of SOC behavior? Turcotte (1999) compared the spread of wars over contiguous areas with people of identical political ideology and to metastable neighbor countries with the forest fire model, where a fire spreads over a contiguous local group of trees, and subsequently to neighboring tree groups, if they are located within a critical distance. The higher the number of trees per area, the larger is the size of the forest fire, similarly to how a local war can spread to a global conflict in case of high population density. However, while some statistical features of human activities can be modeled with a SOC model, we have to be aware that human interactions are far more complex (and arbitrary) than what can be modeled with a lattice model (Chapter 2) with well-defined probabilities.

A selection of references studying SOC behavior in human activities is summarized in Table 1.2.

1.4 SOC in Biophysics

Charles Darwin's theory of evolution explains life on Earth as a chain reaction of mutations, adaptation, survival of the fittest species, and elimination of the least-fit, resulting into a natural selection of the surviving species. Mutations of cells happen relatively rarely, for instance by absorption of ultraviolet light, with a higher probability for long-lived cells (which is the reason for a higher likelihood of cancer development in aging people). Since mutation happens rarely and in episodic steps, rather than with slowly-varying continuity, evolutionary changes occur in episodic bursts, separated by calm periods (similar to the stock market behavior). Evolutionary changes can include creation of new species as well as extinction of old species (like gains and losses on the stock market). The suggestion that evolution takes place in bursts separated by calm periods was made by Gould and Eldredge (1977). Statistics on the extinction of species were clearly found to be episodic at all times (Raup 1986; Sepkoski 1993, see Fig. 1.6), such as the famous Cretaceous-Tertiary event (65 million years ago) when dinosaurs became extinct, which was speculated to be

Table 1.2 SOC behavior in human activities.

Phenomenon	References
urban growth, city sizes	Zipf (1949)
	Zanette (2007)
word frequency in language	Zipf (1949)
cotton prices	Mandelbrot (1963)
stock market	Scheinkman and Woodford (1994)
	Feigenbaum and Freund (1996)
	Sornette et al. (1996)
	Sornette and Johansen (1997)
	Mantegna and Stanley (1997)
	Feigenbaum (2003)
	Bartolozzi et al. (2005)
traffic jam	Bak (1996)
	Nagel and Herrmann (1993)
	Nagel and Paczuski (1995)
	Nagatani (1995a,b,c,d)
war casualties	Richardson (1941, 1960)
	Levy (1983)
	Roberts and Turcotte (1998)
social networks	Newman et al. (2002)
internet traffic	Willinger et al. (2002)

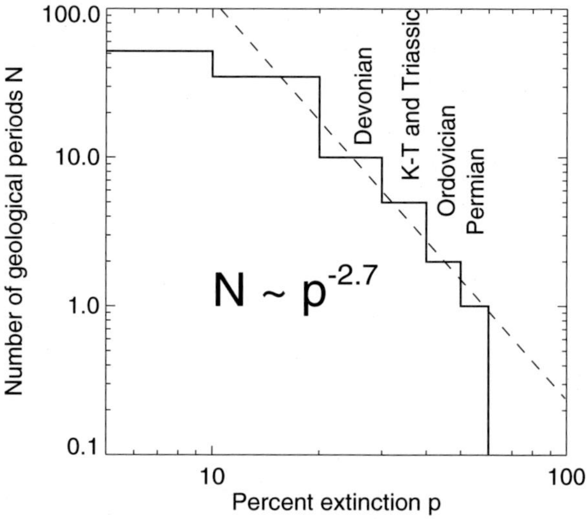

Fig. 1.6 Log-log histogram of the number of biological extinction events over the last 600 million years as recorded by Sepkoski Jr., who estimated from fossil records the percent of families that went extinct within time periods of approximately 4 million years. The dashed line shows a powerlaw fit of $N \propto p_{ext}^{-2.7}$ (adapted from Sepkoski 1993).

a consequence of an asteroid impacting Earth. An even bigger extinction event happened during the Cambrian era (500 million years ago), when up to 95% of all species on Earth disappeared. Raup (1991) discusses the origins of extinctions in his book *"Extinction: Bad Genes or Bad Luck?"* An extinction phase can involve a chain reaction of multiple species, which may depend on each other. On the other hand, the extinction of one species can trigger the growth of another competing species. Dinosaurs and mammals were believed to co-exist for a long period of time, while the number of mammal species grew explosively after the disappearance of dinosaurs, because their degree of fitness increased without the competition of the giant dinosaurs. Episodic changes in the size of species can also be triggered by meteorological changes. Warm-blooded mammals may die at higher rates during ice-ages. Global warming periods increase the overall temperature, which can enhance the probability for the spread of diseases, since chemical reactions generally occur at a rate of a factor of two faster when the temperature rises by 10%, speeding up the spread of pandemic diseases.

The episodic evolution and extinctions of species that leads to the biodiversity at a given time has been characterized with a model called *punctuated equilibria model* (Gould and Eldredge 1977, 1993). The dynamical concept consists of bursty episodes of high activity (i.e., punctuated events, like selected points in a time series), and calm intervening time intervals of low activity (i.e., waiting times; Chapter 5), where the dynamic system settles into a near-equilibrium state. A simple model of biological evolution based on punctuated equilibrium and criticality was developed by Bak and Sneppen (1993), which self-organizes into a critical steady state with intermittent co-evolutionary avalanches of all sizes. This concept involves "collaborative evolution", which is much more efficient than noncooperative scenarios with independent (and thus unlikely) mutation steps. The punctuated or stepwise behavior of evolutionary changes was reproduced by numerical simulations (Maslov et al. 1994). A unified class of systems far from equilibrium processes, including the Bak–Sneppen evolution model, interface depinning models, and invasion percolation models, was combined by Paczuski et al. (1996). A general discussion of criticality and scaling terms in evolutionary ecology is given in Sole et al. (1999). A more recent model on the time-dependent extinction rate and species abundance is the *tangled-nature model* (Hall et al. 2002), which reproduces both a smooth evolution of microscopic fluctuations as well as intermittency of macroscopic fluctuations (punctuated equilibria). Some other applications of SOC models to biophysics are given in Table 1.3.

1.5 SOC in Geophysics

An excellent introduction into the subject of self-organized criticality in Earth systems can be found in the textbook of Hergarten (2002), with extensive coverage on earthquakes, landslides, and drainage networks. Another textbook related to this subject covers fractals and chaos in geology and geophysics (Turcotte 1997). Besides seismology and earthquakes, SOC behavior is also found in a number of other geophysical systems, such as in landslides, turbidites, geological layers, volcanic eruptions, forest fires, lightning, rainfall, hydrology, snow avalanches, cloud formation, climate fluctuations, etc.; see Table 1.4 for a representative selection.

Table 1.3 SOC in biophysics.

Phenomenon	References
evolution and extinctions	Gould and Eldredge (1977)
	Raup (1986)
	Sepkoski (1993)
neuron firing in a brain	Stassinopoulos and Bak (1995)
	Hopfield (1994)
	Rundle et al. (2002)
neural reverberations of spiking nerve cells	Herz and Hopfield (1995)
	da Silva et al. (1998)
learning and memory	Chialvo and Bak (1999)
breathing in lung	Barabasi et al. (1996)
heart rate	Goldberger et al. (2002)
epileptic seizures	Osorio et al. (2009a,b)
spread of diseases	Johansen (1994)
measles epidemics	Rhodes and Anderson (1996)
	Rhodes et al. (1997)
flying formation of birds	Nathan and Barbosa (2006)
termite nest architecture	O'Toole et al. (1999)
phylogenetic (evolutionary) trees	Vandewalle and Ausloos (1995)

Table 1.4 SOC in geophysics.

Phenomenon	Selected references
earthquakes	Gutenberg and Richter (1954)
	Aki (1981)
	Bak et al. (2002)
landslides	Fuyii (1969)
	Hovius et al. (1997, 2000)
	Pelletier et al. (1997)
	Malamud et al. (2001)
turbidite depositions	Rothman et al. (1994)
volcanoclastic turbidite deposits	Hiscott et al. (1992)
volcanic acoustic emission	Diodati et al. (1991, 2000)
volcanic activities	Grasso and Bachelery (1995)
rock texture in craters	Wu and Zhang (1992)
plastic shear bands in rocks	Poliakov and Herrmann (1994)
epizonal mineral deposits	Henley and Berger (2000)
propagating brittle failure	Katz (1986)
snow avalanches	Birkeland and Landry (2002)
river networks	Rinaldo et al. (1996)
drainage networks	Hergarten (2002)
Nile river fluctuations	Hurst (1951)
rainfall	Andrade et al. (1998)
cloud formation	Nagel and Raschke (1992)
climate fluctuations	Grieger (1992)
aerosols in atmosphere	Kopnin et al. (2004)
forest fires	Kasischke and French (1995)
	Malamud et al. (1998)

Earthquakes represent local adjustments to the stressing forces in the upper earth crust (in depths of less than 20 km), which is not static but experiences permanent deformation. The lithosphere is subdivided into several tectonic plates, whose motion is driven by thermal convection in the mantle giving rise to spreading centers ocean ridges and subduction zones at ocean trenches. The tectonic plates behave elastically until the stresses exceed a certain threshold and a displacement occurs as a consequence – like a stick-and-slip motion – reducing the stress. Earthquakes can also strike in the stable crust, far away from earthquake zones at the edges of tectonic plates (Johnston and Kanter 1990) or hidden on "blind" faults under folded terrain (Stein and Yeats 1989). Earthquakes are episodic and intermittent events, separated by long time intervals of quiescence, sometimes preceded by precursors, or followed by aftershocks. Once an earthquake occurs, seismic waves propagate away from the epicenter and cause damage over an extended area. Statistics of earthquakes and measurements of their magnitude became a research focus during the 20th century. The Gutenberg and Richter (1954) law was established already more than a half century ago. It states that the cumulative distribution of earthquakes follows a powerlaw distribution as a function of the magnitude m. A similar relation holds for the earthquake rupture areas A_E, i.e., $N^{cum}(> A_E) \propto A_E^{-1}$, or $N(A_E) \propto A_E^{-2}$ for the differential occurrence frequency distribution (Aki 1981). [Definitions of differential and cumulative frequency distributions are provided in Section 7.1.] The world-wide statistics of earthquakes for the period of 1977–1994 is given in Fig. 1.7, as a function of the Gutenberg–Richter magnitude m (top axis), as well as a function of the earthquake rupture area A_E (bottom axis).

The relationship of earthquakes to self-organized criticality has been considered by Bak and Tang (1989), Sornette and Sornette (1989), Ito and Matsuzaki (1990), Sornette

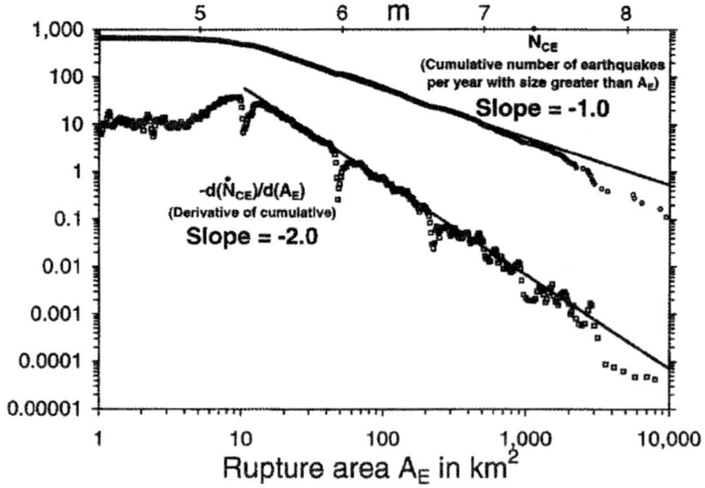

Fig. 1.7 World-wide cumulative (slope = −1) and differential frequency distribution (slope = −2) of earthquakes per year as a function of the rupture area A_E in units of square kilometers, based on the Harvard Centroid-Moment Tensor Data Base (1997) for the years 1997–1994. The conversion into equivalent Gutenberg–Richter magnitudes m is indicated on the top axis (Turcotte 1999).

et al. (1990), Olami et al. (1992), Bak and Chen (1995), and Huang et al. (1998). The fractal dimension of the distribution of earthquake fault gouges was brought into context with the powerlaw behavior of SOC (Sammis et al. 1987). The primary SOC models applied to earthquakes are the Burridge and Knopoff (1967) slider-block model (see reviews by Carlson et al. 1994, Turcotte 1997, 1999, and references therein), while other SOC models include crack propagation (Chen et al. 1991) and interface depinning (Paczuski and Böttcher 1996; Fisher et al. 1997). The waiting times between earthquakes were also found to obey a powerlaw distribution, within a validity range of tens of seconds to tens of years (Bak et al. 2002).

Major natural hazards on Earth include earthquakes, snow avalanches, floods, storms, volcanic eruptions, and landslides, which all exhibit SOC behavior (Hergarten 2002). Landslides occur mostly in mountainous areas (Fig. 1.8), where the inclination angle exceeds a critical slope, similar to the sand and rice piles discussed in Section 1.2. A more physical term for landslides is *gravity-driven mass movements*. Landslides can be triggered by heavy rainfall or earthquakes. After the initial detachment of a certain amount of soil, gravity will accelerate the unstable mass and increase the kinetic energy gradually. The increasing speed will overcome more friction at the front and edges of the sandslide and pull more material along, further increasing the mass and kinetic energy. Landslides and avalanches, therefore, exhibit a multiplicative or exponential-like growth in the time evolution of their area, volume, mass, and energy. Landslides were found to have multifractal properties (Mandelbrot 1985; Feder 1988; Turcotte 1997). Cumulative frequency distributions of landslides as a function of the area (or some other magnitude definition)

Fig. 1.8 A sketch of the mountain "Rossberg" in Arth/Goldau (Switzerland), where a catastrophic rockslide occurred on September 2, 1806 and destroyed the town of Goldau, causing the death of 457 people (drawing by Fritz Morach).

were found to exhibit a powerlaw distribution with slopes of $\beta = 0.96$ in Japan (Fuyii 1969), $\beta = 1.16$ in New Zealand (Hovius et al. 1997), $\beta = 0.70$ in Taiwan (Hovius et al. 2000), $\beta \approx 1.6$–2.0 in Japan, California, and Bolivia (Pelletier et al. 1997), or $\beta = 1.5$ in Italy (Malamud et al. 2001). This powerlaw slope range of $\beta \approx 0.7$–2.0 for cumulative frequency distributions corresponds to $\alpha = \beta + 1 \approx 1.7$–$3.0$ for differential (noncumulative) frequency distributions. Sandslides triggered by earthquakes revealed powerlaw slopes in the range of $\alpha = 2.3$–3.3 for datasets from California, Japan, and Bolivia (see review by Turcotte 1999 and references therein). Frequency distributions of sandslide volumes with powerlaw behavior have been measured over 12 orders of magnitude, on Himalayan roads (Noever 1993). The frequency distributions and fractal properties of landslides seem not to depend on the triggering event (earthquake or rainfall), nor on the steepness of the mountain slope (except for very shallow angles). Theoretical models of sandslides include the classical BTW sandpile model (Bak et al. 1989), the OFC spring-block model (Olami et al. 1992), or physics-based models, e.g., based on partial differential equations that combine the slope stability and mass movements (Hergarten and Neugebauer 1998).

Sediment depositions at the edge of the continental shelf, deposited some 100 million years ago, occurred also as avalanche-like events, called slumps. The resulting sediment layers are called turbidites and have a variety of sizes and thicknesses, which were found to exhibit a powerlaw distribution in the geological layer thicknesses, e.g., as measured in the Death Valley in California (Rothman et al. 1994). Similar powerlaw distributions that indicate SOC behavior have been found in volcanoclastic turbidite deposits (Hiscott et al. 1992), in acoustic emission from volcanic activity of Stromboli (Diodati et al. 1991, 2000), in rock textures and multi-ring structures in the Duolun crater (Wu and Zhang 1992), in plastic shear bands in rocks (Poliakov and Herrmann 1994), in epizonal mineral deposits (Henley and Berger 2000), in the eruptions, volcano-induced earthquakes, dikes, fissures, lava flows, and interflow periods of the *Piton de la Fournaise* volcano (Grasso and Bachelery 1995), in dust grains (aerosols) in the Earth's atmosphere (Kopnin et al. 2004), or in snow avalanches (Birkeland and Landry 2002).

Water plays a fundamental role for life on earth. All parts of the water cycle, from evaporation over the oceans, to formation of clouds, to rainfall, to river formation, were also found to exhibit SOC behavior. Water-related phenomena are studied in the sciences of hydrology and rheology. The formation of small water streams that combine to bigger and bigger rivers exhibit a well-known fractal pattern, also known as Horton's law. Horton's law defines the order of river segments as the number of links to other segments that have to be passed before the river reaches the ocean, which increase as a powerlaw in the order. The most famous manifestation of this principle is the fractal coastline of Norway. River networks and drainage networks have been studied systematically under the aspect of SOC (e.g., Rinaldo et al. 1996; Hergarten 2002). The fluctuations of the water level of the Nile river has been characterized with a Hurst exponent (Hurst 1951), which is related to SOC behavior. SOC behavior was also studied in rainfall (Andrade et al. 1998), cloud formation (Nagel and Raschke 1992), and climate fluctuations (Grieger 1992), which are now in hot debate in the context of the global warming trend.

Forest fires are another phenomenon that exhibit classical SOC behavior. The trigger may be a small accident, such as an out-of-control campfire, a discarded cigarette, or an electric spark of a high-voltage power line, while the outcome can have catastrophic dimensions, depending on the spreading efficiency that can be sped up by dry conditions or wind. The noncumulative frequency distribution of the number of forest fires as a function of burned area was found to have a powerlaw slope of $\alpha = 1.3$ for 4284 fires on US Fish and Wildlife Service land during 1986–1995 (National Interagency Fire Center), $\alpha = 1.3$ for 120 fires in the western US during 1955–1960, calculated from tree ring data (Heyerdahl et al. 1994), $\alpha = 1.4$ for 164 fires in Alaskan Boreal Forests during 1990–1991 (Kasischke and French 1995), or $\alpha = 1.5$ for 298 fires in the Australian Capital Territory during 1926–1991 (ACT Bush Fire Council 1996), as computed by Malamud et al. (1998). Forest fires were one of the first phenomena that have been modeled with a SOC cellular automaton model (Bak et al. 1990; Drossel and Schwabl 1992a,b; Henley 1993). More comprehensive forest-fire models were developed that include also phase transitions, "immune" trees, and applications of the renormalization group theory (see review by Turcotte (1999)).

1.6 SOC in Magnetospheric Physics

Magnetospheric physics deals with the interaction of the Earth's (or some other planet's) magnetic field with the ambient solar wind in the heliosphere, which triggers a host of secondary phenomena, such as ionospheric electric currents, aurorae, magnetic storms, substorms, magnetic reconnection, and turbulence. Some dynamic phenomena occur in the plasma sheet and neutral sheet of the geotail, in the trailing part of the Earth's magnetic field that stretches out past 200 Earth radii away from the Sun. A lot of magnetospheric events (storms) are triggered by solar flares, coronal mass ejections, and the solar wind, which have a highly intermittent and turbulent dynamics, but mostly exhibit SOC behavior. Magnetospheric substorms and auroral activity evolve in response to the solar wind and exhibit distinctly different levels of activity and nonequilibrium phase transitions (Bargatze et al. 1985; Sitnov et al. 2000). Therefore, the bursty nature of magnetospheric phenomena, such as localized current disruptions (Lui et al. 1988), bursty bulk flow events (Angelopoulos et al. 1996, 1999), and the powerlaw magnetic field spectra in the magnetotail (Hoshino et al. 1994), have been interpreted in terms of an open, dissipative nonlinear system near a forced or self-organized critical state (Chang 1992, 1999a,b; Klimas et al. 2000; Chang et al. 2003; Chapman and Watkins 2001; Consolini and Chang 2001). Evidence for the powerlaw characteristics of the probability distribution of energy release events was found in auroral images from Polar/UVI (Lui et al. 2000; Uritsky et al. 2002, 2003, 2006), in ground-based optical auroral observations (Kozelov et al. 2004), in the burst size of the *auroral electron jet index (AE)* (Takalo et al. 1993; Consolini 1997, 2002), or in magnetospheric substorm-related tail current disruptions (Consolini and Lui 1999). An example of an auroral image is shown in Fig. 1.9, and statistics of auroral blob sizes are shown in Fig. 1.10, where a powerlaw distribution over two orders of magnitude is observed for auroral blob areas during quiet time intervals, but not during substorms.

Physical models of the dynamics of the Earth's magnetotail are described in terms of stochastic behavior of a nonlinear dynamical system near forced and/or self-organized criticality. Multi-scale intermittent turbulence of overlapping plasma resonances and current-driven instabilities are believed to lead to the onset and evolution of substorms, which

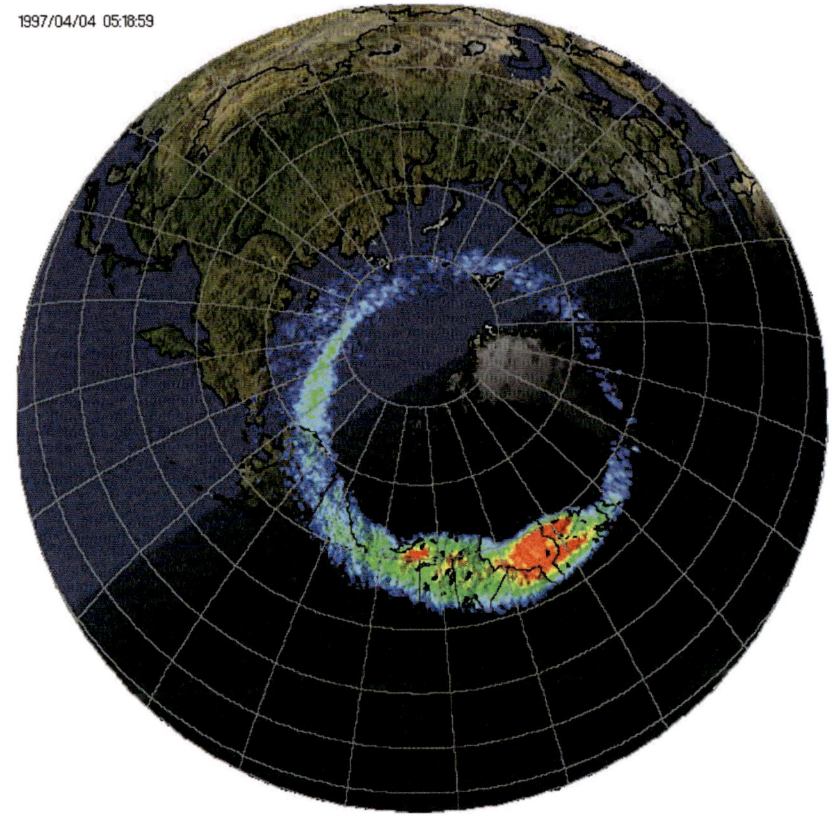

Fig. 1.9 Global image of the auroral oval observed by the Ultraviolet Imager (UVI) onboard the NASA satellite "Polar" on April 4, 1997 at 0519 UT, projected onto an Earth map (credit: NASA, Polar/UVI Team, George Parks).

Table 1.5 SOC in magnetospheric physics.

Phenomenon	Selected references
magnetotail current disruptions	Lui et al. (1988)
substorm current disruptions	Consolini and Lui (1999)
bursty bulk flow events	Angelopoulos et al. (1996, 1999)
magnetotail magnetic field	Hoshino et al. (1994)
auroral UV images	Lui et al. (2000)
	Uritsky et al. (2002, 2003, 2006)
auroral optical images	Kozelov et al. (2004)
auroral electron jet index (AE)	Takalo et al. (1993)
	Consolini (1997, 2002)

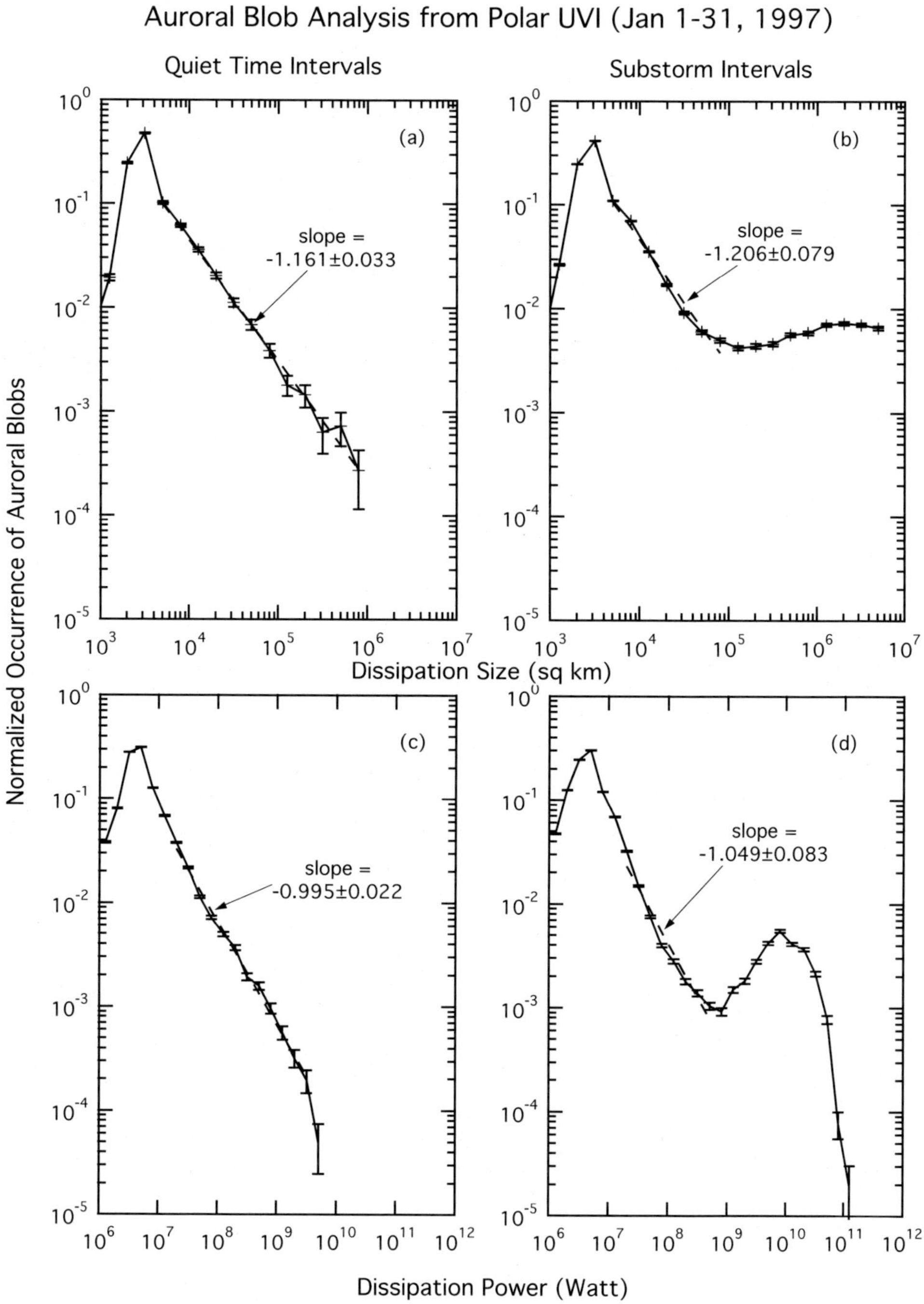

Fig. 1.10 Occurrence rate frequency distributions of auroral blobs (see Fig. 1.9) as a function of the size (top panels) or dissipated energy (bottom panels), during quiet time intervals (left panels) and active time intervals (right panels) (Lui et al. 2000).

explains the localized and sporadic nature of bursty magnetic reconnection and the fractal spectra observed in the magnetotail region (Chang 1999a,b; Klimas et al. 2000, 2004). Extending the original definition of SOC by Bak et al. (1987), Chang (1992) introduced the term *forced criticality*, which differs from Bak's SOC model in the sense that it is critical without self-organization, under the influence of external forcing. Coherent magnetic structures approach each other and merge or scatter under the influence of external forcing, turbulence, or self-organization, until a powerlaw-like spectrum of size scales occurs. It is now argued that the magnetospheric system is driven to a critical or near-critical state as a result of the continuous loading and subsequent unloading above a critical current (Chang 1992, 1999a,b; Consolini and de Michelis 2002; Horton and Doxas 1996). The new model eliminates the older "dripping faucet" model of chaotic loading and unloading (Baker et al. 1990; Klimas et al. 1992). Various cellular automaton SOC models have been developed to model the dynamics of the auroral electron jet (AE) index (Uritsky and Pudovkin 1998; Chapman et al. 1998, 1999; Watkins et al. 1999), the magnetotail current sheet (Takalo et al. 1999a,b; Milovanov et al. 2001), the central plasma sheet (Liu et al. 2006), and extended with renormalization-group analysis (Tam et al. 2000). Some observational references for magnetospheric SOC phenomena are given in Table 1.5.

1.7 SOC in Planetary Physics

A number of SOC phenomena studied in geophysics (earthquakes, volcanic eruptions, landslides, meteorite impacts) are expected to occur also on other rock-like planets, moons, or asteroids. Even water-related phenomena (river networks, fluvial systems, sedimentation) are expected on planets that carried water at some time (e.g., Mars). Planets with atmospheres (Venus, Mars, Jupiter, Saturn, Uranus, Neptune, etc.) are expected to exhibit SOC behavior in climate phenomena (e.g., dust storms, climate changes, transient spots and eddies, latitudinal bands). Planets with magnetic fields (Mercury, Jupiter, Saturn, Uranus, Neptune) are expected to show SOC behavior in magnetospheric phenomena (aurorae, substorms). However, due to the remote location and limited spatial resolution of Earth-bound observations we have very little data on planetary SOC phenomena.

Mars global dust storms do not occur every year, but preferably during late southern spring, when Mars is near its perihelion closest to the Sun. This interannual variability of Mars global dust storms was modeled in terms of a SOC system, where smaller storms ("dust devils") occur between the active years at a lower threshold (Pankine and Ingersoll 2004). SOC models were also applied to the Martian fluvial system (Rosenshein 2003). Self-organized criticality produces avalanches that grow coherently by nearest-neighbor interactions, generally forming size distributions that obey a powerlaw function. In a wider sense, we can even consider accretion, collisions, and scattering of particles as nonlinear dissipative processes possibly governed by SOC behavior. Such nonlinear processes are expected in the formation of planetary systems, planetary rings, asteroids, comets, meteorites, and circumplanetary dust. The distribution of particle sizes in Saturn's ring (Fig. 1.11) indeed follows a powerlaw distribution of $N(L) \propto L^{-3}$ in the range of $1\,\mathrm{mm} < L < 20\,\mathrm{m}$ (Zebker et al. 1985; French and Nicholson 2000). The asteroid size distribution follows a broken powerlaw with $N(L) \propto L^{-2.3}$ for large asteroids (5–50 km) and

Fig. 1.11 Saturns rings observed by Voyager 2 spacecraft (left panel, credit: JPL and NASA), and artistic rendering of saturn ring particles in close-up (right panel).

$N(L) \propto L^{-4}$ for smaller asteroids (0.5–5 km) (Ivezic et al. 2001). The size distribution of lunar craters was found to be approximately $N^{cum}(> L) \propto L^{-2}$ (Cross 1966).

1.8 SOC in Solar Physics

Solar flares are very energetic phenomena where a magnetic reconnection process liberates large amounts of magnetic energy that is dissipated by heating of thermal plasma and by acceleration of high-energy (nonthermal) particles. The high-energy particles propagate along the coronal magnetic field lines and mostly slam into the dense plasma in the chromosphere, while a small fraction escapes upward into interplanetary space. The majority of accelerated particles that precipitate into the chromosphere produce collisional bremsstrahlung in hard X-ray and gamma ray wavelengths, while heated chromospheric plasma "evaporates" up into the postflare loops (Fig 1.12). The hard X-ray emission, which provides a good measure for the total released flare energy, has been recorded for a large number of flares with the *Solar Maximum Mission (SMM)* spacecraft during 1980–1989. When a frequency distribution of these many hard X-ray peak count rates was plotted, an astonishingly straight powerlaw distribution with a slope of $\alpha \approx 1.8$ was found (Dennis 1985), extending over 4 orders of magnitude (Fig. 1.13). Statistics of other flare parameters were calculated, yielding a powerlaw slope of $\alpha = 1.73 \pm 0.01$ for background-subtracted hard X-ray peak rates, $\alpha = 2.54 \pm 0.05$ for flare durations, and $\alpha = 1.53 \pm 0.03$ for nonthermal electron energies above 25 keV (Crosby et al. 1993).

Powerlaw distributions of flare peak intensities were found in virtually all observed wavelengths: in gamma rays, hard X-rays, soft X-rays, extreme ultraviolet, Hα, optical, and radio wavelengths (see Table 1.6 for observational references). Although the values of the powerlaw slope varies over a considerable range ($\alpha \approx 1.1$–2.8) in different wavelengths, most measurements fall in the range of $\alpha \approx 1.5$–1.9. With the advent of hard X-ray detectors with higher sensitivity, the frequency distributions were considerably extended at the lower end, to flares that were smaller up to 3 orders of magnitude, called *microflares*

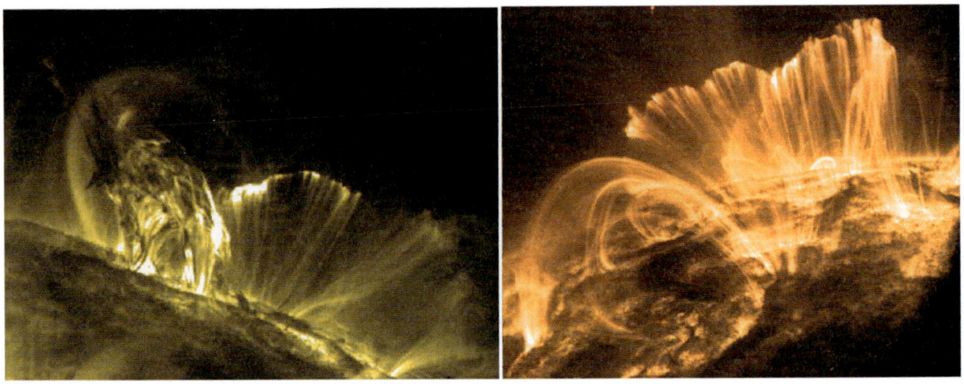

Fig. 1.12 Solar flares observed in EUV with the TRACE spacecraft in 171 Å: The flare of 2001 Apr 15 exhibits an erupting filament in the foreground and a rising postflare arcade behind near the limb (left panel), while the 2000 Nov 9 flare displays the 3-D geometry of the double-ribbon postflare arcade (right panel) (credit: NASA, TRACE).

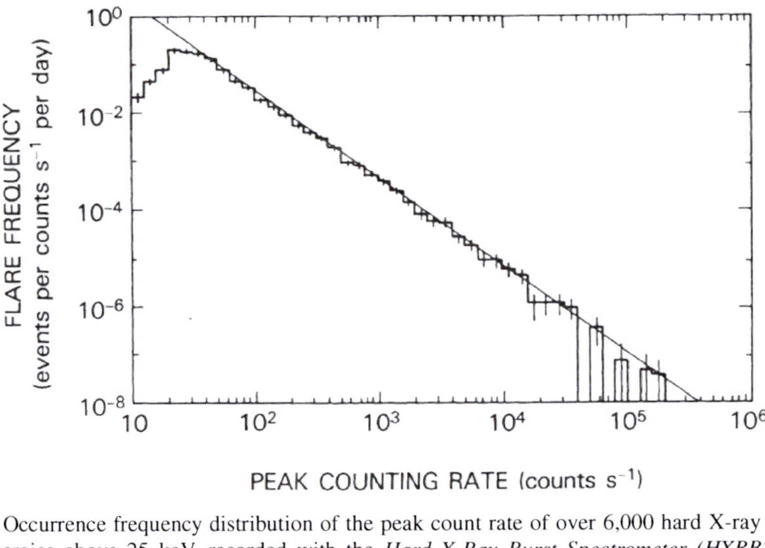

Fig. 1.13 Occurrence frequency distribution of the peak count rate of over 6,000 hard X-ray flares with photon energies above 25 keV, recorded with the *Hard X-Ray Burst Spectrometer (HXRBS)* onboard NASA's *Solar Maximum Mission (SMM)* during 1980–1985. The distribution follows a powerlaw with a slope of $\alpha \approx 1.8$ over 4 orders of magnitude (Dennis 1985).

(e.g., Lin et al. 2001; Christe et al. 2008). A further extension down to 9 orders of magnitude smaller than the largest flares could be observed with high-resolution EUV imagers (SOHO/EIT, TRACE), called *nanoflares* (e.g., Krucker and Benz 1998; Aschwanden and Parnell 2002). Because the powerlaw slope of the frequency distributions was found to be close to $\alpha = 2$ for nanoflares, which is a critical limit where the energy integral diverges at the low or high end, the crucial question came up whether nanoflares significantly contribute to coronal heating (Hudson 1991). A synthesized frequency distribution of flares, microflares, and nanoflares is shown in Fig. 1.14, which exhibits an approximate powerlaw distribution with an overall slope of $\alpha \approx 1.8$ for the flare peak fluxes. Flares with or without coronal mass ejections (CME) were found to have different powerlaw slopes (Yashiro et al. 2006). The exact value of the powerlaw slope depends on event selection, event definition, instrumental sensitivity or flux threshold, instrumental bias, observed wavelengths and temperature regime, geometric models, and energy definitions (e.g., Lee et al. 1995; Isliker and Benz 2001; Aschwanden and Charbonneau 2002; Aschwanden and Parnell 2002; McIntosh and Charbonneau 2001; McIntosh et al. 2002). Another question was how robust the flare frequency distribution is in time. No significant variation of the powerlaw slope was found during the 11-year solar cycle (Bai 1993), although the flare rate varies by orders of magnitude. The flare frequency distributions were also investigated as a function of the spatial size of the associated active regions and some dependencies of the powerlaw slopes and cutoffs were found (Wheatland and Sturrock 1996; Kucera et al. 1997; Sammis 1999; Wheatland 2000c).

While most of the flare emission in hard X-rays, soft X-rays, and EUV is confined to the lower corona, flare-associated phenomena in the upper corona or heliosphere include coronal mass ejections (CMEs), radio bursts, shock waves, and solar energetic particle (SEP) events. Powerlaw distributions of these secondary events in the flare process have been observed for radio bursts and solar energetic particles (see Table 1.6). For CMEs there are probably no suitable data available, because measured observables are the angular width and propagation speed, while CME masses or energies require model-dependent calculations that are not readily available.

An interpretation of the omnipresent powerlaw distributions of flare peak fluxes or energies in terms of SOC models was first proposed by Lu and Hamilton (1991), from which we quote the abstract: *"The solar coronal magnetic field is proposed to be in a self-organized critical state, thus explaining the observed powerlaw dependence of solar-flare-occurrence rate on flare size which extends over more than five orders of magnitude in peak flux. The physical picture that arises is that solar flares are avalanches of many small reconnection events, analogous to avalanches of sand in the models published by Bak and colleagues in 1987 and 1988. Flares of all sizes are manifestations of the same physical processes, where the size of a given flare is determined by the number of elementary reconnection events. The relation between small-scale processes and the statistics of global-flare properties which follows from the self-organized magnetic-field configuration provides a way to learn about the physics of the unobservable small-scale reconnection processes. A simple lattice-reconnection model is presented which is consistent with the observed flare statistics. The implications for coronal heating are discussed and some observational tests of this picture are given."* This seminal paper has been cited over 300 times at the time of writing, which documents what a far-reaching impact this interpretation had in the area of solar physics alone.

Modeling of SOC behavior in solar flares started first with classical Bak–Tang–Wiesenfeld (BTW) and modified cellular automaton models (Lu and Hamilton 1991; Lu et al. 1993; Lu 1995a; Georgoulis et al. 1995, 2001; MacKinnon et al. 1996, 1997; Georgoulis and Vlahos 1998; Macpherson and MacKinnon 1999; Charbonneau et al. 2001; Belanger et al. 2007; Morales and Charbonneau 2008a,b, 2009). Alternatively, also a continuously

Table 1.6 SOC in solar physics: Observations of solar flare phenomena with powerlaw-like event occurrence frequency distributions.

Phenomenon:	Selected references:
flare gamma rays:	Perez-Enriquez and Miroshnichenko (1999)
flare hard X-rays:	Datlowe et al. (1974)
	Dennis (1985)
	Schwartz et al. (1992)
	Crosby et al. (1993, 1998, 1999)
	Lu et al. (1993)
	Lee et al. (1993)
	Bromund et al. (1995)
	Aschwanden et al. (1995, 1998b)
	Kucera et al. (1997)
microflare hard X-rays:	Lin et al. (1984, 2001)
	Biesecker et al. (1993, 1994)
	Su et al. (2006)
	Christe et al. (2008)
flare soft X-rays:	Hudson et al. (1969)
	Drake (1971)
	Shimizu (1995)
	Lee et al. (1995)
	Feldman et al. (1997)
	Veronig et al. (2002a,b)
	Das et al. (2004)
	Yashiro et al. (2006)
microflare soft X-rays:	Shimojo and Shibata (1999)
nanoflare EUV emission:	Krucker and Benz (1998)
	Aschwanden et al. (2000a)
	Aletti et al. (2000)
	Benz and Krucker (2002)
	Aschwanden and Parnell (2002)
flare ultraviolet emission:	Nishizuka et al. (2009)
chromospheric events Hα:	Georgoulis et al. (2002)
flare radio emission:	Akabane (1956)
	Kundu (1965)
	Kakinuma et al. (1969)
	Fitzenreiter et al. (1976)
	Aschwanden et al. (1995, 1998b)
	Mercier and Trottet (1997)
	Das et al. (1997)
	Meszarosova et al. (1999)
	Melendez et al. (1999)
	Nita et al. (2002, 2004)
	Ning et al. (2007)
solar energetic particles events (SEP):	VanHollebeke et al. (1975)
	Cliver et al. (1991)
	Gabriel and Feynman (1996)
	Miroshnichenko et al. (2001)
	Gerontidou et al. (2002)

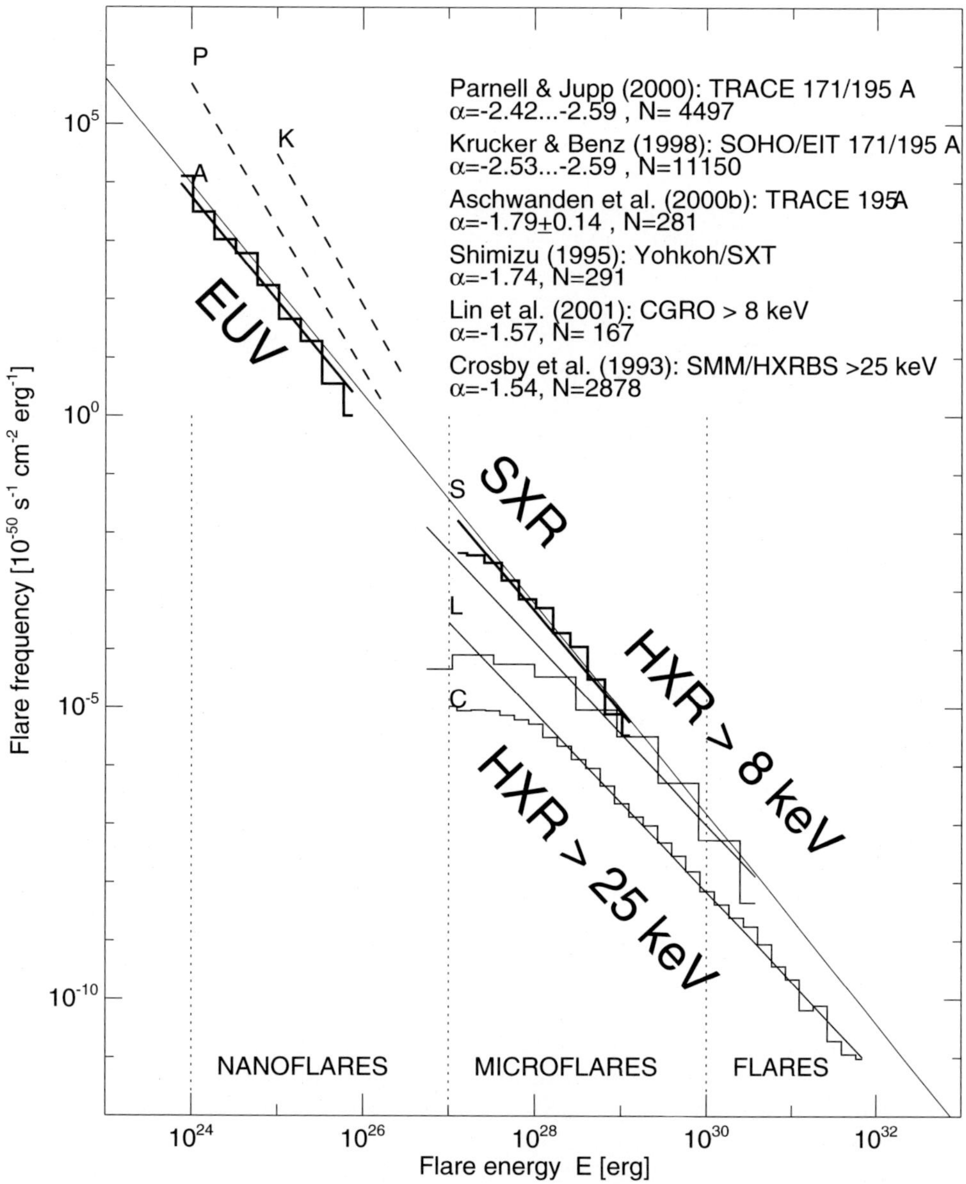

Fig. 1.14 Composite flare frequency distribution in a normalized scale in units of 10^{-50} flares per time unit (s^{-1}), area unit (cm^{-2}), and energy unit (erg^{-1}). The energy is defined in terms of thermal energy $E_{th} = 3n_e k_B T_e V$ for extreme ultraviolet (EUV) and soft X-rays (SXR), and in terms of nonthermal energy in >25 keV (Crosby et al. 1993) or >8 keV electrons (Lin et al. 2001). The slope of -1.8 is extended over the entire energy domain of 10^{24}–10^{32} erg. The offset between the two hard X-ray (HXR) datasets is attributed to different lower energy cutoffs as well as different levels of flare activity during the observed time intervals (adapted from Aschwanden et al. 2000b; Lin et al. 2001).

driven Olami–Feder–Christensen (OFC) model has been applied to solar flares (Hamon et al. 2002). However, although cellular automaton models can reproduce SOC behavior in the form of powerlaw distributions, no particular physics is attached to cellular automaton models, which seem to have some universal characteristics. On the next level, modelers attempted to incorporate physics-based models, such as linking discretized MHD equations to the next-neighbor interactions of cellular automaton models (Vassiliadis et al. 1998; Isliker et al. 2001; Galtier and Pouquet 1998) or MHD simulations (Galsgaard 1996). Along the same line of thought, MHD turbulence and associated heating was linked to SOC models (Walsh et al. 1997; Dmitruk and Gomez 1997; Dmitruk et al. 1998; Boffetta et al. 1999; Liu et al. 2002; Krasnoselskikh et al. 2002). Other physics-based SOC lattice models include particle acceleration in random DC electric fields (Anastasiadis et al. 1997), the scaling of energy release with the magnetic field, i.e., $E \propto B^2$ (Vlahos 2002; Vlahos et al. 2002; Vlahos and Georgoulis 2004), magnetic helicity dissipation (Chou 1999, 2001), chromospheric evaporation (Mitra-Kraev and Benz 2001), annihilation of magnetic elements (Podlazov and Osokin 2002), or cascades of magnetic reconnections (Hughes et al. 2003).

Most existing SOC models are based on lattice grids, where avalanches are created by next-neighbor interactions triggered above a critical threshold, which can be formulated by mathematical rules and rendered by numerical simulations. In contrast to these lattice-type models, the probability distributions of SOC models can also be calculated by analytical means, using parameterized distributions and differential equations to describe the dynamics of a physical process in a SOC avalanche event. Such analytical SOC models have been developed in terms of logistic avalanches (Aschwanden et al. 1998b), master equations (Wheatland and Glukhov 1998; Litvinenko and Wheatland 2001; Wheatland 2009), magnetic separator reconnection models (Litvinenko 1996, 1998a,b; Longcope and Noonan 2000; Litvinenko and Wheatland 2001; Wheatland 2002; Craig and Wheatland 2002; Wheatland and Craig 2003), or torsional Alfvén waves (Wheatland and Uchida 1999). We will discuss physics-based SOC models of solar flares in more detail in Chapter 9.

1.9 SOC in Stellar Physics

Stellar coronae must be governed by similar physics as we observe in the solar corona, especially for stars with similar strong magnetic fields, and similar temperatures and rotation rates. Flares are ubiquitous among coronal stars and have been observed from almost the entire main sequence and giants (e.g., see review by Güdel (2004)). Very energetic flares are frequently seen in ultraviolet, soft X-ray, and radio wavelengths from red dwarf stars of the dMe class, such as from AD Leo, AT Mic, AU Mic, EV Lac, UV Cet, or YZ CMi. Stellar observers started to gather statistics of flare events detected during an observational run and found similar powerlaw distributions as for solar flares, and thus interpreted them in terms of SOC behavior. Powerlaw distributions have been found for stellar flares with the following (noncumulative) slopes: $\alpha = 2.25 \pm 0.1$ for YZ CMi (Robinson et al. 1999), $\alpha \approx 1.6$–2.4 for 12 type F to M stars (Audard et al. 2000), $\alpha \approx 2.0$–2.7 for FK Aqr, V1054 Oph, and AD Leo (Kashyap et al. 2002), or $\alpha \approx 2.3 \pm 0.1$ for AD Leo (Arzner and Güdel 2004).

A synopsis of the dataset of 12 flare stars observed by Audard et al. (2000) is shown in Fig. 1.15 and is compared with the statistics of some 19,000 solar flares observed with RHESSI, all normalized to the same energy definition (in terms of total radiated energy). We see that the largest stellar flares observed from each star have all a larger total energy than the largest solar flares, in excess of up to 3 orders of magnitude. While solar flares have been observed up to maximum energies of $E \lesssim 10^{32}$ ergs, stellar flares range up to $E \lesssim 10^{35}$ ergs (e.g., for HD 2756, a class F2 V type star). This does not mean that stellar flares of the size of solar flares do not exist, but their detection is mainly limited by the instrumental sensitivity of the detectors. The distances to the stars indicated in Fig. 1.15 run from 2.4 pc (CN Leo) to 45 pc (HD 2726). Since 1 pc is about a 200,000 times larger distance than 1 AU, this means that the flux of the largest solar flares would be a factor of 11–14 orders of magnitude fainter at those stellar distances. For the most distant star of this sample, a 10^{12} times larger sensitivity is therefore required to detected the weakest flare on this star shown in Fig. 1.15, which is still a factor of 100 brighter than the largest solar flare.

Another oddity of stellar flare frequency distributions is their systematically steeper slope ($\alpha \gtrsim 2$) than their solar counterparts ($\alpha \lesssim 2$). However, we have to be aware that most of the samples of stellar flares are obtained during a very restricted observing time interval in the order of hours, which typically includes only $n \approx 5$–15 events per star (Audard et al. 2000). The small-number statistics suggests that only the upper cutoff part of a frequency distribution is sampled, which often shows an exponential drop-off, while the powerlaw part is usually manifest when the distribution extends over a larger inertial range (i.e., the powerlaw part of the distribution), say over at least two orders of magnitude (Aschwanden 2007). The distribution of solar flares (from RHESSI) that extends over almost 5 orders of magnitude shown in Fig. 1.15 demonstrates this point: a powerlaw-like part is seen in the lower 4 orders of magnitude in energy, while an exponential drop-off is apparent at the high-energy end, probably limited by the maximum active region size (Kucera et al. 1997). Thus, the systematic difference in observed powerlaw slopes may be mostly an observational limitation, rather than a fundamental difference in the physics of flares on the Sun and stars, and thus may not provide a valid argument for the dominance of coronal heating by nanoflares (which requires a powerlaw slope of steeper than 2; Hudson (1991)).

Theoretical interpretations of stellar flare observations in terms of SOC behavior range from superposition of stochastic flaring and heating events (Arzner and Güdel 2004), to self-regulation of the coronal density driven by chromospheric evaporation and radiative cooling (Uzdensky 2007), or self-driving embedded Sweet–Parker reconnection (Cassak et al. 2008).

Cataclysmic variable stars (CV) are stars that exhibit irregular brightness variations, with quiescent time intervals in between. Over 1,600 CV systems are known today. The interpretation for these episodic bursts is thought to be a mass transfer between close binary stars (Fig. 1.16), consisting of a white dwarf primary star and a secondary star with an orbital period in the range of $P \approx 1$–10 hours. Gravitational disturbances of the primary dwarf star on the secondary star (the donor star) can trigger infall of matter that is accreted in the primary star, leading to an accretion disk around the dwarf star. Occasional instabilities in the accretion disks lead to X-ray bursts whenever an avalanche of unstable massis

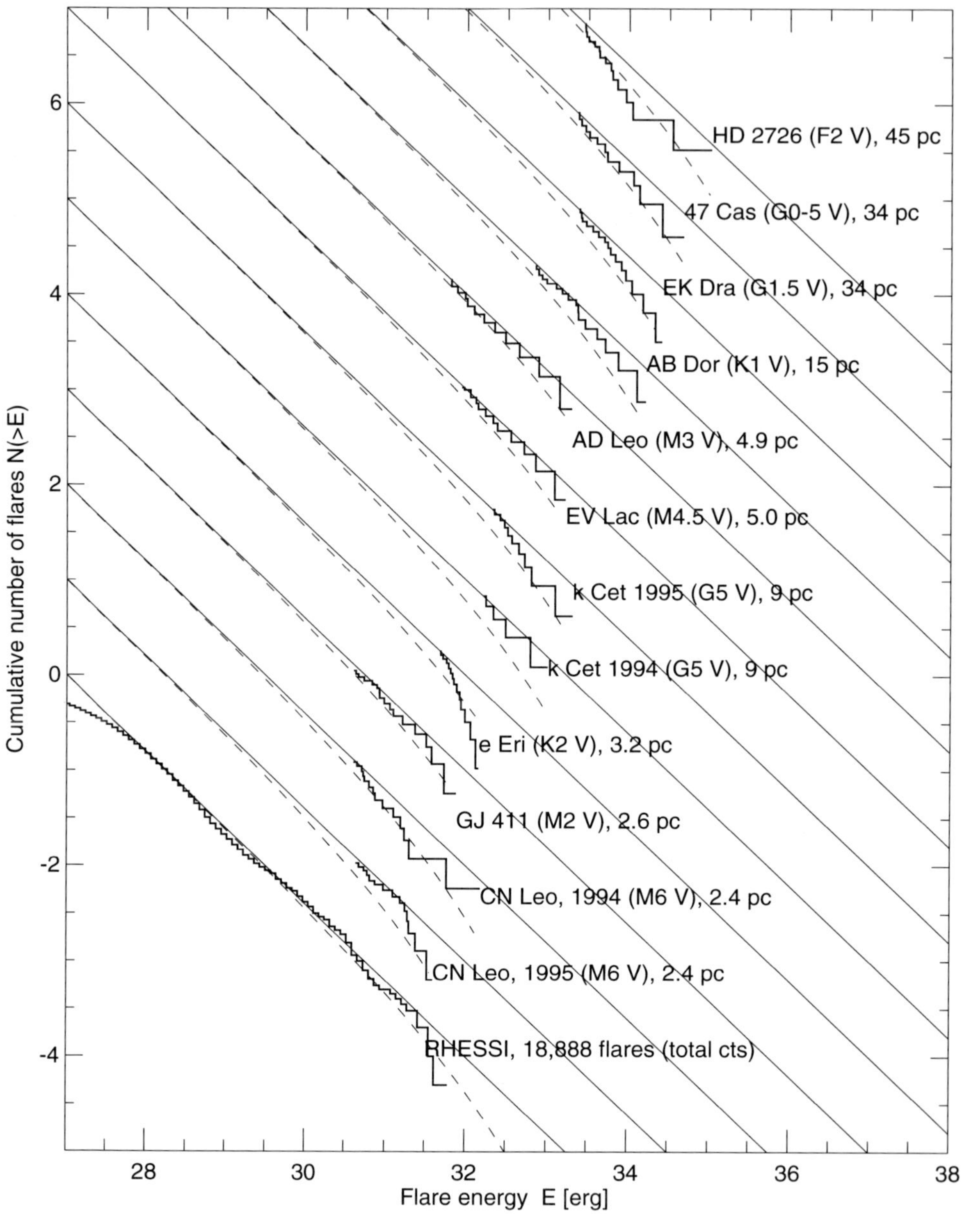

Fig. 1.15 Cumulative occurrence frequency distributions $N(>E)$ of stellar flares (vertically shifted for each star, in order of stellar distance) are shown as a function of the flare energy E, observed from 12 different stars by Audard et al. (2000). For comparison we show also the statistics of total hard X-ray emission of 18,888 solar flares observed with RHESSI, which covers an energy range of $E \approx 10^{27.0}$–$10^{31.7}$ ergs, while stellar flares have an energy range of $E \approx 10^{30.7}$–$10^{35.0}$ ergs. Note that most distributions show a (cumulative) powerlaw part with a slope of $\alpha \approx 0.8$ (diagonal lines) at the low end and an exponential high-energy drop-off. The distributions are incrementally shifted for clarity.

Fig. 1.16 Artistic rendering of the cataclysmic variable star RS Ophiuchi, which exhibits a nova outburst about every 20 years. This binary system contains a white dwarf and a red giant with mass transfer (credit: PPARC, David A. Hardy).

hurled inward. Even larger instabilities can produce nova outbursts, whenever the density and temperature raises above a critical threshold to ignite nuclear fusion reaction.

Optical emission from CVs or accretion disks exhibits short rapid random variability, called "flickering". The power spectral density $S(f)$ as a function of the frequency f was found to have a powerlaw behavior, i.e., $S(f) \propto f^{-1 \ldots -2}$ (Bruch 1992, 1995; Kato et al. 2002), and thus was interpreted in terms of SOC systems (Mineshige et al. 1994a; Dendy et al. 1998; Wiita and Xiong 1998; Mineshige 1999). The waiting-time distribution of rapid fluctuations in X-rays from Cygnus X-1 was found to be exponential (Negoro et al. 1995), implying a random process that can be modeled with a shot-noise model (Negoro et al. 1995; Focke 1998). Massive CVs with accretion disks can form black holes, where the gravitational field is so powerful that it even traps light and precludes escape. The evolution of a black hole thus involves also accretion disks, for which SOC behavior is postulated, similarly to other accretion disk systems (Sivron 1998; Mineshige and Negoro 1999; Xiong et al. 2000). Modeling of the SOC behavior in terms of cellular automaton models was conducted and could reproduce the power spectrum characteristics of the flicker noise of cataclysmic variables (Yonehara et al. 1997; Takeuchi and Mineshige 1997; Pavlidou et al. 2001) or black holes (Takeuchi et al. 1995).

Pulsars are highly magnetized, rotating neutron stars, exhibiting short periods in the range of 1.4 ms to 8.5 s. The most famous one is the Crab pulsar with a period of 33 ms. Although pulsars rotate with such a high precision that they are used as timekeepers, there are some glitches observed occasionally, which are related to the superfluidity and superconductivity inside the star that allows the neutrons to flow without friction (Pizzochero et al. 1997). The probability of pulsar glitches as function of the glitch energy was found to be a powerlaw, i.e., $P(E) \propto E^{-3.5}$ (Argyle and Gower 1972; Lundgren et al. 1995), $P(E) \propto E^{-1.14}$ (Morley and Garcia-Pelayo 1993), $P(E) \propto E^{-2.8}$ (Cognard et al. 1996), which has been related to a SOC system (Young and Kenny 1996). Modeling of pulsar glitches has been conducted in terms of electric fields, for which a powerlaw-like probability distribution of $P(E) \propto E^{-4.6...-9}$ was found (Cairns 2004; Cairns et al. 2004). A cellular automaton model of pulsar glitches was constructed based on the superfluid vortex unpinning paradigm (Warzawski and Melatos 2008; Melatos et al. 2008).

A related class is the soft gamma repeater (SGR), an object that shows a repetitive emission of low-energy gamma-ray bursts. SGRs have been interpreted in terms of starquakes, which are the result of a fracture of the crust of a magnetically powered neutron star or "magnetar" (Duncan and Thompson 1992; Thompson and Duncan 1996). The fluence distribution of SGR bursts was found to have a powerlaw with slopes of $\alpha \approx 1.4$–1.7 (Gogus et al. 1999, 2000), and thus has been associated with SOC behavior like earthquakes, and hence, called starquakes. A compilation of references for SOC phenomena in stellar physics is given in Table 1.7.

1.10 SOC in Galaxies and Cosmology

On galactic scales, X-ray variability has been observed in so-called *active galactic nuclei (AGN)*, which often are radio galaxies or blazars. Physical interpretations attributed to such X-ray variability are X-ray jets, disk-corona models where X-ray emission originates from comptonization of soft UV thermal photons, or black hole models. An X-ray light curve analysis of a radio galaxy was found to be consistent with shot-noise statistics, which lends to the same SOC interpretation as for the black hole candidate Cygnus X-1 (Leighly and O'Brien 1997; Ciprini et al. 2003).

Cosmic structures at large scales, such as galactic spirals and galaxy clusters, display complex and fractal geometries (although the universe turns over to being remarkably homogeneous and isotropic at cosmological scales of the microwave background). Since fractals are scale-free, as self-organized criticality leads to scale-free spatial and temporal structures, SOC behavior could also play a role in the creation of large-scale structures in the universe, not requiring any fine tuning to achieve a critical state. Although a full SOC model has not been developed yet for structures in the early universe, self-organization and fractal scaling (which is SOC without criticality) has already been applied to some large-scale structures, such as to the formation of the interstellar medium (Tainaka et al. 1993), the galactic spiral structure (Nozakura and Ikeuchi 1988), the stellar dynamics in elliptical galaxy formation (Kalapotharakos et al. 2004), or the initial mass function of starbursts (Melnick and Selman 2000), or gravitational structure formation in general on many scales (Da Rocha and Nottale 2003). In numerical N-body simulations of elliptical galaxy formation it is found that the initial chaotic orbits become ordered as the short axis tube type during their evolution, which is a self-organization process (Kalapotharakos et al. 2004),

What about SOC and cosmology? Although the following excerpt is not mainstream cosmology, it contains intriguing thoughts about the possible application of SOC to cos-

Table 1.7 SOC behavior in stellar physics: The references include observational and theoretical modeling papers.

Phenomenon	References
stellar flares	Robinson et al. (1999) Audard et al. (2000) Kashyap et al. (2002) Arzner and Güdel (2004)
cataclysmic variable (CV) stars	Bruch (1992, 1995) Yonehara et al. (1997) Takeuchi and Mineshige (1997) Kato et al. (2002)
accretion disks	Dendy et al. (1998) Wiita and Xiong (1998) Mineshige (1999) Xiong et al. (2000) Pavlidou et al. (2001)
black holes	Mineshige et al. (1994a) Takeuchi et al. (1995) Sivron (1998) Mineshige and Negoro (1999)
pulsar glitches	Argyle and Gower (1972) Morley and Garcia-Pelayo (1993) Cognard et al. (1996) Lundgren et al. (1995) Young and Kenny (1996) Cairns (2004) Cairns et al. (2004) Warzawski and Melatos (2008) Melatos et al. (2008)
soft gamma repeaters (SGR)	Duncan and Thompson (1992) Thompson and Duncan (1996) Gogus et al. (1999, 2000)

mology. We quote from Moffat (1997): *A major problem in modern cosmology is this: How could the universe evolve during more than 10 Gyr and become so close to spatial flatness and avoid the horizon problem? Why is the universe so homogeneous and isotropic? How could such a critical state of the universe come about without a severe fine tuning of parameters. The usual explanation for these questions is based on the idea of inflation. However, inflation is a type of phenomenon that in statistical mechanics corresponds to the existence of an attractor that requires fine tuning of the parameters.* Moffat (1997) then proposes that the universe evolves as a SOC system (in the sense of the Bak-Tang-Wiesenfeld model) with the Hubble expansion undergoing "punctuated equilibria" with energy being dissipated at all scales. Note that punctuated equilibria involve a number of episodic bursts, separated by time intervals of near steady-state, so the implication is that there occurred many inflation phases in the past, not just one. A consequence is

Table 1.8 SOC concepts in galaxies and cosmology.

Phenomenon	References
active galactic nuclei, blazars	Leighly and O'Brien (1997)
	Ciprini et al. (2003)
interstellar medium formation	Tainaka et al. (1993)
galactic spiral structure	Nozakura and Ikeuchi (1988)
elliptical galaxy formation	Kalapotharakos et al. (2004)
IMF of starbursts	Melnick and Selman (2000)
gravitational structure formation	Da Rocha and Nottale (2003)
cosmology, big bang, inflation	Moffat (1997)
quantum gravity	Ansari and Smolin (2008)

that the SOC model predicts a critical value for the density profile, $\Omega = 1$, which is the critical value between an open ($\Omega_c \le 1$) or closed universe, independent of the initial conditions and without fine tuning of the parameters. *The metric fluctuations display 1/f flicker noise, correlations of fluctuations occur at all length scales, and the universe evolves at the "edge of chaos". There is only one possible stable choice (i.e., stable under local perturbations) for the present expanding universe whatever its initial conditions. According to our assumptions the space-time geometry fluctuates randomly at some length scale. If we assume that the metric fluctuations are very intense at the beginning of the universe, and that they smear out the light cones locally, then for a given short duration of time Δt after the big-bang there will be communication of information "instantaneously" throughout the universe. This will resolve the "horizon" problem and explain the high degree of isotropy and homogeneity of the present universe* (Moffat 1997). A related SOC concept has also been applied to quantum gravity (Ansari and Smolin 2008). A few references to SOC concepts applied to galaxies and cosmology are given in Table 1.8.

1.11 Summary

Self-Organized Criticality (SOC) is a theoretical concept that describes the statistics of nonlinear processes. It is a fundamental principle that is common to many nonlinear dissipative systems in the universe. Due to its universality and ubiquity, SOC is a law of nature, for which we derive the theoretical framework and specific physical models in this book. The SOC concept was introduced by Bak, Tang, and Wiesenfeld in 1987 and has been applied to laboratory experiments of sandpiles, to human activities such as population growth, language, economy, traffic jams, or wars, to biophysics, geophysics (earthquakes, landslides, forest fires), magnetospheric physics, solar physics (flares), stellar physics (flares, cataclysmic variables, accretion disks, black holes, pulsar glitches, gamma ray bursts), and to galactic physics and cosmology. From an observational point of view, the hallmark of SOC behavior is the powerlaw shape of occurrence frequency distributions of spatial, temporal, and energy scales, implying scale-free nonlinear processes. Powerlaws are a necessary but not sufficient condition for SOC behavior, because intermittent turbulence also produces powerlaws. While we surveyed the manifestation of SOC behavior from "microscopic" scales (sandpiles) across the universe out to cosmological scales in this introductory chapter, as documented in the literature of some 1,000 research publications, we will provide a more detailed introduction into the theoretical concepts, modeling, and interpretation of SOC behavior in the following chapters.

1.12 Problems

Problem 1.1: Count the words of an English text, either visually or with some computer program (e.g., http://www.hermetic.ch/index.php), and plot the frequency of words in a rank-ordered Zipf plot (similar to the graph in Fig. 1.4 right panel for city sizes). Do you obtain the same order of words as it is found for the English language in general? The first 15 words, in order of frequency, are: *"the"*, *"of"*, *"and"*, *"in"*, *"to"*, *"a"*, *"is"*, *"that"*, *"it"*, *"as"*, *"this"*, *"by"*, *"for"*, *"be"*, *"not"*.

Problem 1.2: How are rank-ordered plots, cumulative frequency distributions, and noncumulative (differential) frequency distributions related to each other? Derive a mathematical proof that a powerlaw in a rank-ordered plot corresponds to a powerlaw of a cumulative frequency distribution. What is the difference in the powerlaw slope between a cumulative and noncumulative frequency distribution? (Check your answers in Section 7.1.)

Problem 1.3: Is there a concentration or preferred range of powerlaw slopes α of (noncumulative) occurrence frequency distributions? (Hint: Plot a histogram of the values quoted in Chapter 1.)

2. Numerical SOC Models

Computers are useless. They can only give you answers.

Pablo Picasso

I do not fear computers. I fear the lack of them.

Isaac Asimov

Self-organized criticality (SOC) is the natural state into which a nonlinear dissipative system evolves into, without fine tuning of the initial conditions. Generally, some external forcing mechanism drives a system into criticality, where energy is dissipated sporadically in avalanche-like events. Such nonlinear dissipative systems are also called complex systems, which are composed of many interconnected parts that interact in a nonlinear way. Complex systems, such as the particles in a fluid that shows Brownian motion, have too many components to be described in terms of an n-body system with an equally large number of differential equations. So it is understandable that we first start with empirical computer simulations, which we call *numerical models*, to study the SOC state. In this chapter we exclusively describe such numerical models, in contrast to *analytical models* that will be developed in Chapter 3, based on approximations of the average behavior of some macroscopic physical parameters. The numerical models that have been used to study SOC behavior are mostly *cellular automaton* models, which essentially consist of an equi-spaced lattice grid and a set of mathematical rules to simulate a time-progressive interaction between next neighbors in the lattice. Hence, they are also called lattice-type simulations. There is no universal cellular automaton algorithm that works for every system in nature, but there is a rather large variety, each one adapted to capture the nonlinear dynamics of a particular phenomenon in nature. Thus we organize this chapter in the same order as Chapter 1, starting from small systems in laboratories out to the largest astrophysical systems in the universe. In addition to cellular automaton models, some other alternative numerical models have been used to study SOC behavior, such as mechanical systems (e.g., coupled pendulums or the block-slider model), n-body simulations (e.g., stellar dynamics), or percolation systems (e.g., solar active regions).

This chapter describes examples of numerical SOC simulations used in astrophysics (Sections 2.5–2.7), which essentially all were first inspired by non-astrophysical applications (Sections 2.1–2.4). Other generic reviews on numerical SOC models can also be found in Bak and Chen (1991), Bak and Paczuski (1995), Bak (1996), Jensen (1998), and Turcotte (1999).

2.1 SOC Simulations of Laboratory Experiments

As we have seen in Section 1.2, laboratory experiments with SOC behavior are difficult to carry out (especially with real sand or rice piles), so it is much easier to simulate these experiments with the computer. A few basic examples of computer models of laboratory experiments are coupled pendulums (Section 2.1.1), sandpiles (Section 2.1.2-3), lattice gases (Section 2.1.4), or coupled slider-block springs (Section 2.4.1), all representing idealized dissipative n-body systems that have a nonlinear coupling between next neighbor elements, from which the SOC behavior can be studied.

2.1.1 Coupled Pendulums

One of the prototypes of mechanical SOC models is a system of coupled oscillators, which mimic a complex system with many degrees of freedom. A relatively simple computer model was conceived by Tang et al. (1987), with a horizontal two-dimensional grid of pendulums that can rotate in a vertical plane and are mutually connected with the next neighbors with springs with force constant k (Fig. 2.1). The 2-D array of n pendulum balls of mass m is driven by a time-periodic square-wave force $F(t)$ and has the equation of motion

$$m\ddot{y} = -\gamma\dot{y}_j + k(y_{j+1} - 2y_j + y_{j-1}) - a\sin(2\pi y_j) + F, \qquad j = 1, 2, ..., n\ , \tag{2.1.1}$$

where y_j is the position of the j-th pendulum ball, γ is the damping constant, and a is the amplitude of the potential. Tang et al. (1987) conducted computer simulations of this dissipative system and found novel patterns in the formation of metastable states. In further computer experiments of the group by Tang, Wiesenfeld, and Bak, energy was pumped into the system by random selection of pendulums that would perform a full initial rotation, which occasionally triggered adjacent pendulums to execute a full rotation (Fig. 2.1), similar to a domino effect. To make the system dissipative, a significant amount of friction was built in (with the damping constant γ). A typical experiment would involve 50×50 gridpoints (or $n = 2{,}500$ pendulums). Every random energy input triggered a different chain reaction of rotating pendulums, whose magnitude or size was counted by the number of pendulums that executed a full rotation during one chain reaction (or avalanche) event. Finally, the group of Tang, Wiesenfeld and Bak discovered that the size distribution $N(S)$ of avalanches as a function of the size S exhibited a powerlaw function in a log-log plot,

$$N(S) \propto S^{-\alpha}\ , \tag{2.1.2}$$

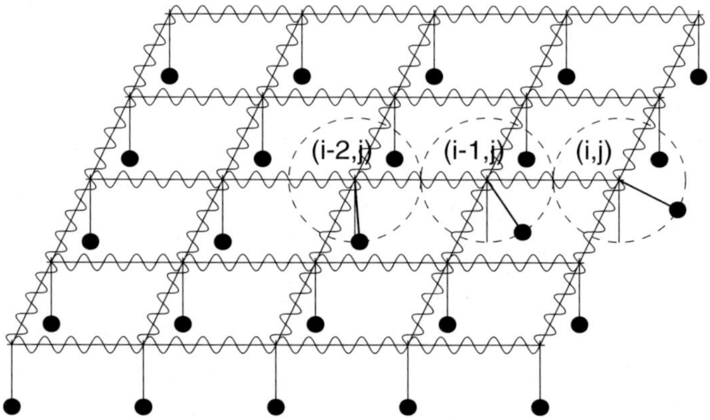

Fig. 2.1 System of coupled pendulums in a two-dimensional grid system as simulated by Tang et al. (1987). Each gridpoint has a pendulum that is coupled to its next neighbors with a spring. At the time shown, the pendulum at position (i,j) is excited and pulls the adjacent pendulums at positions $(i-1,j)$ and $(i-2,j)$ into the same direction through the coupling of the springs.

with a powerlaw slope of $\alpha \approx 1.1$ (see Fig. 11 in Bak 1996, p.47). At the upper end of the powerlaw distribution there is a cutoff given by the maximum number n of the pendulums contained in the system. This is one of the prototypes of nonlinear systems where the state of *self-organized criticality* (SOC) was discovered. The term "self-organizing" refers to the invariant endstate (with the same powerlaw slope) into which the system evolves, without fine tuning of the initial (random) energy input.

2.1.2 The Bak-Tang-Wiesenfeld 1-D Sandpile Model

The most famous paradigm of SOC models, introduced by Per Bak, is the sandpile. Sand grains are randomly dropped on a pile, which trigger avalanches of different sizes and durations. The system is dissipative because of the friction, which provides also the required instability threshold with many metastable states. The nonlinear dynamics comes in from the complicated collisional interactions at various scattering angles, speeds, kinetic, and gravitational energies of each sand grain involved in an avalanche. A complete analytical description of the kinematics of an n-body system that is involved in a sandpile avalanche would be prohibitively complex to solve, even when we include only the simplest mechanical forces occurring in collisions in a gravitational potential. The computer models of sandpiles are much simpler designed, reducing a sandpile just to a regular lattice grid with simple mathematical rules that mimic the interactions between neighbored pixels in a simplified way. In the simplest case, we envision just a one-dimensional sandpile, like the rice pile experiment between two parallel glass plates shown in Fig. 1.3. The 1-D sandpile model is also described in Bak et al. (1988), Bak (1996), and Jensen (1998).

The 1-D computer sandpile model resembles a histogram that has a number of h_i sand grains in each bin i (Fig. 2.2). The slope is defined by the difference $z_i = h_i - h_{i+1}$. The dynamics of sandpile avalanches is simply described by two operators: (a) adding a grain to

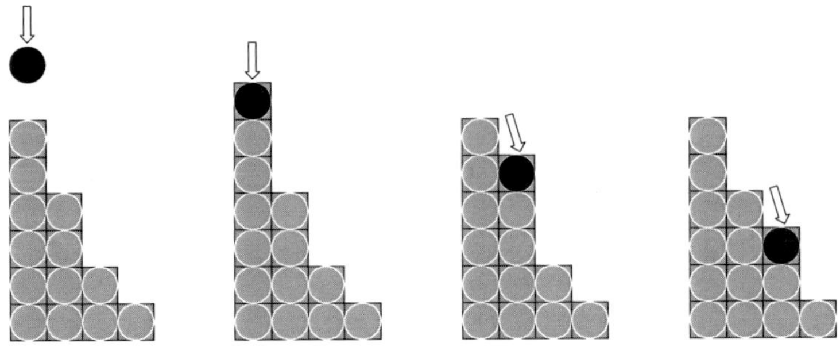

Fig. 2.2 Example of a 1-D Bak–Tang–Wiesenfeld sandpile. The dropped sand grain topples to the next lower level if the local slope $z_i = h_{i+1} - h_i = 2$ is steeper than a critical value of $z_c = 2$, i.e., when $z_i > z_c$. After the third time step the sandpile is stable again.

the pile, and (b) relaxing the slope of the pile wherever the local gradient exceeds a critical threshold for stability, i.e., $z_i > z_c$. In the simple example shown in Fig. 2.2 we have an initial state of $h_i, h_{i+1}, ..., h_{i+3} = [6,4,2,1]$. If we define a stable slope with a critical limit of $z_c = 2$, we see that the sandpile is stable, because the slope nowhere exceeds the critical limit. Now, we drop randomly a grain on the first bin, so that $h_i = 6+1 = 7$, which is unstable to the next bin, since $z_i = h_i - h_{i+1} = 7-4 = 3 > z_c$. Consequently, the unstable sand grain will topple into the next bin, i.e., $h_i = 7-1 = 6$ and $h_{i+1} = 4+1 = 5$, restoring the stable slope above, $z_i = h_{i+1} - h_i = 6-5 = 1$. However, this is unstable towards the next lower bin, since $z_{i+1} = h_{i+1} - h_{i+2} = 5-2 = 3 > z_c$. Consequently, the sand grain will topple into the next bin, i.e., $h_{i+2} = 5-1 = 4$ and $h_{i+3} = 2+1 = 3$, which is now stable everywhere with $h_i, h_{i+1}, ..., h_{i+3} = [6,4,3,1]$, and thus the avalanche stops. Thus, this little avalanche took two topplings. The number of sandgrains is conserved if we combine the previous sandpile plus the input. We note that the additional sandgrain falls on bin i and finally landed on bin $i+2$, while all the other bins stay the same. A system is said to be *"conservative"* when the number of sand grains is invariant during a redistribution rule. Even when sand grains are allowed to fall off the edge of a sandpile, the system can be conservative in the time average, if it balances the time-averaged input.

The dynamical behavior of this 1-D sandpile is easy to predict (Jensen 1998): Sand grains will pile up wherever the slope is less than critical, until the slope becomes critical everywhere. This critical state is also called *global attractor* (in nonlinear system dynamics) and is reached no matter where we start from initially. If we allow the system to exit the toppling sandgrains from the bin with the lowest number, the number of sand grains is conserved in the time average. Avalanches of different sizes are created by the random irregularities of local slopes that are slightly less (or equal) to the critical value. Therefore, the output of the system is highly fluctuating, even for a regular input rate, and thus the energy of the system is conserved in the time average only. The most remarkable property is the robustness of the endstate, which constitutes the concept of *self-organized criticality*.

2.1.3 The Bak–Tang–Wiesenfeld 2-D Sandpile Model

Since the world is not one-dimensional, the next logical step was to study SOC behavior in two dimensions, so our 2-D sandpile model is played now on a checker board, where each field has a 2-D position (i,j) in a cartesian coordinate system. The next-neighbor interactions are now extended to the 4 next neighbors at positions $(i-1,j)$, $(i+1,j)$, $(i,j-1)$, and $(i,j+1)$. One of the simplest SOC games that was initially studied had the simple setup of a critical threshold of $z_c = 4$ and the rule of a redistribution to the next 4 neighbors, whenever the local threshold was exceeded. So, there are two rules: (1) input of one grain at a random position, and (2) relaxation of four sand grains to the next 4 neighbors if a threshold of $z_c = 4$ is exceeded:

$$\begin{aligned} &z(i,j) = z(i,j)+1 && \text{initial input} \\ &z(i,j) = z(i,j)-4 && \text{if } z(i,j) \geq 4, \\ &z(i\pm 1, j\pm 1) = z(i\pm 1, j\pm 1)+1 && \end{aligned} \qquad (2.1.3)$$

Such an algorithm is also called *cellular automaton*, because the dynamic evolution of a system is described by mathematical rules that enable "automated" steps, executed by interactions between the next neighbor cells in a (cellular) lattice grid.

Let us illustrate an avalanche in such a 2-D lattice sandpile model with an example given by Bak (1996; p.53), shown here in Fig. 2.3. Starting from an initially stable sandpile at time $t=1$, a sand grain is dropped at position $z(3,3)=4$, which makes it unstable and causes an avalanche. In the first toppling event between time $t=2$ and $t=3$, the unstable pixel changes its state $z(3,3)=4 \mapsto 0$, while the 4 unstable sand grains get redistributed to the adjacent next neighbors at $z(2,3)=2 \mapsto 3$, $z(3,2)=3 \mapsto 4$, $z(3,4)=2 \mapsto 3$, and $z(4,2)=3 \mapsto 4$. Hence, we have two unstable pixels $z(4,3)=4$ and $z(3,2)=4$ at time $t=3$ and the avalanche continues, until we reach a stable state at time $t=9$ again. The envelope of all unstable sites that ever had an instability with $z(i,j)=4$ entails a total of 8 pixel sites, which represents the area of the avalanche event. The time duration of the avalanche lasted 7 time steps, during which a total of 9 topplings occurred. In this example, a single sand grain is added at the beginning and further input is interrupted until the resulting avalanche comes to a halt, which is also called a *"stop-and-go"* sandpile, in contrast to a *running sandpile*, where the driver continues rregardless of simultaneous avalanching. In the former case, separate timescales control driving and avalanching.

In their seminal papers, Bak, Tang, and Wiesenfeld (1987, 1988) simulated such a cellular automaton algorithm for a 50×50 2-D lattice grid, as well as for a $20\times 20\times 20$ 3-D lattice grid. The 3-D generalization involves 8 nearest neighbors, and hence the rules are:

$$\begin{aligned} &z(i,j,k) = z(i,j,k)+1 && \text{initial input} \\ &z(i,j,k) = z(i,j,k)-8 && \text{if } z(i,j,k) \geq 8, \\ &z(i\pm 1, j\pm 1, k\pm 1) = z(i\pm 1, j\pm 1, k\pm 1)+1 && \end{aligned} \qquad (2.1.4)$$

The avalanche size S was measured by the area of the clusters in the chain reaction of unstable pixels triggered by each single perturbation, and the duration T of an avalanche was measured by the number of toppling time steps that were needed until a stable configuration was reached. Bak et al. (1987, 1988) plotted then the distribution $N(S)$ of avalanche

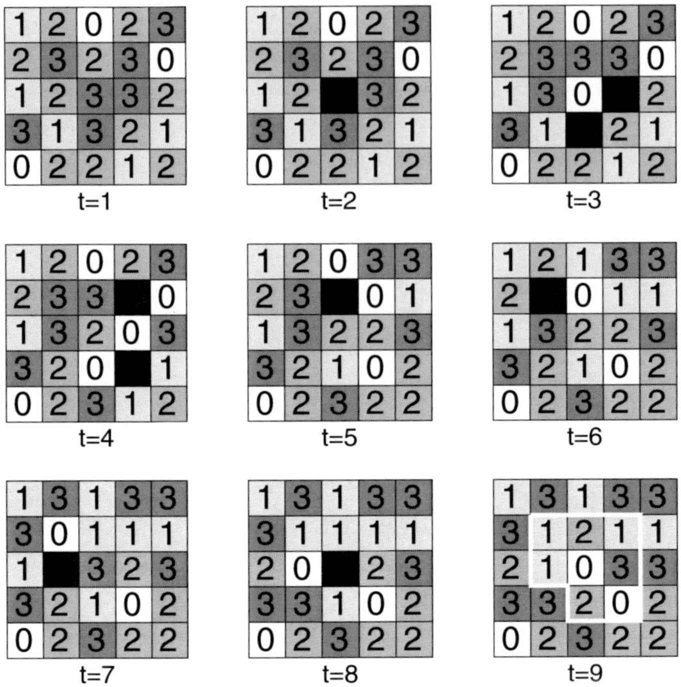

Fig. 2.3 Example of an avalanche in a 2-D Bak–Tang–Wiesenfeld sandpile. The initial state of the sandpile at time $t = 1$ is stable, since none of the states $z(i, j)$ exceeds the critical threshold $z_c = 4$. At time $t = 2$, a sand grain is dropped in the middle of the sandpile, which causes an avalanche of subsequent topplings. At times $t = 3$ and $t = 4$, two topplings occur in the same time step. At time $t = 9$, the sandpile becomes stable again and the total avalanche size is indicated with a white polygon, entailing 8 pixels (adapted from Bak 1996, p.53).

sizes S and durations T on a log-log scale and discovered the famous powerlaw distributions that have since become the hallmark of SOC. The pioneering results are shown in Fig. 2.4, where a powerlaw slope of $\alpha_S = 1.0$ was found for the avalanche sizes in the 2-D lattice over a range of more than two decades, and $\alpha_S = 1.37$ in the 3-D lattice. For the avalanche durations T, a powerlaw was fitted too,

$$N(T) \propto T^{-\alpha_T}, \tag{2.1.5}$$

yielding a slope of $\alpha_T \approx 0.43$ for the 2-D lattice, and $\alpha_T \approx 0.92$ for the 3-D lattice, respectively. All powerlaw distributions show a rollover cutoff at the upper end, which is a finite-size effect, indicating that the largest avalanche events are limited by the system size. These initial simulations were performed for a system with open boundaries $[z(1,y) = z(n+1,y) = z(x,1) = z(x,n-1) = 0]$, for two initial conditions: (a) far from equilibrium, and (b) far from a flat surface. Subsequent simulations were also performed for closed boundaries, but the results were identical for systems with open or closed boundaries, after the process was run for a while.

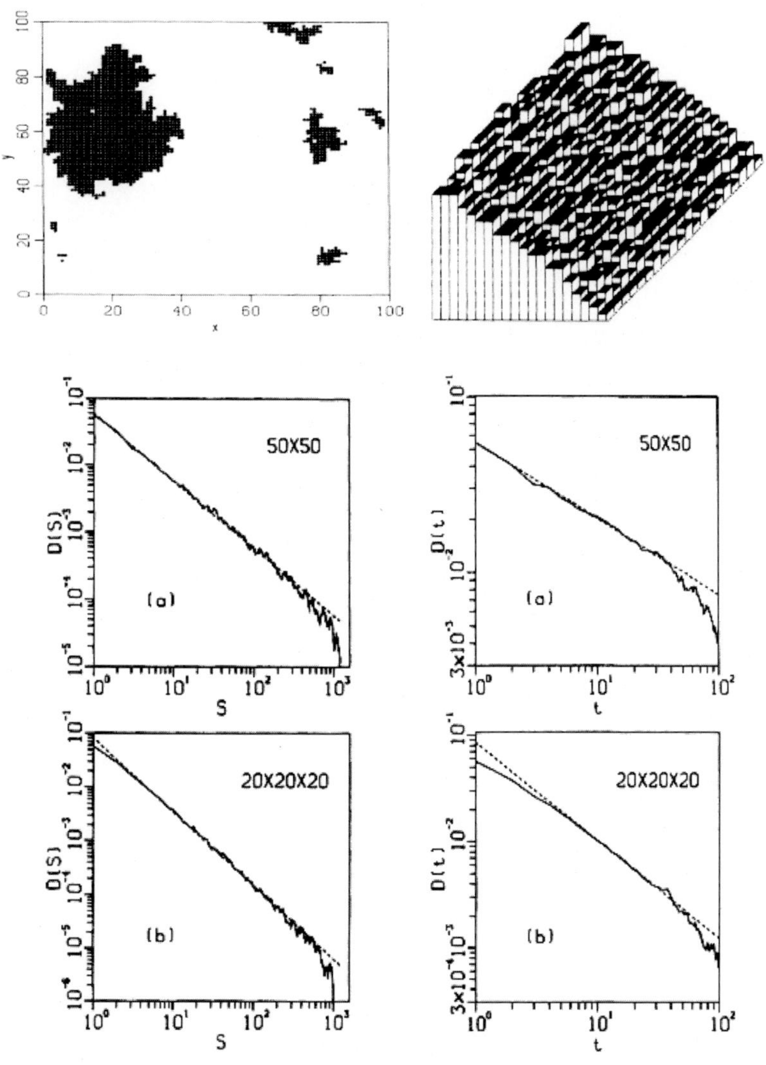

Fig. 2.4 Examples of a fragmented avalanche (top left) occurring in a 2-D (computer) sandpile (top right) and occurrence frequency distribution of avalanche cluster sizes (left panels) and avalanche durations (right panels) of the original BTW sandpile cellular automaton simulation. The simulations have been performed for a 50×50 2-D lattice (middle panels) and for a 20×20×20 3-D lattice grid (bottom panels). The powerlaw slopes are $\alpha_S = 1.0$ and $\alpha_T = 0.42$ for the 2-D grid (middle panels) and $\alpha_S = 1.37$ and $\alpha_T = 0.92$ for the 3-D grid. Reprinted from Bak, Tang, and Wiesenfeld (1987, 1988) with permission; Copyright by American Physical Society.

These initial results raise two interesting questions: (1) what is the physical interpretation of the powerlaw slopes α_S and α_T, and (2) how do the powerlaw slopes depend on the dimensionality (1-D, 2-D, 3-D) of the system? The value of the powerlaw index of time scales near unity, i.e., $\alpha_S \approx 1$, was brought in context of the $1/f$-noise (or white noise), which scales with $P(\nu) \approx \nu^{-1}$. We will discuss the relationship between time scale distributions $N(T)$ and power spectra $P(\nu)$ in Section 4.8.

There is a large number of studies on the original BTW cellular automaton model, which either investigate the properties of the original BTW model or explore variants with modified rules. Some frequently quoted studies deal with: critical exponents and scaling relations (Tang and Bak 1988; Zhang 1989; Manna 1991a), two-state model of SOC (Manna 1991b), mean-field theory (Alstrom 1988; Christensen and Olami 1993; Zapperi et al. 1995; Vespignani and Zapperi 1997, 1998), preferred avalanche directions (Dhar and Ramaswamy 1989), invasion percolation (Roux and Guyon 1989), the *Abelian* property, i.e., the final state is invariant to the time order of the individual events (Dhar and Majumdar 1990; Majumdar and Dhar 1992; Dhar 1999), conservation laws and anisotropy (Grinstein et al. 1990), coexisting periodic attractors (Wiesenfeld et al. 1990), height correlations in sandpiles (Majumdar and Dhar 1991), avalanche dynamics at domain wall boundaries (Carlson et al. 1990), non-conservation in SOC models (Christensen et al. 1992; Socolar et al. 1993), renormalization group methods (Pietronero et al. 1994; Vespignani et al. 1995), the cutoff of the avalanche size distribution (Lise and Jensen 1996), the Landau–Ginzburg theory of SOC (Gil and Sornette 1996), emergent spatial structures (Tadic and Dhar 1997), critical values of driving field and dissipation (Dickman et al. 1998), and the universality of SOC models (Milshtein et al. 1998; Chessa et al. 1999). Reviews on BTW cellular automaton modeling and numerical simulations are given in Bak (1996), Jensen (1998), and Turcotte (1999).

2.1.4 The Lattice-Gas Model

The lattice-gas model is a particular cellular automaton model that has only a dual state per lattice point, either zero or one. One might think of the distribution of table tennis balls in a large egg carton (Fig. 2.5). If the tennis balls are shaken, they will fall back into the next hole, but only one ball can then occupy a hole. The lattice-gas model can be defined in multiple dimensions, but we consider a 2-D model in Fig. 2.5. Each cell contains either one or no particle, with no double occupancy. Neighboring particles repel each other by a unity force. At each timestep, the particles are redistributed according to a displacement vector that adds up the normalized forces,

$$\begin{aligned} x_{i',j'} &\mapsto Integer[x_{i,j} + F_x(i,j)/F] \\ y_{i',j'} &\mapsto Integer[y_{i,j} + F_y(i,j)/F] \end{aligned} \tag{2.1.6}$$

where $F = \sqrt{F_x^2 + F_y^2}$ is the total force, and the function $Integer[...]$ denotes the nearest integer number (i, j), because the positions (x_i, y_j) are defined in a discretized 2-D cartesian (lattice) grid. The rule (2.1.6) can only be applied if the new position $(x_{i',j'}, y_{i',j'})$ is empty. If two particles want to move to the same position, only the particle with the larger force wins and is moved, while the other stays at the old place. If two particles with equal force want to move to the same position, neither is moved. An example of a particle update during one time step is shown in Fig. 2.5. There are setups with closed boundaries, open boundaries, and periodic boundaries, which do not conserve the total number of particles

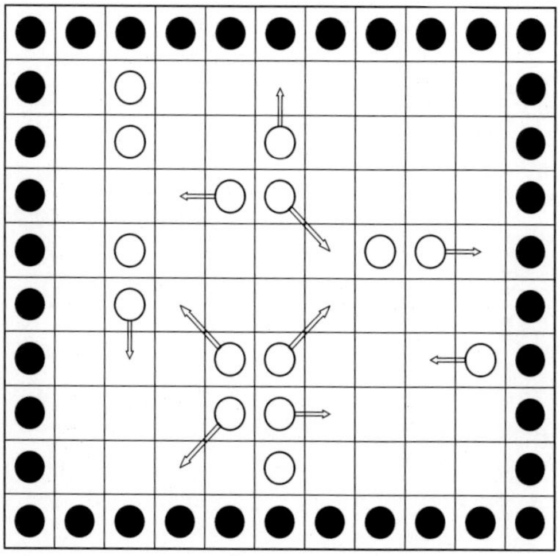

Fig. 2.5 Examples of particle updates in the lattice gas model. The boundaries are represented here with stationary particles (Jensen 1998).

inside the box. The fluctuations $n(t)$ of the total number of particles inside the box is of particular interest in the study of SOC.

Numerical simulations of the lattice-gas model are described in Jensen (1998) and have been studied by Jensen (1990), Fogedby et al. (1991), Andersen et al. (1991), and Fiig and Jensen (1993). The temporal fluctuations of the total number of particles $n(t)$ were found to follow a power spectrum $S(f)$ that is constant below a critical crossover frequency f_c, and a 1/f powerlaw above the crossover frequency,

$$S(f) \propto \begin{cases} \text{constant if } f < f_{cr} \\ 1/f^{\gamma} \quad \text{if } f > f_{cr} \end{cases} , \tag{2.1.7}$$

where the powerlaw slope is $\gamma \approx 1$ and the crossover frequency scales as $f_{cr} \approx 1/L^2$ with the area L^2 of the lattice. Individual particles perform a random walk, as in a diffusion process. The lifetime T, defined by the number of time steps between entering and leaving of the box boundaries for an individual particle, was found to have a powerlaw distribution of

$$N(T) \approx T^{-\alpha_T} , \tag{2.1.8}$$

with $\alpha_T \approx 3/2$ for $T < T_{cr}$, where $T_{cr} \propto L^2$ scales with the area L^2 of the box. The lattice-gas system essentially exhibits SOC behavior, because it organizes itself scale-free with a $1/f$ power spectrum, except for finite-size effects at the upper end of time scales.

The lattice-gas model was found to exhibit the same SOC behavior as the flux noise experiment designed to study the onset of motion of vortices in a superconductor (Yeh and Kao 1984). The scale-invariant SOC behavior in the spatial and temporal scales in the motion of vortices in superconductors was demonstrated by Field et al. (1995).

2.2 SOC Simulations of Human Activities

2.2.1 Conway's Game of Life Model

Cellular automaton models, which are frequently used in numerical SOC models, have originally been devised by the British mathematician John Horton Conway in 1970 (Gardner 1970). One of the first applications was *Conway's Game of Life*, which was inspired by the problem of the mathematician John von Neumann, who tried to create a hypothetical machine (or robot) that could replicate itself, at least in a mathematical world with rectangular grids and mathematical rules.

We describe the original concept of Conway's Game of Life because it is the most basic prototype of a cellular automaton. We have an infinite two-dimensional orthogonal grid of square cells at our disposal, where each cell has the dual state of *live* or *dead* (like the binary memory plate of a computer). The game can be started by creating a particular initial state (of live or dead assignments to each cell), which is the only interaction of the player with the game, while the subsequent evolution is fully determined by mathematical rules. So, the player can just watch the evolution without further interaction or feedback. Each cell interacts with its 8 neighbors, sitting adjacent in horizontal, vertical, or diagonal direction. At each time step (or generation), the following 4 mathematical rules are applied (from *Wikipedia*):

1. Any live cell with fewer than 2 live neighbors dies (underpopulation).
2. Any live cell with more than 3 live neighbors dies (overcrowding).
3. Any live cell with 2 or 3 live neighbors lives on to the next generation.
4. Any dead cell with exactly 3 live neighbors becomes a live cell.

The game can evolve into myriads of unexpected interesting patterns (called "gliders", "blinkers", "F-pentomino" "boat", "pulsar", "diehard", "acorn", and so forth; see examples in Fig. 2.6), similar to a chess game, except that the game evolves automatically without any players interaction. The original motivation of Conway's Game of Life was the study of initial configurations that lead to population growth, extinction, oscillatory, or stable situations. Numerical simulations revealed that specific patterns emerged recurrently, as

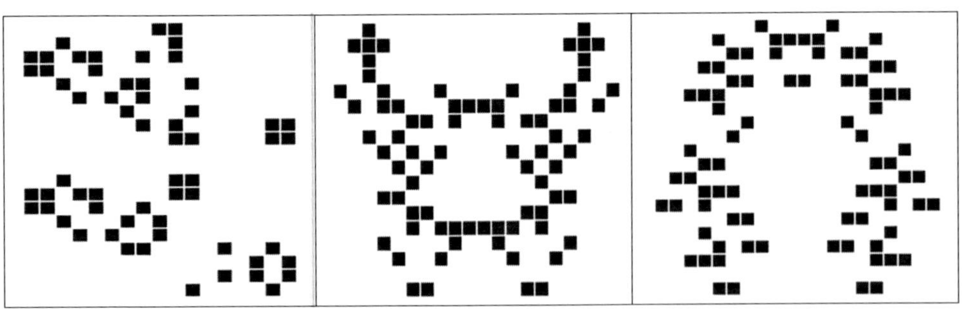

Fig. 2.6 Examples of "glider" structures occurring in Conway's Game of Life. Note the fractal structure of these replicating structures (credit: Eppstein).

well as an infinite variety and richness of self-organizing structures that were thought to mimic aspects of the emergence of complexity in nature.

Bak, Chen, and Creutz (1989) wondered whether the Game of Life was in a critical state with SOC behavior. They simulated a game until it came to a static situation (ending up with static structures and simple oscillatory "blinkers"), and defined this way a time duration T from the number of time steps it took to become static, and a size S from the number of births and deaths occurring during one game, which represents an avalanche event in the SOC terminology. The game was repeated by just changing one random cell, like adding a sandgrain to a stable sandpile, and they sampled the next avalanche, continuing the game ad infinitum. The statistical outcome was indeed a powerlaw distribution with a slope of $\alpha_S = 1.3$ for the size distribution $N(s/L)$, i.e., normalized by the system size L (Bak 1996). A computer simulation of 40,000 games (avalanche events) on a 100×100 lattice grid exhibited powerlaw size distributions with slopes of $\alpha_S = 1.4$ for spatial scales and $\alpha_T = 1.6$ for temporal scales (Bak, Chen, and Creutz 1989), but a rollover at the upper cutoff was found that was interpreted as finite-size effect. This finding of finite-size effects at the upper cutoff was later confirmed with 1024×1024 lattice grid simulations (Alstrom and Leao 1994).

2.2.2 Traffic Jam Simulations

We alluded to the SOC behavior of traffic jams in Section 1.3. Although traffic seems to be a highly complex system with many different vehicles and driver individuals, the basic dynamics of traffic jams can be simulated and understood in a relatively simple way. Nagel and Paczuski (1995) used the following model. The traffic system is reduced to a single-lane freeway, represented by a one-dimensional array of length L. The spatial position of the highway is discretized to x_i, $i = 1, ..., n$, with $x_n = L$. In each position x_i there are $v_{max} + 2$ states, either it is empty (with no car) or occupied (with a car) with an (integer) velocity $v_i = 0, ..., v_{max}$, where they choose $v_{max} = 5$ in their simulations. A basic (cellular automaton) rule is that car movements occur "crash free". To reinforce this rule, the velocity of each car has to be adjusted to the distance of the car in the front, which is defined by the number of pixels n_{gap} between two car-occupied pixels. The detailed mathematical rules for crash-free traffic in this system are:

1. A vehicle is stationary when it travels at maximum velocity v_{max} and has free headway, $n_{gap} \geq v_{max}$, just maintaining its velocity, and is updated by $x(t_{i+1}) = x(t_i) + v_{max}\Delta t$.
2. If a vehicle is not stationary, it is jammed, in which case two possible rules apply:
   (a) *Acceleration of free vehicles:* A vehicle with spacing $n_{gap} = v$ maintains its velocity, while a vehicle with spacing $n_{gap} \geq v + 1$ accelerates to $v(t_{i+1}) = v(t_i) + 1$ with a probability of $1/2$.
   (b) *Deceleration due to other cars:* Each vehicle with an "unsafe" gap of $n_{gap} \leq v_1$ has to slow down to $v(t_{i+1}) = v(t_i) - 1$ in subsequent time steps until it regains an ideally safe distance of $v = n_{gap}$. However, the slow-down manoeuvre is a bit randomized with a probability of $1/2$ to end up in a slightly too small interval $v = n_{gap} + 1$ (for tail-gaters) or $v = n_{gap} - 1$ (for very cautious drivers).

3. *Movement:* In each time step Δt, each vehicle moves to position $x(t_{i+1}) = x(t_i) + v\Delta t$, excluding backward motions ($v \geq 0$).

The outcome of such a one-dimensional traffic simulation is shown in Fig. 2.7, where the spatial position is the horizontal direction and time is progressing downward in vertical direction. Free vehicles with constant speed move along diagonals from top left to bottom right, while jams are visible as concentrations of slower velocities (steeper diagonals) that propagate backward on the street, and eventually dissolve. In SOC parlance, every local jam is called an "avalanche", and the lifetime T of an avalanche is defined by the number of time steps it takes until the number of jammed cars is zero. Furthermore, the spatial extent w of a jam, the number of jammed vehicles n, and the overall space-time size $s \approx nT$ (mass) of the jam can be measured. Fig. 2.8 shows the probability distribution $P(T)$ of jam lifetimes, which in this numerical simulation yielded a perfect powerlaw distribution with a powerlaw slope of $\alpha_T = 1.50 \pm 0.01$ over six orders of magnitude, limited by the maximum time span of the simulation at $t = 10^6$,

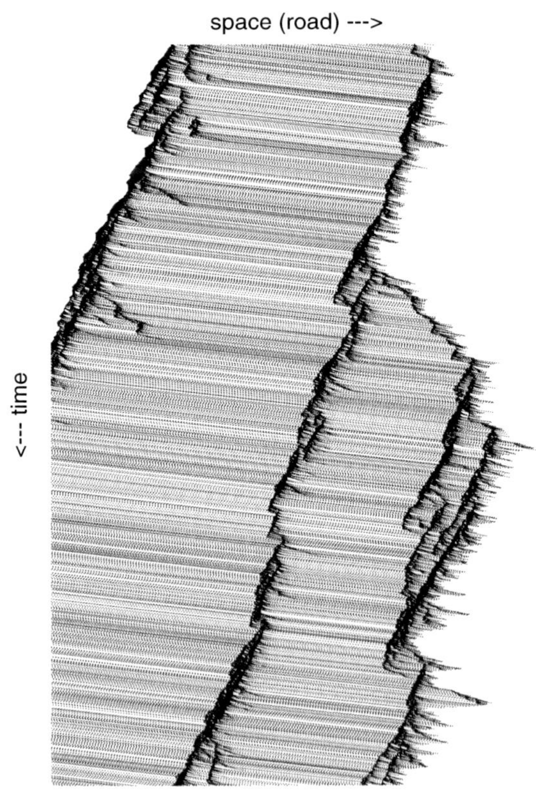

Fig. 2.7 Traffic jam simulations of a 1-D model (with space coordinate in horizontal direction and time running down in vertical direction) showing laminar flows (undisturbed diagonal car traces) and congestions (dark ridges). Reprinted from Nagel and Paczuski (1995) with permission; Copyright by American Physical Society.

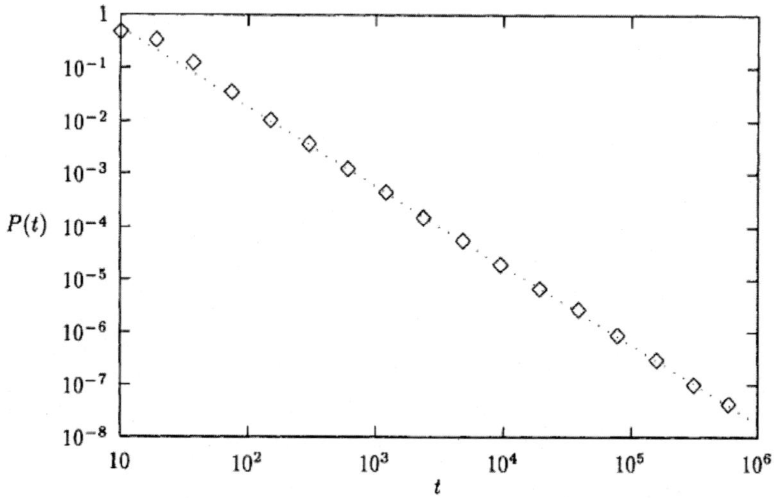

Fig. 2.8 Lifetime distribution $P(T)$ of emergent traffic jams in the outflow region, averaged over more than 65,000 jams. A powerlaw slope of $\alpha_T = 1.50 \pm 0.01$ is found (Nagel and Paczuski 1995).

$$P(T) \propto T^{-\alpha_T} , \quad \alpha_T = 1.50 \pm 0.01 . \tag{2.2.1}$$

It is a quite surprising result that complex traffic behavior can be nailed down by a single number α_T in the framework of the SOC concept. Is this numerical value of $\alpha_T = 3/2$ a magic number with a deeper physical meaning? Nagel and Paczuski (1995) provide a simple analytical model that explains this value in terms of one-dimensional random walk. Since a random walk or diffusion process is a well-understood concept, we outline the analytical connection between diffusion and SOC behavior here, because it gives us a deeper physical insight for this particular SOC system.

The probability distribution $P(n,t)$ for the number n of cars in the jam at time t can be described by the following rate equation,

$$P(n,t+1) = (1 - r_{in} - r_{out})P(n,t) + r_{in}P(n-1,t) + r_{out}P(n+1,t) , \tag{2.2.2}$$

where the rates r_{in} and r_{out} represent empirical quantities that depend on the car densities behind and in front of the jam. For large numbers n the rate equation (2.2.2) transforms into the differential equation,

$$\frac{\partial P}{\partial t} = (r_{out} - r_{in})\frac{\partial P}{\partial n} + \frac{r_{out} + r_{in}}{2}\frac{\partial^2 P}{\partial n^2} , \tag{2.2.3}$$

For sake of simplicity we consider the equilibrium case when the rate of cars entering the jam equals the cars leaving the jam, i.e., $r = r_{in} = r_{out}$, in which case we obtain the one-dimensional diffusion equation,

$$\frac{\partial P}{\partial t} = r\frac{\partial^2 P}{\partial n^2} . \tag{2.2.4}$$

If we plug in the observed probability of lifetimes, $P(t) \propto t^{-3/2}$, into the left side, i.e., $\partial P/\partial t \propto t^{-5/2}$, and use the solution of a diffusion process for one-dimensional random walk,

$$n \propto t^{1/2} , \tag{2.2.5}$$

we infer a probability $P(n)$ for the number n of cars involved in a jam,

$$P(n) \propto (t^{1/2})^{-3/2} = n^{-3} , \tag{2.2.6}$$

which fulfills the diffusion equation (2.2.4), since $\partial^2 P/\partial n^2 \propto n^{-5} \propto t^{-5/2}$. Thus, the evolution of a traffic jam simulation behaves like a one-dimensional diffusion process and explains the observed powerlaw distribution of $P(T) \propto T^{-3/2}$ of jam lifetimes. Remarkably, the simple diffusion process explains not only the powerlaw behavior of this SOC system, but also the value of the powerlaw slope for lifetimes. Note, that the same powerlaw index of $\alpha_T = 3/2$ was also found for lattice gas models (Section 2.1.3), but is different for sand avalanches in the BWT model (Section 2.1.2), where also the dimensionality of the system plays a role.

Some first pioneering Monte-Carlo simulations of 1-D traffic jams that mimic the transition from laminar (uncongested) traffic flow to (congested) start–stop-waves with increasing car density were performed by Nagel and Schreckenberg (1992) and Nagatani (1995a,e,f). Deterministic 1-D traffic jam models with continuous positions and velocities showed that SOC behavior is driven by the slowest car (Nagel and Herrmann 1993). Further SOC studies on traffic jams involved temporary stopping of cars (Nagatani 1995d), entering and exiting of cars on highway ramps (Nagatani 1995b), and 2-D cellular automaton models (with many parallel highway lanes) to avoid jam driving (Nagatani 1995c). A major conclusion of traffic jam simulations is that the critical state of SOC behavior occurs at the maximum throughput rate. Thus, if we want to make our highways most efficient and achieve the highest throughput rate, we have to expect traffic jams at the critical state of SOC. So, we cannot have maximum throughput without traffic jams! The throughput rate is controlled today by metering lights at the entering ramps of some highways, for instance in California.

2.2.3 Financial Market Simulations

One of the first SOC concepts applied to the financial market is the BCSW model of fluctuations in aggregate production (Bak, Chen, Scheinkman, and Woodford 1993). It is a variation of the 2-D BTW sandpile model (Bak, Tang, and Wiesenfeld 1987) in which sites on the lattice represent firms, while the interactions between next-neighbor lattice points are sales and purchases of products. An avalanche starts with the input of a product on the market, which triggers a chain reaction of sales and purchases. High prices produce small avalanches, while low prices can create large avalanches. The SOC behavior explains the large fluctuations in aggregate production.

A related concept is the multiple-strategy agent-based model. For instance, one such model envisions a self-organized network of competing Boolean agents who compete on the market based on actions and information obtained from a small group of other agents

(Paczuski et al. 2000). The agents play a competitive game that rewards those in the minority. Computer simulations of this model show that the network evolves to a stationary but intermittent state where random mutation of the worst strategy can change the behavior of the entire network.

SOC modeling and simulations of the financial market is comprehensively summarized in the review entitled "Financial physics" by Feigenbaum (2003), entailing the BCSW sandpile model (Bak et al. 1993), the percolation model of Cont and Bouchard (2000), multiple-strategy agent-based models (e.g., Paczuski et al. 2000), the minority game model (Challet and Zhang 1997), and log-periodic precursors to financial crashes (Feigenbaum and Freund 1996; Sornette and Johansen 1997). SOC simulations of the financial market may be able to explain some universal scalings that have been observed for widely different economies and different time epochs: (i) the fluctuation of price changes of any stock market is characterized by a probability density function that is a simple power law with an exponent $\alpha = 3$ extending over 8 orders of magnitude (on the y-axis), and (ii) for a wide range of economic organizations, the size of organizations is inversely correlated to the fluctuations in size with an exponent of $\beta \approx 0.2$ (Stanley et al. 2002).

2.3 SOC Simulations in Biophysics

2.3.1 The Punctuated Equilibrium (Bak–Sneppen Model)

Biological evolution is not a gradual process. Genetic mutations that lead to the origin of new species, or natural disasters that lead to mass extinctions, happen very intermittently, separated by long quiescent periods of near-equilibrium. This mixed state of near-equilibrium and rare loss-of-equilibrium events is called the theory of punctuated equilibrium. A numerical model with punctuated equilibrium and criticality applied to evolution was developed by Bak and Sneppen (1993). It is a cellular automaton model with a one-dimensional set of n sites, where each site represents a species and its fitness ρ_i is characterized with a random number between zero and unity, i.e. $0 \le \rho_i \le 1$. At each time step, the site with the smallest value of ρ_i is replaced by a new random value from the range $0 \le \rho_i \le 1$, which corresponds to a random mutation of the least-fit species (or dissipation of the least-fit organism in a food chain). After many steps, a quasi-stationary state is established where all values end up in a restricted range of $\rho_c \le \rho_i \le 1$, where a critical value of $\rho_c = 0.6670$ is found. This state corresponds to a near-equilibrium situation where each species has a minimal fitness of ρ_c, essentially capturing Charles Darwin's notion of the evolution theory where only the fittest species survive. Continuing the random mutations, either the equilibrium is maintained if a new assigned state is larger than the critical value, $\rho_i \ge \rho_c$, or an avalanche is started when $\rho_i < \rho_c$. Thus, an avalanche of evolutionary changes propagates to the adjacent sites, until a new equilibrium is obtained with all fitness values $\rho_i \ge \rho_c$. Counting the number of species that temporarily have a fitness value of $\rho_i < \rho_c$ defines the size of a (mutation) avalanche. Bak and Sneppen (1993) found a powerlaw frequency distribution for the avalanche sizes with a slope of $\alpha \approx 1$ for a 1-D model, and $\alpha \approx 1.26$ for a 2-D model. Tracking a single species i, there are long time periods when the fitness parameter $\rho_i(t)$ does not change, but once it or its next neighbor becomes the species with the lowest fitness, it is part of a chain-reaction of mutations until the whole neighborhood is restored into quasi-equilibrium. Consequently, a plot of the number of mutations as a function of time has the appearance of a sequence of irregular step functions (i.e., a "Devil's staircase"), which is the hallmark of punctuated equilibrium theory (Bak and Paczuski 1995). A visualization of the "fractal evolution" of the Bak–Sneppen model is shown in Fig. 2.9. The time intervals of quasi-equilibrium can be considered as lifetimes of a species, while the "punctuations" or episodes of mutations represent a transition, equivalent to an extinction and replacement of a species. Such a simulated evolution of extinctions could mimic the observed powerlaw distribution of some 19,000 species extinctions observed in fossil history by Raup (1986) and Sepkoski (1993), see Fig. 1.6 and discussions in Bak and Paczuski (1995), Sneppen et al. (1995), and Bak (1996).

This simple cellular automaton model for evolution could also be treated analytically in terms of mean field theory. In a one-dimensional model, the dynamics can be described in terms of a "repetitious random walker" and anomalous diffusion diffusion with exponent 0.4 (Flyvbjerg et al. 1993). The distribution of avalanche durations (or co-evolutionary mutations) has a mean field exponent of $\alpha_T = 3/2$ (de Boer et al. 1994, 1995). A conjecture

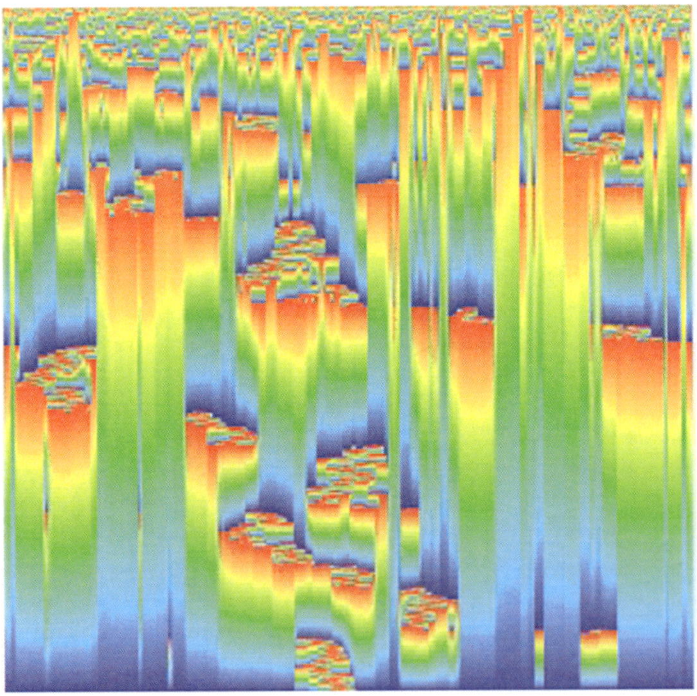

Fig. 2.9 Sample of the Bak–Sneppen model evolution. The population status (or species number) is shown on the x-axis and the history (or time) is on the y-axis from top to bottom. Each "floating iceberg" structure (visualized with a shading along the z-axis) represents a population episode of evolution or extinction (credit: Claudi Rocchini).

that the same model applies to Reggeon field theory (Grassberger and de la Torre 1979) in high-energy particle physics was made by Paczuski et al. (1994), and Ray and Jan (1994), and the analytical derivation in terms of the "gap equation" and the "gamma equation" is discussed in Bak and Paczuski (1995). Scaling laws of SOC powerlaw coefficients and power spectra are studied in Maslov et al. (1994), Paczuski et al. (1996), Böttcher and Paczuski (1996, 1997), and Sole and Manrubia (1996). Reviews of the Bak–Sneppen SOC model applied to biological evolution can also be found in Turcotte (1999) and Sole et al. (1999). The mean field theory applied to the Bak–Sneppen model is summarized in Jensen (1998).

2.4 SOC Simulations in Geophysics

2.4.1 Slider-Block Spring Model

The slider-block spring model (Fig. 2.10) is a mechanical model that was especially designed to mimic the forces between tectonic plates in the Earth's crust, where each stick-and-slip motion is manifested as an earthquake. Alternatively, one can also think of pulling sandpaper across a carpet. This mechanical model consists of two plates and a set of mass elements that all are elastically coupled between neighboring elements, similar to the coupled pendulums described in Section 2.1. All elements experience the same driving force from the upper driver plate, but the individual elements, which are elastically coupled to both the next neighbors as well as to the driver plate, execute a stick-and-slip motion due to the friction on the rough surface. A block moves when the pulling force exceeds the static friction F_s. Once a block moves, it experiences a smaller amount of friction (i.e., dynamic friction, with $F_d < F_s$) than in rest. The elastic coupling can trigger a number of neighbored blocks to slip in a chain reaction, which produces an avalanche, and thus SOC behavior is expected in a similar way as for the coupled pendulums or sandpiles. This type of slider-block spring model has been originally introduced by Burridge and Knopoff (1967).

The mechanical force acting on a block in the slider-block spring model is composed of the driver (leaf) spring constant k_p and the connector spring constants k_c (see Fig. 2.10),

$$F_i(t) = m\frac{d^2x_{i,j}}{dt^2} + k_p\, x_{i,j} + k_c(x_{i+1,j} + x_{i-1,j} + x_{i,j+1} + x_{i,j-1} - 4x_{i,j}) = F_d \ , \qquad (2.4.1)$$

where $x_{i,j}$ is the position of block (i,j), t is the time, and F_d is the dynamic frictional force during motion. When the block is at rest, we have the inequality condition,

$$k_p x_{i,j} + k_c(x_{i+1,j} + x_{i-1,j} + x_{i,j+1} + x_{i,j-1} - 4x_{i,j}) < F_s \ , \qquad (2.4.2)$$

where F_s is the static resisting force. The dynamical behavior of the system can be characterized by the free parameters of the ratio of the static to dynamic friction (F_d) and the stiffness of the system ($q_k = k_c/k_p$). For soft systems, $q_k \mapsto 0$, the blocks exhibit stick-and-slip behavior independently, while stiff systems, $q_k \mapsto \infty$ show more coherent motions as a

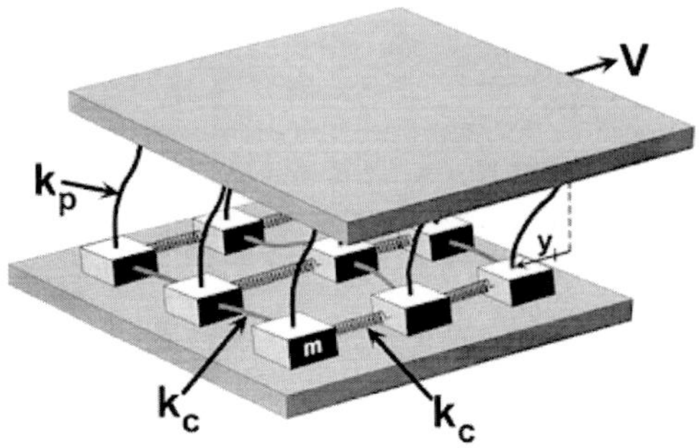

Fig. 2.10 The slider-block model consists of an array of blocks, each with mass m, which are pulled across a surface by a driver plate at a constant velocity v. Each block is coupled to the adjacent blocks with either leaf or coil springs (with spring constant k_c), and to the driver plate with leaf springs (with spring constant k_p) (Turcotte 1999). (Reprinted with permission of the American Physical Society)

whole. Stiff systems produce an over-abundance of coherent large-scale slips (with peaks at the high end of the occurrence frequency distribution), while soft systems exhibit powerlaws for small events, but a lack of large events (Huang et al. 1992). Typical powerlaw distributions with a slope of $\alpha \approx 1.0$–1.3 are found for such systems (Turcotte 1999).

Early computer simulations with 2,000 slider blocks revealed a powerlaw distribution for the occurrence frequency of events as a function of the number of blocks that slip (event size), and thus evidence for SOC behavior was found (Otsuka 1972). Simulations of a 1-D chain of 400 blocks using a velocity-weakening (nonlinear) friction law also exhibited chaotic behavior and powerlaw distributions of slip avalanches, which are characteristic of SOC behavior (Carlson and Langer 1989a,b). Using a cellular-automaton model of threshold elements, instead of solving the differential equations (2.4.1–2.4.2), greatly simplified the calculations (Nakanishi 1990, 1991; Brown et al. 1991). A large number of studies on the slider-block spring model have been conducted and variants of it have been modeled and simulated, such as the following frequently quoted studies: Smalley et al. (1985), Sornette and Sornette (1989), Rundle and Klein (1989), Carlson (1991a,b), Carlson et al. (1991), Feder and Feder (1991), the Olami–Feder–Christensen model (Olami et al. 1992), Christensen and Olami (1992a,b), Shaw et al. (1992), Cowie et al. (1993), de Sousa Vieira et al. (1993), Grassberger (1994), Shaw (1995), or Middleton and Tang (1995). The slider-block spring model is the primary basis for associating earthquakes with SOC (Turcotte 1999), but alternative models also have been proposed, such as the crack propagation model (Chen et al. 1991) and interface depinning (Paczuski and Böttcher 1996; Fisher et al. 1997). Reviews on the slider-block spring model can be found in Carlson et al. (1994), Bak (1996), Turcotte (1999), and Hergarten (2002).

2.4.2 The Forest-Fire Model

The forest-fire model is a cellular automaton model that exhibits SOC behavior and mimics the spread of fire in a forest, but is also applied to the spread of biological diseases. We describe the basic model in two dimensions, with a square grid with integer positions $x_{i,j}$.

At every time step a tree (seed) is dropped onto a random location, which corresponds to the planting of a tree if the site is unoccupied. In addition, a match is dropped at a random position every n_s-th time step, so with a "sparking frequency" of $f_s = 1/n_s$. If the site of the dropped match is empty, nothing happens, while a forest fire is started if is occupied by a tree. The spread of the forest fire consumes the tree at the local site and propagates to all adjacent (non-diagonal) neighbor sites, if occupied by a tree. So, the forest fire will spread over a small or large area, depending on the tree density, and will stop once burning finds no further next tree. In the long-term time average, there will be a stable tree density at an intermediate value that is not too small (so that fires cannot spread) and not too high (that would lead to a catastrophic total burn-down). The long-term average tree density will be a critical state that is independent of the initial condition and does not require any fine-tuning of how the initial forest configuration is built up. In several models, a necessary condition for SOC behavior is the double limit of (1) a very low growth probability ($\ll 1$) and (2) an ignition probability much lower than the growth probability. An example of a numerical simulation of this forest-fire model is shown in Fig. 2.11, showing 4 cases from a small fire (with 5 burned trees) to a large forest fire (with over 5,000 burned trees).

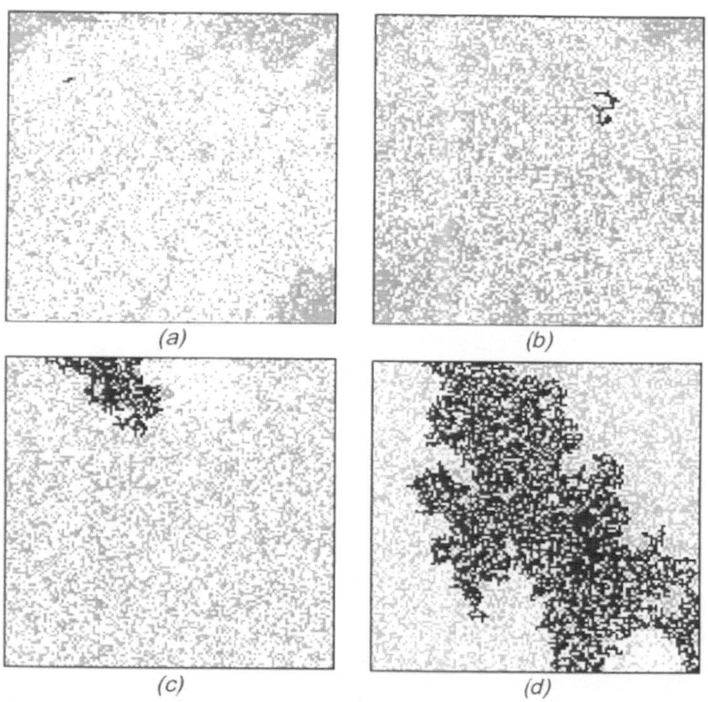

Fig. 2.11 Four examples of typical model forest fires are given for a 128×128 grid with a sparking frequency of $f_s = 1/2{,}000$. The black regions mark the sizes of forest fires, while the grey regions indicate the green (unburned) forest, and white regions correspond to sites without trees. The areas of the 4 forest fires are (a) 5, (b) 51, (c) 505, and (d) 5,327 trees, spanning the entire grid (Turcotte 1999). (Reprinted with permission of the American Physical Society)

In numerical simulations of this forest-fire algorithm, the area A_f of each forest fire is measured and the frequency distributions exhibit powerlaws with slopes in the range of $\alpha \approx 1.0$–1.2 (Turcotte 1999). This is somewhat lower than the observed distribution with a powerlaw of $\alpha \approx 1.3$–1.4 (Section 1.5). The results of three runs are shown in Fig. 2.12. For the smallest sparking frequency ($f_s = 1/2{,}000$), the tree density grows so high in between two forest fires, that the next fire consumes almost the entire forest area (or grid size), which shows up as a peak at the upper limit in Fig. 2.12. This deviation from a straight powerlaw function at the upper end is called a *finite grid size effect*.

Initial numerical simulations of the forest-fire SOC model have been pioneered by Bak et al. (1990), Drossel and Schwabl (1992a,b), Mossner et al. (1992), Henley (1993), and Grassberger (1993). Further model simulations spanned from one to six dimensions (Christensen et al. 1993), were compared with analytical solutions (Drossel et al. 1993), interpreted as a turbulent cascade process (Paczuski and Bak 1993), applied a renormalization group theory (Loreto et al. 1995), or derived scaling laws of critical exponents (Clar et al. 1994). Reviews on the forest-fire model can be found in Mossner et al. (1992), Clar et al. (1996, 1999), Jensen (1998), Hergarten (1998), and Turcotte (1999). Descriptions of forest-fire models can be found in the textbooks of (Jensen 1998, p.65) and Hergarten (1998).

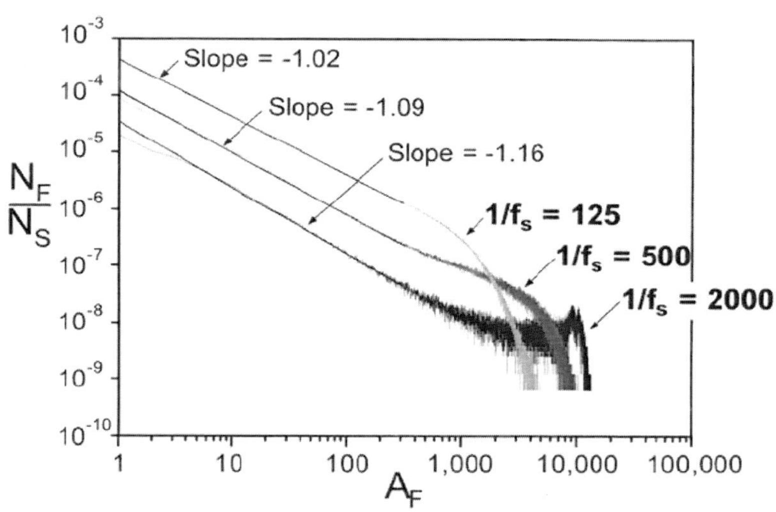

Fig. 2.12 Frequency distributions of forest-fire areas A_f for three different setups, with different sparking frequencies of $f_s = 125$, 500, 2,000. Each run contains $\approx 10^9$ time steps run on a 128×128 grid. The powerlaw slopes are similar (in the range of $\alpha = 1.02$–1.16, but the upper cutoff shows finite grid size effects, approaching the upper limit of the grid $A_F \leq 128^2$ for the smallest sparking frequency (Turcotte 1999). (Reprinted with permission of the American Physical Society)

2.5 SOC Simulations in Magnetospheric Physics

2.5.1 SOC Model with Finite System Size

Some magnetospheric phenomena were found to exhibit powerlaw distributions, such as the burst area of the auroral electro-jet index (AE) or substorm-related tail current disruptions (see observational references in Section 1.6), which seem therefore to be scale-free and consistent with SOC models. On the other side, some other parameters, such as the intensity or time intervals between substorm events seem to have well-defined probability distributions with characteristic scales (see Fig. 1.10). SOC models thus can not explain both statistics, or an additional physical reason may modify the SOC behavior significantly. This dilemma led Chapman et al. (1998) to a dual SOC model where constant energy input (inflow) generates (1) scale-free internal energy discharges with a SOC-like powerlaw behavior, whereas (2) system-wide discharges with a well-defined size distribution do not exhibit SOC. In analogy to the Bak–Tang–Wiesenfeld 2-D sandpile model, the first group of events would include small avalanches confined to the surface of the sandpile that do not reach the base, while the second group would include the largest avalanches that propagate all the way down to the base of the sandpile, where they reach their maximum size, and then would roll over the edge, if the sandpile is mounted on a circular plate (like the IBM experiment with a high-precision scale by Held et al. 1990, see Section 1.2).

Chapman et al. (1998) conducted numerical simulations of such a dual SOC model as follows. A cellular automaton is represented by a 1-D grid of n equally-spaced cells, each one with sand at height h_j and local gradient $z_j = h_j - h_{j-1}$, which is assumed to be stable below a critical angle of repose z. The selection rule for the critical gradients on the n nodes is a random number uniformly distributed in the range $[0,1]$, parameterized by an exponential function for the probability $P(z)$ in the interval $[z, z+dz]$,

$$P(z) = z^y \exp\left(\frac{-z^{1+y}}{1+y}\right) , \tag{2.5.1}$$

which has the cumulative probability distribution $F(>z)$

$$F^{cum}(>z) = \int_z^\infty P(z) dz = 1 - \exp\left(-\frac{z^{1+y}}{1+y}\right) , \tag{2.5.2}$$

and the normalization

$$\int_{z=0}^\infty P(z)\, dz = P^{cum}(>0) = 1 . \tag{2.5.3}$$

Sand is then added at cell 1 at a constant rate $r \ll 1$, until the critical gradient is exceeded at cell 1, triggering a redistribution to neighboring cells until the slope is "flattened" back to the angle of repose ($z_j = 0$). From time to time an avalanche reaches the end of the 1-D grid (at $j = n$), causing a system-wide discharge, in which case the entire sandpile is emptied and returns to the angle of repose. For the statistics of dissipated energies E per avalanche, a nonlinear scaling with a quadratic function is assumed,

$$E = \sum_{j=1}^{n} h_j^2(t_{before}) - \sum_{j=1}^{n} h_j^2(t_{after}) , \tag{2.5.4}$$

similar to the model of Dendy and Helander (1997). The quadratic energy scaling can be understood in terms of an area-like spreading for 2-D avalanches, which grows quadratically with the length scale h_j, assuming a constant energy dissipation rate per unit area. Fig. 2.13 (left panel) shows the evolution of dissipated energy as a function of time. Internal avalanches reduce the energy by a partial amount, while system-wide avalanches dump the entire energy of the system to zero, which corresponds to the energy state of the sandpile at the angle of repose. The probability distribution $P(E)$ of the energy dissipated per avalanche exhibits two different components: (1) a powerlaw distribution for internal events, and (2) a peaked distribution for system-wide avalanches (Fig. 2.13, right panel).

This particular set-up mimics the bimodal distribution of magnetospheric substorms (Fig. 1.10) observed by Lui et al. (2000), leading to the interpretation in terms of a nonlinear dissipative system that is driven by a constant inflow, but exhibits scale-free SOC behavior only for internal energy dissipation events, while it exhibits a particular scale for system-wide energy dumps. Such a deviation from a straight powerlaw was also found in system-wide forest-fire events in the simulations shown in (Fig. 2.12), where it was interpreted as a finite-size effect also. Numerical simulations of system-wide avalanches were performed for various (low and high) input rates, in order to mimic the variable loading rates of the magnetosphere, and it was found that the the powerlaw signature of large-scale internal events persisted (Chapman et al. 1999, 2001). Alternatively, a bimodal SOC behavior was also found in a 1-D cellular automaton simulation of the central plasma sheet with two different local instability criteria (Liu et al. 2006).

2.5.2 Cellular Automaton Model with Discretized MHD

SOC behavior in the magnetotail plasma has been inferred early on from the powerlaw distribution of lifetimes found in magnetospheric disturbances based on the auroral electron jet (AE) index (Takalo et al. 1993; Consolini 1997), from near-Earth magnetotail current disruptions (Lui et al. 1988), substorm current disruptions (Consolini and Lui 1999), or bursty bulk flow events (Angelopoulos et al. 1996, 1999). An early attempt to simulate the statistics of these observed features in terms of a cellular automaton SOC model was carried out by Takalo et al. (1999a), as described in the following.

Maxwell's equations in MHD applications generally assume the nonrelativistic limit of plasma motion ($v \ll c$) and are expressed in terms of the current density $\mathbf{j}$ according to Ampère's law (in cgs units),

$$\mathbf{j} = \frac{1}{4\pi}(\nabla \times \mathbf{B}) , \tag{2.5.5}$$

yielding together with Ohm's law (with electric conductivity σ) the so-called induction equation,

$$\frac{\partial \mathbf{B}}{\partial t} = \nabla \times (\mathbf{v} \times \mathbf{B}) + \eta \nabla^2 \mathbf{B} , \tag{2.5.6}$$

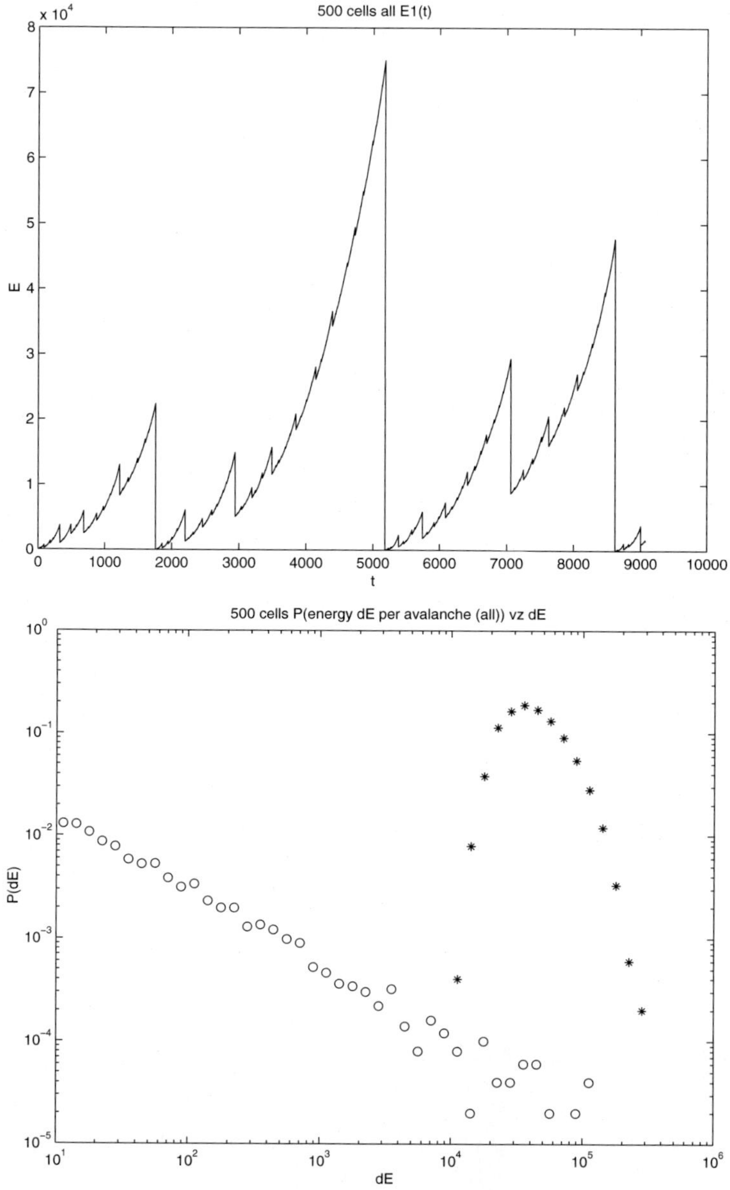

Fig. 2.13 *Left:* Time evolution of a sandpile SOC system with quadratically growing energy and random energy releases. Zero energy corresponds to the energy of the sandpile at the angle of repose. Internal avalanches reduce the total energy to values > 0, while system-wide avalanches drop the energy to values $= 0$. *Right:* A log-log plot of the probability distributions of the energy dissipated for 50,000 internal avalanches (O symbols) and for 10,000 system-wide avalanches ($*$ symbols). Note that the former frequency distribution has a powerlaw shape, while the latter forms a peaked distribution with a peak at $E \approx 10^{4.5}$ (Chapman et al. 1998). (Reprinted with the permission of the American Geophysical Union)

which contains a convective and a magnetic diffusion term (with a magnetic diffusity $\eta = c^2/4\pi\sigma$), and fulfill the divergence-free condition for the magnetic field,

$$\nabla \cdot \mathbf{B} = 0 \ . \tag{2.5.7}$$

Using the vector identity,

$$\nabla \times (\nabla \times \mathbf{B}) = \nabla(\nabla \cdot \mathbf{B}) - (\nabla \cdot \nabla)\mathbf{B} \ , \tag{2.5.8}$$

and Maxwell's equations (Eqs. 2.5.5 and 2.5.7) we have the relation

$$4\pi \ (\nabla \times \mathbf{j}) = -\nabla^2 \mathbf{B} \ . \tag{2.5.9}$$

Making use of Stokes' theorem for a vector field $\mathbf{j}$,

$$\int_S (\nabla \times \mathbf{j}) \cdot \mathbf{n} \ d\mathbf{S} = \oint_C \mathbf{j} \cdot d\mathbf{l} \tag{2.5.10}$$

we have the relationship,

$$\int \frac{\nabla^2 \mathbf{B}}{4\pi} \cdot \mathbf{n} \cdot d\mathbf{S} = - \oint_C \mathbf{j} \cdot d\mathbf{l} \ . \tag{2.5.11}$$

Applying this functions to a discretized 2-D cellular grid in the xy plane (with cell size $\Delta x = \Delta y$ and area Δx^2) and magnetic field vectors in orthogonal z-direction, $\mathbf{B} = (0, 0, B_z)$, the current density $\mathbf{j}$ has non-zero components only in the xy plane,

$$\mathbf{j} = \frac{1}{4\pi}(\nabla \times \mathbf{B}) = \left(\frac{dB_z}{dy}, -\frac{dB_z}{dx}, 0 \right) = (j_x, j_y, 0) \ , \tag{2.5.12}$$

which can be computed from the the magnetic field $\mathbf{B}(x,y,z)$ at the midpoints $x \pm \frac{1}{2}$ and $y \pm \frac{1}{2}$ between the cell boundaries (Fig. 2.14),

$$j_x \left(x, y \pm \frac{1}{2} \right) = + \frac{1}{4\pi} \frac{dB_z}{dy} = \pm \frac{1}{4\pi} \frac{B_z(x, y \pm 1) - B_z(x,y)}{\Delta y} \ , \tag{2.5.13}$$

$$j_y \left(x \pm \frac{1}{2}, y \right) = - \frac{1}{4\pi} \frac{dB_z}{dx} = \pm \frac{1}{4\pi} \frac{B_z(x,y) - B_z(x \pm 1, y)}{\Delta x} \ . \tag{2.5.14}$$

Summing the currents (j_x, j_y) along the four boundaries of a quadratic cell (Fig. 2.14) leads according to Eq. (2.5.11) to the Laplacian L,

$$\begin{aligned} L\,\Delta x^2 &= \nabla^2 \mathbf{B}\, \Delta x^2 = 4\pi \oint_C \mathbf{j} \cdot d\mathbf{l} \\ &= [j_x(x, y+\tfrac{1}{2}) - j_y(x+\tfrac{1}{2}, y) - j_x(x, y-\tfrac{1}{2}) + j_y(x-\tfrac{1}{2}, y)]\Delta x \\ &= [(B_z(x, y+1) + B_z(x+1, y) + B_z(x, y-1) + B_z(x-1, y) - 4B_z(x,y)] \ . \end{aligned} \tag{2.5.15}$$

The diffusive term in the induction equation (2.5.6) is then replaced by this Laplacian L, and the conductive term is replaced by a general source term $S(x,y,t)$, representing some

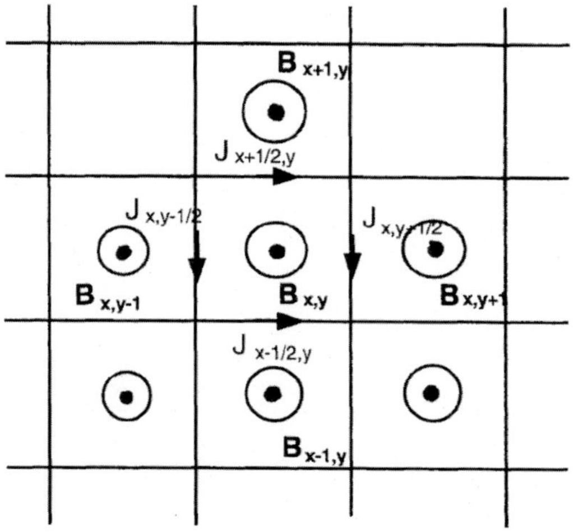

Fig. 2.14 Cellular automaton model containing magnetic fluxtubes, each one characterized by a magnetic field $B_z(x \pm 1, y \pm 1)$ and by four segments of currents $J(x, y \pm \frac{1}{2})$ and $J(x \pm \frac{1}{2}, y)$ at the cell boundaries (Takalo et al. 1999a). (Reprinted with the permission of the American Geophysical Union)

external driving, such as loading of the magnetospheric plasma,

$$\frac{d\mathbf{B}}{dt} = S(x,y,t) + \eta L \, . \tag{2.5.16}$$

The critical threshold level that initiates an avalanche in a cellular automaton model is controlled by the resistivity η of the plasma, which is realized here by a critical Laplacian L_{cr},

$$\eta(x,y) = \eta\theta(L - L_{cr}) \, , \tag{2.5.17}$$

where $\theta(...)$ is the Heaviside step function. The Laplacian represents the net current around the cell (x,y). The bigger the resistivity η, the more flux is distributed to neighboring cells during an avalanche. When an avalanche occurs, the excess field,

$$\Delta B = \eta(x,y)(L - L_{cr}) \, , \tag{2.5.18}$$

is evenly redistributed over the next four neighboring cells, one fourth to each of them, similar to the Bak–Tang–Wiesenfeld model (Section 2.1.3). The total energy contained in the magnetic field of the model is then calculated as

$$E(t) = \frac{1}{8\pi} \sum_{x,y=1}^{n} B(x,y,t)^2 \, , \tag{2.5.19}$$

for the cellular grid $(x = 1, ..., n, y = 1, ..., n)$. This 2-D cellular automaton model was run with a grid of $N^2 = 50^2$ cells. The system was continuously driven by a random input with small emerging fields that appear as fluctuations in $B(x,y)$. The evolution of the system in SOC state exhibited frequency distributions of avalanche durations and sizes that are close

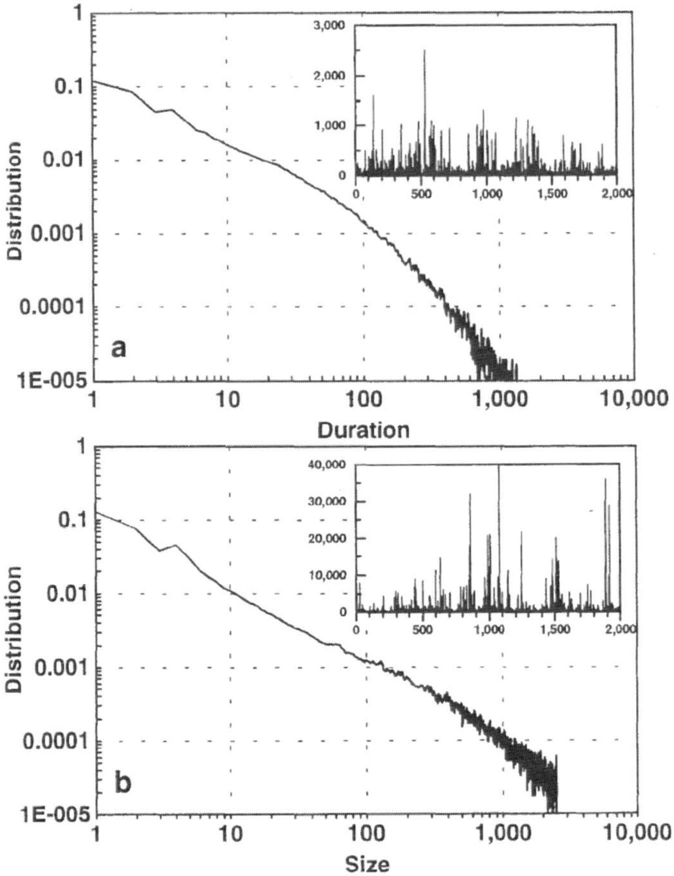

Fig. 2.15 Frequency distributions and time series (inserts) of avalanche durations (top) and sizes (bottom) of 100,000 avalanche events simulated with the cellular automaton model of Takalo et al. (1999a). (Reprinted with the permission of the American Geophysical Union)

to powerlaws at lower values (with a slope of $\alpha \approx 1$), but with a knee at the upper end (Fig. 2.15).

Takalo et al. (1999a) also explored sudden changes of the input, to mimic the southward turning and subsequent northward turning of the interplanetary magnetic field (IMF), by feeding the input of the model with occasional stronger pulses. These intermittent stronger disturbances pushed the model out of the SOC state and triggered larger avalanches during a while, but the system always returned to the SOC state, which demonstrates the robustness of the SOC state. This simulation thus can mimic two different components of magnetospheric substorms: (i) externally triggered substorms caused by perturbations in the solar wind, and (ii) internal substorms caused by the intrinsic dynamics of the magnetosphere near SOC state.

More elaborate numerical simulations of SOC systems with discretized MHD and current-driven kinetic instabilities have been conducted by Klimas et al. (2004), which we will discuss in a selection of physical SOC models (Chapter 9).

2.6 SOC Simulations in Solar Physics

There is a growing industry of SOC avalanche models in solar physics, which is best summarized in the review of Charbonneau et al. (2001). Most SOC models of solar flares are based on the basic idea that a flare is produced by an avalanche of small-scale magnetic reconnection events that cascade in a highly stressed coronal magnetic field, ultimately driven by random photospheric magneto-convection and the solar dynamo.

2.6.1 Isotropic Cellular Automaton Models

Solar flares exhibit a strikingly perfect powerlaw distribution of their peak count rates in hard X-rays, extending over 3–4 orders of magnitude (Fig. 1.13; Dennis 1985). In analogy to the BTW sandpile model, the coronal magnetic field is thought to play the same role of a nonlinear dissipative system in SOC state, where the photospheric convective motion plays the role of the random input of sand grains, the critical angle between misaligned magnetic field lines that leads to magnetic reconnection plays the role of the critical angle of repose in the sandpile, and solar flares play the role of avalanches in SOC state. This interpretation in terms of sandpile SOC systems was first proposed by Lu and Hamilton (1991).

A cellular automaton model was conceived by Lu and Hamilton (1991), where a 3-D grid represents the solar corona, characterized by a magnetic field strength $\mathbf{B_{ij}}$ in each volume element, with a local magnetic field gradient $\Delta\mathbf{B}$ defined by the difference between the local magnetic field and the average of its six nearest neighbors ($\mathbf{B}_{nn}$),

$$\Delta\mathbf{B} = \mathbf{B}_{ij} - \frac{1}{6}\sum_{nn}\mathbf{B}_{nn} \ . \qquad (2.6.1)$$

The magnetic field structure is unstable to magnetic reconnection when the difference exceeds a critical field value, $|\Delta\mathbf{B}| > B_c$, similar to the magnetic discontinuity angle proposed by Parker (1988). When a reconnection instability occurs, the redistribution rule is set up in a way that the average magnetic field gradient (Eq. 2.6.1) vanishes, $\Delta\mathbf{B} \mapsto 0$,

$$\mathbf{B}_{ij} \mapsto \mathbf{B}_{ij} - \frac{6}{7}\Delta\mathbf{B} \ , \quad \mathbf{B}_{nn} \mapsto \mathbf{B}_{nn} + \frac{1}{7}\Delta\mathbf{B} \ . \qquad (2.6.2)$$

The simulation starts with a uniform magnetic field and is driven by adding random vectors $\delta\mathbf{B}$ at random positions of the field. Once a site gets unstable, the redistribution rule (2.6.2) is applied, the magnetic field gradient is recalculated, and the redistribution rule is applied iteratively until it becomes stable. Each magnetic reconnection event releases a magnetic energy of

$$E_m = \Delta \sum_k B_k^2 = \left(\frac{6}{7}\right) \Delta \mathbf{B}^2 \ , \tag{2.6.3}$$

where k sums over the unstable cell (i,j) and its next neighbors $(n,n) = (i \pm 1, j \pm 1)$. A new disturbance of the magnetic field is added after the old instability has relaxed.

Note that this redistribution rule is *conservative*, in the sense that the quantity B is conserved after every redistribution step, because the same amount is transferred to the next neighbors that is taken away from the central cell. This conservative property is also genuine to the original sandpile model, where avalanches redistribute sand grains without creating or destroying any. However, although the field quantity B is conserved, the energy B^2 is not conserved after a redistribution step, because of the nonlinear (quadratic) dependence. In fact, every redistribution of $|\Delta B| > B_c$ dissipates energy from the system.

Let us quantify the amount of released energy in every redistribution step more generally for a D-dimensional lattice (e.g., $D = 2$ or 3), with B_{ij} (or B_{nn}) being a scalar variable, and requiring a critical threshold of $|\Delta B| > B_c$ for redistribution. If we denote the new values of the magnetic field with B'_{ij} and B'_{nn}, the average field difference to the next neighbors is then (Eq. 2.6.1),

$$\Delta B = B_{ij} - \frac{1}{2D} \sum_{nn=1}^{2D} B_{nn} \ , \quad |\Delta B| > B_c \ , \tag{2.6.4}$$

and the redistribution rule (2.6.2) reads as

$$B_{ij} \mapsto B'_{ij} = B_{ij} - \frac{2D}{2D+1} \Delta B \ , \tag{2.6.5}$$

$$B_{nn} \mapsto B'_{nn} = B_{nn} + \frac{1}{2D+1} \Delta B \ . \tag{2.6.6}$$

This specific choice of redistribution rule leads to $\Delta B = 0$ after the redistribution, in contrast to the redistribution rules used by Lu et al. (1993). Calculating now the total magnetic energy before (E_m) and after (E'_m) the redistribution step,

$$E_m = B_{ij}^2 + \sum_{nn} B_{nn}^2 \ , \tag{2.6.7}$$

$$E'_m = (B'_{ij})^2 + \sum_{nn} (B'_{nn})^2 \ , \tag{2.6.8}$$

we find the following energy difference, inserting the magnetic field values from Eqs. (2.6.5) and (2.6.6),

$$\begin{aligned} \Delta E &= E'_m - E_m \\ &= \left(B_{ij} - \tfrac{2D}{2D+1} \Delta B\right)^2 + \textstyle\sum_{nn} \left(B_{nn} + \tfrac{1}{2D+1} \Delta B\right)^2 \\ &\quad - B_{ij}^2 - \textstyle\sum_{nn} B_{nn}^2 \\ &= -2B_{ij} \left(\tfrac{2D}{2D+1}\right) \Delta B + \left(\tfrac{2D}{2D+1} \Delta B\right)^2 \\ &\quad + 2\textstyle\sum_{nn} B_{nn} \left(\tfrac{1}{2D+1}\right) \Delta B + \textstyle\sum_{nn} \left(\tfrac{1}{2D+1}\right)^2 \Delta B^2 \end{aligned} \ . \tag{2.6.9}$$

From Eq. (2.6.4) we can express

$$\sum_{nn=1}^{2D} B_{nn} = 2D(B_{ij} - \Delta B) \ . \tag{2.6.10}$$

and, together with $\sum_{nn} = 2D$ for summation over constants, inserting into Eq. (2.6.9), we find the final result,

$$\Delta E_m = -\frac{2D}{2D+1}\Delta B^2 \ , \tag{2.6.11}$$

which is the minimum energy quantum that can be released in a redistribution step. This amounts to $\Delta E_m \approx (4/5)\Delta B^2$ for $D = 2$, or $\Delta E_m \approx (6/7)\Delta B^2$ for $D = 3$, respectively, as used in Eq. (2.6.3) by Lu and Hamilton (1991).

A slightly different treatment is given in Charbonneau et al. (2001), where an amount of B_c is redistributed, rather than ΔB,

$$B_{ij} \mapsto B'_{ij} = B_{ij} - \frac{2D}{2D+1} B_c \ , \tag{2.6.12}$$

$$B_{nn} \mapsto B'_{nn} = B_{nn} + \frac{1}{2D+1} B_c \ . \tag{2.6.13}$$

which leads to the following amount for the change of energy (Eq. 5 in Charbonneau et al. 2001),

$$\Delta E_m = -\frac{2D}{2D+1}\left(2\frac{|\Delta B|}{B_c} - 1\right) B_c^2 \ , \tag{2.6.14}$$

but is identical with the result (Eq. 2.6.3 or 2.6.11) of Lu and Hamilton (1991) in the limit of weak driving, i.e., $\Delta B \gtrsim B_c$. In the weak driving limit, only small increments $\delta B \ll B_{ij}$ are added as input to maintain the SOC state. The treatment of Charbonneau et al. (2001) is identical to that of Lu et al. (1993), while the original redistribution rule of Lu and Hamilton (1991) turned out to numerically unstable, as demonstrated by Liu et al. (2002).

A new disturbance of the magnetic field is added after the old instability has relaxed. Lu and Hamilton (1991) find approximate powerlaw distributions for the released energies, i.e., $N(E) \propto E^{-1.4}$, the peak fluxes P, i.e., $N(P) \propto P^{-1.8}$, and the time durations T, similar to those from the observed distributions of hard X-ray flares (Dennis 1985; Crosby et al. 1993),

$$\begin{aligned} N(P) &\propto P^{-1.67\pm0.04} \\ N(E) &\propto E^{-1.53\pm0.02} \ , \\ N(T) &\propto T^{-2.17\pm0.05} \end{aligned} \tag{2.6.15}$$

where the peak energy flux P and total energy E is calculated from the nonthermal spectrum of electrons above a lower cutoff of ≥ 25 keV. The numerically simulated distributions of this SOC model are shown in Fig. 2.16, extending over approximately two orders of magnitude.

The same cellular automaton model for solar flares was repeated for different system sizes, using 3-D grids with length sizes of $L = 10, ..., 50$ (Lu et al. 1993). It was found that the value of the powerlaw slope was invariant to the system size, but the upper cutoffs

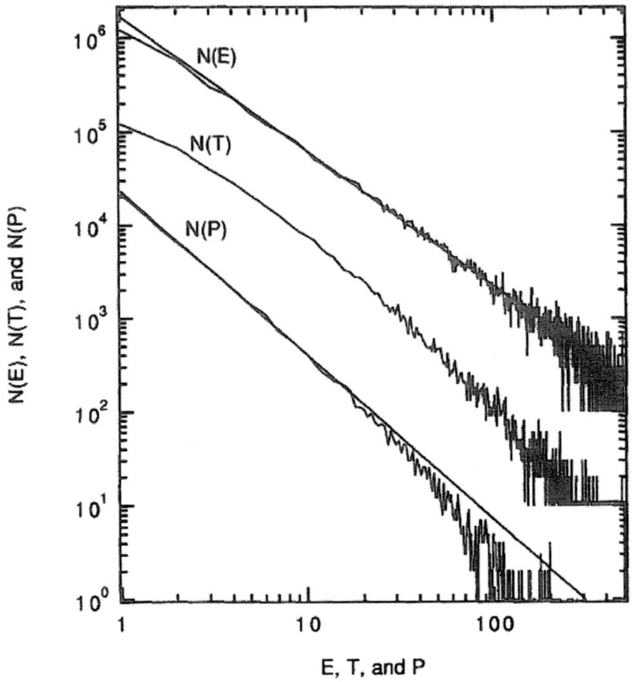

Fig. 2.16 Frequency distributions of event energy release $N(E)$, peak flux $N(P)$, and duration $N(T)$ for avalanches in a $30 \times 30 \times 30$ grid according to the SOC model of Lu and Hamilton (1991). The distributions are offset by 1 and 2 orders of magnitude, and slopes of $N(E) \propto E^{-1.4}$ and $N(P) \propto P^{-1.8}$ are indicated with straight lines (reproduced by permission of the AAS).

moved to larger sizes for larger grids, and thus clearly reflect a system-size effect. The rollover at the upper cutoff was found to be close to an exponential function, so that the total distribution can be fitted by

$$N(X) \propto X^{-\alpha} \exp(-X/X_c) , \qquad (2.6.16)$$

where X represents (E,P,T) and X_c is the exponential cutoff value. The relationship between the three powerlaw slopes could be explained by a simple physical model of a magnetic reconnection event,

$$\Delta P = \frac{\Delta E}{\Delta T} = \Delta L^2 \langle B_\perp^2/8\pi \rangle \frac{v_A}{\zeta} , \qquad (2.6.17)$$

where v_A is the Alfvén speed and ζ a constant. The relations between the powerlaw slopes of E,P,T reflect correlations that can approximately be quantified by the physical relationship given in Eq. (2.6.15). Scatterplots between these three parameters and the resulting powerlaw distributions fitted to solar hard X-ray flare data from the ISEE-3 spacecraft are shown in Fig. 2.17. We will discuss the functional relationship between powerlaw slopes

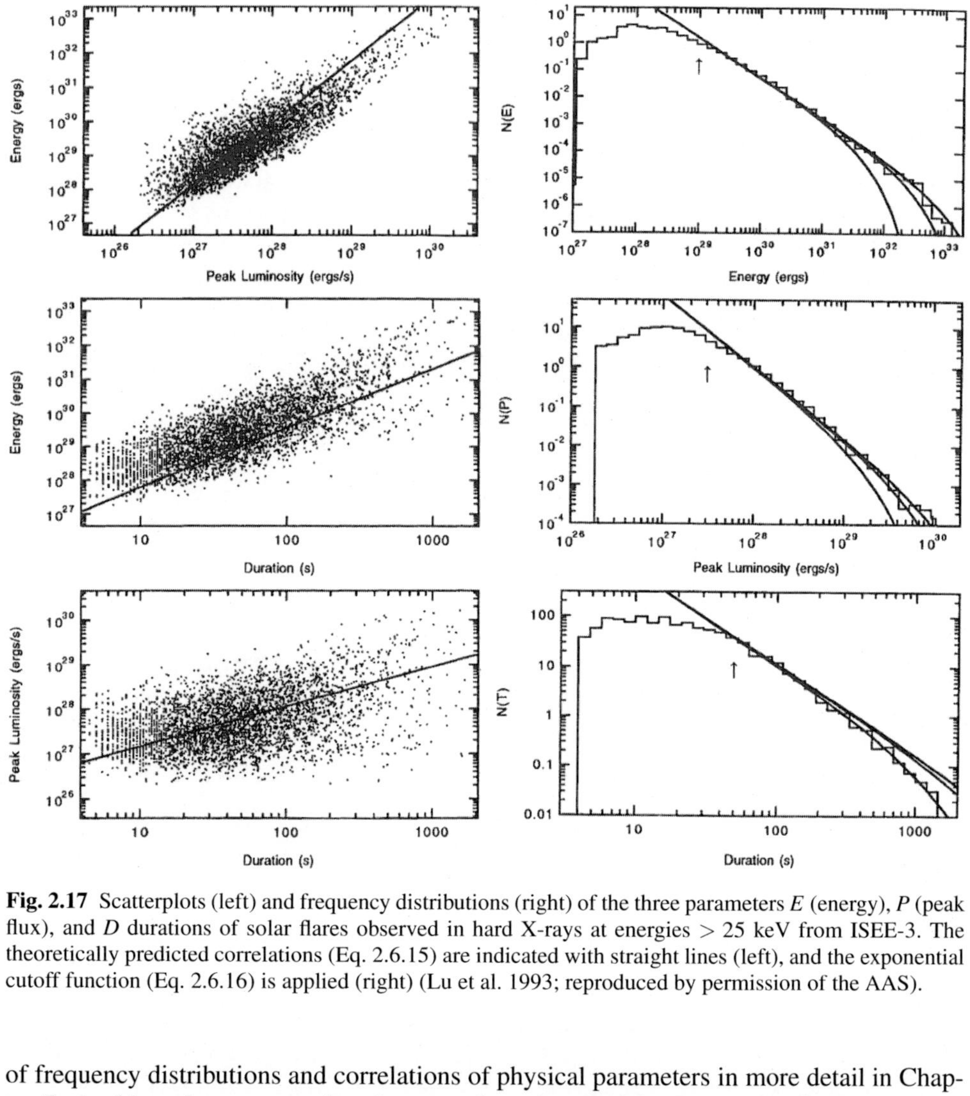

Fig. 2.17 Scatterplots (left) and frequency distributions (right) of the three parameters E (energy), P (peak flux), and D durations of solar flares observed in hard X-rays at energies > 25 keV from ISEE-3. The theoretically predicted correlations (Eq. 2.6.15) are indicated with straight lines (left), and the exponential cutoff function (Eq. 2.6.16) is applied (right) (Lu et al. 1993; reproduced by permission of the AAS).

of frequency distributions and correlations of physical parameters in more detail in Chapter 7. At this point we note that the powerlaw slopes of frequency distributions can be explained by physical relationships that describe the functional dependence between the three parameters E, P, and T. So, it is clearly demonstrated that the powerlaw slope of a specific SOC system is not an universal constant, but depends on the chosen parameters and can be explained by physical laws.

2.6.2 Anisotropic Cellular Automaton Models

In the classic BTW sandpile model, next-neighbor interactions are isotropic, which means that there is an equal probability for the propagation of avalanches or redistribution of energies in every direction, in 2-D or 3-D lattice models (see Eqs. 2.1.3 and 2.1.4). This isotropic dissipation of energy was also assumed for the magnetic field relaxation in the cellular automaton models first applied to solar flares by Lu and Hamilton (1991) (see Eq. 2.6.2). However, since the solar corona is permeated by magnetic fields, which introduce a preeminent structuring into one-dimensional flux tubes, energy dissipation is not expected to propagate isotropically. Energy release in solar flares is believed to take place in the process of magnetic reconnection, which produces a central current sheet with an X-point that demarcates an ion and electron diffusion region with anisotropic particle transport and energy dissipation. This characteristic of anisotropic energy dissipation was incorporated in 3-D cellular automaton models by defining 6 magnetic field gradients to the next neighbor cells (Vlahos et al. 1995),

$$\begin{aligned}
\Delta B^1_{i,j,k} &= B_{i,j,k} - B_{i+1,j,k} \\
\Delta B^2_{i,j,k} &= B_{i,j,k} - B_{i-1,j,k} \\
\Delta B^3_{i,j,k} &= B_{i,j,k} - B_{i,j-1,k} \\
\Delta B^4_{i,j,k} &= B_{i,j,k} - B_{i,j+1,k} \\
\Delta B^5_{i,j,k} &= B_{i,j,k} - B_{i,j,k-1} \\
\Delta B^6_{i,j,k} &= B_{i,j,k} - B_{i,j,k+1}
\end{aligned} \qquad (2.6.18)\ .$$

instead of one single slope averaged over the 6 next neighbors (Eq. 2.6.1) as introduced by Lu and Hamilton (1991). A redistribution or energy release is then applied when a critical value B_{cr} is exceeded by any of the 6 slopes, say in direction $[i, i+1]$,

$$B_{i,j,k} \mapsto B_{i,j,k} - \frac{6}{7} B_{cr}\ , \quad B_{i+1,j,k} \mapsto B_{i+1,j,k} + \frac{6}{7} B_{cr}\ . \qquad (2.6.19)$$

If more than one direction exceeds the critical threshold B_{cr}, the redistribution of energy is applied to all unstable directions $a \le 6$, weighted by the relative magnetic field gradient,

$$B_{i\pm1,j\pm1,k\pm1} \mapsto B_{i\pm1,j\pm1,k\pm1} + \frac{6}{7} B_{cr} \frac{\Delta B^a_{i,j,k}}{\sum_a \Delta B^a_{i,j,k}}\ . \qquad (2.6.20)$$

In contrast to the vector field $\mathbf{B}(\mathbf{x})$ used by Lu and Hamilton (1991), a scalar field $B(\mathbf{x})$ is used in Vlahos et al. (1995), which implies that no energy is placed in twisting magnetic fields. Using this anisotropic cellular automaton model, Vlahos et al. (1995) find frequency distributions that are quite different from isotropic cellular automaton models, as shown in Fig. 2.18. The distributions show a plateau with a nearly constant occurrence rate at low energies, and a steep powerlaw cutoff at high energies, with an approximate powerlaw index of $\alpha_E = 3.4 \pm 0.1$ for energies E, $\alpha_P \approx 3.7 \pm 0.1$ for peak luminosities P, and $\alpha_D \approx 8.5 \pm 1.5$ for durations D. It is also found that the extension of the plateau is shorter the lower the critical threshold B_{cr} is set in the numerical simulations. This dual behavior is found to be robust, even when the energy input (i.e., loading by photospheric turbulence) is driven at different rates (Georgoulis and Vlahos 1998).

In a hybrid model, the anisotropic and isotropic model were added together, yielding a synthesized distribution (Fig. 2.19) with a steep powerlaw slope ($\alpha \approx 3.5$) at low energies

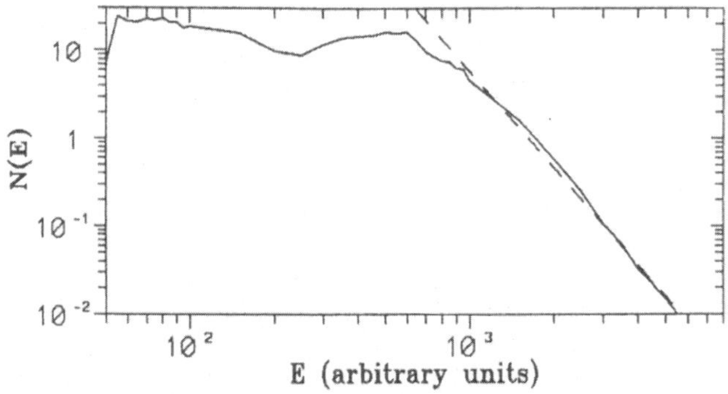

Fig. 2.18 Frequency distribution of energies simulated in a $100 \times 100 \times 100$ grid using a SOC model with anisotropic energy dissipation. The distribution has a flat plateau at low energies and a steep powerlaw slope of $\alpha \approx 3.7$ at high energies (Vlahos et al. 1995).

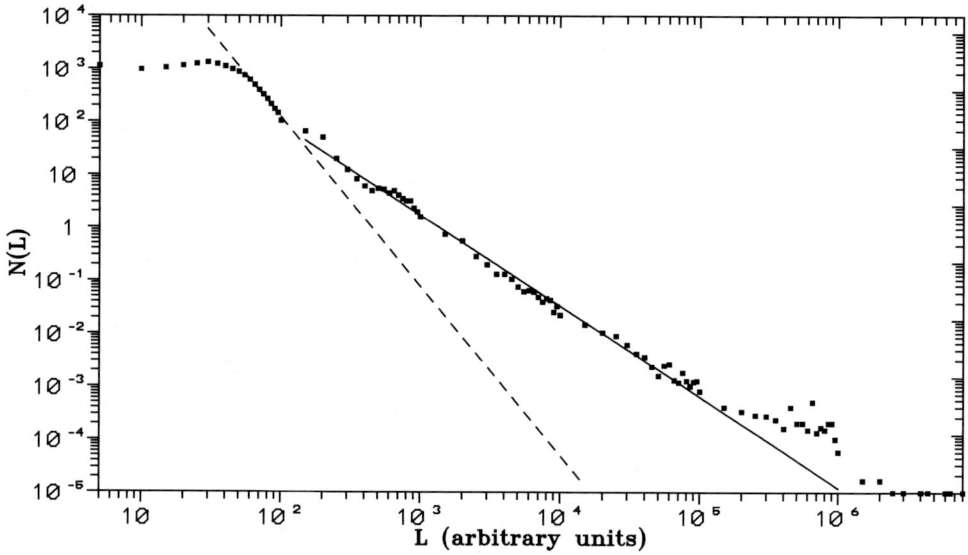

Fig. 2.19 Frequency distribution of the peak luminosity L in a hybrid model with anisotropic energy dissipation for nanoflares and isotropic dissipation for large flares, simulated in a $150 \times 150 \times 150$ grid. The distribution exhibits a steep powerlaw slope of $\alpha \approx 3.3$ at low energies (nanoflares) and a flatter powerlaw slope of $\alpha \approx 1.7$ at high energies for large flares (Georgoulis and Vlahos 1996).

for small flares, and flatter slopes ($\alpha \approx 1.8$) for larger flares (Vlahos et al. 1995; Georgoulis et al. 1995; Georgoulis and Vlahos 1996). This model was also applied to model the bimodal distributions of flare data observed with the WATCH spacecraft (Georgoulis et al. 2001).

This result was interpreted in terms of a unified two-component model of (i) normal flares that release energy in an active region, and (ii) nanoflare events that heat the overall corona. Since the singular value $\alpha = 2$ of the powerlaw slope in energy demarcates the limit where the integral of the frequency distribution diverges, either at the upper end (for $\alpha < 2$) or lower end (for $\alpha > 2$), the numerical result of the steep slope of nanoflares beyond the critical limit seems to imply that nanoflares could have an unlimited amount of undetected energy available at the lowest energies that could account for coronal heating. However, although some observations suggested a steeper slope for nanoflares (see Fig. 1.14), this result could also be explained by instrumental biases and remained controversial. Sampling nanoflares with comprehensive temperature coverage yielded the same powerlaw slope for nanoflares and large flares (Fig. 1.14), and thus did not support the hybrid model of anisotropic energy dissipation in nanoflares and isotropic dissipation in large flares. However, since anisotropic energy dissipation during magnetic reconnection is more likely based on physical models, it probably applies to both nanoflares and large flares.

2.6.3 Discretized MHD Cellular Automaton Models

The magnetohydrodynamic (MHD) evolution of the coronal or solar flare plasma can be simulated with numerical simulations that solve the ideal or resistive MHD equations, which are based on the basic Maxwell equations known in classical electrodynamics. Such numerical MHD simulations are usually coded in a discretized 2-D or 3-D lattice grid, and thus entail next-neighbor interactions in discretized grids as in the (physics-free) cellular automaton algorithms. So, it is a natural desire to derive a physics-based discretization of the MHD equations in order to understand SOC models in terms of physical models, such as magnetic reconnection that drives solar flares. Some studies have been conducted to derive physics-based cellular automaton models from the discretization of the MHD equations (Vassiliadis et al. 1998; Isliker et al. 1998a, 2000, 2001).

The basic approach of Vassiliadis et al. (1998) has already been described in Section 2.5.2 applied to magnetospheric substorms by Takalo et al. (1999a). Each cell of a 2-D lattice is associated with the cross-section of flux tubes aligned with a magnetic field $\mathbf{B}(x,y) = (0,0,B_z)$ in perpendicular direction (z) to the lattice plane (x,y). The currents $\mathbf{j}(x,y,0) = (j_x, j_y, 0)$ along the four sides of each cell boundary are then computed from the induction equation quantified in terms of discretized magnetic field gradients $(dB/dx, dB/dy)$ between the next-neighbor cells. The induction equation $d\mathbf{B}/dt = \eta \nabla^2 \mathbf{B} + S(x,y,t)$, consisting of magnetic diffusion (with magnetic diffusivity constant η) and a convective term, which is represented here with a source term $S(x,y,t)$ that randomly disturbs the magnetic field (like the dropping sand grains on a sandpile), describe the nonlinear dynamics of the system. Magnetic diffusion with classical resistivity η is very slow, and thus nonlinear resistivity $\eta(\mathbf{j})$ is required (for anisotropic cellular automaton models) or hyper-resistivity $\eta(\nabla^2 \mathbf{j})$ (for isotropic ones) in order to enable a rapid dissipation and relaxation mechanism (Vassiliadis et al. 1998). The discrete MHD equations satisfy the necessary conditions for a SOC state in terms of local conservation of the magnetic flux and rapid energy dissipation and relaxation by nonlinear resistivity.

Isliker et al. (1988) calculate a discretization of the MHD equation by applying the discrete redistribution rules (Eqs. 2.6.1 and 2.6.2) of Lu and Hamilton (1991) to a 3-D magnetic field and obtain partial differential equations that exhibit a mathematical discontinuity at the cell boundaries that is not consistent with the smooth spatial function expected for a magnetic diffusion process.

A new approach was attempted by Isliker et al. (2000), by choosing a vector potential **A** as the independent variable to characterize the physical state of each lattice point, instead of using the magnetic field **B**, which generally does not fulfill Maxwell's equation of a divergence-free field ($\nabla \cdot \mathbf{B} = 0$) when disturbed randomly. The magnetic field is then defined by

$$\mathbf{B} = \nabla \times \mathbf{A} \ , \tag{2.6.21}$$

which automatically fulfills the divergence-free condition,

$$\nabla \cdot \mathbf{B} = \nabla \cdot (\nabla \times \mathbf{A}) = 0 \ . \tag{2.6.22}$$

With this vector potential approach, the redistribution rules (Eqs. 2.6.1 and 2.6.2) of Lu and Hamilton (1991) are then expressed as,

$$\Delta \mathbf{A}_{ij} = \mathbf{A}_{ij} - \frac{1}{6} \sum_{nn} \mathbf{A}_{nn} \ . \tag{2.6.23}$$

$$\mathbf{A}_{ij} \mapsto \mathbf{A}_{ij} - \frac{6}{7} \Delta \mathbf{A}_{ij} \ , \quad \mathbf{A}_{nn} \mapsto \mathbf{A}_{nn} + \frac{1}{7} \Delta \mathbf{A}_{ij} \ . \tag{2.6.24}$$

The changes in the magnetic field variables are then calculated with Eq. (2.6.20), and the current changes with Ampère's Law,

$$\mathbf{j} = \frac{1}{4\pi} (\nabla \times \mathbf{B}) \ . \tag{2.6.25}$$

Having the resulting magnetic field **B** defined this way in every cell (x,y,z) and the current densities **j** at the cell boundaries $(x \pm \frac{1}{2}, y \pm \frac{1}{2}, z \pm \frac{1}{2})$, the nonlinear system dynamics controlled by the induction equation can then be calculated in the same was as in the approach of Vassiliadis et al. (1998), (see Section 2.5.2 for details). The only difference is that the energy input to the SOC system occurs by random disturbances of the vector potential **A**, rather than of the magnetic field **B**. Similar energy distributions are obtained when the energy quantity in each cell is derived from a scalar field $B_k(\mathbf{x})$ or from a vector field $\mathbf{B}(\mathbf{x}) = \nabla \times \mathbf{A}$ (Isliker et al. 2000, 2001). An example of such a simulation is shown in Fig. 2.20, which exhibits similar powerlaw-like frequency distributions as simulated by Lu and Hamilton (1991) or Lu et al. (1993), see Fig. 2.16 and 2.17.

Note that the definition of the released energy based on the square of the magnitude of the vector field as a measure of energy density, i.e. $E \propto \sum B_j^2$ (Eq. 2.6.3) is physically sound when the vector field is assumed to correspond to **B**, but not for **A** (Galsgaard 1996). A modified redistribution rule that removes a constant amount of tension ΔA_{crit}, rather than the full amount of the total field gradient $\Delta A_{i,j,k}$, was suggested by Galsgaard (1996),

$$\mathbf{A}_{i,j,k} \mapsto \mathbf{A}_{i,j,k} - \frac{6}{7} \Delta A_{crit} \frac{\Delta \mathbf{A}_{i,j,k}}{|\Delta \mathbf{A}_{i,j,k}|} \ , \tag{2.6.26}$$

$$\mathbf{A}_{i\pm1,j\pm1,k\pm1} \mapsto \mathbf{A}_{i\pm1,j\pm,k\pm1} + \frac{1}{7} \Delta A_{crit} \frac{\Delta \mathbf{A}_{i,j,k}}{|\Delta \mathbf{A}_{i,j,k}|} \ . \tag{2.6.27}$$

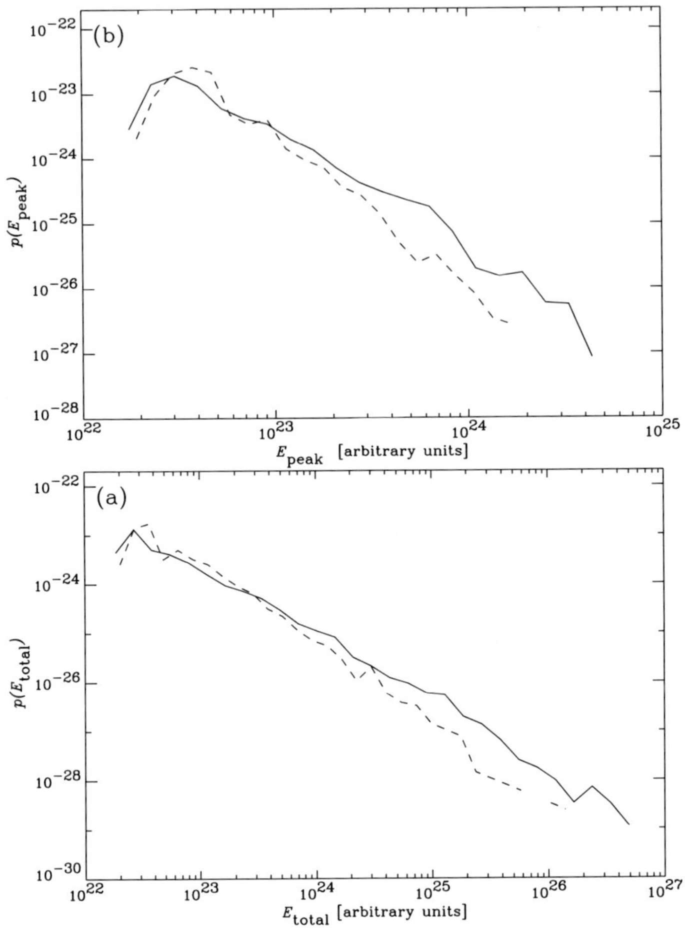

Fig. 2.20 Probability distributions of total energy (top) and peak flux (bottom) of a cellular automaton model that produces current instabilities, mimicking energy releases in solar flares, simulated in terms of discretized MHD equations. The disturbance of the nonlinear system and the threshold for instability is quantified in terms of the vector potential **A** (solid) or current density **j** (dashed) (Isliker et al. 2001).

With some algebra (see Eqs. 2.6.4–2.6.11) it can be shown that the energy released during one redistribution step is (Galsgaard 1996),

$$E_m = \frac{6}{7}\Delta A_{crit}^2 \left(2\frac{|\Delta \mathbf{A}_{i,j,k}|}{\Delta A_{crit}} - 1 \right) , \tag{2.6.28}$$

or more generally for arbitrary dimensions $D = 1, 2, 3, ...$ (Charbonneau et al. 2001),

$$E_m = \frac{2D}{2D+1}\Delta A_{crit}^2 \left(2\frac{|\Delta \mathbf{A}_{i,j,k}|}{\Delta A_{crit}} - 1 \right) . \tag{2.6.29}$$

Thus, the smallest "quantum" of energy that can be released by a lattice is (Charbonneau et al. 2001),

$$E_{min} = \frac{2D}{2D+1} \Delta A_{crit}^2 \ . \quad (2.6.30)$$

Comparisons of numerical SOC simulations with full energy redistributions (Lu et al. 1993) and partial energy redistributions (Eqs. 2.6.26–2.6.28) demonstrated that the functional shape of frequency distributions of peak fluxes and energies sensitively depends on the details of the redistribution rules. Galsgaard (1996) found two criteria to be necessary to obtain powerlaw distributions for the energy release: (1) the field must be systematically driven, so that large-scale regions with coherent tension are obtained, and (2) only a fraction of the field quantity triggering the instability must be removed from the local redistribution procedure.

2.6.4 Divergence-Free Field Braiding Models

The solar corona is envisioned to be a system of one-dimensional flux tubes that are subject to a variety of dynamical forces, such as buoyancy forces that make flux tubes emerge from the photosphere and rise into the corona, twisting and braiding caused by photospheric magneto-convection, as well as impulsive pressure forces during magnetic reconnection processes and coronal mass ejections that kick adjacent field lines and cause damped oscillations. The previously considered 2-D or 3-D lattice cellular automaton models capture dynamic processes as time-variable fluctuations of some physical parameters in the coordinate system of a rigid grid, but cannot follow the dynamic motion of an identical magnetic field line or flux tube. A new approach of a cellular automaton model that consists of a lattice grid of deformable magnetic field lines has been developed by Morales and Charbonneau (2008a,b, 2009), similar to the *coronal field braiding model* postulated by Parker (1988). The model, moreover, ensures that the magnetic field stays divergence-free ($\nabla \cdot \mathbf{B} = 0$) during random disturbances, so we call it a "divergence-free field braiding model".

The basic lattice structure of the cellular automaton model of Morales and Charbonneau (2008a) is shown in Fig. 2.21, consisting of initially parallel field lines that have their footpoints rooted in a horizontal 1-D lattice grid (x), while their lengths extend over the vertical axis (y), so that every field line position has a unique coordinate (x, y). This initial setup is periodically disturbed by a random displacement of node (x, y) to position $(x + \Delta x, y)$, which is measured by a misalignment angle $\theta(x, y)$. If the misalignment angle exceeds a critical angle θ_c, the intersecting field lines can reconnect into an alternative configuration by a "cut-and-splice" operation, until a sub-critical misalignment angle is reached. The redistribution rule involves a new connectivity for crossing field lines that has a lower misalignment angle. The example shown in Fig. 2.21 shows a two-step redistribution scheme where field lines 1 and 3 exchange their connectivity in two time steps and end up with a sub-critical angle everywhere. Multiple reconnection steps are often required to reach a new stable state. Such chains of reconnection steps represent the avalanches in sandpiles. Note that the avalanches are highly anisotropic (similar to the models described in Section 2.6.2), in contrast to isotropic avalanches in the BTW sandpile model (Section 2.6.1).

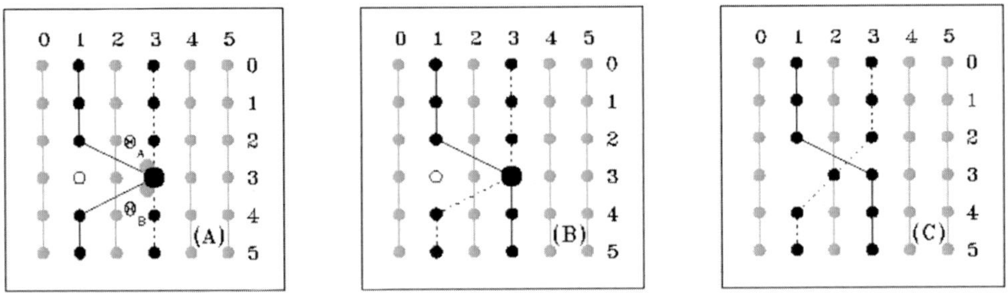

Fig. 2.21 Sequence of a two-step redistribution scheme in the divergence-free field braiding cellular automaton model of Morales and Charbonneau (2008a). Node $j = 3$ of field line $i = 1$ is displaced by two units, causing misalignment angles θ_A and θ_B above a critical value (left frame A), triggering a new connectivity between field line $i = 1$ and $i = 3$ (middle frame B), which is still not stable and triggers another new connectivity between field line 1 and 3 (dashed; right frame C) (reproduced by permission of the AAS).

This braiding model is designed to maintain divergence-freeness during each time step. Initially, every node has a distance Δy to the next neighbor. Let us consider field lines with an initial length of $L = n \times \Delta y$. If a node is displaced by a transverse disturbance of δx in an intermediate position, the new length of the deformed field line is $L' = L - 2\Delta y + 2\sqrt{\Delta y^2 + \delta x^2}$, and the misalignment angle is $\theta = \delta x/\Delta y$. Each field line has a cross-section of $A = \Delta x^2$. In order to ensure mass conservation, the cross-section A' at the deformed location has to reduce (i.e., thinning) to compensate for the stretching of the flux tube mass M,

$$M = \rho AL = \rho A'L' , \tag{2.6.31}$$

if we assume a constant density ρ in an incompressible fluid. At the same time, the magnetic flux $\Phi = AB$ has to be conserved

$$\Phi = AB = A'B' . \tag{2.6.32}$$

These two conditions yield a scaling of the magnetic field B' with the deformed length L' of the field line,

$$B' = B\left(\frac{L'}{L}\right) . \tag{2.6.33}$$

The total magnetic energy of the magnetic field line is then

$$E_m(t) = \frac{1}{8\pi}\int_V \mathbf{B}(t)^2 dV = \frac{V_0}{8\pi}B_0^2 \sum_{i=1}^{n-1}\left(\frac{l_i(t)}{l_0}\right)^2 \tag{2.6.34}$$

where V_0 and B_0 are the initial volume and field strength of a field strand, and $l_i(t)$ is the length of strand i at time t. Every deformation of a flux tube corresponds to a lengthening of $l_i(t)$ and thus represents an energy input into the system. Once the threshold is exceeded, an avalanche of reconnection occurs which shortens the loop segments, corresponding to a decrease or dissipation of the energy according to Eq. (2.6.34). This procedure to vary the magnetic field strength of a flux strand, in response to stretching its length, conserves the magnetic flux $\Phi = AB$ of a strand by design, and thus the model automatically satisfies the flux conservation constraint or divergence-free condition $\nabla \cdot \mathbf{B} = 0$.

Morales and Charbonneau (2008a) perform numerical simulations with this model and find that the system transitions into a SOC state after an initial stressing phase, where avalanches of all sizes occur and produce powerlaw-like frequency distributions with slopes of $\alpha_E \approx 1.63$–1.72 for energies, $\alpha_P \approx 1.73$–1.84 for peak fluxes, and $\alpha_T \approx 1.79$–1.95 for durations, which are close to the observed values for solar flares (Eq. 2.6.15). This model can be scaled to coronal loops with lengths of $L_0 = 100$ Mm and diameters of $w = 1$ Mm, which yields flare energies in the range of $E_m \approx 10^{23}$–10^{29} erg, for an instability threshold angle of $\theta_{cr} = 11^\circ$, which is similar to the nanoflare model of Parker (1988).

Statistics of avalanche peak areas A with the same model yields powerlaw distributions also,

$$N(A) \propto A^{-\alpha_A} , \tag{2.6.35}$$

with a powerlaw slope of $\alpha_A \approx 2.45$ (Morales and Charbonneau 2008b), which indeed also agree with flare areas observed in EUV ($\alpha_A \approx 2.3$–2.7) and soft X-rays ($\alpha_A \approx 1.7$–2.1) (Aschwanden and Parnell 2002). In order to obtain more realistic geometries, the 2-D lattice model was also wrapped onto a cylinder, stretched along the cylinder axis, bent into a semi-circular loop structure, rotated into an arbitrary line-of-sight direction, and projected into the plane of the sky, which mimics an observed image of a flare loop. These transformations changed the frequency distributions of the observed flare areas A slightly, with a powerlaw index of $\alpha_A \approx 2.37$ (Morales and Charbonneau 2009). In addition, the simulated images allow also to determine the fractal Hausdorff dimension D_2,

$$A(L) \propto L^{D_2} , \tag{2.6.36}$$

for which a value of $D_2 = 1.17$–1.24 was found (Morales and Charbonneau 2009), which is significantly smaller than the Euclidean limit $D_2 \leq 2$ of solid filling. These simulated values are somewhat lower than observed at the times of peak flux in nanoflares ($D_2 \approx 1.5$–1.9; Aschwanden and Parnell 2002) and in large flares ($D_2 \approx 1.0$–1.9; Aschwanden and Aschwanden 2008a). However, we have to be aware that the fractal structure of observed flare areas generally appears smoothed by insufficient instrumental resolution, temperature discrimination, and background confusion, while numerical simulations produce sharper and crisper images, and thus can measure lower values for the fractal dimension. The inclusion of the fractal dimension in volume modeling affects also energy models of flares, and thus the powerlaw slope of energy frequency distributions (McIntosh and Charbonneau 2001; McIntosh et al. 2002; Mitra-Kraev and Benz 2001; Aschwanden and Parnell 2002). More details about the fractal dimension of SOC structures will be discussed in Chapter 8.

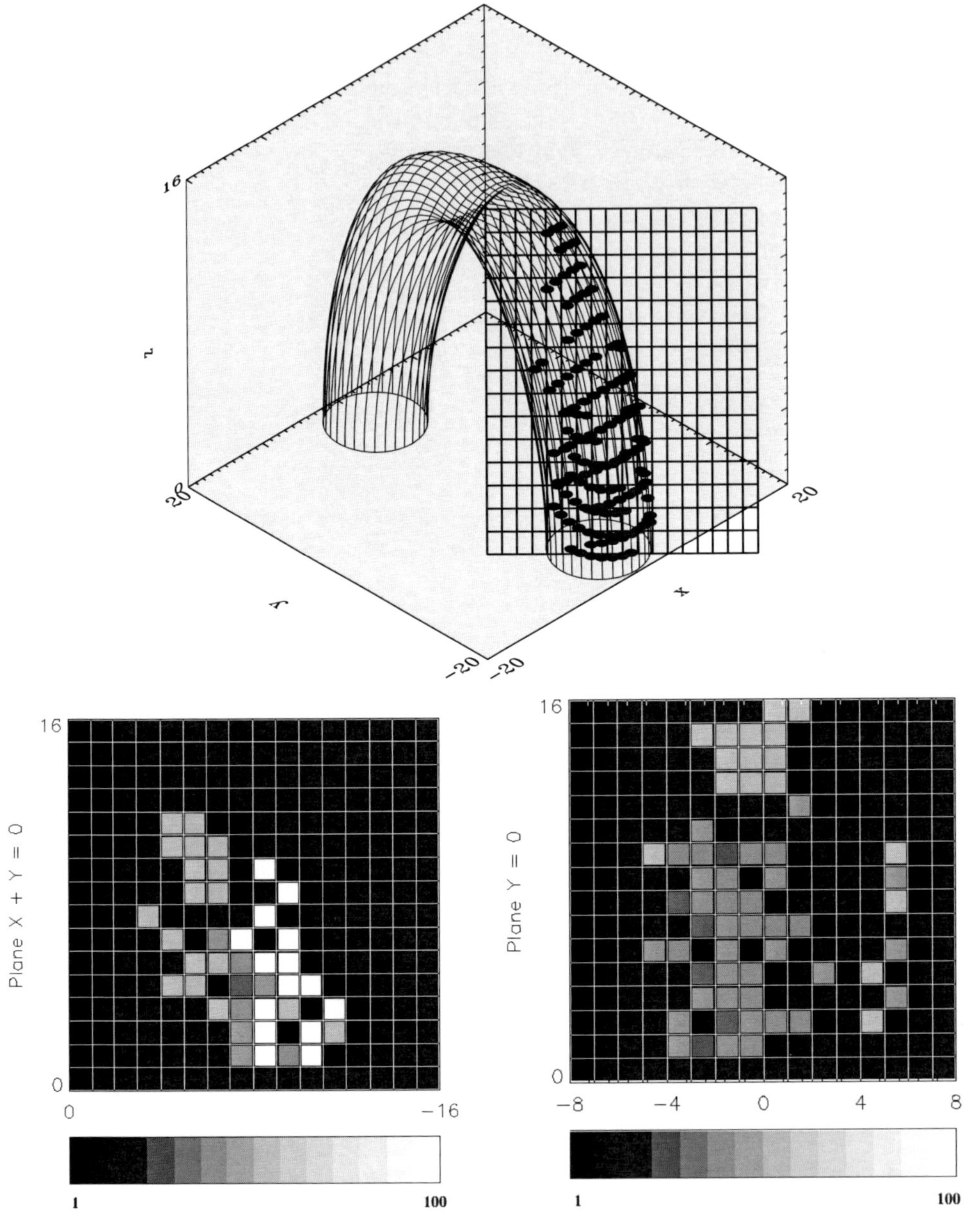

Fig. 2.22 Transformation of the flat 2-D lattice geometry of the divergence-free braiding model onto a pseudo 3-D loop envelope and plane-of-sky for an arbitrary observer's line-of-sight direction (top panel). The fractal area of the projected avalanches are shown for two different directions, for the planes $X+Y=0$ (left) and $Y=0$ (right) (Morales and Charbonneau 2009b; reproduced by permission of the AAS).

2.6.5 Branching Process Models

The previously discussed cellular automaton models all involve a redistribution rule to the next neighbor cells of a lattice grid, which operates whenever a critical threshold is exceeded. A similar concept is a *branching process* in probability theory, which expresses next-neighbor interactions in terms of probabilities.

A simple 1-D cellular automaton model using a branching process applied to solar flares is described in MacKinnon et al. (1996). A lattice site has three states: (1) quiescent, (2) flaring, and (3) flared, so each cell can have three values, $x_i = 0, 1, 2$. The system starts with all sites in state $x_i = 0$. At any time $t = 0, 1, 2, \ldots$, the i-th site may change from state 0 to state 1 with probability p_0 (representing the driving of the system from outside) if none of its neighboring sites were in state 1 in the previous time step, but with probability $p_1 (> p_0)$ if either of the neighboring sites $i-1$ or $i+1$ were in state 1 in the previous time step, similar to the forest-fire model (Section 2.4.2). Applying these rules iteratively over many time steps leads (by combinational arguments) to the probability distribution $P(n)$ of an event of size n as,

$$P(n) = n p_1^{(n-1)} (1 - p_1)^2 \ . \tag{2.6.37}$$

Averaging these probabilities produces then a size distribution

$$\langle P(n) \rangle = \int_0^1 P(n) d p_1 = \frac{2}{(n+1)(n+2)} \propto n^{-2} \ , \tag{2.6.38}$$

in the limit of $n \mapsto \infty$. Thus, if we associate the size n of an event with the energy E of a flare and the probability $\langle P(n) \rangle$ with the occurrence frequency distribution $N(E)$, this branching process model predicts a size distribution of $N(E) \propto E^{-2}$. Numerical simulations of a branching process with modified rules can produce flatter power law slopes, as demonstrated in Macpherson and MacKinnon (1999).

In an attempt to generalize this 1-D branching process to higher dimensions, Litvinenko (1998a) points out a result from a tree branching process, for which an asymptotic limit was found (Otter 1949),

$$\langle P(n) \rangle \propto n^{-3/2} \exp\left(-\frac{n}{n_0} \right) \ , \tag{2.6.39}$$

which is close to the observed frequency distributions of flare energies.

2.7 SOC Simulations in Astrophysics

The most striking powerlaw distributions observed in astrophysics are the spectra of ultra-high energy cosmic rays that extend over 11 orders of magnitude, or giant pulses from black hole accretion disks, both being the possible result of SOC processes. In the following we summarize a SOC cellular automaton model that has been applied to model the giant pulses from accretion disk systems.

2.7.1 Cellular Automaton Model of Accretion Disk Fluctuations

The X-ray light curves of black-hole candidates, such as Cygnus X-1, were found to have quiescent periods with Planckian power spectra (soft state), as well as active periods with powerlaw-like spectra (hard state), i.e., f^{-p} with $p \approx 1$, which were brought into context of $1/f$-fluctuations or flicker noise of SOC systems (Mineshige et al. 1994a). The rapid variability of X-rays with $1/f$ spectra were discovered not only in black-hole objects, for binary sources, and active galactic nuclei, but also in neutron stars and cataclysmic stars.

Inspired by the BTW model (Bak et al. 1988), the following basic cellular automaton model was set up by Mineshige et al. (1994a). The accretion disk is divided into an outer portion where material smoothly drifts inward, and an inner portion where it suffers an instability and tends to form blobs. Using a 2-D cylindrical coordinate system (r, φ) for the the disk plane (Fig. 2.23), cells are labeled with a radial coordinate $r_i, i = 1, ..., n_i$ and an azimuthal coordinate $\varphi_j, j = 1, ..., n_j$, where r_1 marks the outermost ring of the inner zone. A particle with mass m is put into the outermost cell at r_1. The mass density $M_{i,j}$ is defined as the number of mass particles in cell r_i and φ_j. When the mass $M_{i,j}$ exceeds some critical mass density $M_{crit} \propto r$, a part of the accumulated mass in cell (r_i, φ_j) falls into the 3 adjacent cells of the adjacent inner ring, as a consequence of an unknown instability (such as magnetic reconnection or flares),

$$\begin{array}{ll} M_{i,j} & \mapsto M_{i,j} - 3m \\ M_{i,j-1} & \mapsto M_{i,j-1} + m \\ M_{i,j} & \mapsto M_{i,j} + m \\ M_{i,j+1} & \mapsto M_{i,j+1} + m \end{array} , \qquad (2.7.1)$$

After this single redistribution step, further redistributions are executed if one of the newly filled cells exceeds the critical mass M_{crit}, until the whole system is stable again. The

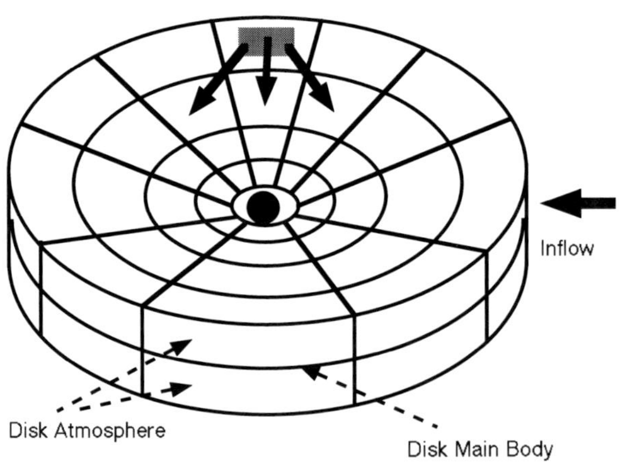

Fig. 2.23 Schematic view of a cellular-automaton model applied to an accretion disk with a black hole in the center. Mass flow from outside randomly triggers avalanches from the outermost ring to inner rings (Yonehara et al. 1997; reproduced by permission of the AAS).

motion of each redistributed k-th particle is tracked from its initial orbit r_i to the final orbit r_j and the the total change in gravitational energy is summed together. The X-ray luminosity L_X resulting from such a mass transfer event is assumed to be approximately proportional to the change in gravitational energy ΔE_{grav},

$$L_X \propto \Delta E_{grav} = G M_{BH} m \sum_k \left(\frac{1}{r_j^k} - \frac{1}{r_i^k} \right) , \tag{2.7.2}$$

where G is Newton's gravitational constant and M_{BH} is the mass of the central black hole. This basic cellular automaton model applies to an accretion disk with rigid rotation. However, a more realistic model should include the effects of differential rotation, which introduces an azimuthal shift j' in the position of the transferred blob, since the orbital time scale in a black-hole accretion disk is much smaller than the blob drift time to an inner radius. Hence, the cellular automaton model should be modified to incorporate this azimuthal shift j' by an amount that corresponds to the differential rotation rate (Mineshige et al. 1994a),

$$\begin{array}{ll} M_{i,j} & \mapsto M_{i,j} - 3m \\ M_{i,j+j'-1} & \mapsto M_{i,j+j'-1} + m \\ M_{i,j+j'} & \mapsto M_{i,j+j'} + m \\ M_{i,j+j'+1} & \mapsto M_{i,j+j'+1} + m \end{array} , \tag{2.7.3}$$

Using this simple cellular automaton model, Mineshige et al. (1994a) are able to produce a powerlaw distribution of energies $N(E) \propto E^{-1.35}$ and a powerlaw distribution of pulse time scales $N(T) \propto T^{-1.7}$, as well as a power spectrum of $S(f) \propto f^{-1.8}$ of the simulated time series, which are all close to the observed values of X-ray pulses from black-hole candidates.

In a slightly modified model, Mineshige et al. (1994b) incorporate a viscous diffusion process, which changes the power spectrum to $S(f) \propto f^{-1.6}$, which is closer to the observed power spectra of $f^{-1.7}$ (Makishima 1988) or $f^{-1.5}$ (Negoro 1992) at $f \gtrsim 1$ Hz. The authors point out that they can obtain different power spectra ($f^{-1.3}$, ..., $f^{-1.1}$) or time scale distributions (from powerlaw-like to exponential) if they change the redistribution rule slightly.

New observations (Negoro et al. 1995) indicated that the peak intensities of X-ray fluctuations from Cygnus X-1 exhibited exponential distributions, rather than powerlaws as expected for SOC, which triggered more modifications of the SOC model in terms of enhancing the effects of gradual diffusion. This model contains additional mass transfer (besides the avalanches specified with Eq. (2.7.3)),

$$\begin{array}{ll} M_{i,j} & \mapsto M_{i,j} - m' \\ M_{i+1,j} & \mapsto M_{i+1,j} + m' \end{array} , \tag{2.7.4}$$

with $m' = m/100,\ m/10,\ m/5$, which flattens the power spectrum and deforms the powerlaw distribution of time scales into exponential distributions (Takeuchi et al. 1995). Further modeling was performed by describing the X-ray fluctuations with fluid dynamics in an advection-dominated disk with critical behavior, which also could reproduce the $1/f$ power spectrum (Takeuchi and Mineshige 1997). More extensions of the model included relativistic effects, which causes substantial differences if the disk is viewed from directions far from the accretion disk axis (Xiong et al. 2000).

Besides the X-ray fluctuations from X-ray binaries or black-hole accretion disks, which is emitted in an optically thin medium, also the UV radiation from cataclysmic stars (such as Canes Venatici [CV] types), which is emitted from an optically thick medium, was modeled with the same SOC cellular automaton model, which was able to reproduce the observed power spectra that are flat at lower frequencies and have a powerlaw slope of $p \approx 1$–2 at high frequencies (Yonehara et al. 1997).

More advanced SOC cellular automaton models of accretion disks address also the physics of the magnetic field, which entails effects such as the Balbus–Hawley instability (driving a disk dynamo), buoyancy of magnetic fields, magnetic flux emergence, disappearance, and flaring, by including non-local transport of angular momentum in terms of the kinematic viscosity of the magnetic loops in the disk corona (Pavlidou et al. 2001). A snapshot of the magnetic loops in the modeled accretion disk are shown in Fig. 2.24.

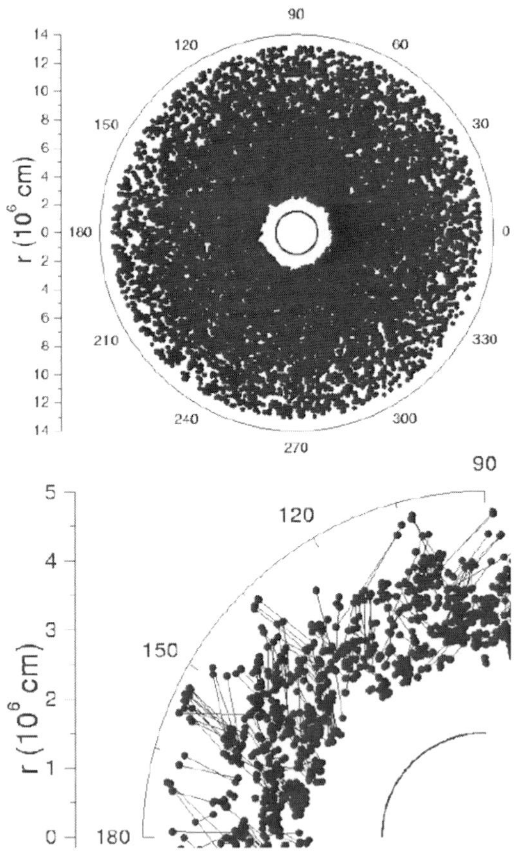

Fig. 2.24 A snapshot of the magnetic loops in a modeled accretion disk surrounding a compact object, shown in full view face-on (top) and as an enlarged segment (bottom), according to simulations with a numerical SOC cellular automaton model by Pavlidou et al. (2001).

2.8 Summary

The theoretical understanding of nonlinear dissipative systems in the state of self-organized criticality started with numerical simulations of a cellular automaton model, which is driven by slow external forcing. The essential framework of a cellular automaton model is a mathematical algorithm that consists of (1) a critical threshold, (2) a next-neighbor redistribution rule that is applied when a local threshold is exceeded, and (3) subsequent iterations of the redistribution rules, until a new equilibrium is reached. Cellular automaton models can simulate the dynamics and outcome of complex multi-element systems, which is often not achievable by analytical theories, but are essentially physics-free models. A great success of cellular automaton models is the reproduction of powerlaw-like distributions for the peak fluxes, energies, and time scales of avalanche events, which emerge as robust characteristics without fine-tuning of the initial conditions. The most influential cellular automaton model in SOC processes is the Bak–Tang–Wiesenfeld (BTW) sandpile model, as well as the slider-block and the forest-fire model in geophysics. After 1990, these cellular automaton models have also been applied in astrophysics, such as in magnetospheric substorms, in solar flares, and in accretion disks, but the meaning of the energy quantity in the BTW sandpile model is often replaced by magnetic energies. Some attempts at reverse engineering have been made to translate the mathematical redistribution rules in SOC sandpile models to physical, discretized differential equations in terms of magneto-hydrodynamics (MHD), which also imply anisotropic transport processes and additional constraints from Maxwell's equations (such as divergence-free magnetic fields). The reduction of physical models of nonlinear processes to discretized cellular automaton algorithms is still in its infancy for astrophysical systems.

2.9 Problems

Problem 2.1: Derive the MHD induction equation (Eq. 2.5.6) from Maxwell's equations and the discretization of the MHD equations outlined in the Eqs. (2.5.5–2.5.19).

Problem 2.2: Derive the amount of released energy in a D-dimensional lattice vector field $\mathbf{A}$ based on the redistribution rule given in Eqs. (2.6.1–2.6.3) and prove the result given in Eqs. (2.6.4–2.6.14).

Problem 2.3: Generalize the divergence-free field braiding model from a 2-D flux tube surface geometry to a 3-D (solidly filled) cylindric geometry. What are the equivalent equations of (2.6.31–2.6.34)?

3. Analytical SOC Models

Every theory should be as simple as possible, but not over-simplified.

Albert Einstein

There is something universal about SOC systems that does not depend on some particular physical parameters, because they all exhibit powerlaw distributions, a necessary but not sufficient condition. There is also something universal about a fractal dimension, because it is manifested in many different physical systems. In the previous chapter we reviewed cellular automaton models, which are based on mathematical redistribution rules and do not require any specific physical model. In this chapter we develop an analytical theory of SOC phenomena that is "physics-free", and thus can be applied to sandpiles, earthquakes, or solar flares equally. Essentially, our analytical approach provides an understanding of SOC phenomena in terms of *the universal statistics of nonlinear systems*, which in the limit of linear system behavior degenerates to *the statistics of random processes*. The powerlaw distributions of SOC phenomena always represent the statistics of relatively rare events, the high-end or fat tails of probabilistic distributions, which are discernible in a log-log representation only. At the low-end, events are most frequent and can often be described by standard probability theory, such as Poisson statistics or Gaussian normal distributions. Monte-Carlo simulations can simulate such event statistics for any arbitrary physical process, in form of discretized event parameters that can be sampled in binned histograms. In the continuum limit, such distributions can sometimes be described by analytical functions. In analogy, we develop an analytical SOC theory that describes the continuum limit of numerical SOC simulations, as they have been reviewed in the previous chapter. The power of analytical models is the prediction of exact distributions, which can significantly deviate from exponential or powerlaw functions, often not evident from observations due to the limited statistics of rare events. Analytical models thus can then be forward-fitted to the statistical distributions of observed SOC events. The analytical theory will provide us also a rigorous mathematical definition of SOC processes.

3.1 The Exponential-Growth Model

Avalanches occurring in the state of self-organized criticality represent local instabilities that grow explosively for some time interval. The released energy grows in a nonlinear way above some energy threshold, which can be parameterized by some nonlinear function, for instance by an exponential growth function. We define the time evolution of the energy release rate $W(t)$ of a nonlinear process that starts at a threshold energy of W_0 by

$$W(t) = W_0 \exp\left(\frac{t}{\tau_G}\right), \qquad 0 \le t \le \tau , \tag{3.1.1}$$

where τ_G represents the exponential growth time. The process grows exponentially until it saturates at time $t = \tau$ with a saturation energy W_S,

$$W_S = W(t = \tau) = W_0 \exp\left(\frac{\tau}{\tau_G}\right) . \tag{3.1.2}$$

We define a peak energy release rate P that represents the maximum energy release rate W_S, after subtraction of the threshold energy W_0, that corresponds to the steady-state energy level before the nonlinear growth phase,

$$P = W_S - W_0 = W_0 \left[\exp\left(\frac{\tau}{\tau_G}\right) - 1\right] . \tag{3.1.3}$$

In the following, we will refer to the peak energy release rate P also briefly as "peak energy". For the saturation times τ, which we also call "rise times", we assume a random probability distribution, approximated by an exponential function $N(\tau)$ with e-folding time constant t_S,

$$N(\tau) d\tau = \frac{N_0}{t_S} \exp\left(-\frac{\tau}{t_S}\right) d\tau . \tag{3.1.4}$$

This probability distribution is normalized to the total number of N_0 events,

$$\int_0^\infty N(\tau)\, d\tau = N_0 . \tag{3.1.5}$$

In order to derive the probability distribution $N(P)$ of peak energy release rates P, we have to substitute the variable of the peak energy, P, into the function of the rise time $\tau(P)$,

$$N(P)\, dP = N(\tau)\, d\tau = N[\tau(P)] \left|\frac{d\tau}{dP}\right| dP . \tag{3.1.6}$$

This requires the inversion of the evolution function $P(\tau)$ (Eq. 3.1.3),

$$\tau(P) = \tau_G \ln\left(\frac{P}{W_0} + 1\right) , \tag{3.1.7}$$

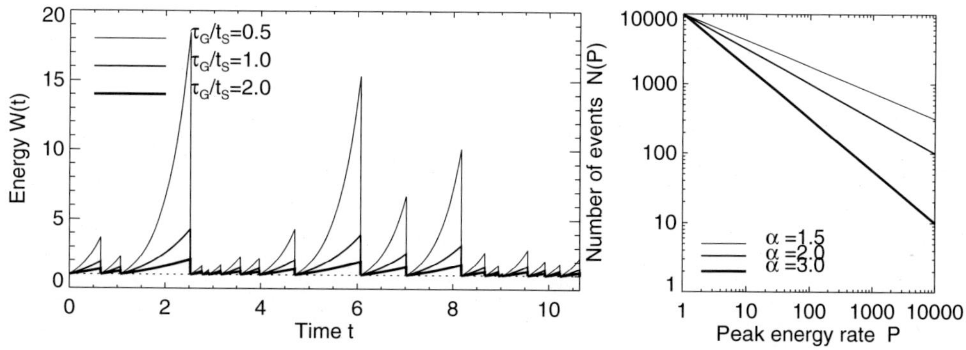

Fig. 3.1 Time evolution of energy release rate $W(t)$ for 3 different ratios of growth times to saturation times, $\tau_G/t_S = (0.5, 1.0, 2.0)$ (left) and the corresponding powerlaw distributions of the peak energy release rate P. Note that the event set with the shortest growth time ($\tau_G/t_S = 0.5$) reaches the highest energies and thus produces the flattest powerlaw slope ($\alpha = 1 + \tau_G/t_S = 1.5$).

and the calculation of its derivative $d\tau/dP$, which is

$$\frac{d\tau}{dP} = \frac{\tau_G}{W_0} \left(\frac{P}{W_0} + 1 \right)^{-1} . \tag{3.1.8}$$

Inserting the probability distribution of saturation times $N(\tau)$ (Eq. 3.1.4), the inverted evolution function $\tau(P)$ (Eq. 3.1.7) and its time derivative $(d\tau/dP)$ from Eq. (3.1.8) into the frequency distribution $N(P)$ in Eq. (3.1.6) yields then,

$$N(P)\, dP = \frac{N_0(\alpha_P - 1)}{W_0} \left(\frac{P}{W_0} + 1 \right)^{-\alpha_P} dP , \tag{3.1.9}$$

which is an exact powerlaw distribution for large peak energies ($P \gg W_0$) with a powerlaw slope α_P of

$$\alpha_P = \left(1 + \frac{\tau_G}{t_S} \right) . \tag{3.1.10}$$

The powerlaw slope thus depends on the ratio of the growth time to the e-folding saturation time, which is essentially the average number of growth times. Examples of time series with avalanches of different growth times ($\tau_G/t_S = 0.5, 1.0, 2.0$) are shown in Fig. 3.1, along with the corresponding powerlaw distributions of peak energies P. Note that the fastest growing events produce the flattest powerlaw distribution of peak energies.

Once an instability has released a maximum amount W_S of energy, say when an avalanche reaches its largest velocity on a sandpile, the energy release gradually slows down until the avalanche comes to rest. For sake of simplicity we assume a constant energy decay rate η after the peak of the energy release, lasting for a time interval D until it drops to the threshold level W_0,

$$\eta = \frac{W_S - W_0}{D} = \frac{W_0}{\tau_D} , \tag{3.1.11}$$

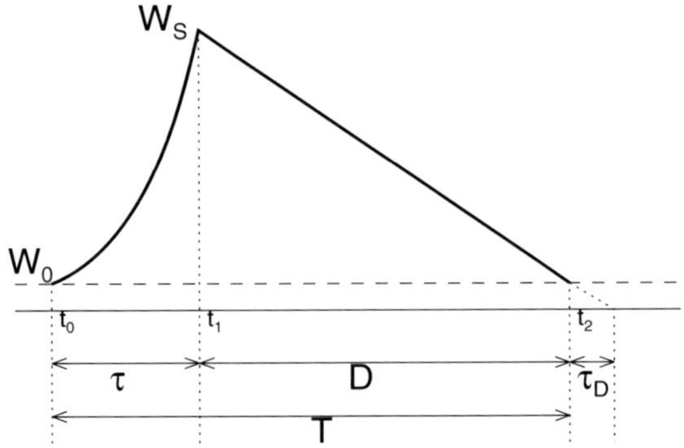

Fig. 3.2 Schematic of the time evolution of an avalanche event, consisting of (i) a rise time (τ) with exponential growth of the energy release $W(t)$ from a threshold level W_0 to the saturation level W_S, and (ii) a decay time (D) with a constant decay rate $\eta = dW/dt = W_0/\tau_D$.

which produces a linear decay of the released energy (see Fig. 3.2),

$$W(t) = W_0 + (W_S - W_0)\left(1 - \frac{(t-t_1)}{D}\right) \quad t_1 < t < t_2 \ , \tag{3.1.12}$$

where t_2 is the end time of the process at $t_2 = t_1 + D$. The time interval D of the energy decay thus depends on the peak energy release rate P as (with Eqs. 3.1.11 and 3.1.3),

$$D = \tau_D\left(\frac{W_S}{W_0} - 1\right) = \tau_D\left(\frac{P}{W_0}\right) \ . \tag{3.1.13}$$

We define now the time interval T of the total duration of the avalanche process as the sum of the exponential rise phase τ (Eq. 3.1.7) and the linear decay phase D (Eq. 3.1.13), as illustrated in Fig. 3.2,

$$T = \tau + D = \tau_G \ln\left(\frac{P}{W_0} + 1\right) + \tau_D \frac{P}{W_0} \ . \tag{3.1.14}$$

We see that this relationship predicts an approximate proportionality of $T \propto P$ for large avalanches, since the second term, which is linear to P, becomes far greater than the first term with a logarithmic dependence ($\propto \ln P$).

For the calculation of the distribution $N(\tau)$ we express the total duration T in terms of the rise time τ (with Eqs. 3.1.5, 3.1.13, and 3.1.14),

$$T(\tau) = \tau + \tau_D\left[\exp\left(\frac{\tau}{\tau_G}\right) - 1\right] \tag{3.1.15}$$

Since the exponential term $\exp(\tau/\tau_G)$ becomes much greater than the linear term for large avalanches, we can neglect the first term,

$$T(\tau) \approx \tau_D\left[\exp\left(\frac{\tau}{\tau_G}\right) - 1\right] \tag{3.1.16}$$

With this approximation we can invert $T(\tau)$,

$$\tau(T) = \tau_G \ln\left[\frac{T}{\tau_D} + 1\right] , \tag{3.1.17}$$

and calculate the derivative,

$$\frac{d\tau}{dT} = \frac{\tau_G}{\tau_D}\left[\frac{T}{\tau_D} + 1\right]^{-1} , \tag{3.1.18}$$

which allows us to calculate the frequency distribution $N(T)$ of total durations T by substituting Eqs. (3.1.17) and (3.1.18) into the distribution $N(\tau)$ Eqs. (3.1.4) of rise times,

$$N(T)\, dT = N[\tau(T)] \left|\frac{d\tau}{dT}\right| dT = \frac{N_0(\alpha_P - 1)}{\tau_D}\left(\frac{T}{\tau_D} + 1\right)^{-\alpha_P} , \tag{3.1.19}$$

which is a powerlaw function for large durations T with the same slope as the peak energy rate P (Eq. 3.1.9).

We define also the total released energy E by the time integral of the energy release rate $W(t)$ during the event duration T, but neglect the rise time τ (i.e., $T \approx D$), using Eq. (3.1.12) and subtract the threshold level W_0 before the avalanche,

$$E = \int_0^T [W(t) - W_0]\, dt \approx \int_\tau^{\tau+D} [W(t) - W_0]\, dt = \frac{1}{2} PD\, . \tag{3.1.20}$$

Inserting the relations $D(P)$ (Eq. 3.1.13) we obtain the dependence of the total energy $E(P)$ on the peak energy P

$$E(P) = \frac{\tau_D}{2W_0} P^2\, . \tag{3.1.21}$$

Defining a reference energy E_0,

$$E_0 = \frac{W_0 \tau_D}{2} , \tag{3.1.22}$$

we have the inverted function $P(E)$ of $E(P)$ from Eq. (3.1.21),

$$P(E) = W_0\left(\frac{E}{E_0}\right)^{1/2} , \tag{3.1.23}$$

and the derivative dP/dE,

$$\frac{dP}{dE} = \frac{W_0}{2E_0}\left(\frac{E}{E_0}\right)^{-1/2} , \tag{3.1.24}$$

and we can then derive the frequency distribution $N(E)$ of energies by inserting $P(E)$ (Eqs. 3.1.23) and dP/dE (3.1.24) in the distribution $N(P)$ of peak energies (Eq. 3.1.9),

$$N(E)\, dE = N[P(E)] \left|\frac{dP}{dE}\right| dE = \frac{N_0(\alpha_P - 1)}{2E_0}\left[\sqrt{\frac{E}{E_0}} + 1\right]^{-\alpha_P}\left[\frac{E}{E_0}\right]^{-1/2} \tag{3.1.25}$$

The resulting frequency distribution $N(E)$ of energies is close to a powerlaw distribution and converges to the slope $\alpha_E = (\alpha_P + 1)/2$ for large energies,

$$N(E)\, dE \approx \frac{N_0(\alpha_P - 1)}{2E_0} \left(\frac{E}{E_0} \right)^{-(\alpha_P+1)/2} . \tag{3.1.26}$$

We show the frequency distributions of the total energy E, peak energy P, rise time τ, and total duration T in Fig. 3.3, for three different ratios of the growth rate to the average saturation time t_S, i.e., $\tau_G/t_S = 0.5$, 1, and 2. We see that this model can accommodate a range of powerlaw slopes in the upper energy range and predicts particular correlations between the three parameters E, P and T (from Eqs. 3.1.14 and 3.1.21),

$$\begin{aligned} (E/E_0) &\approx (P/W_0)^2 \\ (T/\tau_D) &\approx (P/W_0)^1 \\ (E/E_0) &\approx (T/\tau_D)^2 \end{aligned} \tag{3.1.27}$$

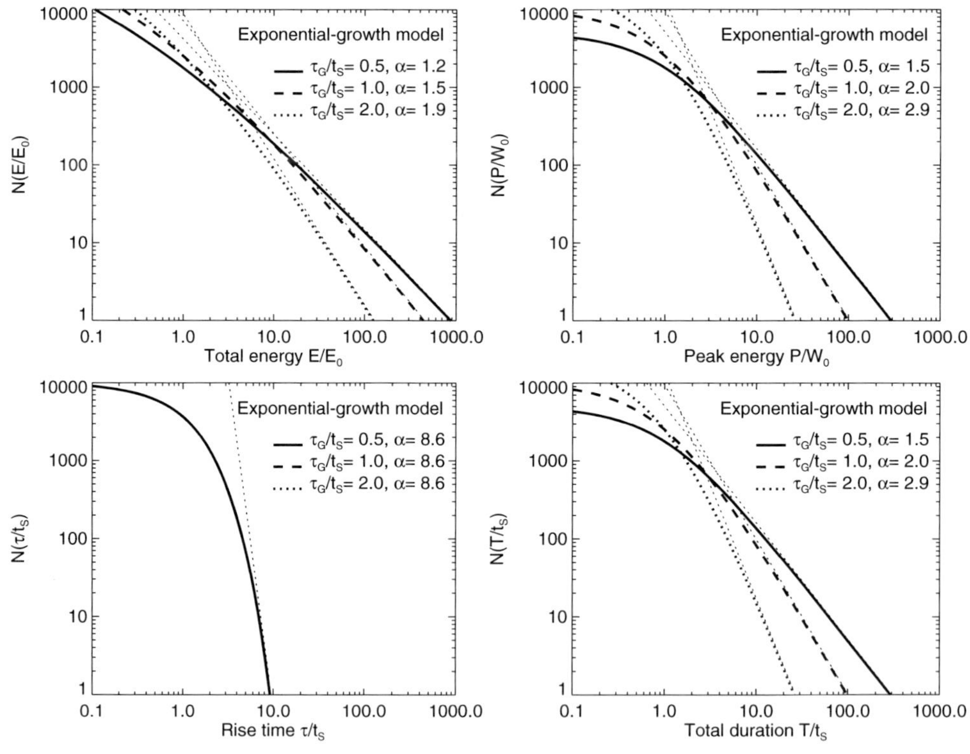

Fig. 3.3 Occurrence frequency distribution of total energies E (top left), peak energies P (top right), energy release times τ (bottom left), and total time durations T (bottom right) for the exponential-growth model, for τ_G/t_S=0.5, 1, and 2. Powerlaw fits are performed at the upper end of the distributions (dotted thin lines), with the slopes α_P indicated in each panel. The distributions contain $N_0 = 10^4$ events. The probability density functions can be obtained by dividing the y-axis by $N_0 = 10^4$.

while the powerlaw slopes are related to each other by

$$\begin{aligned} \alpha_P &= 1 + \tau_G / t_S \\ \alpha_T &= \alpha_P \\ \alpha_E &= (\alpha_P + 1)/2 \end{aligned} \quad (3.1.28)$$

We will compare these predictions with numerical SOC simulations and observed distributions in the following.

This simple analytical model in terms of an exponential growth phase with saturation after a random time interval goes back to Willis and Yule (1922) who applied it to geographical distributions of plants and animals. Yule's model was applied to cosmic rays (Fermi 1949), to cosmic transients and solar flares (Rosner and Vaiana 1978; Aschwanden et al. 1998b), to the growth dynamics of the world-wide web (Huberman and Adamic 1999), as well as to the distribution of the sizes of incomes, cities, internet files, biological taxa, and in gene family and protein family frequencies (Reed and Hughes 2002). For the application to solar flares, Rosner and Vaiana (1978) interpreted the time interval between two subsequent events in terms of an energy storage time. However, it was argued that there is no observational evidence for a correlation between the energy storage time and the magnitude of a solar flare peak (Lu 1995b; Crosby 1996; Wheatland 2000b; Georgoulis et al. 2001). In contrast, the exponential rise time τ in our model here corresponds to the energy release time during an instability (Aschwanden et al. 1998b).

3.2 The Powerlaw-Growth Model

The exponential-growth model we discussed in the previous section is most suitable for multiplicative avalanche processes, where the increase per time step during the rise phase is based on a multiplicative factor, such as it occurs in nuclear chain reactions, population growth, or urban growth (e.g., see Zanette 2007). Alternatively, avalanche processes that continuously expand in space may show an energy increase that scales with the area or volume, i.e., with a powerlaw function of $W(t) \propto A(t) \propto r(t)^2 \propto t^2$ or $W(t) \propto V(t) \propto r(t)^3 \propto t^3$, or with any smaller power index for fractal structures. In the following analytical model we derive the resulting frequency distributions for such avalanche processes with a powerlaw-growth behavior (Fig. 3.4).

We assume that the avalanche process grows with a nonlinear power p, typically $1 \le p \le 3$ in 1-D to 3-D geometrical space,

$$W(t) = W_0 \left[1 + \left(\frac{t}{\tau_G} \right)^p \right] , \quad (3.2.1)$$

where τ_G represents a time constant that corresponds to the first doubling of energy. Coherent growth occurs during a time τ and the process saturates at energy W_S,

$$W_S = W(t = \tau) = W_0 \left[1 + \left(\frac{\tau}{\tau_G} \right)^p \right] . \quad (3.2.2)$$

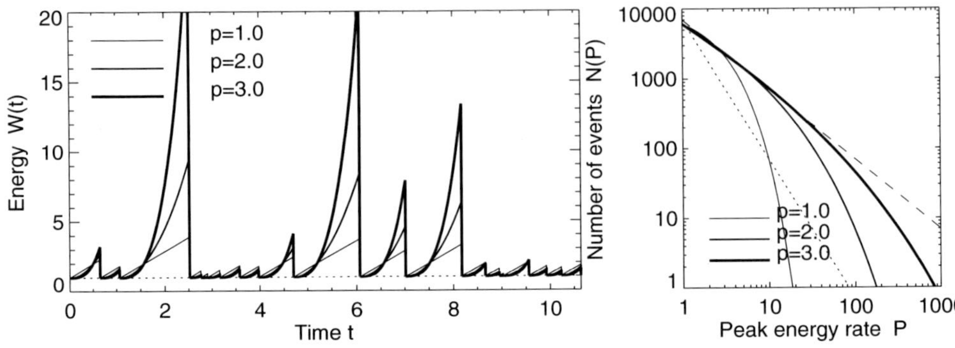

Fig. 3.4 Time evolution of energy release rate $W(t)$ according to the powerlaw-growth model for power indices of $p=1$ (thin solid curves), $p=2$ (thick solid curves), and $p=3$ (very thick solid curves), for a ratio of growth time to saturation time $\tau_G/t_S = 0.5$. The linear growth model ($p=1$) produces an exponential frequency distribution of energies, while nonlinear power indices ($p=2,3$) produce a powerlaw-like distribution with an exponential high-energy cutoff, approaching a powerlaw with a slope of $\alpha_P \mapsto 1$ for the asymptotic limit $p \mapsto \infty$. Powerlaw slopes of $\alpha_P = 1$ (dashed) and $\alpha_P = 2$ (dotted) are indicated for comparison.

We define the peak energy release reate P that represents the maximum energy release rate W_S minus the threshold energy W_0 in the same way as previously for the exponential-growth model (Eq. 3.1.3),

$$P = W_S - W_0 = W_0 \left(\frac{\tau}{\tau_G}\right)^p . \tag{3.2.3}$$

In order to derive the probability distribution for the peak energies P we have to invert the function $P(\tau)$ (Eq. 3.2.3),

$$\tau(P) = \tau_G \left(\frac{P}{W_0}\right)^{1/p} , \tag{3.2.4}$$

and calculate its derivative,

$$\frac{d\tau}{dP} = \frac{\tau_G}{pW_0} \left(\frac{P}{W_0}\right)^{1/p-1} . \tag{3.2.5}$$

Again we assume that the time scales τ of coherent avalanche growth are governed by a random process, approximated by an exponential function (Eq. 3.1.4). Substituting the relations of the time scales on the peak energy (Eqs. 3.2.4 and 3.2.5) into the time frequency distribution $N(\tau)$ of time scales (Eq. 3.1.4) we obtain the frequency distribution of peak energies via Eq. (3.1.6),

$$N(P)\, dP = \frac{N_0(\alpha_P - 1)}{pW_0} \exp\left[-(\alpha_P - 1)\left(\frac{P}{W_0}\right)^{1/p}\right] \left(\frac{P}{W_0}\right)^{1/p-1} . \tag{3.2.6}$$

For the energy dissipation process we assume a constant dissipation rate $\eta = W_0/\tau_D$ as in the previous model (Eq. 3.1.11), which yields a proportionality of the duration D to the peak energy P as in the previous model (Eq. 3.1.13), and thus a frequency distribution with the same functional dependence as $N(P)$ (Eq. 3.2.6). For the calculation of the distribution $N(T)$ we express the total duration T in terms of the peak energy P (with Eqs. 3.2.4 and 3.1.13),

$$T(P) = \tau + D = \tau_G \left(\frac{P}{W_0} \right)^{1/p} + \tau_D \frac{P}{W_0} \ . \tag{3.2.7}$$

Since the linear term in P becomes much greater than the term with $P^{1/p}$ for large avalanches and positive power indices p, we can neglect the first term (i.e., the rise time) and obtain with the definition of $P(\tau)$ Eq. (3.2.3)

$$T(P) \approx D = \tau_D \frac{P}{W_0} \ . \tag{3.2.8}$$

With this approximation we can invert $T(P)$,

$$P(T) = W_0 \left[\frac{T}{\tau_D} \right] \ , \tag{3.2.9}$$

and calculate the derivative,

$$\frac{dP}{dT} = \frac{W_0}{\tau_D} \ , \tag{3.2.10}$$

which allows us to calculate the frequency distribution $N(T)$ of total times T by substituting into the distribution $N(P)$ Eq. (3.2.6) of peak energies,

$$N(T)\, dT = N[P(T)] \left| \frac{dP}{dT} \right| = \frac{N_0(\alpha_P - 1)}{p\ \tau_D} \exp\left[-(\alpha_P - 1) \left(\frac{T}{\tau_D} \right)^{1/p} \right] \left(\frac{T}{\tau_D} \right)^{1/p-1} \tag{3.2.11}$$

We define the total released energy E by the time integral of the energy release rate $W(t)$ during the event duration T, but neglect the rise time τ ($T \approx D$), using Eq. (3.1.12) and subtract the threshold level W_0 before the avalanche,

$$E = \int_0^T [W(t) - W_0]\ dt \approx \int_\tau^{\tau+D} [W(t) - W_0]\ dt = \frac{1}{2} PD \tag{3.2.12}$$

Inserting the relations $D(P)$ (Eq. 3.2.8) we obtain the dependence of the total energy $E(P)$ on the peak energy E,

$$E(P) = \frac{\tau_D}{2W_0} P^2 \ . \tag{3.2.13}$$

Defining a reference energy E_0,

$$E_0 = \frac{W_0 \tau_D}{2} \ , \tag{3.2.14}$$

we have the inverted function $P(E)$ of $E(P)$ from Eq. (3.2.13),

$$P(E) = W_0 \left(\frac{E}{E_0} \right)^{1/2} \ , \tag{3.2.15}$$

and the derivative dP/dE,

$$\frac{dP}{dE} = \frac{W_0}{2E_0} \left(\frac{E}{E_0} \right)^{-1/2} , \tag{3.2.16}$$

we can then derive the frequency distribution $N(E)$ of energies by inserting Eqs. (3.2.15) and (3.2.16) in the distribution $N(P)$ of peak energies (Eq. 3.2.6),

$$N(E)\,dE = N[P(E)] \left| \frac{dP}{dE} \right| dE$$
$$= \frac{N_0(\alpha_P - 1)}{2pE_0} \exp \left(-(\alpha_P - 1) \left(\frac{E}{E_0} \right)^{1/2p} \right) \left(\frac{E}{E_0} \right)^{1/2p-1} \tag{3.2.17}$$

We show the frequency distributions of the total energy E, peak energy P, rise time τ, and total duration T in Fig. 3.5, for the area-like, quadratic ($p = 2$) powerlaw-growth model, for different ratios of the growth time to saturation time, i.e., $\tau_G/t_S = 0.5$, 1, and 2.

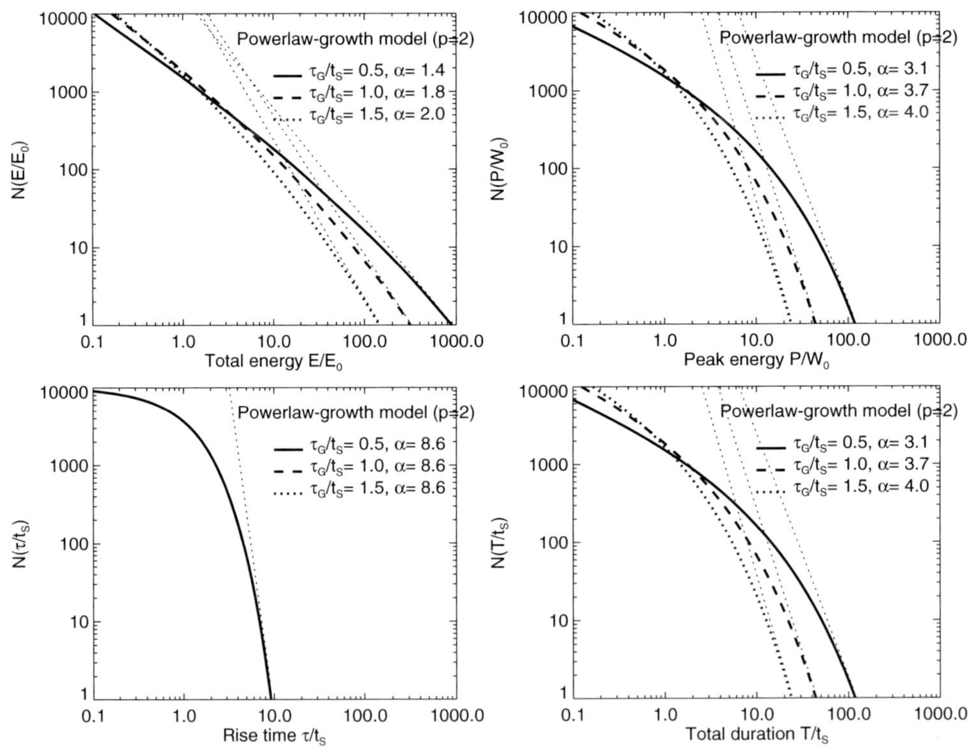

Fig. 3.5 Occurrence frequency distribution of total energies E (top left), peak energies P (top right), energy release times τ (bottom left), and total time durations D (bottom right) for the powerlaw-growth model, for τ_G/t_S=0.5, 1, and 2. Powerlaw fits are performed at the upper end of the distribution (dotted thin lines), with the slopes α_P indicated in each panel.

The corresponding plots are also shown for a volume-like, cubic ($p = 3$) powerlaw-growth model in Fig. 3.6. The frequency distributions are powerlaw-like, but have a gradual steepening at higher energies. This model predicts the following approximate correlations between the three parameters E, P and T, based on Eqs. (3.2.4) and (3.2.13),

$$\begin{aligned} E &\propto P^2 \\ T &\propto P^1 \\ E &\propto T^2 \end{aligned} \qquad (3.2.18)$$

These correlations are identical with the exponential-growth model, because we approximated the total event duration T with the decay time (which has a linear decay and identical scaling $D \propto P$ in both models), and neglected the rise time τ, which would have a slightly different scaling in both models.

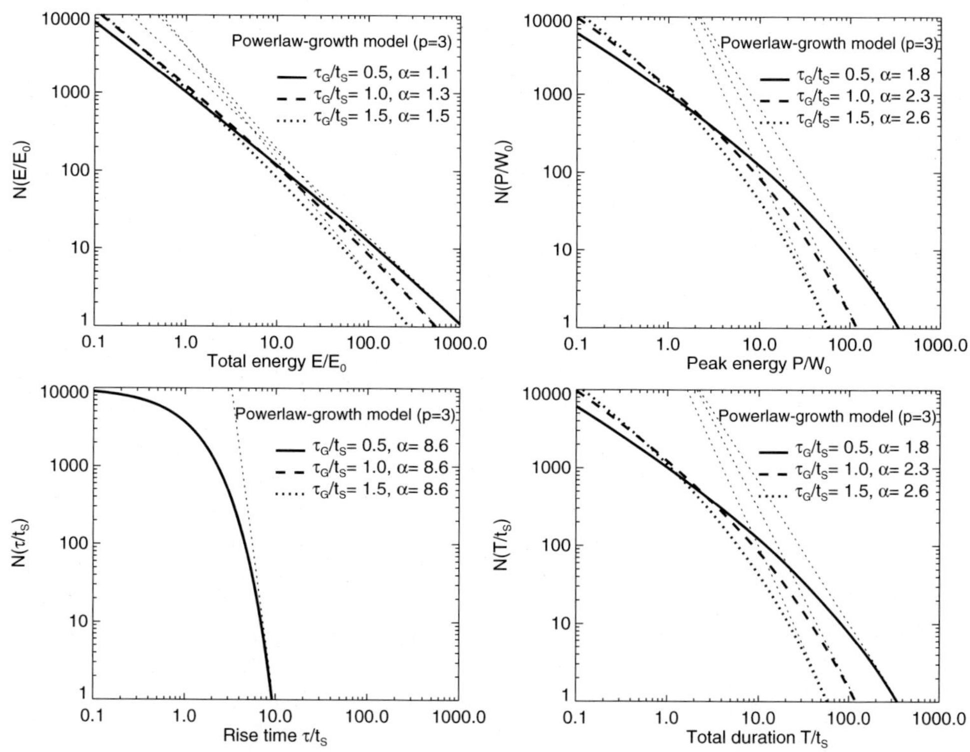

Fig. 3.6 Similar representation as Fig. 3.5, but for a powerlaw-growth model with volumetric or cubic growth ($p = 3$).

3.3 The Logistic-Growth Model

Many energy dissipation processes can be understood in terms of nonlinear instabilities that exhibit an initial exponential growth phase and saturate after a number of e-folding growth times; this is essentially our definition of an *avalanche*. For exponentially growing instabilities, the rate of released energy dE/dt is proportional to the already released energy $E(t)$,

$$\frac{dE(t)}{dt} = \Gamma E(t) \ , \tag{3.3.1}$$

with Γ denoting a constant growth rate (or reciprocally defined as *growth time*, $\tau_G = 1/\Gamma$). Such a behavior of exponential growth has been observed, e.g., in avalanches, nuclear chain reactions, or population growth. The differential equation (3.3.1) has the exponential function as the solution, i.e., $E(t) \propto \exp(\Gamma t)$ (Fig. 3.7 left).

The exponential model suffers from an unrealistic description of the saturation phase, because the energy release rate $dE(t)/dt$ exhibits an unphysical discontinuity at saturation time $t = t_1$ (Fig. 3.7 left). A more realistic approach is to describe the saturation phase of a nonlinear growth process with the so-called *logistic equation*, which has been widely used in ecologic applications, but has universal validity for nonlinear systems with limited free energy. The logistic equation is defined by a simple first-order differential equation (discovered by Pierre François Verhulst in 1845; see textbooks on nonlinear dynamics, e.g., May 1974; Beltrami 1987 (p.61-64); Jackson 1989 (p.75)),

$$\frac{dE(t)}{dt} = \Gamma \ E(t) \cdot \left[1 - \frac{E(t)}{E_\infty} \right] \tag{3.3.2}$$

which just represents a generalization of the exponential growth curve (Eq. 3.3.1), by including the growth limitation at the asymptotic saturation level E_∞. The solution $E(t)$ of this differential equation is initially exponential (because the term $[1 - E(t)/E_\infty]$ is close to unity for $E(t) \ll E_\infty$, and approaches asymptotically the value E_∞ for large times (because $dE(t)/dt \mapsto 0$ for $E(t) \mapsto E_\infty$). The physical basis of the *logistic term* $[1 - E(t)/E_\infty]$ is the constraint of an uppermost limit E_∞ (also called *"carrying capacity"* in ecologic applications), which slows down the exponential growth proportionally to the shrinking free energy, i.e. $dE/dt \approx \Gamma[E_\infty - E(t)]$ near the asymptotic value $E(t) \mapsto E_\infty$. The logistic equation can directly be integrated (by variable separation $dE/[E(1 - E/E_\infty)] = \Gamma \ dt$) and has the solution

$$E(t) = \frac{E_\infty}{1 + \exp(-\frac{t - t_1}{\tau_G})} \ , \tag{3.3.3}$$

where $\tau_G = 1/\Gamma$ is the exponential growth time, and we call the integration constant t_1 the saturation time. This analytical solution $E(t)$ is shown in Fig. 3.7 (top right).

We define a threshold energy E_0 at time $t = 0$, where the instability sets in, in the same way as for the exponential model (Eq. 3.1.1),

$$E_0 = E(t = 0) = \frac{E_\infty}{[1 + \exp(\frac{t_1}{\tau_G})]} \ . \tag{3.3.4}$$

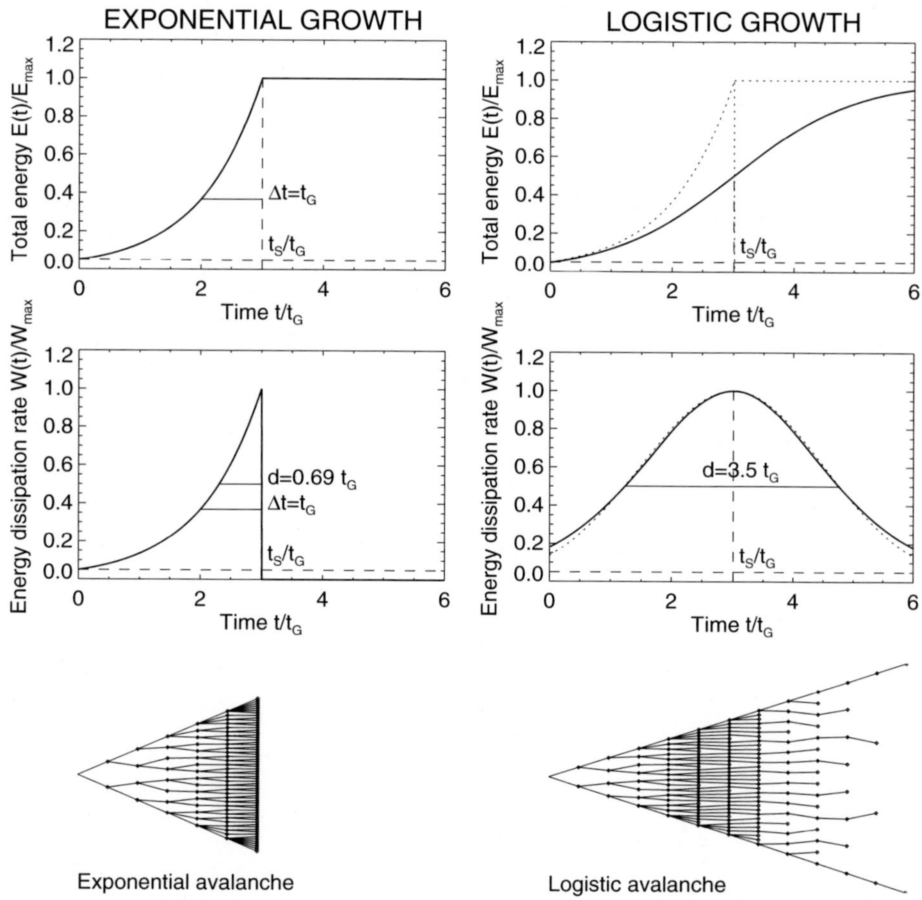

Fig. 3.7 Time evolution of total released energy $E(t)$ (top panels), the energy release rate $W(t) = dE(t)/dt$ (middle panels), and binary representation of avalanche growth rate (bottom panels), for both the exponential (left panels) and the logistic growth model (right panels). An exponential curve (right top) and a Gaussian curve (right middle) are drawn (with dotted lines) onto the logistic curves for comparison (Aschwanden et al. 1998b).

The maximum energy E_∞ is reached only in the asymptotic limit $t \mapsto \infty$, while the energy level E_1 at the time t_1 corresponds to half of the asymptotic limit E_∞,

$$E_1 = E(t = t_1) = \frac{E_\infty}{2} = \frac{E_0}{2}\left[1 + \exp\left(\frac{t_1}{\tau_G}\right)\right] . \quad (3.3.5)$$

The meaning of the saturation time t_1 becomes clearer when we calculate the energy release rate $W(t) = dE/dt$ of the logistic equation, by taking the time derivative of Eq. (3.3.5),

$$W(t) = \frac{dE(t)}{dt} = \frac{E_\infty}{\tau_G} \frac{\exp(-\frac{t-t_1}{\tau_G})}{[1 + \exp(-\frac{t-t_1}{\tau_G})]^2} . \quad (3.3.6)$$

The temporal evolution of $W(t)$ has a maximum at the saturation time $t = t_1$, the time of the maximum energy release rate,

$$W_1 = W(t = t_1) = \frac{E_\infty}{4\tau_G} = \frac{E_1}{2\tau_G} , \tag{3.3.7}$$

and is symmetric in time with respect to the maximum time t_1. The energy release rate $W(t)$ closely resembles to a Gaussian curve, as shown in Fig. 3.7 (middle right).

We quantify also the duration of the energy release function $W(t)$ by the full width at half maximum (FWHM). From the definition of the FWHM duration d, i.e. $W_1/2 = W(t = t_1 - FWHM/2)$, we find a quadratic equation which has the solution

$$FWHM = \tau_G \cdot 2\ln(3+\sqrt{8}) \approx \tau_G \cdot 3.53 . \tag{3.3.8}$$

This FWHM duration d depends only on the *growth time* τ_G, but does not depend on the saturation time t_1, the maximum energy release rate W_1, or the saturation energy W_S. The FWHM in the exponential-growth model is $FWHM = (\ln 2)\tau_G \approx 0.7\tau_G$, so about a factor of 5 shorter.

After we have defined the time evolution of the energy release rate $W(t)$ (Eq. 3.3.6) of a nonlinear process with logistic growth, we want to explore the statistics of total energies E, peak energy rates P, and total durations T as in the previous models (Sections 3.1 and 3.2). Again we assume that the probability distribution of saturation times τ, which is the time interval from a threshold level W_0 to the maximum release rate W_1, i.e., $\tau = t_1$, is governed by a random process, which we approximate with an exponential distribution as in Eq. (3.1.4),

$$N(\tau)\, d\tau = \frac{N_0}{t_S} \exp\left(-\frac{\tau}{t_S}\right) d\tau , \tag{3.3.9}$$

where t_S is the e-folding time constant of the distribution of saturation times, or the mean saturation time. With Eq. (3.3.4) we can express the dependence of the total energy $E(\tau)$ as a function of the saturation time $\tau = t_1$,

$$E(\tau) = E_\infty - E_0 = E_0 \exp\left(\frac{\tau}{\tau_G}\right) , \tag{3.3.10}$$

which can easily be inverted to obtain $\tau(E)$ and the derivative $d\tau/dE$, and hence the frequency distribution of total energies $N(E)$ by substituting $\tau(E)$ into the distribution $N(\tau)$ of Eq. (3.3.9),

$$N(E)\, dE = N[\tau(E)] \left| \frac{d\tau}{dE} \right| dE = \frac{N_0(\alpha_P - 1)}{E_0} \left(\frac{E}{E_0}\right)^{-\alpha_P} dE , \tag{3.3.11}$$

which is a powerlaw distribution with slope $\alpha_P = (1 + \tau_G/t_S)$. The peak energy release rate $P = W_1$ is proportional to the total energy E in the logistic model (according to Eq. 3.3.7),

$$P(E) = \frac{E}{4\tau_G} , \tag{3.3.12}$$

and thus the frequency distribution $(N(P))$ of peak energies is,

$$N(P)\, dP = N[E(P)] \left| \frac{dP}{dE} \right| dP = \frac{N_0(\alpha_P - 1)}{W_0} \left(\frac{P}{W_0} \right)^{-\alpha_P} dP \ , \qquad (3.3.13)$$

if we define the threshold peak rate as $W_0 = E_0/4\tau_G$. The total duration of an avalanche with a logistic time profile, which has equally long rise and decay times, is simply the double rise time, $T = 2\tau$, and thus the frequency distribution $N(T)$ is an exponential distribution like the distribution $N(\tau)$ of rise times (Eq. 3.3.9),

$$N(T)\, dT = \frac{N_0}{2t_S} \exp\left(-\frac{T}{t_S} \right) dT \ , \qquad (3.3.14)$$

We show the frequency distributions of the total energy E, peak energy P, rise time τ, and total duration T in Fig. 3.8, for three different ratios of the growth rate to the average saturation time t_S, i.e., $\tau_G/t_S = 0.5$, 1, and 2. We see that this model predicts equal pow-

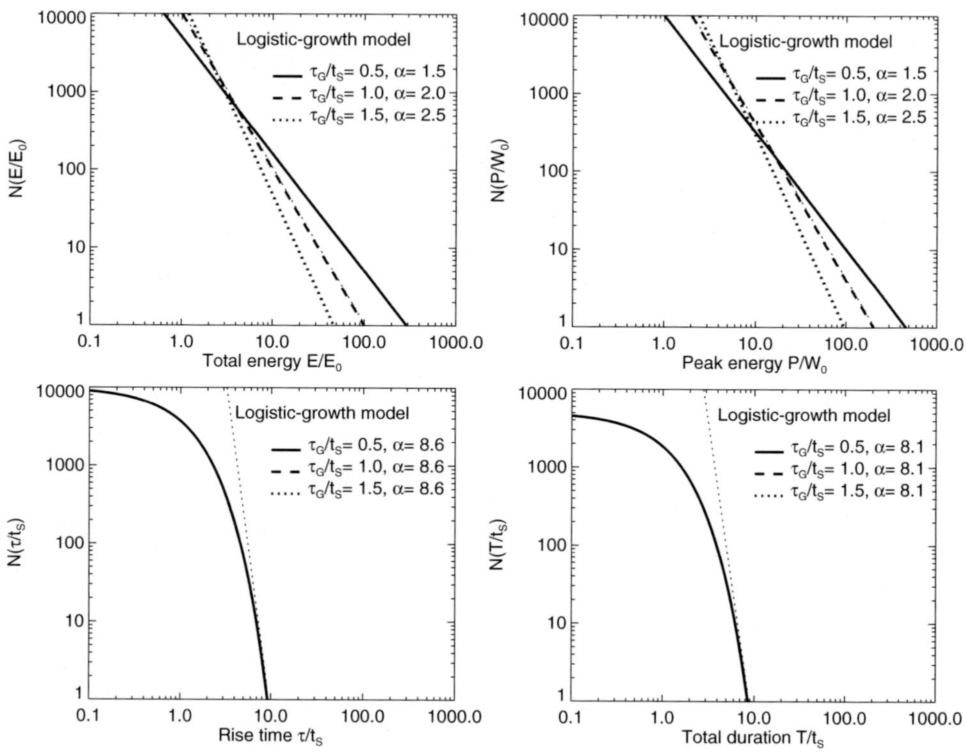

Fig. 3.8 Occurrence frequency distribution of total energies E (top left), peak energies P (top right), rise times τ (bottom left), and total time durations T (bottom right) for the logistic-growth model, for $\tau_G/t_S = 0.5$, 1, and 2. Powerlaw fits are performed at the upper end of the distribution and the values are indicated in each panel.

erlaw distributions for the total energy E and peak energy rate P with a common slope of $\alpha_P = 1 + \tau_G/t_S$, but exponential distributions for both the rise times τ and total durations T, which is an intermediate behavior between the exponential-growth model (Section 3.1) and the powerlaw-growth model (Section 3.2). The logistic-growth model has been applied to the energy release in solar flares (Aschwanden et al. 1998b) or to magnetic energy storage in the solar corona (Wang et al. 2009).

3.4 Analytical Fit to Numerical SOC Simulations

Numerical SOC models require iterative applications of mathematical redistribution rules that mimic complex behavior of next-neighbor interactions. There is an infinite variety of mathematical redistribution rules that can be applied in cellular automaton models, and their iterative simulation can lead to SOC behavior or to other unexpected nonlinear dynamics. A glimpse of the infinite variety of complex patterns that can result from simple mathematical redistribution rules can be gleaned from the very illustrative book *A New Kind of Science* by Stephen Wolfram (2002), the founder of *Mathematica*. Here we study only a subgroup of mathematical redistribution rules that mimic the dynamics of a nonlinear system in a critical state, the so-called SOC state. The question here is whether the dynamics resulting from iterative applications of mathematical redistribution rules can be approximated by a simple analytical function of the time evolution convolved with the statistics of random time scales. In the previous three sections we have demonstrated that powerlaw distributions can be produced by nonlinear functions subject to random statistics of time scales. In order to validate the usefulness of these analytical models we have to compare the analytically formulated distributions of SOC parameters (energy, peak energy rate, total duration) with the results of numerically generated distributions of SOC systems. Since we calculated our analytical distributions in explicit form, we can easily make this comparison by simple forward-fitting of the analytical distributions to the numerically generated ones. In a first step we compare only computer-generated SOC models, because they represent the most direct parameters of SOC avalanche processes, regarding spatial, temporal, and energy scales, while observed parameters in astrophysics require additional modeling to quantify the energy content of an avalanche.

We consider the numerical SOC simulation of Lu and Hamilton (1991), shown in Fig. 2.16, where the frequency distributions of energies, $N(E) \propto E^{-1.4}$, peak fluxes $N(P) \propto P^{-1.8}$, and time durations, $N(T)$, were accumulated from a cellular automaton model, with the energy release rate P defined by the redistribution rule Eq. (2.6.3). We fit our analytical expressions of the exponential-growth model (Section 3.1) onto these 3 frequency distributions, which are shown in Fig. 3.9. We vary the only free parameter and find a best fit with $\tau_G/t_S = 1.0$, which produces a powerlaw slope of $\alpha_P = 1 + \tau_G/t_S = 2.0$ for the peaks P and durations T at the upper end of the distribution, and $\alpha_E = (\alpha_P + 1)/2 = 1.5$ for energy. This value of $\tau_G/t_S = 1.0$ provides frequency distributions that are fully consistent with the numerical simulations, with an average powerlaw slope of $\alpha_P \approx 1.8$ and $\alpha_E = (\alpha_P + 1)/2 = 1.4$ in the medium range of the distributions. Also the rollovers at the lower end can be fitted, which particularly constrains the reference time scale $T/t_s = 1$ of the durations. The other analytical models (i.e., the powerlaw-growth or logistic-

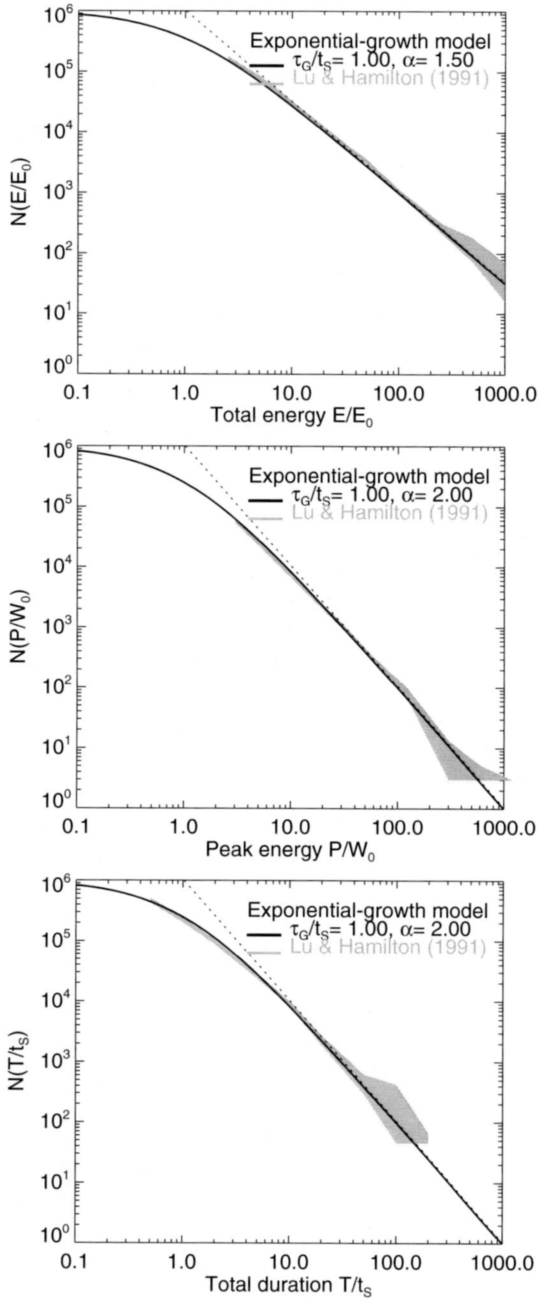

Fig. 3.9 Frequency distributions of total energy E (top), peak energy P (middle), and total duration T (bottom) of the numerical simulations of the SOC model of Lu and Hamilton (1991) (gray curves), fitted with the exponential-growth model (Section 3.1). The best fit yields a parameter of $\tau_G/t_S = 1.0$. Powerlaw functions are fitted at the upper end of the analytical distributions (dotted lines), with slopes of $\alpha_E = 1.5$ and $\alpha_P = \alpha_T = 2.0$.

growth model) do not provide such excellent fits in all three distributions. Therefore, the exponential-growth model seems to be the most suitable analytical model that is consistent with numerical SOC simulations.

Our next comparison is with the SOC simulations shown in Fig. 6 of Charbonneau et al. (2001), produced by a numerical SOC simulation of avalanches in a 10^7 iteration run carried out on a $N^D = 32^3$ lattice. Again we fit the exponential-growth model (Section 3.1) and find a best fit with the same parameter that fitted the simulations of Lu and Hamilton (1991), namely $\tau_G/t_S = 1.0$, which produces a powerlaw slope of $\alpha_E = 1.5$ for energies E and $\alpha_P = \alpha_t = 2.0$ for peak energies P and time durations T (Fig. 3.10). The same model fits also the correlations between these three parameters well, as predicted by this model, $E \propto P^2$, $P \propto T^1$, and $E \propto T^2$. Thus, the exponential-growth model seems to be the most suitable analytical model among the considered ones to reproduce the statistics of sandpile avalanches in a SOC state.

How does the analytical model translate into a numerical SOC model, for instance in the case of the 2-D lattice cellular automaton model of BTW, applied to solar flares by Lu and Hamilton (1991). The analytical coherent-growth model (Section 3.1) has an initial phase of exponential growth with a rise time τ_G, which has been modeled with a random distribution. A numerical avalanche spreads to the next 4 nearest neighbors in a cellular automaton model, and thus the maximum spreading corresponds to a circular area with a speed of one cell per time step. Thus, the areas of coherent growth correspond to circular patches with random radius r_i, as indicated in Fig. 3.11 (left frame). In the cellular automaton model, those areas need to have large fluctuations of the lattice parameter B_k (Fig. 3.11, right frame), which corresponds to the magnetic field in the solar case. If the field fluctuations are large and close to the threshold, they are in the state of SOC criticality, where a small disturbance or instability can quickly spread and amplify. In the most unstable situation, every next neighbor becomes unstable after every redistribution step, and thus the number of unstable nodes increases by a multiplicative factor, i.e., by a factor of 4 in 2-D or a factor of 6 in 3-D. This corresponds to an average exponential growth time of $\tau_G/t_S = 1/\ln(2D)$, i.e., $\tau_G/t_S = 0.72$ in 2-D or $\tau_G/t_S = 0.56$ in 3-D. Thus, the number of unstable cells grows near-exponentially during this phase, as does the released energy E, since the released energy per time step is approximately proportional to the squared field gradient $|\Delta B|^2$ (Eq. 2.6.11), which is modeled with an exponential function in our exponential-growth model (Eq. 3.1.1). Once the boundary of the unstable area is reached, the fluctuations become more random and decay gradually, which we modeled with a linear function in our model (Eq. 3.1.12). This gradual decay of fluctuations is essentially a random process that decouples adjacent field fluctuations and makes them more incoherent with each redistribution process. Thus, our analytical model implies a differential equation that is proportional to the number of nodes n during the coherent growth phase, but is proportional to the time t during the incoherent decay phase,

$$\begin{aligned} dn(t)/dt &\propto n(t) \quad \text{(rise time)} \\ dn(t)/dt &\propto -t \quad \text{(decay time)} \end{aligned} \tag{3.4.1}$$

which has the analytical solutions of an exponential function during the rise time and a linear function during the decay time (Fig. 3.2).

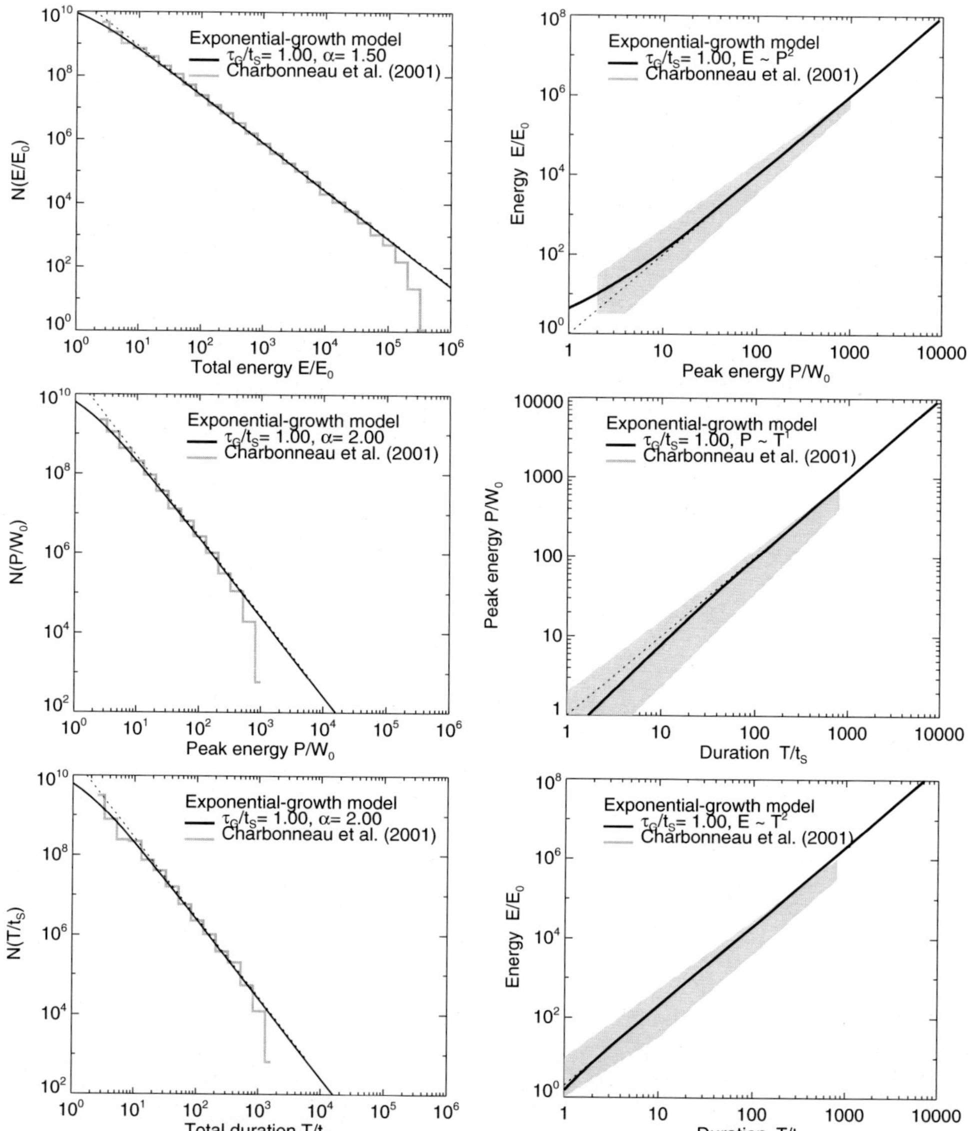

Fig. 3.10 Occurrence frequency distributions of total energy E (top left), peak energy P (middle left), and total duration T (bottom left) of the numerical simulations of the SOC model of Charbonneau et al. (2001) (gray histograms), fitted with the exponential-growth model (Section 3.1). The best fit yields a parameter of $\tau_G/t_S = 1.0$. Powerlaw functions are fitted at the upper end of the analytical distributions (dotted lines), with slopes of $\alpha_E = 1.5$ and $\alpha_P = \alpha_T = 2.0$. The correlations between the three parameters E, P, and T are shown in the right-hand panels (data points from Charbonneau et al. 2001), compared with the predicted correlations according to Eqs. (3.1.28)

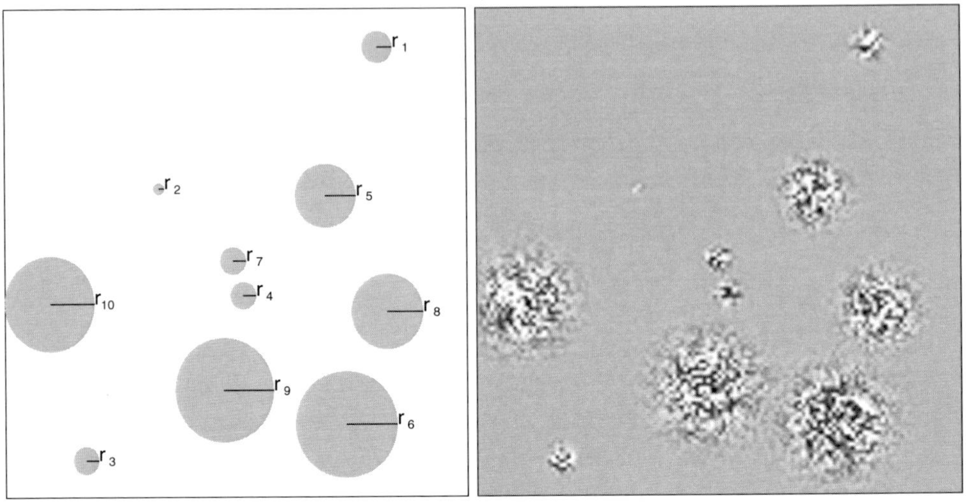

Fig. 3.11 Analytical (left) and numerical schematic representation (right) of areas that are susceptible to coherent growth, for a set of 10 avalanches. The circles with random sizes (with radius r_i) in the left panel indicate areas of possible coherent growth, which correspond to locations with large fluctuations B_k near the threshold level in 2-D lattice cellular automaton models.

3.5 Inertial Range, Lower and Upper Cutoff

We discuss now the lower and upper bound of frequency distributions, which define the so-called *inertial range* in between the two bounds, which often can be characterized with a single powerlaw function in SOC statistics (Fig. 3.12).

In the exponential-growth model (Section 3.1) we find a flattening of the frequency distributions $N(E)$, $N(P)$, and $N(T)$ at the low bound, at the threshold energy $E \lesssim E_0 = W_0 \tau_D/2$ (Eq. 3.1.22), peak energy $P \lesssim W_0$, and $T \lesssim \tau_D = W_0/\eta$ (Eq. 3.1.11). The threshold energy rate W_0 essentially represents a watershed between incoherent (random) and coherent energy fluctuations. A nonlinear instability can grow coherently only above this critical energy level. This threshold energy release rate W_0 constitutes also a lower bound on the minimum time scale, which is given by $\tau_D = W_0/\eta$, where η represents the average energy decay rate (Eq. 3.1.11). Taking the two limits together, we have also a limit of the lowest total energy $E_0 = W_0 \tau_D/2$, which is just the product of the average energy $W_0/2$ and the minimum duration τ_D (Fig. 3.2). If small events near the lower bound can be sampled we can indeed determine these threshold parameters W_0, E_0 and τ_D, such as τ_D in the SOC simulations of Lu and Hamilton (1991); see fit in Fig. 3.9 (bottom panel).

In astrophysical observations, the lower bound of the sampled frequency distribution is additionally affected by the detection threshold, which can appear as a sharp cutoff when an observable is directly proportional to an avalanche parameter (such as the hard X-ray count rate and peak energy release rate P), but can result in smooth rollovers for avalanche parameters that have a dependence on multiple physical parameters (e.g., the total energy E in a solar flare depends on the temperature and electron density with weakly correlated thresholds).

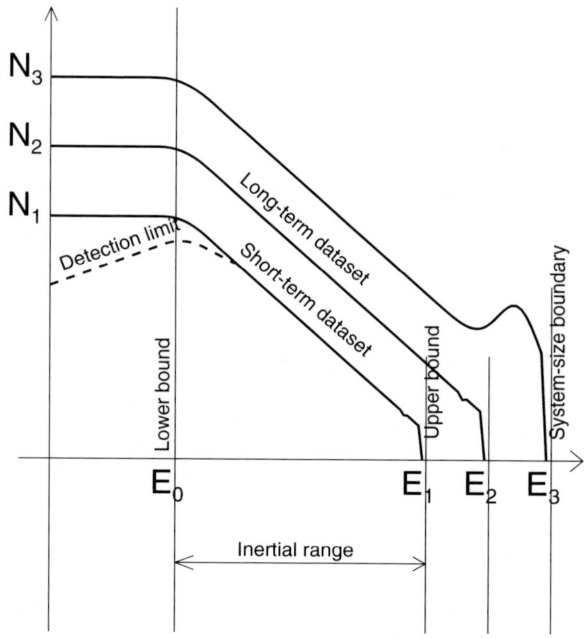

Fig. 3.12 Schematic frequency distribution with nomenclature definitions. The inertial range $[E_0, E_1]$ is the powerlaw part of the distribution between a lower and upper bound. The lower bound corresponds to the threshold energy E_0 in theoretical models. Real observations commonly show a rollover due to undersampling below the detection limit. Short-term datasets (N_1 events) have a lower upper cutoff E_1 than long-term datasets (N_2 events). System-wide avalanches show a hump at the upper bound (at E_3).

What is the upper bound of a frequency distribution? Most frequency distributions are monotonically dropping off towards higher values (of E, P, and T), so we can define an upper bound when the probability distribution drops to 1 event. All our analytical models are based on the random distribution $N(\tau)$ of energy release saturation times τ (Eq. 3.1.4), which is normalized to a total number of N_0 detected events (Eq. 3.1.5). Thus, setting the cumulative frequency distribution to one event,

$$\int_{\tau_{max}}^{\infty} N(\tau)\, d\tau = N_0 \exp\left(-\frac{\tau_{max}}{t_S}\right) = 1 \tag{3.5.1}$$

we obtain the maximum time scale τ_{max},

$$\tau_{max} = t_S \ln N_0 \ . \tag{3.5.2}$$

Say for a dataset with $N_0 = 10^4$ events we expect a maximum saturation time of $\tau_{max}/t_S \approx 9$.

The maximum values of E, P, and T can simply be estimated by setting the respective frequency distributions to a value of one event, which yields, e.g., for the exponential-growth model (with Eqs. 3.1.9, 3.1.19, and 3.1.26),

$$\left(\frac{E_{max}}{E_0}\right) \approx \left(\frac{N_0(\alpha_P - 1)}{2E_0}\right)^{2/(\alpha_P+1)} , \tag{3.5.3}$$

$$\left(\frac{P_{max}}{W_0}\right) \approx \left(\frac{N_0(\alpha_P - 1)}{W_0}\right)^{1/\alpha_P} , \tag{3.5.4}$$

$$\left(\frac{T_{max}}{t_D}\right) \approx \left(\frac{N_0(\alpha_P - 1)}{t_D}\right)^{1/\alpha_P} . \tag{3.5.5}$$

For a dataset with $N_0 = 10^4$ events and a ratio $\tau_G/t_S = 1.0$ (corresponding to $\alpha_P = 1 + \tau_G/t_S = 2$) and normalization $W_0 = 1$, $t_D = 1$, and $E_0 = W_0 t_D/2 = 1/2$ thus we expect a maximum energy of $E_{max}/E_0 = 300$, a maximum peak energy rate of $P_{max}/W_0 = 100$, and a maximum total duration of $T_{max}/t_D = 100$. Note that these maximum event values X_{max} correspond to the minimum event occurrence frequency number $N(X_{max}) = 1$ at the bottom of the graphs in Fig. 3.3. Thus the inertial range is about two orders of magnitude for a dataset of $N_0 = 10^4$ events. For models with powerlaw slopes of $\alpha_P \approx 2$, the inertial range spans about half of the logarithmic range of the number of events.

The dependence of the inertial range X_{max}/X_0 on the number of events N_0, for $X = E, P, T$,

$$\frac{X_{max}}{X_0} \approx N_0^{1/\alpha_P} , \tag{3.5.6}$$

has the implication that the inertial range grows with increasing time with a power of $1/\alpha_P$ (Fig. 3.12). If events are sampled with a mean rate of $dN_0/dt = 1/\langle\Delta t\rangle$, the inertial range grows as,

$$\frac{X_{max}}{X_0} \approx \left(\frac{t}{\langle\Delta t\rangle}\right)^{1/\alpha_P} , \tag{3.5.7}$$

where $< \Delta t >$ is the mean waiting time between two subsequent events. For example, for a powerlaw index of $\alpha_P = 2$, the observing time has to be increased by a factor of $10^{\alpha_P} = 10^2$ to gain one more decade in inertial range. For example, the average rate of solar flares during the solar maximum is about 15 events per day for hard X-ray flare events detected with HXRBS/SMM (Crosby et al. 1993). This means that we need about one week to obtain an inertial range of 1 decade for the frequency distribution of the peak count rate P, or 3 months for an inertial range of 2 decades, or 25 years for an inertial range of 3 decades. The inertial range for energies E, however, would extend over 4 decades during the same time range, because the powerlaw index is flatter ($\alpha_E \approx 1.5$).

While we considered only the temporal cutoff at the upper bound of frequency distributions, there could also be a spatial boundary that prevents avalanches to propagate to their maximum size. Such spatial boundaries could be the finite lattice size in computer simulations, the edge or base of a sandpile, the vegetation borders in forest fires, the polar cusp in magnetospheric substorms, active regions in solar flares, or the size of the accretion disk in gamma-ray bursts. If system-wide avalanches encounter such spatial boundaries, the frequency distributions can be strongly modified at the upper bounds (Fig. 3.12), e.g., see the size of dissipation power in auroral blobs (Fig. 1.10), which could successfully be reproduced with numerical SOC simulations (Fig. 2.13), as well as for the area distributions of forest fires (Fig. 2.12). Essentially, the spatial boundary stops avalanches from further growing and produces an overabundance of event parameters at the particular spatial scale that corresponds to the system size. Investigations of the effect of boundary conditions have been studied by Galsgaard (1996), who finds that the frequency distributions sensitively depend on the boundary conditions (for his chosen stability criterion).

3.6 Continuum Limit of Cellular Automaton Model

We try now to understand the analytical exponential-growth model in terms of the numerical BTW sandpile SOC model, which is formulated as a mathematical redistribution rule involving discretized changes among next neighbors. We attempt to map the analytical model onto the numerical SOC model in order to derive a continuum limit of the discretized numerical redistribution rules.

The spatial pattern of a large avalanche produced by such a 2-D redistribution rule (Eqs. 2.6.1–2.6.2 or 2.6.23–2.6.27) is shown in Fig. 3.13, as it would appear in the most unstable situation. Essentially, the spatial size of the spreading avalanche grows by one cell in each direction every time step, because only next-neighbor interactions are allowed. This means that the size of the avalanche grows linearly in time, and the area grows quadratically in a 2-D lattice grid, but the number of nearest-neighbor interactions grows exponentially. The unaffected cells that have random fluctuations below the threshold are marked with black in Fig. 3.13, while the unstable cells to which the redistribution rule is applied

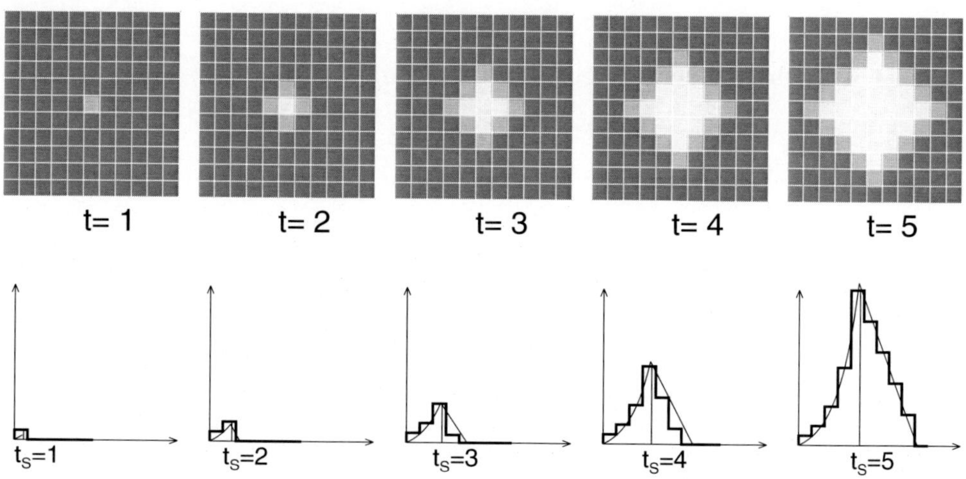

Fig. 3.13 Spatial patterns of a propagating avalanche in subsequent time steps in a 2-D cellular automaton model with a next-neighbor redistribution rule (top) and time profiles of energy release rate (bottom), for saturation times of $t_S = 1, 2, ..., 5\Delta t$. The black cells represent cells with random fluctuations below the threshold, $z_k < z_c$, the gray cells contain possibly unstable cells with fluctuation $z_k \geq z_c$ that are subject to a first redistribution, while the white cells have already been affected by a redistribution rule before. Most avalanches die out after step t $\gtrsim$ 2.

for the first time are marked in gray, and the cells that were already redistributed earlier are marked in white. The propagation pattern shown in Fig. 3.13 is a "worst-case scenario" for a large avalanche with the most unstable cell configuration, but most avalanches die out after a few time steps due to the stabilizing effect of the redistribution rule.

Let us consider now the mapping of our analytical model onto the numerical SOC model. The smallest avalanche consists of one single cell that becomes redistributed. If the environment is sufficiently unstable the avalanche proceeds and will affect a maximum of 4 (or 6) nearest neighbors in a 2-D (or 3-D) lattice. If the avalanche grows further, it will affect a maximum of 4^2 (or 6^2) cells, and so forth. This multiplicative behavior is consistent with the exponential function in our analytical model. Even if the average multiplication factor is smaller, say $1 < q \le 4$, the number of cells affected by nearest-neighbor interactions will grow exponentially with $q(t) \propto q^t$. In the numerical SOC model we have discretized time steps, so we can express the time scale t, the growth time τ_G, and the saturation time t_S, with integer numbers n_t, n_G and n_S,

$$t = n_t \Delta t \ , \quad \tau_G = n_G \Delta t \ , \quad \tau_S = n_S \Delta t \ , \tag{3.6.1}$$

leading to a discretized exponential-growth function of (Eq. 3.1.1), with discretized time intervals $t = n_t \Delta t$,

$$W(n_t) = W_0 \exp\left(\frac{n_t}{n_G}\right) . \tag{3.6.2}$$

The released energy during the redistribution of one discretized time step Δt is for $\Delta B \gtrsim B_c$ according to Eq. (2.6.11),

$$|\Delta E_m| = \frac{2D}{2D+1} \Delta B^2 \approx \frac{2D}{2D+1} B_c^2 \ . \tag{3.6.3}$$

Thus, choosing the threshold energy $W_0 = B_c^2$ and attributing the energy difference $|\Delta E_m|$ to $[W(n_t) - W_0]$, we have after one redistribution time step $n_t = 1$,

$$\Delta E_1 = W(n_t = 1) - W_0 = W_0 \left(\exp \frac{1}{n_G} - 1 \right) = W_0 \frac{2D}{2D+1} \ , \tag{3.6.4}$$

which leads to a (discretized) growth time $n_G = \tau_G / \Delta t$ of,

$$n_G = \frac{1}{\ln\left[\frac{2D}{(2D+1)} + 1\right]} \ , \tag{3.6.5}$$

amounting to $n_G \approx 1.70$ for $D = 2$, or $n_G = 1.62$ for $D = 3$. Thus, the released energy grows by an exponential factor of $e \approx 2.7$ within less than two redistribution time steps. Examples of saturation after $n_S = t_S / \Delta t = 1, 2, ..., 5$ time steps are shown in Fig. 3.13.

We can now calculate the resulting powerlaw slopes of the peak energies P and total time duration T,

$$\alpha_P = \alpha_T = \left(1 + \frac{\tau_G}{t_S}\right) = \left(1 + \frac{n_G}{n_S}\right) = 1 + \frac{1}{n_S \ln\left[\frac{2D}{(2D+1)} + 1\right]} \ , \tag{3.6.6}$$

Table 3.1 Powerlaw slopes α_P, α_T, and α_E of the exponential-growth model predicted for the numerical BTW simulations as as function of the mean discretized saturation time $n_S = t_S/\Delta t = 1, 2, ..., 5$ and dimension $D = 2, 3$ of the lattice. The values in parentheses were obtained from numerical lattice simulations by Charbonneau et al. (2001).

Saturation time $n_S = t_S/\Delta t$	Dimension D	Peak energy powerlaw slope $\alpha_P = \alpha_T$	Energy powerlaw slope α_E
1	2	2.70	1.85
2	2	1.85 (1.73,1.72)	1.42 (1.42)
3	2	1.57	1.28
4	2	1.42	1.21
5	2	1.34	1.17
1	3	2.61	1.81
2	3	1.81 (1.92,1.79)	1.40 (1.49)
3	3	1.54	1.27
4	3	1.40	1.20
5	3	1.32	1.16

and total released energies E with Eq. (3.1.28), i.e., $\alpha_E = (\alpha_P + 1)/2$, as a function of the discretized saturation time $n_s = t_S/\Delta t$ in the framework of our analytical exponential-growth model. We tabulate the values obtained for $n_S = 1, 2, ..., 5$ and $D = 2, 3$ in Table 3.1. Comparing these theoretical values with the numerically obtained powerlaw slopes in Charbonneau et al. (2001), indicated in parentheses in Table 3.1, we find that a mean saturation time of $n_t = 2$ is most consistent with the numerical SOC simulations. Since the time scale t_S represents the mean saturation time of the exponential distribution (defined in Eq. (3.1.4)), this means that the avalanches typically saturate after two time steps.

Thus, the choice of $n_G = \tau_G/\Delta t \approx 1.7$ and $n_S = t_S/\Delta t \approx 2.0$ represents the most consistent mapping of the continuous analytical functions in our exponential-growth model to the discretized numerical avalanche simulations. It allows us to predict the statistically averaged time profiles of the energy release rate of avalanches and the frequency distributions of their peak energy P, total energy E, and total duration T in agreement with the numerical lattice simulations. There are only two parameters (n_G, n_S) that constrain the transformation from the numerical to the analytical model, which we determined for the simulations by Charbonneau et al. (2001). For other SOC simulations that use different mathematical redistribution rules or a different level $\Delta B/B_c$ to drive the SOC state, we expect slightly different transformation parameters.

The analytical model gives us also new insights into the concept of self-organized criticality. We may ask what defines the criticality in the analytical model? A subcritical state means a growth time that is much longer than the mean saturation time, in which case almost no avalanches occur. This is also the case in the initial phase of numerical SOC simulations, when the field fluctuations ΔB are small compared with the critical threshold B_c. Thus, the buildup phase of a numerical SOC simulation corresponds to a gradual decrease of the growth time τ_G from an initially infinite value down to a critical value that is commensurable with the mean saturation time, $\tau_G \approx t_S$, constrained by the amount of released energy per redistribution process among the next neighbors (Eq. 3.6.4). What warrants the powerlaw of the energy distribution is the constancy of the growth time τ_G in the critical SOC state, which can be understood in terms of two conditions: (1) a fixed probability for next-neighbor interactions, and (2) a fixed energy quantum that is released per redistribution step. The first condition of a fixed probability for next-neighbor interactions is automatically guaranteed because there is a constant number of next neighbors per definition (4 in 2-D or 6 in 3-D), which would not be the case in models with non-local communication (e.g., MacKinnon and Macpherson 1997). The second condition of a fixed energy quantum released per time step is warranted in the weak-driving limit, where the triggering field fluctuation exceeds the threshold only by a small amount, e.g., $\Delta B \gtrsim B_c$, which is ultimately guaranteed by the constant threshold value B_c and the limit of weak driving. What happens when the limit of weak driving is violated, i.e., $|\delta B|/\langle B\rangle \ll 1$, is that mid-size avalanches are favored and the scale-free powerlaw behavior disappears, according to numerical SOC simulations in the strong-driving limit (Charbonneau et al. 2001). In short, since our analytical exponential-growth model strictly predicts a powerlaw, it implies also the weak-driving limit. The evolution of the growth time $\tau_G(t)$ during the build-up phase reflects the approach towards the critical state, and the growth time τ_G becomes a fixed constant when reaching the SOC state. The critical value τ_G (expressed in units of mean saturation times) can easily be derived from the powerlaw slope of numerical SOC simulations, but can in principle also be calculated from the probability of field fluctuations $\Delta B = B_{ij} - B_{nn}$ and the threshold value B_c.

Our approach of mapping an analytical model with an explicit function of the energy release $W(t)$ onto numerical simulations of SOC avalanches provides a statistically averaged continuum limit of the BTW cellular automaton redistribution rule. There are a number of other approaches to derive a continuum limit from the BTW cellular automaton rules.

Lu (1995c) envisions avalanches in a continuum-driven dissipative system, which is characterized by a coupled equation system of a one-dimensional diffusion process,

$$\frac{\partial B(x,t)}{\partial t} = \frac{\partial}{\partial x}\left[D(x,t)\frac{\partial B}{\partial x}\right] + S(x,t) , \tag{3.6.7}$$

$$\frac{\partial D(x,t)}{\partial t} = \frac{Q(|\partial B/\partial x])}{\tau} - \frac{D(x,t)}{\tau} , \tag{3.6.8}$$

where $B(x,t)$ is a scalar field, $D(x,t)$ is a spatially and temporally varying diffusion term, $S(x,t)$ is a source term, $Q(|\partial B/\partial x|)$ is a double-valued Heaviside function that has a low or high state that depend on the time history and an instability threshold. Lu (1995c) demonstrated that the complex dynamic behavior of this differential equation can be approximated by a much simpler cellular automaton simulation.

Isliker et al. (1998a) discretize the 3-D cellular automaton redistribution rule into a differential equation that represents a diffusion process,

$$\frac{\partial \mathbf{B}(\mathbf{x},t)}{\partial t} = \eta \nabla^2 \mathbf{B}(\mathbf{x},t) + \mathbf{S}(\mathbf{x},t) , \tag{3.6.9}$$

with a source term $\mathbf{S}(\mathbf{x},t)$ and a diffusion coefficient $\eta = 1/7(\Delta h^2/\Delta t)$. This differential equation contains a continuous function $\mathbf{B}(\mathbf{x},t)$ that behaves the same way as the nearest neighbors during one redistribution step, but a singularity occurs at the center location at $\Delta h \mapsto 0$, which requires a modification of the cellular automaton rule.

Liu et al. (2002) and Charbonneau et al. (2001) transform the cellular automaton rule of Lu and Hamilton (1991) into a finite difference equation,

$$\frac{\partial B}{\partial t} = -\frac{\partial^2}{\partial x^2}\kappa(B_x x^2)\frac{\partial^2 B}{\partial x^2} , \qquad (3.6.10)$$

where $\kappa(B_x x^2)$ is a diffusion coefficient that depends on the local curvature B_{xx}^2. This is a fourth-order nonlinear hyperdiffusion equation, which is interpreted as continuum limit of the cellular automaton rule, compatible with MHD in the regime of strong magnetic field and strong MHD turbulence (with high effective magnetic diffusity).

3.7 Summary

We developed several analytical SOC models that consist of: (1) an equation that describes (in explicit form) the nonlinear (explosive) energy release $W(t)$ during the rise time ($t \le t_S$) of an instability, (2) an equation that characterizes the (linear) decay rate after saturation, and (3) a probability distribution $N(\tau)$ of random time scales for the instability rise times t_S. We quantify the explosive phase with three different parameterizations: (i) exponential-growth, (ii) powerlaw-growth (of area or volume), and (iii) logistic growth phases. For each of the scenarios we derive the frequency distributions of the peak energy release rate P, the total released energy E, and the total time duration T. Each model predicts a specific analytical function for the occurrence frequency distributions, which range from powerlaw-like functions to exponential-like functions. These analytical models predict also correlations between the avalanche parameters P, E, and T. We find that the exponential-growth model fits the numerically simulated SOC avalanches most suitably regarding their frequency distributions and parameter correlations. The analytical models quantify also the inertial range with its lower and upper bound, which are a function of the avalanching threshold and the total avalanche sampling time. Our analytical models can be considered as a continuum limit of numerical cellular automaton simulations. Alternative approaches derive a continuum limit by discretizing the cellular automaton redistribution rules into differential equations with anomalous or hyper-diffusion. The class of analytical SOC models we discuss in this chapter are "physics-free" in the sense that they can be generally applied to any arbitrary avalanching SOC system, such as to solar flares as well as to earthquakes.

3.8 Problems

Problem 3.1: Verify the derivation of the powerlaw energy distribution (Eqs. 3.1.7–3.1.9) in the exponential-growth model (Eqs. 3.1.1–3.1.2).

Problem 3.2: Derive the corresponding time evolution of the energy release decay phase $W(t)$ for observed correlations of $P \propto D^q$ between the peak energy release rate P and decay phase duration D (see solution 3.1.12 for a linear decay rate $q = 1$).

Problem 3.3: How accurate are the approximations of the parameter correlations given in Eq. (3.1.27)? For what parameters E, P, and T do you find the largest discrepancy?

Problem 3.4: Plot the accurate expression for the correlation between the total energy $E(\tau)$ and rise time τ for the powerlaw-growth model (Eq. 3.2.12) and find another suitable approximation function for the cases of $p = 2, 3$.

Problem 3.5: Determine the size of the sample plotted in the frequency distribution $N(P)$ of the solar flare peak count rate shown in Fig. 1.13 from the inertial range relation (Eq. 3.5.4).

4. Statistics of Random Processes

Although dice have been mostly used in gambling, and in recent times as "randomizing" elements in games (e.g., role playing games), the Victorian scientist Francis Galton described a way to use dice to explicitly generate random numbers for scientific purposes, in 1890.

Wikipedia 2009, *Hardware Random Number Generator*

The phenomenon of *self-organized criticality (SOC)* can be identified from many observations in the universe, by sampling statistical distributions of physical parameters, such as the distributions of time scales, spatial scales, or energies, for a set of events. SOC manifests itself in the statistics of nonlinear processes. The powerlaw shape of the occurrence frequency distribution of events is one of the telling indicators of SOC. By observing a single event it would be impossible to establish whether the system is in a SOC state or not. Statistics is therefore of paramount importance for modeling and interpretation of SOC phenomena. Of course, statistics always implies random deviations from smooth distributions, as they are often defined by analytical functions. Only for strictly deterministic systems can one accurately predict the outcome of an event based on its initial conditions. In reality, however, initial conditions are never known exactly and many random disturbances occur during the evolution of an event, which prevents us from making accurate predictions. The most accurate statements we can make about almost any physical system is of statistical nature. In our Chapter 3 on analytical models of SOC phenomena, the statistics of random time scales was one of the primary assumptions in the derivation of occurrence frequency distributions, which needs to be rigorously defined and quantified. In this Chapter 4 we deal with the most common statistical probability distributions of random processes, such as the binomial distribution, the Gaussian distribution, the Poisson distribution, and the exponential distribution. Random processes produce various types of noise, such as white noise, pink, flicker or $1/f$ noise, Brownian or red noise, or black noise. The time scales of SOC phenomena, such as the durations of SOC avalanches, are often attributed to $1/f$ noise, which we study in this chapter. The definition of power spectra of these various types of noise enable us to construct a variety of analytical SOC models that are needed to understand and identify SOC and non-SOC phenomena according to their intrinsic noise characteristics, from earthquakes to starquakes.

4.1 Binomial Distribution

We mostly deal with random processes when studying the behavior of SOC, so we have to start with the statistical probability distributions. Basic introductions into probability distributions can be found in many textbooks, e.g., see Section 2 of Bevington and Robinson (1969). The binomial distribution is generally used in experiments with a small number of different final states, such as coin tosses, card games, casino games, particle physics experiments, or quantum mechanics. In our focus on SOC, we might describe the probability of next-neighbor interactions above a threshold with binomial statistics.

Starting from first principles, the most fundamental probability distribution is the *binomial distribution*. It can be derived from the probabilities of tossing coins or rolling dice. If we toss n coins and consider the probability that x coins end in a particular outcome (either head or tail), the number of possibilities is the number of permutations $Pm(n,x)$, say for $n=4$ and $x=1,2,3,4$ we have,

$$\begin{aligned}
Pm(n=4,x=1) &= n &&= 4 \\
Pm(n=4,x=2) &= n(n-1) &&= 4\times 3 = 12 \\
Pm(n=4,x=3) &= n(n-1)(n-2) &&= 4\times 3\times 2 = 24 \\
Pm(n=4,x=4) &= n(n-1)(n-2)(n-3) &&= 4\times 3\times 2\times 1 = 24
\end{aligned} , \tag{4.1.1}$$

which can be expressed more generally in terms of factorials,

$$Pm(n,x) = n(n-1)(n-2)\dots(n-x+1) = \frac{n!}{(n-x)!} . \tag{4.1.2}$$

However, the outcomes of different permutations for one state x has $x!$ possible combinations, so the number of different combinations $C(n,x)$ has to be divided by this factor, if we do not distinguish between identical cases. The resulting fractions are also called *binomial coefficients* $\binom{n}{x}$,

$$C(n,x) = \binom{n}{x} = \frac{Pm(n,x)}{x!} = \frac{n!}{x!(n-x)!} , \tag{4.1.3}$$

which yields the following number of different possible combinations,

$$\begin{aligned}
C(n=4,x=0) &= 1 &&= 1 \\
C(n=4,x=1) &= n/1 &&= 4 \\
C(n=4,x=2) &= n(n-1)/2! &&= 6 \\
C(n=4,x=3) &= n(n-1)(n-2)/3! &&= 4 \\
C(n=4,x=4) &= n(n-1)(n-2)(n-3)/4! &&= 1
\end{aligned} . \tag{4.1.4}$$

The name *"binomial coefficients"* stems from the algebraic binomial equation,

$$(a+b)^n = \sum_{x=0}^{n} \binom{n}{x} a^x b^{(n-x)} , \tag{4.1.5}$$

which reads, e.g., for $n=4$, as

$$(a+b)^4 = 1 \cdot a^4 + 4 \cdot a^3 b + 6 \cdot a^2 b^2 + 4 \cdot ab^3 + 1 \cdot b^4 . \tag{4.1.6}$$

Now, to obtain the probability for each state x we have to normalize to unity. If p is the basic probability for each state, say $p = 1/2$ for tossing coins (head or tails), the probability for n coins to be in a particular state x is p^x, and the probability for the other $(n-x)$ coins to be in the other state is $(1-p)^{n-x}$, while the product of these two parts is the probability $P_B(x;n,p)$ of the combination,

$$P_B(x;n,p) = \binom{n}{x} p^x (1-p)^{n-x} = \frac{n!}{x!(n-x)!} p^x (1-p)^{n-x} , \tag{4.1.7}$$

which is called the *binomial distribution*, expressed in terms of factorials. If the probability is $p = (1-p) = 1/2$, the normalization factor becomes trivially $p^x(1-p)^{n-x} = p^n$, which is just the reciprocal number of the total number of combinations $2^n = 16$ for $n = 4$.

As a practical example let us consider a 2-D lattice where every cell has 4 next neighbors (Fig. 4.1) and the probability that a neighboring cell becomes unstable is $p = 1/2$. There are $n = 2^4 = 16$ possible outcomes, but there are only 5 non-distinguishable classes (labeled with the number of states $x = 0,...4$), if one only cares about the total number of unstable states. We show all 16 different possibilities in Fig. 4.1 and find the probabilities $P_B(x = [0,1,2,3,4]; n = 4, p = 1/2) = [1,4,6,4,1]/16$. Thus, there is a six time higher probability to have only 2 neighboring cells triggered than all 4 next neighbors together. In Section 3.6 we calculated the growth time τ_G of an avalanche for maximum unstable conditions, where the probability for a next-neighbor interaction is $p = 1$. However, if the lattice is somewhat subcritical, so that the probability for one next-neighbor interaction is $p = 1/2$ for instance, there is only a combined probability of $p = 1/16$ that all 4 next neighbors are triggered together and the avalanche propagates with the maximum growth factor. A related probabilistic SOC model was also conceived by MacKinnon et al. (1996) and Macpherson and MacKinnon (1999), which we discussed in Section 2.6.5 on branching processes.

In Fig. 4.2 we show binomial distributions $P_B(x;n,p = 1/2)$ for $n = 2$ to $n = 20$, all displayed on a normalized axis of states x/n. It can clearly be seen that the binomial distribution turns into a Gaussian distribution for large number of states $n \mapsto \infty$. However, factorials are not practicable to calculate for large values of n, say for a gas that has $n \approx 10^{26}$ atoms per cubic centimeter, so it is more useful to approximate Eq. (4.1.7) with an analytical function, which turns out to be the Gaussian function.

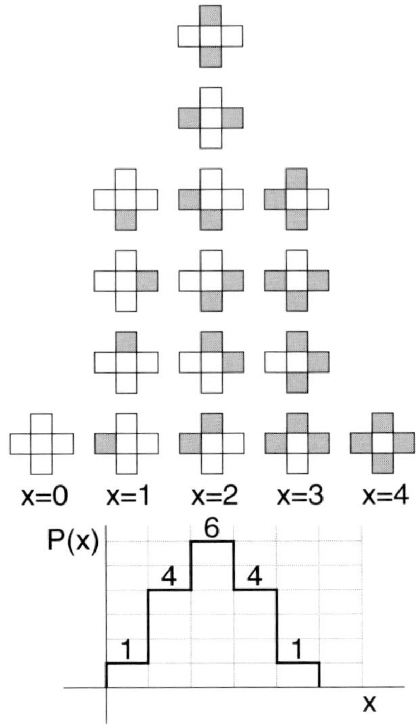

Fig. 4.1 Binomial distributions of all possible combinations of next-neighbor interactions in a 2-D lattice. There are $n = 4$ next neighbors and 5 possible states, $x = 0, 1, ..., 4$, and the probabilities $P(x;n) = 1, 4, 6, 4, 1$ are given in the histogram at the bottom of the figure. The number of all combinations amounts to 16 cases, while the number of distinguishable combinations amounts to 5 different cases only.

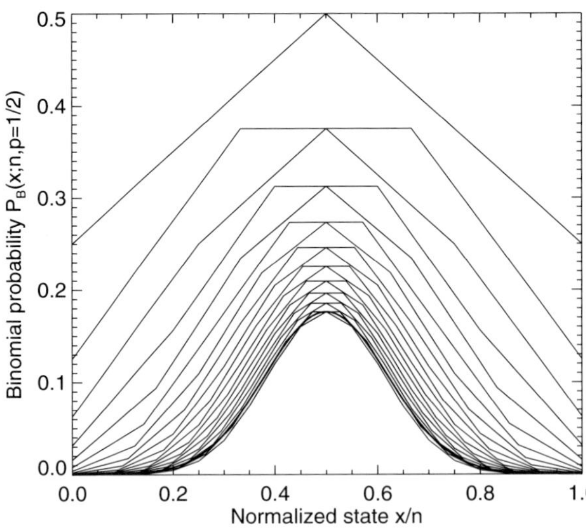

Fig. 4.2 Binomial distributions $P_B(x;n,p = 1/2)$ for $n = 2, 20$, overlaid on the same normalized state x/n. Note that the binomial distribution turns into a Gaussian function for $n \mapsto \infty$ (thick curve).

4.2 Gaussian Distribution

A very accurate approximation to the binomial distribution is the *Gaussian distribution* function, which represents the special case when the number of possible observations n becomes infinitely large and the probability for each state is finitely large, so that $np \gg 1$. In this case it is more convenient to parameterize the function in terms of the the mean μ and standard deviation σ, rather than in terms of the states x and number n items of the sample (i.e., coins). The Gaussian probability function P_G is defined as,

$$P_G(x;\mu,\sigma) = \frac{1}{\sigma\sqrt{2\pi}} \exp\left[-\frac{1}{2}\left(\frac{x-\mu}{\sigma}\right)^2\right] . \tag{4.2.1}$$

It can be shown that the mean of this Gaussian or normal distribution is $\mu = np$ and the standard deviation fulfills $\sigma^2 = np(1-p)$, and the distribution is normalized to $\sum_{x=0}^{\infty} P_G(x,\mu) = 1$. For instance, if we approximate the probability of next-neighbor interactions in a 2-D lattice (Fig. 4.1) with a Gaussian, using $n = 4$ and $p = 1/2$, the mean would be $\mu = np = 2$ and the standard deviation $\sigma = \sqrt{np(1-p)} = 1$, which agrees with the histogram of the binomial distribution shown in Fig. 4.1.

In astrophysical applications, random intensity fluctuations from a steady source are expected to obey a Gaussian distribution function to first order. For instance, simple histograms of soft X-ray intensity fluctuations from the solar corona observed with the *Soft X-ray Telescope (SXT)* onboard the *Yohkoh* spacecraft were sampled by Katsukawa and Tsuneta (2001). They used 25 different image sequences, each one consisting of about 20 images with a size of 128×128 pixels and a pixel size of $2.45''$. The data were processed to remove spacecraft pointing jitter using onboard attitude sensors as well as cross-correlation techniques, because the pointing jitter broadens the Gaussian noise distribution. In addition, little bursts in each time series from each pixel were removed, in order to have a clean separation of true photon noise from small soft X-ray bursts. Since the soft X-ray brightness level $I_0(x,y,t)$ varies across an image (x,y), they constructed histograms for three separate levels, around $I_0 \approx 10^{1.5}$, $10^{2.3}$, and $10^{3.0}$. The standard deviation σ_p of the photon noise for these three intensity levels is shown in Fig. 4.3 (top row), which is found to fit the core part of the distributions as,

$$\sigma_p = (1.5 \pm 0.3) I_0^{0.51 \pm 0.03} , \tag{4.2.2}$$

close to the theoretical expectation of $\sigma_p \propto \sqrt{I_0}$. Although the core of the distributions closely fit a Gaussian, the wing component apparently contains some other contributions, such as transient brightenings or time-dependent gradual variations of the mean intensity $I_0(t)$. Identifying the locations (x,y) in the images that contribute to the wing components, Katsukawa and Tsuneta (2001) did a second pass through the original data cube and removed those time intervals, which led to a clean Gaussian core component without wings (Fig. 4.3, bottom row). Although these cleaned time series produced a perfect Gaussian distribution that can be interpreted as pure random noise of emitted photons with Gaussian width σ_p, a slight excess was noted relative to the theoretical expectation of the instrumental noise characteristics, which was attributed to unresolved nanoflares with a hypothetical

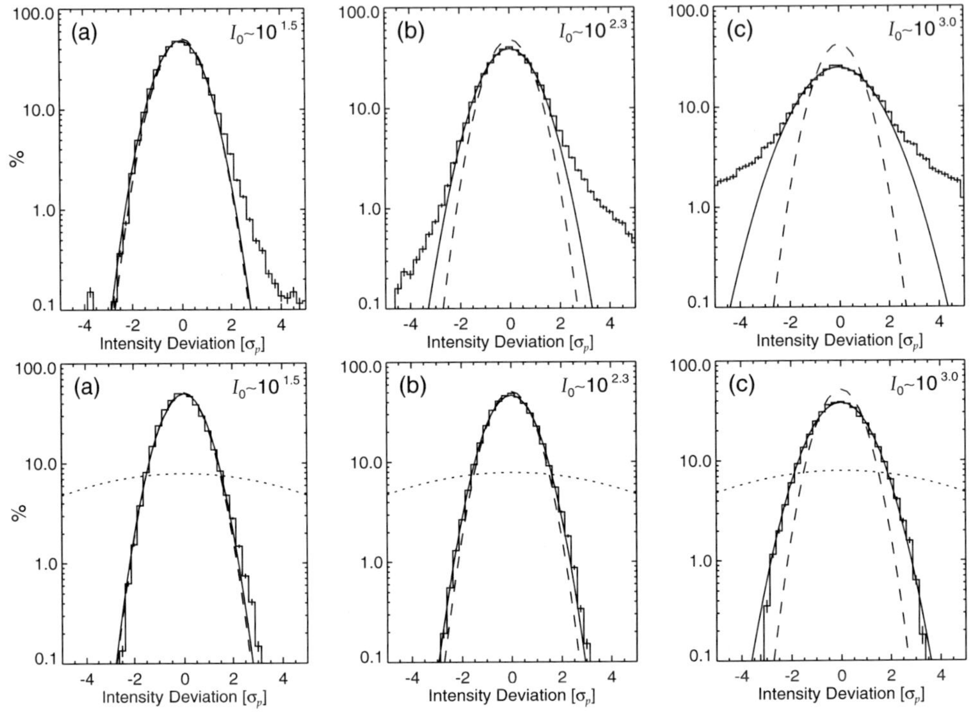

Fig. 4.3 *Top:* Histogram of soft X-ray intensity fluctuations around the mean intensity I_0 for three intensity levels, measured from a series of images observed with SXT/Yohkoh. The units of the horizontal axis are the standard deviations σ_p of the photon noise. The solid curves represent Gaussian fits to the core parts, with the pure photon noise distribution indicated by dashed curves. *Bottom:* Histogram of soft X-ray intensity fluctuations around the mean intensity I_0 for three intensity levels after removing the wing component originating from transients or time-dependent background variations. The solid curves represent Gaussian fits to the core parts, the theoretical photon noise distribution is indicated with dashed curves, and a 5-times wider Gaussian with dotted curves (Katsukawa and Tsuneta 2001; reproduced by permission of the AAS).

Gaussian distribution width σ_n, which adds in quadrature to the photon noise width σ_p to a slightly broadened observed width σ_{obs},

$$\sigma_{obs} = \sqrt{\sigma_p^2 + \sigma_n^2} \ . \tag{4.2.3}$$

Of course, because of the quadratic dependence of the Gaussian function on the width σ (Eq. 4.2.1), the convolution of two Gaussians is a Gaussian distribution again, with the Gaussian widths σ_i added in quadrature,

$$
\begin{aligned}
&P_G(x;\mu,\sigma_1) \times P_G(x;\mu,\sigma_2) \\
&= \frac{1}{\sigma_1\sqrt{2\pi}} \exp\left[-\frac{1}{2}\left(\frac{x-\mu}{\sigma_1}\right)^2\right] \times \frac{1}{\sigma_2\sqrt{2\pi}} \exp\left[-\frac{1}{2}\left(\frac{x-\mu}{\sigma_2}\right)^2\right] . \\
&= \frac{1}{\sqrt{2\pi(\sigma_1^2+\sigma_2^2)}} \exp\left[-\frac{1}{2}\frac{(x-\mu)^2}{(\sigma_1^2+\sigma_2^2)}\right]
\end{aligned}
\qquad (4.2.4)
$$

In the case of the soft X-ray fluctuations observed by Katsukawa and Tsuneta (2001), the excess of the Gaussian component was very small ($\sigma_n/\sigma_p \approx 0.05 \pm 0.02$), and thus the interpretation in terms of nanoflares versus unknown instrumental effects remained debatable. However, the study demonstrates that the observed random photon noise from the solar soft X-ray corona fulfills a Gaussian distribution with high precision, allowing us to determine additional non-noise components down to a relative brightness level of a few percent. Soft X-ray emission from stellar sources have of course much poorer statistics due to the large distances, which requires observations with significantly longer sampling time intervals to discriminate between photon noise and extraneous transient brightenings.

4.3 Poisson Distribution

In astrophysical observations, count statistics is often limited, either due to the large (stellar) distances or due to the paucity of photons at high energies. It is therefore common that we deal with count rates of less than one photon per second in observations of solar gamma-ray flares or soft X-rays from black hole accretion disks. Such low count rates do not fulfill the condition $np \gg 1$ required for the Gaussian approximation (Eq. 4.2.1), and thus another approximation to the binomial distribution has to be found in this low count-rate limit. Such an approximation was first derived by Siméon Denis Poisson (1781–1840), published in his work *Research on the Probability of Judgements in Criminal and Civil Matters*. A concise derivation can be found, e.g., in Bevington and Robinson (1969).

The Poisson distribution is an approximation to the binomial distribution (Eq. 4.1.7) in the limit of a small number of observed outcomes x (with a mean of μ) with respect to the number n of items because of very small probabilities, $p \ll 1$, and thus $\mu = np \ll n$. For instance, the probability of a photon from a faint stellar source to hit the aperture of a telescope on Earth can be extremely small, i.e., $p \ll 1$, and thus also the mean detection rate $\mu = np \ll n$ is very small compared with the number n of emitted photons at the source. Although the binomial statistics correctly describes the probability $P_B(x;n,p)$ of detected events, i.e., x photons per second, the enormous large (and unknown) number n (of emitted photons) makes it impossible to calculate the n factorials in Eq. (4.1.7). However, we can detect an average counting rate μ, and thus it is more convenient to express the Poisson approximation as a function of the count rate x and the mean μ, i.e., $P_P(x;\mu)$, rather than as a function of the unknown numbers n and p. Going back to the original expression of the binomial distribution (Eq. 4.1.7), the factorial $n!/(n-x)!$ has x factors that are all close to n for $x \ll n$, so we can approximate it with the product n^x,

$$\frac{n!}{(n-x)!} = n(n-1)(n-2)...(n-x-1) \approx n^x . \qquad (4.3.1)$$

The approximated second term (n^x) together with the third term p^x in Eq. (4.1.7) becomes then $(np)^x = \mu^x$. The fourth term, $(1-p)^{n-x}$ can be split into two factors, where one term is close to unity, i.e., $(1-p)^{-x} \approx 1$ for $p \ll 1$, and the remaining term $(1-p)^n$ can be rearranged by substituting $n = \mu/p$ to show that it converges towards $e^{-\mu}$,

$$\lim_{p \mapsto 0} (1-p)^n = \lim_{p \mapsto 0} \left[(1-p)^{1/p} \right]^{\mu} = \frac{1}{e}^{\mu} = e^{-\mu} \ . \tag{4.3.2}$$

Combining these approximations in the binomial distribution we arrive at the *Poisson distribution*,

$$P_P(x;\mu) = \lim_{p \mapsto 0} P_B(x;n,p) = \frac{\mu^x}{x!} e^{-\mu} \ , \tag{4.3.3}$$

where $\mu = np$ is the mean value and $\sigma = \sqrt{\mu}$ is the standard deviation of the probability distribution. The Poisson distribution is normalized so that $\sum_{x=0}^{\infty} P_P(x;\mu) = 1$. In Fig. 4.4 we show 10 Poisson distributions for means of $\mu = 1, 2, ..., 10$ within the range of $x = 0, ..., 20$. Note that the Poisson distribution is a discrete distribution at integer values of x. The Poisson distribution is strongly asymmetric for small means μ, but becomes more symmetric for larger μ and asymptotically approaches the Gaussian distribution.

The numerical calculation of the Poisson distribution can be simplified by the following recursive relationship (avoiding the factorials in the denominator),

$$P_P(0;\mu) = e^{-\mu} \ ,$$

$$P_P(x;\mu) = \frac{\mu}{x} P_P(x-1;\mu) \ . \tag{4.3.4}$$

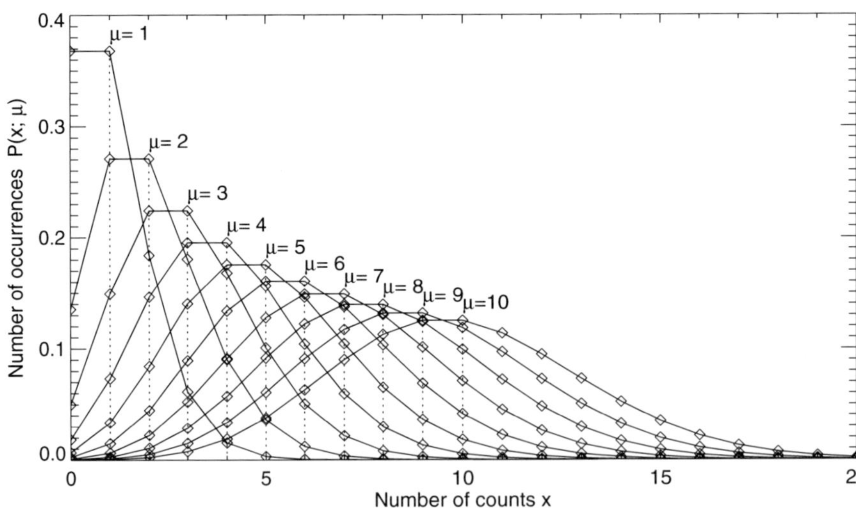

Fig. 4.4 Ten Poisson probability distributions $P_P(x;\mu)$ for $\mu = 1, 2, ..., 10$. Note that the distributions are only defined at discrete integer values $x = 0, 1, 2, ...$, while the smooth curves serve only to indicate the connections for each curve.

The Poisson probability distribution is one of the most common statistics applied to random processes. Regarding SOC phenomena, the waiting time between two subsequent avalanche events is generally assumed to be a random process, which can be tested by fitting the distribution of waiting times with a Poisson distribution (Eq. 4.3.3). We will deal with the statistics of waiting-time distributions in Chapter 5 and with the statistics of observed time scales in Chapter 7.

4.4 Exponential Distribution

In the limit of rare events ($x \ll n$) and small probabilities ($p \ll 1$), the discrete Poisson distribution $P_P(x;\mu)$ (Eq. 4.3.3) can simply be approximated by an exponential distribution. For the rarest events, say in the range of $0 \le x \le 1$, the factorial ($x! = 0! = 1! = 1$) is unity in the expression for the Poisson distribution (Eq. 4.3.3), and the exponential $\exp^{-\mu} \approx 1$ is also near unity when the mean value $\mu = np \ll 1$ is much smaller than unity. In this case the Poisson probability is only proportional to the function μ^x, which can be written as,

$$P_e(x;\mu) \approx \mu^x = (\exp^{\ln \mu})^x = \exp^{-x\ \ln(1/\mu)} , \tag{4.4.1}$$

which is a pure exponential function, i.e., $P_e(x) \approx \exp^{-ax}$, with $a = \ln(1/\mu)$. We show the comparison of a (discrete) Poisson distribution (Eq. 4.3.3) with a (continuous) exponential distribution (Eq. 4.4.1) in Fig. 4.5 for $x = 0,...,5$ and for means $\mu = np = 10^{-1}, 10^{-2}, 10^{-3}$. The approximation is almost exact in the range of $x = [0,1]$, but overestimates the Poisson distribution progressively for larger numbers of x by factors $\approx x!$, i.e., by a factor $2! = 2$ for $x = 2$. Thus, the exponential approximation to the Poisson statistics should only be applied for $x \lesssim 1$.

The exponential distribution is a continuous probability function, while the Poisson distribution is discretized by integer values of x. Since the approximation should only be used for $\mu \ll x$, the coefficient $a = \ln(1/\mu)$ in the exponent (Eq. 4.4.1) is according to the Taylor expansion of the natural logarithm,

$$\ln(1/\mu) \approx \left(\frac{1}{\mu} - 1\right) + ... \approx \frac{1}{\mu} \qquad \text{for } \mu \ll 1 , \tag{4.4.2}$$

and we can express the exponential distribution simply by,

$$P_e(x;\mu) \approx \frac{1}{\mu} \exp\left(-\frac{x}{\mu}\right) , \tag{4.4.3}$$

where the factor $1/\mu$ results from the normalization to $\int_0^\infty P_e(x;\mu)\, dx = 1$.

For instance, let us consider a random process for the growth phase of a nonlinear instability as we introduced it in our first analytical SOC model (Section 3.1). Since the coherent growth phase of a nonlinear instability is subject to many random factors, the rise times were assumed to be produced by a random process and were characterized by an exponential distribution. The nonlinear instability grows coherently during this rise time until it becomes quenched by some saturation mechanism, after a duration that we call

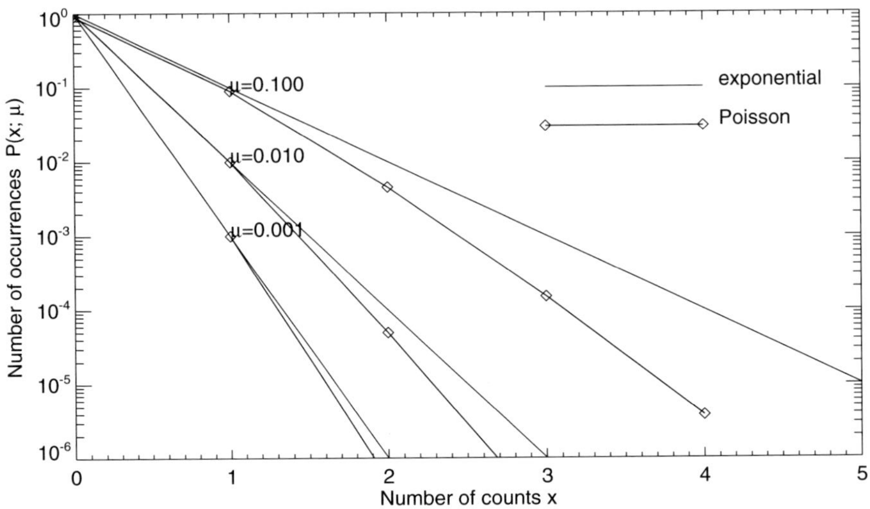

Fig. 4.5 Comparison of a discrete Poisson distribution (thick curves with diamonds; Eq. 4.3.3) with a (continuous) exponential approximation (solid line; Eq. 4.4.1) for means of $\mu = 10^{-1}, 10^{-2}, 10^{-3}$ in the range of $x = 0, ..., 5$. Note that the deviations are only significant for $x \gtrsim 1$.

the *saturation time* t_S. If we sample many such events of the same nonlinear process with different random conditions, the distribution $N(t_S)$ of saturation times t_S is expected to follow approximately an exponential distribution function, with an e-folding time constant t_{Se},

$$N(t_S)\, dt_S = \frac{N_0}{t_{Se}} \exp\left(-\frac{t_S}{t_{Se}}\right) dt_S \ , \tag{4.4.4}$$

where N_0 is the total number of events. This distribution is normalized so that the integral over the entire distribution in the range $[0, \infty]$ yields the total number of events N_0,

$$\int_0^\infty N(t_S)\, dt_S = N_0 \ . \tag{4.4.5}$$

If we normalize to unity, i.e., $N_0 = 1$, the event distribution $N(t_S)$ turns into a *differential probability distribution* $dP(t_S)$ (Fig. 4.6, solid curve),

$$dP(t_S)\, dt_S = \frac{1}{t_{Se}} \exp\left(-\frac{t_S}{t_{Se}}\right) dt_S \ . \tag{4.4.6}$$

The integral in the range $[0, t_s]$ yields the total probability $P(t_S)$ that an event occurs at time $t = t_S$ (Fig. 4.6, dashed curve),

$$P(t_S) = \int_0^{t_S} dP(t'_s) dt'_S = (1 - e^{-t_S/t_{Se}}) \ , \tag{4.4.7}$$

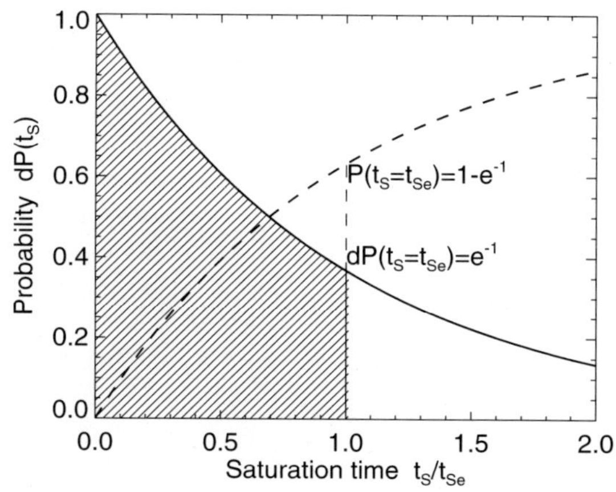

Fig. 4.6 The differential probability function $dP(t_S)$ (solid line) and the total probability function $P(t_S)$ (dashed line) that an event occurs after time t_S is shown for a random process.

which has a minimum probability of $P(t=0)=0$ at $t=0$ and a maximum probability of $P(t=\infty)=1$ at the asymptotic limit $t_S \mapsto \infty$. The probability after an e-folding time scale is $P(t_S=t_{Se})=(1-e^{-1}) \approx 0.63$.

The mean saturation time $\langle t_S \rangle$ in an exponential distribution is actually exactly the e-folding saturation time t_{Se},

$$\langle t_S \rangle = \int_0^\infty t_S \, dP(t_S) \, dt_S = t_{Se} \ , \tag{4.4.8}$$

as it can be shown by using the integral $\int x e^{ax} \, dx = (e^{ax}/a^2)(ax-1)$ with $x = t_S/t_{Se}$. So, the e-folding time scale t_{Se} is also a good characterization of the typical saturation time for random processes.

A mathematically generalized family of probability distribution functions was proposed by Karl Pearson (1895), which consists of a classification of distributions according to their first four moments. This unified formulation of distribution functions contains 12 different types, containing the Gaussian, exponential, β-, or γ-distribution function as special cases. Pearson's system was originally devised for modeling the observed skewed distributions in biometrics, but recent applications to model astrophysical SOC phenomena such as solar nanoflares have also been tackled (Podladchikova 2002).

Astrophysical observations of time scale distributions will be discussed in Chapter 7, where we find numerous examples of solar flare related time scale distributions observed in gamma rays, hard X-rays, or radio wavelengths to be consistent with the exponential distribution of a random process.

4.5 Count Rate Statistics

The statistics of events is usually defined by unique time points, say by n event times $t_i, i = 1,...,n$. For waiting-time distributions, the event times t_i have to be sorted in time and we can then sample the time intervals $\Delta t_i = (t_{i+1} - t_i)$, for $i = 1,...,n-1$. Mathematically, such discrete events localized at unique time points are called *point processes*. A point process is a random element whose values are *"point patterns"* on a mathematical set. A physical example is the series of arrival times of photons (or particles) from an astrophysical source that are so rare that each single photon (or particle) can be counted individually.

Random events occurring at low rates can be counted individually, which yields a discrete time series $t_i, i = 1,...,n$. If a random process produces a high rate of events, or if the temporal resolution of a detector is insufficient to separate individual events, we may be able to count the number of events in time intervals of length Δt and can produce a time series $f(t_i)$ that contains the counts f_i in each time bin $[t_i, t_i + \Delta t]$. For instance, for most astrophysical sources we detect a count rate, which quantifies the number of counts per time interval Δt, so we observe a time series $f_i = f(t_i)$ that can be represented as a binned time profile of a continuous function $f(t)$.

Transitioning from a discrete point process of event times to a continuous time series changes also our analysis technique of the temporal behavior. We can analyze discrete point processes by means of waiting-time statistics (Chapter 5), while (equidistant) continuous time series can conveniently be studied by means of auto-correlation, Fourier transforms, power spectra, or wavelet analysis. The fundamental aspect of random or Poisson processes, however, is manifested in both point processes and continuous time series in a very similar way.

Let us demonstrate the random behavior by constructing time series of random events that are sampled with low and high rates. In Fig. 4.7 we show time series of random events with mean rates from $\langle C(t) \rangle = 10^{-1}$ to 10^3 per time interval Δt. Stationary random processes with very low probabilities exhibit a Poisson distribution of count rates (Eq. 4.4.3), which can be approximated with an exponential function (Eq. 4.3.3), while random processes with high probabilities can be characterized by a Gaussian distribution (Eq. 4.2.1) of count rates. The Gaussian distributions with a mean of C have a standard deviation of $\sigma_C = \sqrt{C}$. However, despite the different analytical distribution functions, all examples of low and high count rate time profiles shown in Fig. 4.7 are consistent with the statistics of random noise. The diagnostics of random processes play a fundamental role in the statistical discrimination of SOC phenomena, which exhibit powerlaw-like distributions (e.g., of count rates C), rather than binomial, Gaussian, Poisson, or exponential distributions.

4.6 White Noise

We have generated time profiles of random processes in Fig. 4.7. Time series $f(t)$ are often analyzed with the Fourier transform $P(\nu)$, which decomposes a time profile into a sum of harmonic functions, i.e., $\exp(-i2\pi\nu t/n) = \cos(2\pi\nu t/n) + i\sin(2\pi\nu t/n)$, where the amplitude for each frequency ν is specified with a power spectrum $P(\nu)$ in frequency

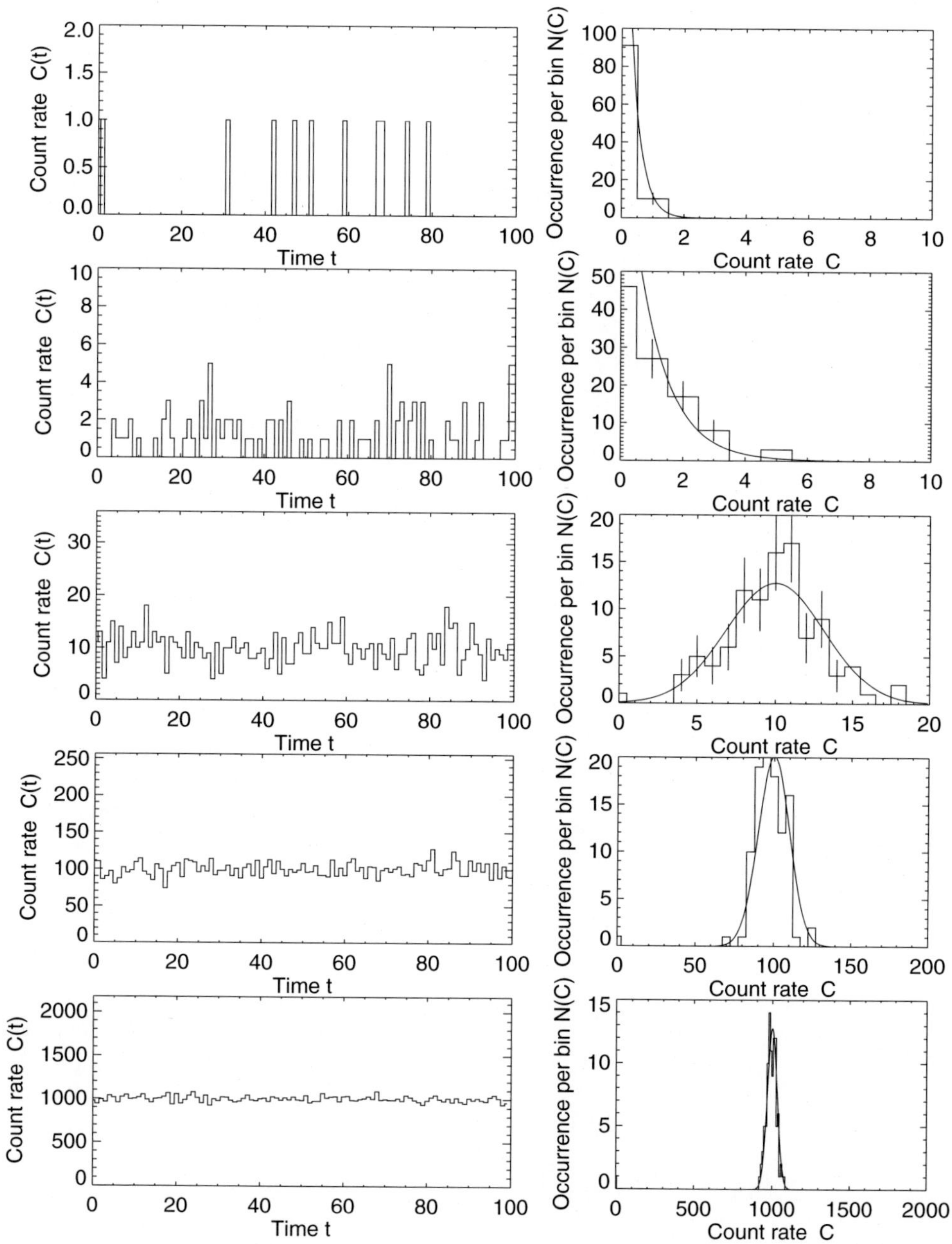

Fig. 4.7 Time series of random processes with average count rates of $C = 10^{-1}$ (top) to 10^3 counts per time interval (bottom). The binned time series is shown on the left side, and the histogram of counts per time bin on the right side. Note that the distributions of count rates with low rates can be approximated with an exponential function [for $C = 0.1$ (top) and $C = 1$ (second row)], while the distributions with high count rates can be approximated with a Gaussian function, which have a mean and standard deviation of $C = 10$, $\sigma_C \approx 3$ (third row), $C = 10^2$, $\sigma_C = 10$ (fourth row), and $C = 10^3$, $\sigma_C \approx 30$ (bottom row).

space, i.e. (in complex form),

$$P(\nu) = \frac{1}{n}\sum_{t=0}^{n-1} f(t) \exp\left(-\frac{i2\pi\nu t}{n}\right) . \quad (4.6.1)$$

The power spectral density is usually expressed with a real number, by calculating the absolute value of the complex power spectrum, i.e., $|P(\nu)|$, which discards the phase information that is contained in the complex number of the power spectrum $P(\nu)$. The Fourier transform is particularly useful to extract periodic pulses with a particular period in a time series, even in the presence of heavy noise. If there are multiple periodic fluctuations present in a time series, the power spectrum will reveal each one with a peak in the power spectrum at the particular period or frequency.

However, what does the power spectrum of a random process look like? In Fig. 4.8 we show a time series $f(t)$ of a random process. Calculating the power spectral density $P(\nu)$ with the *Fast Fourier Transform*, we find a completely flat power spectrum from the minimum frequency $\nu_{min} = 1/(n\Delta t)$ to the maximum frequency $\nu_{max} = 1/(2\Delta t)$ (i.e., the half sampling frequency which is also called *Nyquist frequency* or *cutoff frequency*). Since this constant power at all frequencies is similar to white light, consisting of all colors in the visible wavelength range, such a flat power spectrum is also called *white noise spectrum*.

Another characterization of time profiles is the auto-correlation function, which is useful to evaluate the distribution of time scales of pulses that occur in a time series. The

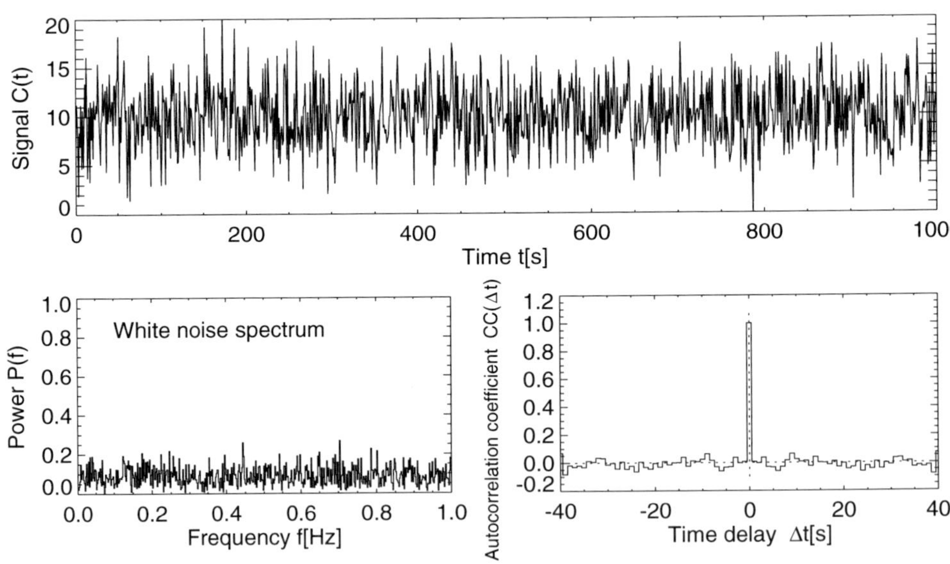

Fig. 4.8 Random time series with $n = 1\,000$ time points and time interval $dt = 1.0$ s, with a mean count rate of $C = 10$ cts s^{-1} and a standard deviation of $\sigma_C \approx 3$ cts s^{-1} (top frame). The Fourier power spectrum is flat, called a *white noise spectrum* (bottom left). The auto-correlation function is zero everywhere except for a delta-function peak $CC(\Delta t = 0) = 1$ at $\Delta t = 0$ (bottom right).

auto-correlation function $f_{AC}(\Delta t)$ is simply defined by the normalized product of a time series with the time-shifted or delayed time series as a function of the delay Δt, i.e.,

$$f_{AC}(\Delta t) = \frac{\sum[f(t+\Delta t) - f_0)][f(t) - f_0]}{\sum[f(t) - f_0]^2} , \tag{4.6.2}$$

where $f_0 = \langle f(t) \rangle$ is the average value of the time series. If pulses with duration $\tau_p = n_p \Delta t$ exist in a time series, the auto-correlation coefficient will be high for delays $\Delta t \leq \tau_p$, while it will be low for larger delays, since the product $f(t+\Delta t)f(t)$ will largely cancel out for random correlations. Therefore, if $f(t)$ is a random time series, the auto-correlation function $f_{AC}(\Delta t)$ has only a delta-function peak at zero delay, $f_{AC}(\Delta t = 0) = 1$, while it is near zero everywhere else (Fig. 4.8, bottom right frame).

We show a few examples of observed time series $f(t)$ and their auto-correlation function $f_{AC}(\Delta t)$ in Fig. 4.9, observed with a solar radio spectrometer (Aschwanden et

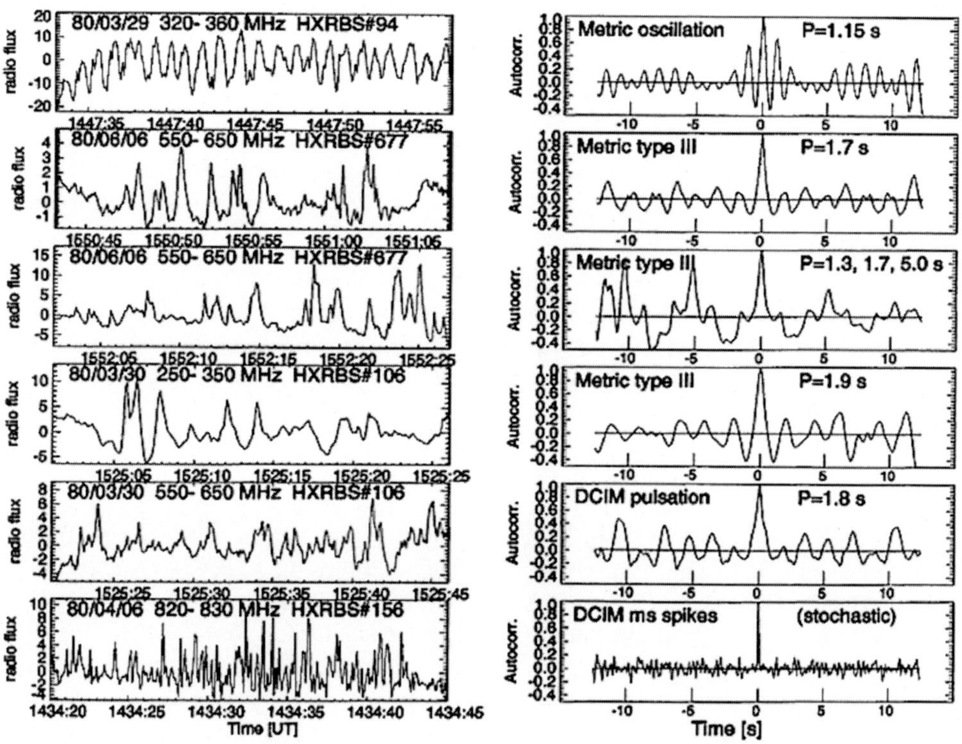

Fig. 4.9 Six time series (left-hand panels) of solar radio burst emission recorded at decimetric radio frequencies in the range of 250–830 MHz with the radio spectrometer of ETH Zurich, Switzerland. The auto-correlation functions of these time profiles is shown in the right-hand panels, with significant time periods P identified from FFT power spectra indicated. Note that the upper five examples show significant periodicities, while the sixt example (bottom panels) reveals truly stochastic pulses with a δ-function in the auto-correlation function at a time lag of $\Delta t = 0$ (Aschwanden et al. 1994).

al. 1994). The six examples shown in Fig. 4.9 contain different types of decimetric radio bursts, all occurring during solar flares. The first case shows a very periodic time profile in the frequency range of 320–360 MHz, called a *metric oscillation* burst, which produces also a smoothed oscillatory pattern with a period of $P = 1.15$ s in the auto-correlation function. The next three cases are *metric type III* radio bursts, which seem to have quite erratically fluctuating time profiles, but the auto-correlation function reveals some periodicity with periods in the range of $P \approx 1$–5 s. The fifth case appears to be more periodic and is called *decimetric pulsation* event, which reveals a periodicity of $P = 1.8$ s in the auto-correlation function. The last type consists of thousands of unresolved *decimetric millisecond spikes*, which are randomly distributed in time and frequency, and indeed produce an auto-correlation function that has only a δ-function peak at $\Delta t = 0$, perfectly consistent with a white noise spectrum as shown in Fig. 4.8. These examples demonstrate that the time scales and periodicity or randomness of time structures can be diagnosed with the auto-correlation function $f_{AC}(\Delta t)$, even when it is not evident from the time series $f(t)$. The discrimination of random pulses (with finite duration) from the (white) noise floor is an important capability in the statistics of SOC phenomena.

4.7 1/*f* Power Spectra Nomenclature

Besides the *white noise spectrum* there exists a more general class of noise spectra that all have in common that the power spectral density $P(\nu)$ is proportional to a negative powerlaw of the frequency ν,

$$P(\nu) \propto \nu^{-p} , \tag{4.7.1}$$

with the power index p most frequently found in the range of $0 < p < 2$. Since most noise spectra found in nature and technology have a value near $p \approx 1$, this class of noise spectrum is also called $1/f$ *noise*, where f means the frequency, and $1/f$ corresponds to ν^{-1} in our notation. $1/f$-noise spectra occur most commonly in nature and technology, because they contain a balance of short and long fluctuations from different processes, such as occur in semiconductors, diodes, transistors, or films, but also in the Earth rotation, highway traffic, or nerve membranes (e.g., see Schuster 1988, p.92; Press 1978).

The nomenclature of noise spectra borrows from the analogy to color spectra. White light is defined as the sum of all visible wavelengths from ultraviolet ($\lambda \approx 2{,}000$ Å) to infrared ($\lambda \approx 8{,}000$ Å), and thus a flat noise spectrum is called *white noise spectrum* (Section 4.6). A color spectrum that has more red color has an overabundance of long wavelengths λ, or low frequencies ($\nu = c/\lambda$), and therefore falls off with higher frequencies, e.g., as a powerlaw spectrum $P(\nu) \propto \nu^{-p}$ with a positive power index p. A noise spectrum that falls off with the second power, $P(\nu) \propto \nu^{-2}$, has therefore been named *red noise*, more commonly known as *Brownian noise*, since it occurs in Brownian molecular motion. Noise spectra that are intermediate between the white ($p = 0$) and the red noise spectrum ($p = 2$), say with $p \approx 1$, have therefore been named as *pink noise spectrum*, to indicate the mixture of white and red colors. The *Brownian noise* is also called briefly *Brown noise*, but it does not refer to the color, but rather to Robert Brown, the discoverer of the Brownian motion. Since this type of $1/f$ noise is very common in electric signals, it is also called *flicker noise*. The powerlaw range goes even steeper in some phenomena up to $p \approx 3$, in which case it is called *black noise spectrum*, in analogy to extending the color spectrum to the invisible beyond red. Black-noise phenomena govern natural and unnatural catastrophes like floods, droughts, bear markets, or power outages (e.g., see Schroeder 1991, Section 5 therein). In Table 4.1 we summarize this nomenclature of noise spectra.

Table 4.1 Nomenclature of noise spectra.

Power spectrum	Power index	Spectrum nomenclature
$P(\nu) \propto \nu^0$	$p = 0$	white noise
$P(\nu) \propto \nu^{-1}$	$p = 1$	pink noise, flicker noise, $1/f$ noise
$P(\nu) \propto \nu^{-2}$	$p = 2$	red noise, Brown(ian) noise
$P(\nu) \propto \nu^{-3}$	$p = 3$	black noise

Let us visualize an example of each of these spectra in Fig. 4.10: a white noise spectrum ($p = 0$), a pink noise spectrum ($p = 1$), a red noise spectrum ($p = 2$), and a black noise spectrum ($p = 3$). We multiply a white noise spectrum simply by the appropriate powerlaw function $P(\nu) \propto \nu^{-p}$ and construct the corresponding time series $f(t)$ by the inverse Fourier transform (with the forward Fourier transform defined in Eq. 4.6.1),

$$f(t) = \frac{1}{n} \sum_{t=0}^{n-1} P(\nu) \exp\left(+\frac{i2\pi\nu t}{n}\right) . \tag{4.7.2}$$

The white noise spectrum shown in the top panel of Fig. 4.10 is identical to that shown in Fig. 4.8, and is multiplied with a powerlaw function ν^{-p} with $p = 1, 2, 3$ in the other examples shown in Fig. 4.10. The resulting time profiles show a mixture of short and longer pulses for the case of the pink noise spectrum (second row in Fig. 4.10), but are completely dominated by long-duration pulses for the cases of the *red noise spectrum* (third row in Fig. 4.10) and the *black noise spectrum* (bottom row in Fig. 4.10), since the high-frequency noise is strongly suppressed for $p \gtrsim 2$. A natural way to produce $1/f$ noise spectra is to apply a high-pass filter to a white-noise signal.

In astrophysical time series, random processes with $1/f$ noise have been studied extensively, in order to discriminate between photon noise and significant signals from solar flares, flare stars, cataclysmic variables, neutron stars, pulsars, and black hole candidates. Data analysis techniques range from filtering out noise in time series analyzed with the Fast Fourier Transform (Brault and White 1971; Scargle 1989), modeling random processes in the time domain (Scargle 1981), spectral analysis of unevenly spaced data (Scargle 1982), auto-correlation and cross-correlation methods (Scargle 1989), modeling chaotic and random processes with linear filters (Scargle 1990), to Bayesian blocks in photon counting data (Scargle 1998). A correlation dimension and wavelet analysis of time series of solar radio bursts observed in microwaves revealed that the intensity profiles $I(t)$ could be decomposed into a sum of white noise (or thermal) component $I_T(t)$ and a flicker-type (or nonthermal) random component for pulses $I_{NT}(t)$,

$$I(t) = I_T(t) + I_{NT}(t) , \tag{4.7.3}$$

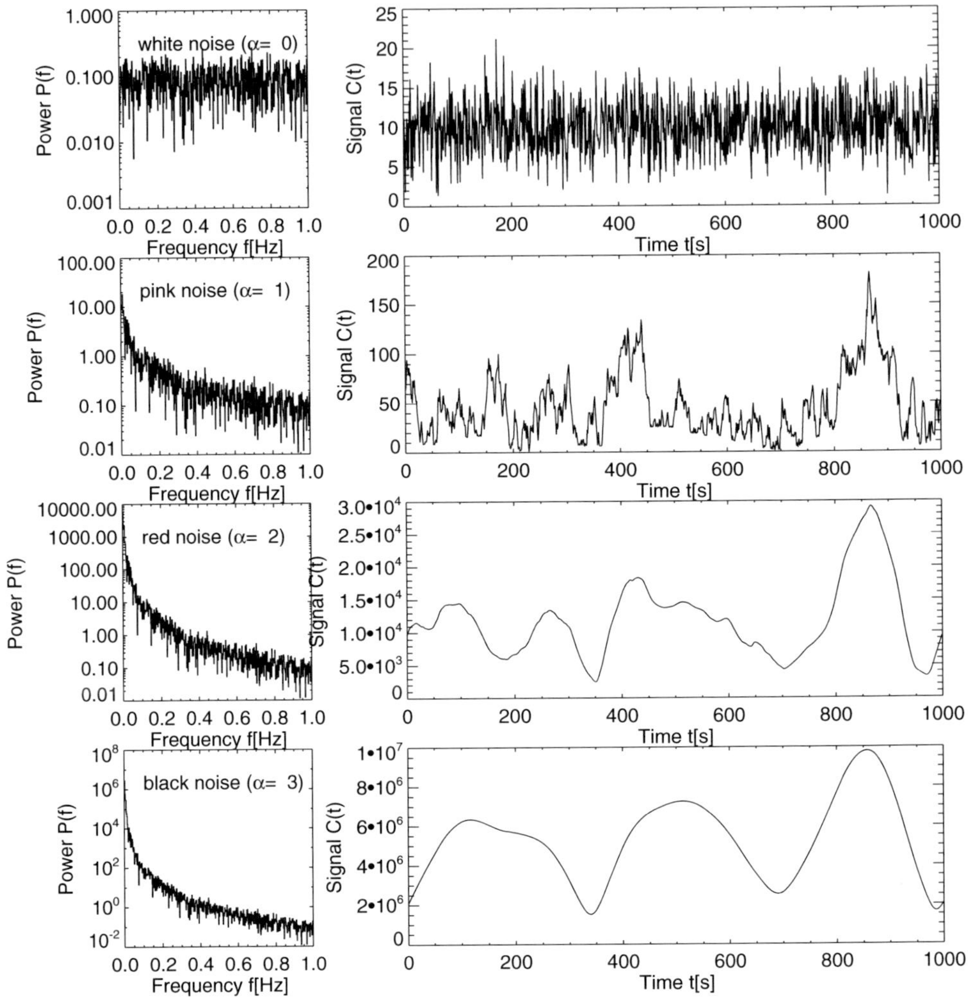

Fig. 4.10 Noise power spectra (left panels) and corresponding time series (right panels) for power spectral indices $p = 0$ (top row: white noise spectrum), $p = 1$ (second row: pink noise spectrum), $p = 2$ (third row: red noise spectrum), and $p = 3$ (bottom row: black noise spectrum). The white noise spectrum is identical to Fig. 4.10 and is multiplied with ν^{-p} in the other cases. The time series are reconstructed with the inverse Fast Fourier Transform.

where the flicker-type process has a power spectrum $P(\nu) \propto \nu^{-p}$ with $p \approx 0.8, ..., 1.8$ (Ryabov et al. 1997). Soft X-ray observations of stellar black-hole candidates were found to exhibit $1/f$ power spectra, such as the stellar black-hole candidate Cygnus X-1 in the hard state (Tanaka 1989; Makishima 1988), the black-hole candidate GX 339-4 in its very high state (Miyamoto et al. 1991), the low-luminosity type I Seyfert galaxy NGC4051 (Lawrence et al. 1987), the active galactic nuclei (AGN) Seyfert galaxy NGC5506 (McHardy and Czerny 1987), or the X-ray binary pulsar GX 301-1 (Tashiro et al. 1991). Soft X-ray emission from black-hole candidates or accretion disk sources always exhibit two distinctly different spectral states (Tanaka 1989; Mineshige 1994a): in the soft (or high) state the emergent spectra are approximately thermal (Planck spectrum), whereas in the hard (or low) state, the spectra are powerlaw-like ($1/f$ spectra type). Thus, the power spectra $P(\nu)$ of astrophysical light curves can often be decomposed into these two spectral types of thermal (Planck) and nonthermal (powerlaw) emission,

$$P(\nu) = P_T \left(\frac{2h\nu^3}{c^2} \right) \frac{1}{\exp(h\nu/k_B T) - 1} + P_{NT} \left(\frac{\nu}{\nu_0} \right)^{-p} . \tag{4.7.4}$$

corresponding to the thermal or white-noise component $I_T(t)$ and the flicker-type (or nonthermal) random component manifested in superimposed pulses $I_{NT}(t)$ that are detectable in the intensity time profiles $I(t)$ (Eq. 4.7.3). These flicker-type pulses $I_{NT}(t)$ will be identified with individual SOC avalanches in the following.

4.8 Shot Noise or Flicker Noise

Historically, the term *shot noise* (or *flicker noise*) was used to characterize the discreteness of particle transport, such as the DC current of charged particles (electrons and/or holes) in a conductor, first discovered by Walter Schottky in 1918. In essence, electrons in a conductor do not flow uniformly like water in a laminar state, but rather move intermittently with an average rate (driven by the applied voltage) and exhibit random fluctuations around this average rate, which is called "shot noise". A typical system is a *pn* junction diode, where each carrier randomly passes across the depletion region of the junction. Since the electrons pass randomly and independently, their number can be described by Poisson statistics (Section 4.3).

4.8.1 Derivation of Schottky's Theorem

The shot noise concept is a physical model that produces $1/f$ type power spectra and powerlaw distributions of time scales, and thus is highly relevant for SOC models. We start with the original derivation of the shot noise statistics shown by Van der Ziel (1950). The current I is defined by the number n of electrons passing through a point during a time interval T,

$$I = \frac{ne}{T} , \tag{4.8.1}$$

and the time-averaged current $\langle I \rangle$ relates to the time-averaged number $\langle n \rangle$ as

$$\langle I \rangle = \frac{\langle n \rangle e}{T} . \tag{4.8.2}$$

If we assume that electron transport is a random Poisson process, we have a variance of $\sigma^2 = \langle n \rangle$ (or a standard deviation of $\sigma = \sqrt{\langle n \rangle}$),

$$\sigma^2 = \langle n - \langle n \rangle \rangle^2 = \langle n^2 \rangle - \langle n \rangle^2 = \langle n \rangle , \tag{4.8.3}$$

which yields a relationship between the mean quadratic fluctuations $\langle n^2 \rangle$ and the mean number $\langle n \rangle$. To describe the shot noise we are interested in the mean current fluctuations,

$$\langle \Delta I^2 \rangle = \langle I^2 \rangle - \langle I \rangle^2 = \left\langle \left(\frac{ne}{T} \right)^2 \right\rangle - \left(\frac{\langle n \rangle e}{T} \right)^2 . \tag{4.8.4}$$

Inserting the quadratic fluctuations $\langle n^2 \rangle$ (Eq. 4.8.3) and the mean current (Eq. 4.8.2) into Eq. (4.8.4) we find,

$$\langle \Delta I^2 \rangle = \frac{\langle n^2 \rangle e^2}{T^2} - \frac{\langle n \rangle^2 e^2}{T^2} = \frac{\langle n \rangle e^2}{T^2} = \langle I \rangle \frac{e}{T} , \tag{4.8.5}$$

which means that the root-mean-square (rms) fluctuations i_{rms} of the current (i.e., shot noise) is proportional to the square root of the mean current $\langle I \rangle$,

$$i_{rms} = \sqrt{\langle \Delta I^2 \rangle} = \sqrt{\frac{e \langle I \rangle}{T}} . \tag{4.8.6}$$

In order to derive the power spectrum $S(\nu)$ of shot noise, we have to relate the time scale T to the frequency ν. We can define the current in an RCL-circuit as a superposition of current spikes (parameterized with δ-functions) resulting from individual electrons that arrive at random times t_j,

$$I(t) = \sum_j q \delta(t - t_j) . \tag{4.8.7}$$

This time profile $I(t)$ contains random fluctuations of very short pulses (δ-functions). The duration T of a current pulse can be defined by the *auto-correlation function* $R_I(t)$,

$$R_I(t') = \lim_{T \mapsto \infty} \frac{1}{T} \int_{-T/2}^{+T/2} I(t) I(t+t') \, dt . \tag{4.8.8}$$

The *Wiener–Khinchin theorem* (also called *Wiener–Khintchine, Wiener–Khinchin–Einstein*, or *Khinchin–Kolmogorov theorem*) states that the power spectral density $P(\nu)$ of a stationary random process is the Fourier transform of the corresponding auto-correlation function $R(t)$,

$$P_I(\nu) = 2 \int_{-\infty}^{\infty} R_I(t') e^{-i 2\pi \nu t'} \, dt' , \tag{4.8.9}$$

where $P_I(\nu)$ is the one-sided power spectral density (leading to a factor 2 in front of the integral).

Applying now the auto-correlation function $R_I(t)$ (Eq. 4.8.8) to the current pulses with δ-function shapes (Eq. 4.8.7) we have,

$$\begin{aligned} R_I(t') &= \lim_{T \mapsto \infty} \frac{q^2}{T} \sum_k \sum_{k'} \int_{-T/2}^{T/2} \delta(t - t_k) \delta(t - t_{k'} + t') \, dt \\ &= \lim_{T \mapsto \infty} \frac{q^2}{T} \sum_k \sum_{k'} \delta(t - t_{k'} + t') . \end{aligned} \tag{4.8.10}$$

The summation over $t_k = t_{k'}$ contributes $\delta(t')$ for N values of t_k in the range of $-T/2 < t_k < T/2$, while the contributions of $t_k \neq t_{k'}$ will vanish for randomly distributed values, which yields with $N/T = \langle I \rangle / q$,

$$R_I(t') = q\langle I \rangle \delta(t') \ . \tag{4.8.11}$$

An example of an auto-correlation function of δ-function random pulses is shown in Fig. 4.8. The Fourier transform of this auto-correlation function $R_I(t')$ yields then the power spectrum $P_I(\nu)$ according to the Wiener-Khinchin theorem (Eq. 4.8.9),

$$P_I(\nu) = 2q\langle I \rangle \ , \tag{4.8.12}$$

which is the *Schottky theorem*, stating that the shot noise spectrum is a constant and extends uniformly over all frequencies, also called white noise spectrum. Note that this result applies to very short pulses characterized with δ-functions, as shown in Section 4.6. If we set $q = e$ and relate the time interval T to the Nyquist sampling frequency, i.e., $\Delta\nu = 1/2T$, we see (with Eq. 4.8.6) that the power spectral density corresponds to the rms current fluctuations i_{rms}^2 per unity bandwidth $\Delta\nu$,

$$P_I(\nu) = 2e\langle I \rangle = 2 i_{rms}^2 T = \frac{i_{rms}^2}{\Delta\nu} \ . \tag{4.8.13}$$

4.8.2 Shot Noise Spectrum for Rectangular Pulses

Let us now consider current pulses with a significant duration T, for instance a square current pulse $f(t)$ with duration T, as shown in Fig. 4.11 (top). The auto-correlation function $R(t)$ of this rectangular pulse shape $f(t)$ can be computed with Eq. (4.8.8) and yields a single triangle (Fig. 4.11 bottom left),

$$R_I(t') = \lim_{T \mapsto \infty} \frac{1}{T} \int_{-T/2}^{+T/2} I(t) I(t+t')\, dt = q\langle I \rangle \begin{cases} 1 - |t'|/T & \text{for } |t'| \le T \\ 0 & \text{for } |t'| > T \end{cases} \tag{4.8.14}$$

and the Fourier transform of a single triangle yields with the Wiener–Khinchin theorem (Eq. 4.8.8) the power spectrum $P_I(\nu)$,

$$P_I(\nu) = 2q\langle I \rangle \left[\frac{\sin(\pi\nu T)}{\pi\nu T} \right]^2 , \tag{4.8.15}$$

where T is the averaged pulse duration. The resulting power spectrum is a white spectrum at $\nu = 0$, with the value $P_I(\nu = 0) = 2q\langle I \rangle$ (Eq. 4.8.13) of Schottky's theorem, but it falls off with a Gaussian-like function and has a cutoff at $\nu = 1/T$ (Fig. 4.11, right panel).

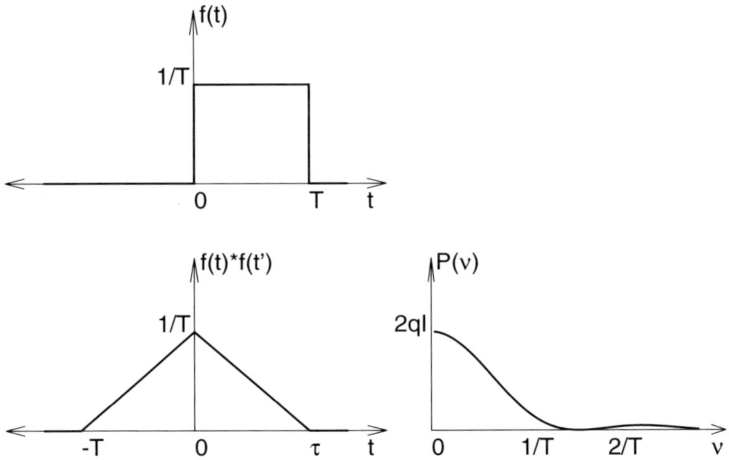

Fig. 4.11 Rectangular pulse shape $f(t)$ with duration T (top left), auto-correlation function $R(t) = f(t) * f(t')$ (bottom left), and corresponding Fourier power spectrum $P(\nu)$ (bottom right).

4.8.3 Shot Noise Spectrum for Exponential-Decay Pulses

Assuming that the generated pulses are subject to a linear relaxation process, a time profile with an exponentially decaying function can be assumed,

$$f(t) = \frac{1}{T} \exp\left(-\frac{t}{T}\right) . \tag{4.8.16}$$

Calculating the correlation function $R(t)$ (Eq. 4.8.8) and the Fourier transform of it (Eq. 4.8.9), the following power spectrum $P(\nu)$ is obtained,

$$P(\nu) = P_0 \frac{1}{1+(2\pi\nu T)^2} , \tag{4.8.17}$$

which is essentially constant at $\nu \lesssim 1/T$ and a powerlaw spectrum $P(\nu) \propto \nu^{-2}$ above the frequency $\nu \gtrsim 1/T$.

An example of a power spectrum $P(\nu) \propto \nu^{-2}$ is shown in Fig. 4.10 (third row), called *"red noise"*. Pulses with some finite time scale T appear in this time profile, which strongly dominate the white noise background fluctuations. Such pulses are also said to have a high *signal-to-noise* ratio. Note that the meaning of the terms *shot noise* or *flicker noise* applies now to significant pulses with a finite duration, which stand out of the white-noise background, which consists of random fluctuations with unresolved (δ-functions) time scales, as originally defined in Schottky's theorem. In other words, the powerlaw slope of the power spectrum tells us whether we deal with random fluctuations of unresolved time scales (i.e., δ- functions in time and white noise spectra with a powerlaw slope of $p \approx 0$) or with random pulses with resolved time scales (i.e., $1/f$ or flicker noise with powerlaw slopes of $p \gtrsim 1$).

4.8.4 Shot Noise Spectrum and Distribution of Pulse Durations

In the previous derivation we assumed a single duration T for a pulse. Anticipating an application of shot noise pulses to SOC simulations or observations, we have to deal with a distribution $N(T)$ of pulse durations T. The derivation of a power spectrum $P(\nu)$ from the frequency duration $N(T)$ of pulses or avalanches has been mentioned in the original article of Bak, Tang, and Wiesenfeld (1987), entitled "Self-Organized Criticality: An Explanation of 1/f Noise", and has been outlined in Bak et al. (1988), or Mineshige et al. (1994a). Essentially, the power spectra $P_T(\nu)$ that are characteristic for a particular pulse duration T are added up linearly for all time scales T, normalized by their number per unit time interval, which is given by the occurrence frequency distribution $N(T)$,

$$P(\nu) = \sum_T N(T) P_T(\nu) . \tag{4.8.18}$$

For the case of avalanches with exponential decay with duration T we have according to Eq. (4.8.17),

$$P_T(\nu) \propto E \frac{1}{1+(2\pi\nu T)^2} , \tag{4.8.19}$$

where E represents the total energy of the avalanche. If we observe a powerlaw-like frequency distribution of energies E,

$$N(E) \propto E^{-\alpha_E} , \tag{4.8.20}$$

and assume a statistical correlation between the total energy E and total duration T of an avalanche event, say a powerlaw relation with coefficient $1+\gamma$ (where γ needs to be determined from numerical simulations, analytical models, or observations),

$$E(T) \propto T^{1+\gamma} , \tag{4.8.21}$$

we can calculate the derivative $|dE/dT| = T^{\gamma}$ and express the frequency distribution $N(T)$ of time scales by substituting $E(T)$ (Eq. 4.8.21) into the energy distribution $N(E)$ (Eq. 4.8.20),

$$N(T) = N(E[T]) \left| \frac{dE}{dT} \right| \approx T^{-\alpha_E(1+\gamma)+\gamma} . \tag{4.8.22}$$

We can insert now the expressions for $E(T)$ (Eq. 4.8.21) and $N(T)$ (Eq. 4.8.22) into the partial power spectrum $P_T(\nu)$ (Eq. 4.8.19) and the total power spectrum $P(\nu)$ (Eq. 4.8.18),

$$P(\nu) \propto \sum_T \frac{T^{-\alpha_E(1+\gamma)+\gamma} T^{1+\gamma}}{1+(2\pi\nu T)^2} . \tag{4.8.23}$$

Each partial power spectrum for a time scale T has a cutoff above $T \gtrsim 1/\nu$ due to the quadratic term $(2\pi\nu T)^2$ in the denominator. Thus, we can integrate each power spectrum from $T_1 = 0$ to $T_2 = 1/\nu$ and replace the summation in Eq. (4.8.23) by an integral over all time ranges $[T_1, T_2]$,

$$P(\nu) \propto \sum_T \int_0^\infty \frac{T^{1+2\gamma-\alpha_E(1+\gamma)}}{1+(2\pi T\nu)^2} \approx \int_0^{1/\nu} T^{1+2\gamma-\alpha_E(1+\gamma)}\, dT \ , \tag{4.8.24}$$

which can straightforwardly be integrated,

$$P(\nu) \propto \int_0^{1/\nu} T^{1+2\gamma-\alpha_E(1+\gamma)}\, dT = \left[T^{2+2\gamma-\alpha_E(1+\gamma)}\right|_0^{1/\nu} = \left(\frac{1}{\nu}\right)^{(2-\alpha_E)(1+\gamma)} \ . \tag{4.8.25}$$

which is a powerlaw spectrum,

$$P(\nu) \propto \nu^{-p} \ . \tag{4.8.26}$$

where the powerlaw index p of the power spectrum is related to the powerlaw coefficient α_E of the energy frequency distribution by

$$p = (2-\alpha_E)(1+\gamma) \ . \tag{4.8.27}$$

We illustrate the superposition of individual power spectra $P_{T_i}(\nu)$ for a series of time scales $T_i = 1/\nu_i$ of frequencies $\nu_i = 10^{1+0.1*i}$ with $i = 1, 30$ in the frequency range of $\nu = 1$–1000 in Fig. 4.12. Each individual power spectrum has a flat part at $\nu < \nu_i$ and falls off with the second power according to Eq. (4.8.19) above this cutoff frequency $\nu \geq \nu_i$. The summation of these power spectra with a relative weighting of $N(T) \propto N(E) \propto E^{\alpha_E}$ with $\alpha_E = 1.1$ and $\gamma = 0$ yields a total power spectrum $P(\nu)$ with an approximate powerlaw $P(\nu) \propto \nu^{0.9}$, in agreement wtih the derived relationship $p = (2-\alpha_E)(1+\gamma) = (2-1.1) = 0.9$ (Eq. 4.8.27).

In the numerical simulations of Bak et al. (1988) a power spectrum of $P(\nu) \propto \nu^{-1.57}$ was found in a 2-D lattice, and of $P(\nu) \propto \nu^{-1.08}$ in a 3-D lattice, respectively, which is close to the value of $P(\nu) \propto \nu^{-1}$ expected for $1/f$ flicker noise.

In astrophysical observations, the term *shot noise* has been used to characterize the statistics of random bursts (shots) that appear superimposed on the background radiation. In the nomenclature of our previous section, these "shots" correspond to the nonthermal component of random pulses appearing superimposed on the thermal background component, such as an elementary flare spikes or small flares in soft X-rays (e.g., Frontera and Fuligni 1979; Ueno et al. 1997), or pulses in X-ray time profiles from accretion or black-hole candidates (in high state) (e.g., Sutherland et al. 1978; Negoro et al. 1995; Vaughan and Nowak 1997; Takeuchi and Mineshige 1997; Negoro et al. 2001; Uttley and McHardy 2001; Li and Muraki 2002; Focke et al. 2005).

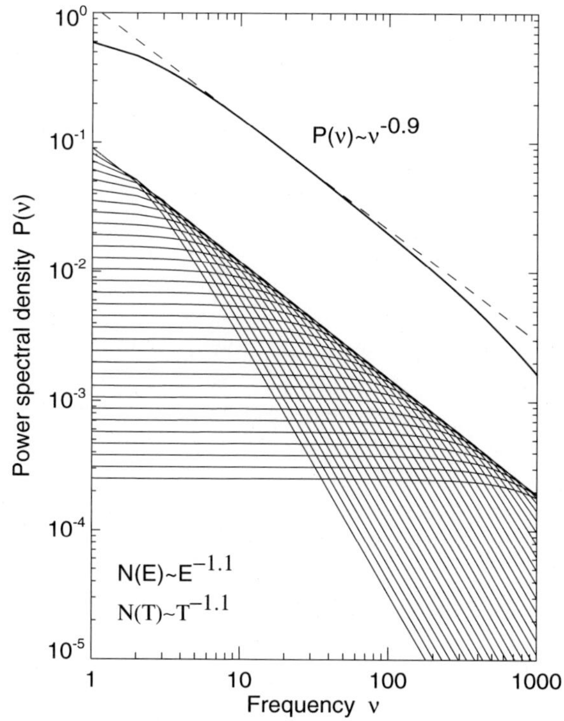

Fig. 4.12 Superposition of power spectra $P_T(\nu)$ of time profiles containing exponentially-decaying pulses with time scales T (thin curves). The sum of the individual power spectra is indicated with a thick curve and fitted with a powerlaw, which has a slope of $P(\nu) \propto \nu^{-0.9}$. The distribution of individual time scales is $T_i = 1/\nu_i$ of frequencies $\nu_i = 10^{1+0.1*i}$ with $i = 1,30$ in the frequency range of $\nu = 1-1{,}000$. The powerlaw distribution of energies is $N(E) \propto E^{1.1}$, which constrains the relative weighting of the time scales $T(E)$ in the summation.

4.9 Log-Normal Distribution

In the shot noise model (Section 4.8), a time profile is produced by a sum of independently occurring shots, i.e., an astrophysical light curve is composed of a superposition of many independent flares, and thus both the time profiles and the power spectra are additive regarding the time scales of individual events (shots). This leads to binomial, Poissonian, or Gaussian distribution of time scales. However, sometimes processes are found in nature that have the logarithm $\ln(X)$ of the random variable X normally distributed, which is referred to as *log-normal distribution* or *Galton distribution*. The log-normal distribution can be thought as the analog of the normal or Gaussian distribution for multiplicative processes, rather than for additive processes. For instance, SOC avalanches are subject to exponential growth (Section 3.1 and 3.3), so if the time scales of saturation τ are normally distributed, we expect that the resulting energies $W_S = exp(\tau/t_G)$ (Eq. 3.1.2) are log-normally distributed. The exponentiation represents a particular multiplication factor of $e^2 = 2.718$ for a growth time t_G, but can more generally be described as a product function,

$$X = \Pi_{i=1}^{N} x_i \ . \tag{4.9.1}$$

Thus, the definition of a log-normal distribution is essentially the same as the Gaussian distribution function (Eq. 4.2.1), except with the variable X replaced by its logarithm $log(X)$. One general univariate form of the log-normal distribution is the 3-parameter definition (e.g., Uttley et al. 2005),

$$f(x;\tau,\mu,\sigma) = \frac{1}{\sigma\sqrt{2\pi}(x-\tau)} \exp\left(-\frac{[\log(x-\tau)-\mu]^2}{2\sigma^2}\right) , \tag{4.9.2}$$

where τ is a threshold parameter representing a lower limit on x, while μ and σ are the means and standard deviation of the log-normal distribution.

Historically, the log-normal distribution was found to apply to phenomena in economy, population statistics, clouds, sand grains (for a review see, e.g., Crow and Shimizu (1988)). Since our basic SOC model (Section 3.1) envisions a multiplicative process we expect that it also applies to most SOC phenomena described in this book. In the astrophysical context, log-normal statistics has been found to apply to solar wind plasma fluctuations (Burlaga and Lazarus 2000), gamma-ray bursts and X-ray binary variability data (Negoro and Mineshige 2002; Quilligan et al. 2002), the extremely variable narrow-line Seyfert 1 Galaxy IRAS 13224-3809 (Gaskell 2004), or Cygnus X-1 (Uttley et al. 2005). We show an example of such a measurement of Cygnus X-1 (Uttley et al. 2005) in Fig. 4.13.

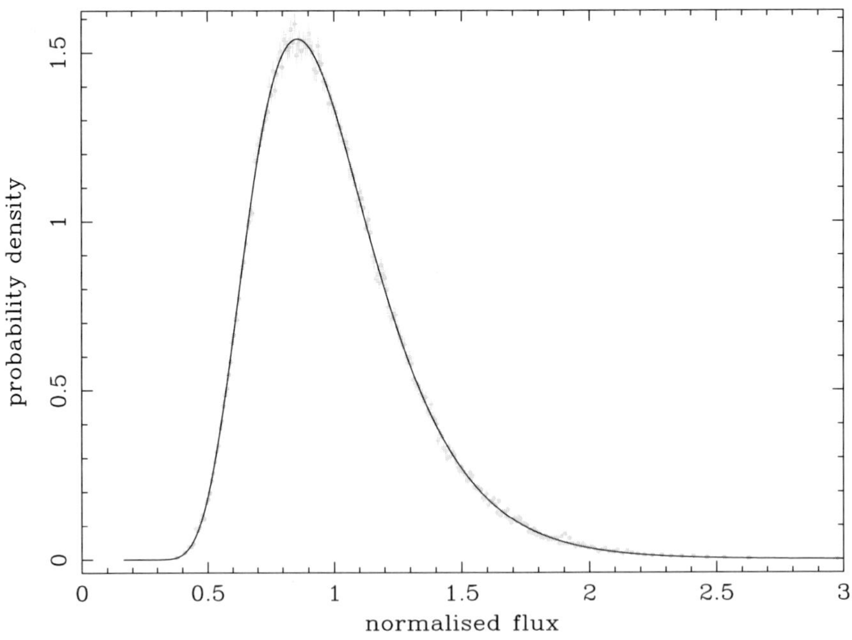

Fig. 4.13 Flux distribution of Cygnus X-1 in December 1996, expressed as a probability density function (gray data points), fitted with a log-normal distribution as defined in Eq. 4.9.2 (Uttley et al. 2005). (Uttley et al. 2005; reprinted with permission of the author)

4.10 Summary

Self-organized criticality (SOC) is a phenomenon that manifests itself in the statistics of nonlinear processes. Basic physical parameters that are used in the statistics of nonlinear processes are time scales, spatial scales, and energies. In this chapter we introduced some basic concepts of the statistics of time scales in random processes that are useful to model and understand SOC behavior. The most basic mathematical distributions of random processes are the binomial (Section 4.1), the Gaussian (Section 4.2), and the Poisson distribution (Section 4.3). The Poisson distribution in the limit of rare events is often approximated with an exponential distribution (Section 4.4), which can be used to describe the distribution of time scales in SOC processes, e.g., event durations, or waiting times between subsequent events. The variability of an astrophysical source is often studied from the count rate statistics of binned time series (Section 4.5) or from Fourier power spectral density distributions. Power spectra are classified into white-noise (Section 4.6), $1/f$ noise, flicker noise, pink noise, red (Brownian) noise, and black noise spectra, depending on their mean spectral powerlaw slope (Section 4.7). The variability of random pulses is most commonly described with the shot noise or flicker noise model (Section 4.8), which produces powerlaw-type spectra. We derived the power spectra of random current fluctuations that leads to a white-noise spectrum (Schottky theorem), as well as power spectra of rectangular and exponential-decay pulses, and derived their relation to time scale distributions as they are measured for SOC avalanches (Section 4.8.4). While incoherent random processes are additive and produce Gaussian or normal distributions, random processes with coherent growth are multiplicative and produce log-normal distributions (Section 4.9). This chapter is an introduction into the statistics of random processes, covering the most relevant tools that are used to diagnose SOC phenomena in time series, power spectra, time scale distributions, and waiting-time distributions, which follow in Chapters 5 and 7.

4.11 Problems

Problem 4.1: Show the relationship between the "Pascal triangle" of binomial coefficients (Eqs. 4.1.5 and 4.1.6) and the combinatorial derivation of the binomial coefficients (Eq. 4.1.3). Calculate the binomial coefficients with both methods for $n = 1, 2, ..., 10$.

Problem 4.2: Generate time profiles with mean count rates of $C = 10^{-1}$, 1, 10, 10^2, 10^3 counts per time interval using a random number generator, similar to those shown in Fig. 4.7 (left). Calculate histograms of the count rates (similar to Fig. 4.7 right) and fit them with Poisson distributions (Eq. 4.3.3). In what cases do you notice deviations from the exponential and Gaussian approximations (as fitted in Fig. 4.7 right) ?

Problem 4.3: Produce a random time series using a random number generator (as shown in Fig. 4.10 top right), verify that it has a white-noise spectrum using a FFT transform (Fig. 4.10 top left), and then multiply the white-noise spectrum with $\nu^{-1}, \nu^{-2}, \nu^{-3}$, (Fig. 4.10 left) to produce the corresponding time series (Fig. 4.10 right) and apply the auto-correlation function (Eq. 4.6.2) to determine the dominant pulse time scale T in each case.

Problem 4.4: Calculate the Fourier transform $P(\nu)$ of the triangular auto-correlation function (Eq. 4.8.14) and verify the power spectrum given in Eq. (4.8.15).

Problem 4.5: Calculate the auto-correlation function of the pulse shape of an exponentially-decaying time profile (Eq. 4.8.16), calculate the Fourier transform $P(\nu)$ and verify the power spectrum given in Eq. 4.8.17).

Problem 4.6: Verify the derivation of the shot-noise power spectrum for exponentially-decaying pulses from Eq. (4.8.18) through Eq. (4.8.27).

Problem 4.7: Determine the powerlaw coefficient γ in the correlation $E \propto T^{1+\gamma}$ (Eq. 4.8.21) from the numerical simulations given in Bak et al. (1988) and compare with the discussion of this result in Mineshige et al. (1994a).

5. Waiting-Time Distributions

Somewhere, something incredible is **waiting** *to be known.*

Carl Sagan

The universe is full of magical things, patiently **waiting** *for our wits to grow sharper.*

Eden Phillpotts

Why are waiting times interesting? While waiting times in our everyday life are usually associated with boredom, the statistics of waiting times contains scientifically interesting information about (1) the mean rates of event occurrence and (2) the randomness and uncorrelatedness of events, in contrast to events that are causally related in a system with local correlations or long-range interactions. Especially in astrophysical observations, event statistics of stellar flares or gamma-ray bursts from black-hole candidates are gathered without any spatial information, and thus we cannot decide whether individual events originate from the same source, or from spatially independent sources, which can only be discriminated by means of waiting-time statistics. Similar issues arise for solar flares and magnetospheric substorms, where spatial localization has often limited accuracy and mutual interactions are hidden by the invisible magnetic field. The classical SOC model of sandpile avalanches predicts no correlation between the input (of dripping sand grains) and the output (of sand avalanches) in the state of self-organized criticality, which implies that the statistics of waiting times (between subsequent avalanches) should be strictly random and consistent with Poisson statistics (e.g., for numerical simulations of cellular automatons see, e.g., waiting-time distribution in Fig. 6 of Charbonneau et al. 2001). There are physical processes that produce similar powerlaw distributions as SOC processes do, such as intermittent turbulence, but the two processes can be discriminated from their different waiting-time distributions, because SOC processes are uncorrelated, while turbulence entails spatial correlations. Waiting times are also called *elapsed times*, *inter-occurrence times*, or *laminar times*, in analogy to hydrodynamic flows in a laminar or turbulent state. Earthquakes were found to have a different waiting-time statistics than those of aftershocks, because earthquakes from different regions and continents are uncorrelated, while aftershocks occur in shorter time intervals due to some local correlation. Earthquake insurance agents might event make money by using the knowledge of waiting-time statistics wisely. Waiting times in the stock market were found to exhibit powerlaws (e.g., Bartolozzi et al. 2005), in contrast to the exponential Poisson process found in sandpile SOC systems with a stationary input rate, so the stock market may be a SOC system with a nonstationary driver. In this chapter we deal with the systematic data analysis of waiting times in astrophysical time series.

5.1 Waiting Times

A *Poisson process* is a point process, i.e., a sequence of events at points on an axis, taken here to be the time axis, which occur randomly and independently of one another (e.g., Cox and Isham 1980). The randomness can be tested by examining whether an observed sample of time scales or time intervals is consistent with a Poisson process, which has an exponential distribution for stationary processes. The sample may consist of time scales or time intervals observed in statistically independent events, sampled at different locations and/or different times. A particular case is the sampling of time intervals between subsequent events in a time series, which is called the *waiting time* Δt. In this case we talk about a *waiting-time distribution* $P(\Delta t)$, which is expected to exhibit an exponential function for a stationary random process,

$$P(\Delta t) = \lambda\, e^{-\lambda \Delta t} \,, \tag{5.1.1}$$

where λ represents the mean event occurrence rate. This probability distribution is formally identical to an exponential distribution of time scales (Eq. 4.4.4), if the time scale is replaced by the waiting time ($t_S \mapsto \Delta t$) and the mean (or e-folding) time scale is expressed with the mean occurrence rate $1/t_{Se} \mapsto \lambda$. Such an event occurrence rate is a Poisson process if subsequent events are statistically independent and not affected by some physical coupling. Especially if subsequent events occur at different locations without communication in between, they are expected to be statistically independent.

For instance, subsequent earthquake events on different continents are expected to occur in statistically independent time intervals, while time intervals between aftershock events in the same region may be clustered and not independent of each other. For another example, solar flares from different active regions on the solar surface are believed to be statistically independent, while subsequent flares that are triggered by the same magnetic connection event are not statistically independent and are called *sympathetic flares*.

It is important to understand that the duration of an event is generally different from the waiting time between two subsequent events, and thus the statistical distribution of event durations and waiting time intervals are not identical. However, both time scales may follow an exponential distribution if they are controlled by random processes, but they may be uncorrelated and have different time ranges, the range of event durations generally being shorter than the range of waiting-time intervals (since an event duration cannot be longer than the time interval to the next event without overlapping). The concept of event durations τ and waiting-time intervals Δt is visualized in Fig. 5.1, both assumed to be subject of random processes. Of course, if event durations are not much smaller

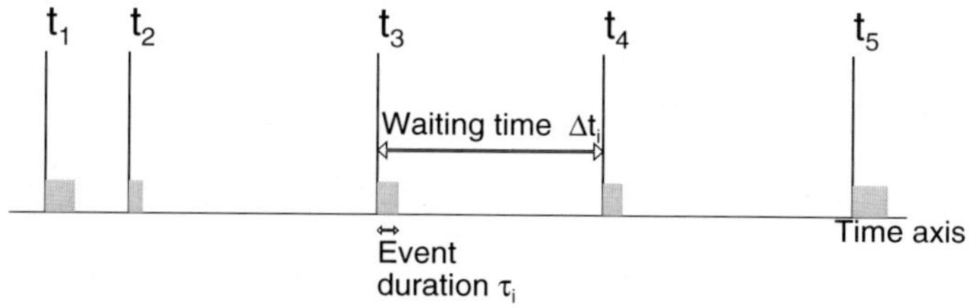

Fig. 5.1 The definition of time durations τ_i of events and waiting times Δt_i in a time series of events t_i. The statistics of event durations is generally independent of the statistics of waiting times, unless subsequent events are triggered by a process with some physical coupling.

than the typical time interval between subsequent events, they could overlap with the next event, which constitutes a tricky measurement problem.

The distribution of waiting times measured in a global system loses the timing information from individual local regions and can be entirely different from the waiting-time distributions of individual local regions. This is illustrated in the example shown in Fig. 5.2, where we combine the times of three periodic processes (with periods of 5, 6, and 7 time units) and sample the waiting times of the combined time series, which shows a continuous distribution of intervals $\Delta t = 0, 1, ..., 5$. The waiting-time distribution of the combined time series is close to a random process, while the underlying individual time series are strictly periodic and not random at all. So, we can never conclude from the waiting times of a global system whether the waiting times in a local region is a random process or not. However, the opposite is true and can be mathematically proven, i.e., that the combination of time series with random time intervals produces a combined time series that has also random time intervals. This property is also called the *superposition theorem of Palm and Khinchin* (e.g., Cox and Isham 1980; Craig and Wheatland 2002) and is analogous to the

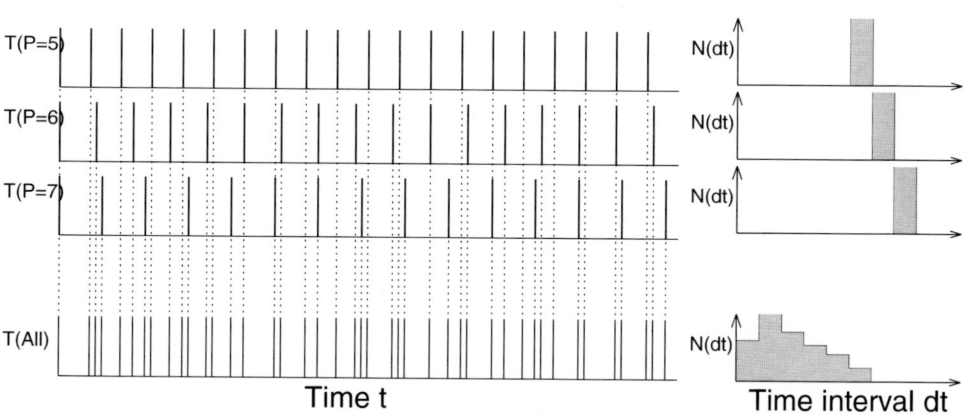

Fig. 5.2 Three time series with pulses at regular intervals with periods of $P = 5, 6, 7$ and the superposition of these three time series. The statistics of time intervals shows single periods for each of the three individual time series (histograms on right-hand side), but a continuous distribution of almost random time intervals for the combined time series.

central limit theorem (Rice 1995). Another example that waiting times in local regions can be completely different from those of the global system was confirmed in earthquake statistics, where aftershocks (occurring in the same local region) exhibit an excess of short waiting times (Omori's law; Omori 1895), compared with the overall statistics of (spatially) independent earthquakes.

5.2 Nonstationary Waiting-Time Statistics

Waiting-time distributions resulting from a random process exhibit an exponential distribution only when the average event rate is time-independent, which is called a *stationary Poisson process*. If the average rate varies with time, the waiting-time distribution reflects a superposition of multiple exponential distributions with different e-folding time scales, and may even resemble powerlaw-like distributions. A classical example is the solar flare rate, which varies by orders of magnitude during an 11-year solar cycle, and thus is a nonstationary process. The statistics of such *inhomogeneous* or *nonstationary Poisson processes* can be characterized with *Bayesian statistics* (e.g., Jaynes 2003; Sivia and Skilling 2006), where a time-dependent Poisson process is decomposed into piecewise constant Poisson processes. In the following derivation we follow the framework and notation used in Wheatland et al. (1998) and Wheatland and Litvinenko (2002).

In a nonstationary Poisson process, the time-varying occurrence rate can be defined with a time-dependent function $\lambda(t)$ and the probability function of waiting times becomes itself a function of time (e.g., Cox and Isham 1980; Wheatland et al. 1998),

$$P(t,\Delta t) = \lambda(t+\Delta t)\exp\left[-\int_t^{t+\Delta t} \lambda(t')\, dt'\right] . \qquad (5.2.1)$$

An approximation of this general expression is a subdivision into discrete time intervals where the occurrence rate is constant within the fluctuations expected from Poisson statistics, so it consists of piecewise stationary processes with occurrence rates $\lambda_1, \lambda_2, ..., \lambda_n$,

$$P(\Delta t) = \begin{cases} \lambda_1 e^{-\lambda_1 \Delta t} & \text{for } t_1 \le t \le t_2 \\ \lambda_2 e^{-\lambda_2 \Delta t} & \text{for } t_2 \le t \le t_3 \\ \dots\dots\dots\dots & \dots\dots \\ \lambda_n e^{-\lambda_n \Delta t} & \text{for } t_n \le t \le t_{n+1} \end{cases} \qquad (5.2.2)$$

where the occurrence rate λ_i is stationary during a time interval $[t_i, t_{i+1}]$, but has different values in subsequent time intervals. The time intervals $[t_i, t_{i+1}]$ where the occurrence rate is stationary are called *Bayesian blocks*, a special application of *Bayesian statistics* (e.g., see Scargle (1998) for astrophysical applications).

If observations of a nonstationary Poisson process are made for the time interval $[0, T]$, then the distribution of waiting times for that time interval will be weighted by the number of events $\lambda(t)dt$ in each time interval $(t, t+dt)$,

$$P(\Delta t) = \frac{1}{N}\int_0^T \lambda(t)\, P(t,\Delta t)\, dt , \qquad (5.2.3)$$

where the rate is zero after the time interval $t > T$, i.e., $\lambda(t > T) = 0$, and $N = \int_0^T \lambda(t)\, dt$. If the rate is slowly varying, so that it can be subdivided into piecewise stationary Poisson processes (into Bayesian blocks), then the distribution of waiting times will be

$$P(\Delta t) \approx \sum_i q(\lambda_i) \lambda_i e^{-\lambda_i \Delta t} , \tag{5.2.4}$$

where

$$q(\lambda_i) = \frac{\lambda_i t_i}{\sum_j \lambda_j t_j} , \tag{5.2.5}$$

is the fraction of events associated with a given rate λ_i and t_i is the piecewise time interval or Bayesian block over which the constant rate λ_i is observed. If we make the transition from the summation over discrete time intervals t_i (Eqs. 5.2.4 and 5.2.5) to a continuous integral function over the time interval $[0 < t < T]$, we obtain,

$$P(\Delta t) = \frac{\int_0^T \lambda(t)^2 e^{-\lambda(t)\Delta t}\, dt}{\int_0^T \lambda(t)\, dt} . \tag{5.2.6}$$

When the occurrence rate $\lambda(t)$ is not a simple function, the integral Eq. (5.2.6) becomes untractable, in which case it is more suitable to substitute the integration variable t with the variable λ. Defining $f(\lambda) = (1/T) dt(\lambda)/d\lambda$ as the fraction of time that the flaring rate is in the range $(\lambda, \lambda + d\lambda)$, or $f(\lambda) d\lambda = dt/T$, we can express Eq. (5.2.6) as an integral of the variable λ,

$$P(\Delta t) = \frac{\int_0^\infty f(\lambda) \lambda^2 e^{-\lambda \Delta t} d\lambda}{\int_0^\infty \lambda f(\lambda)\, d\lambda} , \tag{5.2.7}$$

where the denominator $\lambda_0 = \int_0^\infty \lambda f(\lambda)\, d\lambda$ is the mean rate of flaring.

Let us consider some examples. In Fig. 5.3 we show five cases: (1) a stationary Poisson process with a constant rate λ_0; (2) a two-step process with two different occurrence rates λ_1 and λ_2; (3) a nonstationary Poisson process with a linearly increasing occurrence rate $\lambda(t) = \lambda_0 t/T$, varying like a triangular function for each cycle, (4) a piecewise constant Poisson process with an exponentially varying rate distribution, and (5) a piecewise constant Poisson process with an exponentially varying rate distribution steepened by a reciprocal factor. For each case we show the time-dependent occurrence rate $\lambda(t)$ and the resulting probability distribution $P(\Delta t)$ of events. We see that a stationary Poisson process produces an exponential waiting-time distribution, while nonstationary Poisson processes with a discrete number of occurrence rates λ_i produce a superposition of exponential distributions, and continuous occurrence rate functions $\lambda(t)$ generate powerlaw-like waiting-time distributions at the upper end.

We can calculate the analytical functions for the waiting-time distributions for these five cases. The first case is simply an exponential function as given in Eq. (5.1.1) because of the constant rate $\lambda(t) = \lambda_0$,

$$P(\Delta t) = \lambda_0 e^{-\lambda_0 \Delta t} . \tag{5.2.8}$$

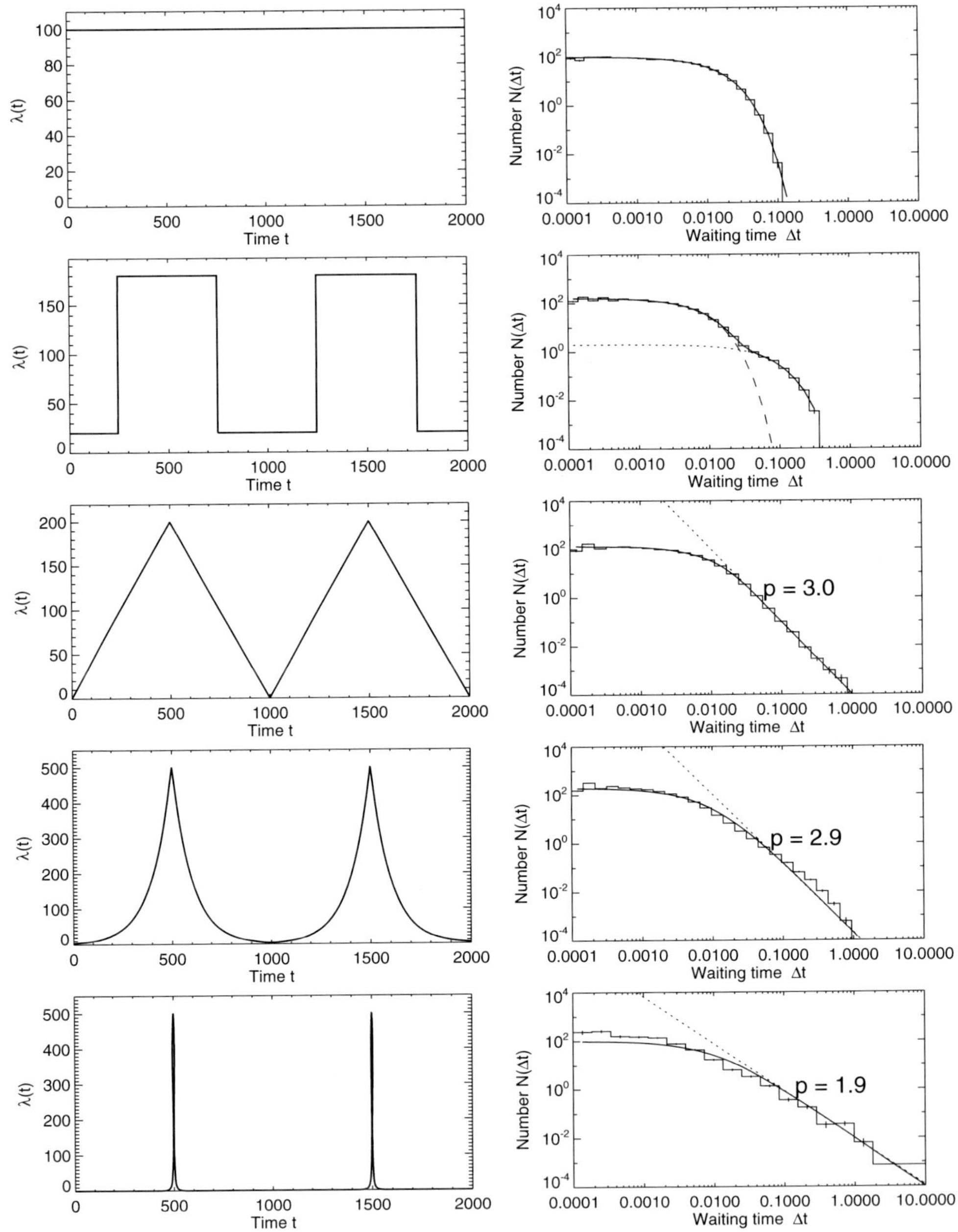

Fig. 5.3 One case of a stationary Poisson process (top) and four cases of nonstationary Poisson processes with two-step, linear-increasing, exponentially varying, and δ-function like variations of the occurrence rate $\lambda(t)$. The time-dependent occurrence rates $\lambda(t)$ are shown on the left side, while the waiting-time distributions are shown in the right-hand panels, in the form of histograms sampled from Monte-Carlo simulations, as well as in the form of the analytical solutions (given in Eqs. 5.2.8–5.2.18). Powerlaw fits $N(\Delta t) \propto \Delta t^{-p}$ are indicated with a dotted line and labeled with the slope p (Aschwanden and McTiernan 2010).

The second case follows from Eq. (5.2.4) and yields

$$P(\Delta t) = \frac{1}{10}\lambda_1 e^{-\lambda_1 \Delta t} + \frac{9}{10}\lambda_2 e^{-\lambda_2 \Delta t} . \tag{5.2.9}$$

The third case can be integrated with Eq. (5.2.6). The time-dependent flare rate grows linearly with time to a maximum rate of $\lambda_m = 2\lambda_0$ over a time interval T, with a mean rate of λ_0,

$$\lambda(t) = \lambda_m \frac{t}{T} = 2\lambda_0 \frac{t}{T} \tag{5.2.10}$$

Defining the constant $a = -\lambda_0 \Delta t/T$ the integral of Eq. (5.2.6) reads as $P(\Delta t) = (-a/\Delta t) \int_0^T t^2 e^{at}\, dt / \int_0^T t\, dt$. The integral $\int x^2 e^{ax}\, dx = e^{ax}(x^2/a - 2x/a^2 + 2/a^3)$ can be gleaned from an integral table. The analytical function of the waiting-time distribution for a linearly increasing occurrence rate is then

$$P(\Delta t) = 2\lambda_0 \left[\frac{2}{(\lambda_0 \Delta t)^3} - e^{-\lambda_0 \Delta t} \left(\frac{1}{(\lambda_0 \Delta t)} + \frac{2}{(\lambda_0 \Delta t)^2} + \frac{2}{(\lambda_0 \Delta t)^3} \right) \right] , \tag{5.2.11}$$

which is a flat distribution for small waiting times and approaches a powerlaw function with a slope of $p = 3$ at large waiting times, i.e., $P(\Delta t) \propto \Delta t^{-3}$ (Fig. 5.3, third case). The distribution is the same for a single linear ramp or for a cyclic triangular variation, because the total time spent at each rate $[\lambda, \lambda + d\lambda]$ is the same.

The fourth case, which mimics the solar cycle, has an exponentially growing (or decaying) occurrence rate, i.e.,

$$f(\lambda) = \left(\frac{1}{\lambda_0} \right) \exp\left(-\frac{\lambda}{\lambda_0} \right) , \tag{5.2.12}$$

defined in the range of $[0 < \lambda < \infty]$, and has a mean of λ_0. The waiting-time distribution can therefore be written with Eq. (5.2.7) as

$$P(\Delta t) = \int_0^\infty \left(\frac{\lambda}{\lambda_0} \right)^2 \exp\left(-\frac{\lambda}{\lambda_0} [1 + \lambda_0 \Delta t] \right) d\lambda , \tag{5.2.13}$$

which corresponds to the integral $\int_0^\infty x^2 e^{ax}\, dx = -2/a^3$ using $a = -(1 + \lambda_0 \Delta t)/\lambda_0$ and thus has the solution $P(\Delta t) = -2/(a^3 \lambda_0^2)$, i.e.,

$$P(\Delta t) = \frac{2\lambda_0}{(1 + \lambda_0 \Delta t)^3} . \tag{5.2.14}$$

For very large waiting times ($\Delta t \gg 1/\lambda_0$) the distribution Eq. (5.2.14) approaches the powerlaw limit $P(\Delta t) \approx (2/\lambda_0^2)(\Delta t)^{-3}$ (see Fig. 5.3 fourth panel). The asymptotic behavior for large waiting times ($\Delta t \mapsto \infty$) can also be evaluated from a Taylor expansion at $\lambda = 0$, which leads (by integrating Eq. 5.2.7 term by term) to

$$P(\Delta t) = \frac{2f(0)}{\lambda_0} (\Delta t)^{-3} + \frac{6f'(0)}{\lambda_0} (\Delta t)^{-4} + \dots , \tag{5.2.15}$$

and analytically demonstrates the convergence to a powerlaw $P(\Delta t) \propto (\Delta t)^{-3}$ with a slope of -3 for large waiting times $\Delta t \gg 1/\lambda_0$. This fourth case defined by Eqs. (5.2.12–5.2.14) corresponds to the model of Wheatland (2000), which yields a somewhat too steep powerlaw slope when compared with observations (Fig. 5.12).

The fifth case has an exponentially growing occurrence rate, multiplied with a reciprocal factor, i.e.,

$$f(\lambda) = \lambda^{-1} \exp\left(-\frac{\lambda}{\lambda_0}\right) , \tag{5.2.16}$$

and fulfills the normalization $\int_0^\infty \lambda f(\lambda)\, d\lambda = \lambda_0$. The waiting-time distribution can then be written with Eq. (5.2.7) as

$$P(\Delta t) = \int_0^\infty \left(\frac{\lambda}{\lambda_0}\right) \exp\left(-\frac{\lambda}{\lambda_0}[1+\lambda_0 \Delta t]\right) d\lambda , \tag{5.2.17}$$

which, with defining $a = -(1+\lambda_0\Delta t)/\lambda_0$, corresponds to the integral $\int x e^{ax}\, dx = (e^{ax}/a^2)(ax-1)$ and becomes $\int_0^\infty x e^{ax}\, dx = 1/a^2$ when integrated over $[0 < x < \infty]$, yielding the solution $P(\Delta t) = 1/(a^2\lambda_0^2)$, i.e.,

$$P(\Delta t) = \frac{\lambda_0}{(1+\lambda_0\Delta t)^2} . \tag{5.2.18}$$

For very large waiting times $(\Delta t \gg 1/\lambda_0)$, the equation Eq. (5.2.8) approaches the powerlaw limit $P(\Delta t) \approx \lambda_0^{-1}(\Delta t)^{-2}$ (see Fig. 5.3, bottom panel), which seems to fit the observations (Fig. 5.12) better than the fourth case used in Wheatland (2000).

Thus we learn from the last four examples that most continuously changing occurrence rates produce powerlaw-like waiting-time distributions $\propto(\Delta t)^{-p}$ with slopes of $p \lesssim 2,...,3$ at large waiting times, despite the intrinsic exponential distribution that is characteristic to stationary Poisson processes. If the variability of the flare rate is gradual (third and fourth case in Fig. 5.3), the powerlaw slope of the waiting-time distribution is close to $p \lesssim 3$. However, if the variability of the flare rate shows spikes like δ-functions (Fig. 5.3, bottom), which is highly intermittent with short clusters of flares, the distribution of waiting times has a slope closer to $p \approx 2$. This phenomenon is also called *clusterization* and has analogs in earthquake statistics, where aftershocks appear in clusters after a main shock (Omori's law; Omori 1895). Thus the powerlaw slope of waiting times contains essential information whether the flare rate is constant, varies gradually, or in form of intermittent clusters.

5.3 Measurement of Waiting Times

Statistics of events and waiting times in astrophysical time series are not trivial, because often multiple physical processes are present that contribute to the observed emission of pulses and backgrounds, which cause confusion, deconvolution, event separation, and background-subtraction problems. Especially small-amplitude fluctuations are ambiguous because they could constitute small nonthermal events or fluctuations of the thermal background. Ideally, we want to sample statistics of SOC avalanche events from one single common physical mechanism and separate out other secondary physical mechanisms that are manifested as a convolution of the primary signal. For instance, the thermalization process after an impulsive plasma heating event can obscure subsequent small impulsive heating events in the cooling phase of a large heating event. The definition of an event includes at least the time t_i of the start or peak of the event. The definition of the waiting time (or *quiescent time*) is the time interval between subsequent events, i.e., $\Delta t_i = t_i - t_{i-1}$, which entirely depends on the definition of the event times t_i.

In the following we discuss some biases of waiting-time statistics that result from particular event definitions. Three different definitions of event times are tested in Buchlin et al. (2005): (i) the peak method, (ii) the threshold method, and (iii) the wavelet method. Examples of the three methods are illustrated in Fig. 5.4 for an artificial time series. The peak method (Fig. 5.4 left) requires a relatively noise-free smoothed time profile, so that noise fluctuations do not contaminate the statistics with multiple peaks per time structure, leading to an excess of short waiting times. The threshold method (Fig. 5.4 middle) requires that the time profiles return to a sub-threshold background level for each event, otherwise events in the decaying tail of a pulse time profile are ignored. The wavelet method (Fig. 5.4 right) has the ability to detect simultaneous pulses with different time scales, which would be impossible with the peak or threshold method. In the presence of a $1/f$ noise spectrum, the *Mexican hat* mother wavelet, which is well adapted to the second derivative of a Gaussian profile, appears to be most suitable to find enhanced structures (Sanz et al. 2001). Besides the detection of time structures, also the definition of the event time t_i within the duration of the pulse structure (e.g., start time, peak time,

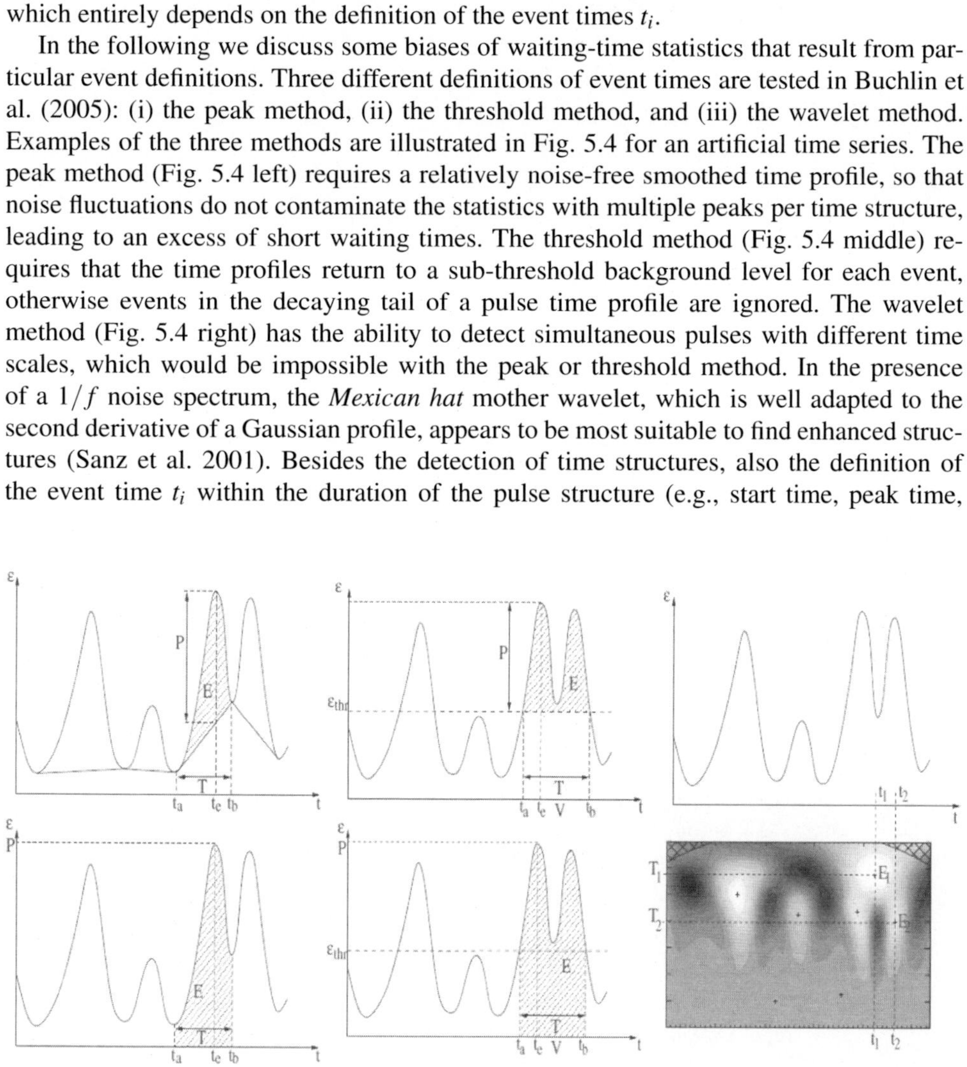

Fig. 5.4 Three definitions of event characteristics (peak energy P, total energy E, event time t_i (labeled as t_a for the start or t_e for the peak), and event duration T: with the peak method (left), the threshold method (middle), and the wavelet method (right), with (top) and without (bottom) background subtraction (Buchlin et al. 2005).

midtime) affects the waiting-time statistics to some degree. The peak times are generally more delayed with respect to the start time in large events, due to the longer rise times, but this effect is probably of secondary importance. In the study of Buchlin et al. (2005), three different numerical simulations are performed based on a shell-model of MHD turbulence (Giuliani and Carbone 1998), and the resulting waiting-time distributions based on the three different event definition criteria are shown, which we reproduce in Fig. 5.5. Interestingly, the three methods reveal quite different waiting-time distributions in each case. The threshold-based method seems to produce powerlaw-like distributions, while the peak-based and wavelet-based methods produce exponential-like distributions, at least in the regime of large waiting times. This result imposes some ambiguity in the interpretation of waiting-time distributions, at least for turbulence-type time series.

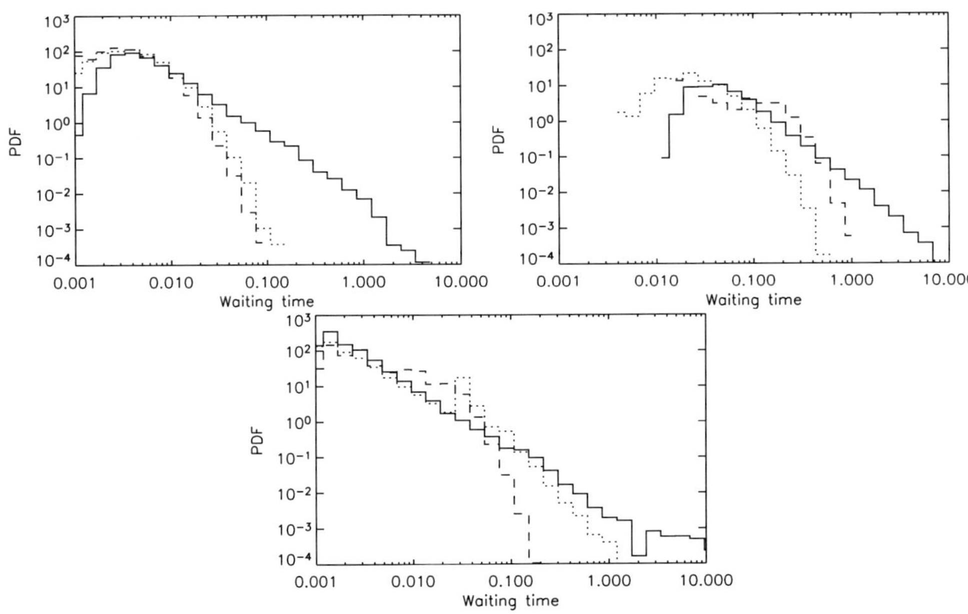

Fig. 5.5 Waiting-time distributions for events determined by peaks (dotted histograms), thresholds (plain histograms), and a wavelet method (dashed histograms), for three different time series (a, b, c) based on numerical simulations of MHD turbulence (Buchlin et al. 2005).

The effect of event definition on the distribution of waiting times has also been numerically simulated with the continuously driven Olami–Feder–Christensen (OFC) model (Olami et al. 1992) by Hamon et al. (2002). Three different definitions of waiting times have been explored: (i) peak-to-peak, (ii) a finite flux threshold, and (iii) no flux threshold (Fig. 5.6 top left). The case (iii) with no flux threshold should be the most complete sample that includes the weakest avalanche events and this was found to have a stretched exponential shape (Fig. 5.6 bottom left), similar to the (strictly) exponential shape expected in the original BTW model (e.g., Charbonneau et al. 2001; Wheatland 2009). However, when the threshold is enhanced, small avalanches become filtered out and an excess of

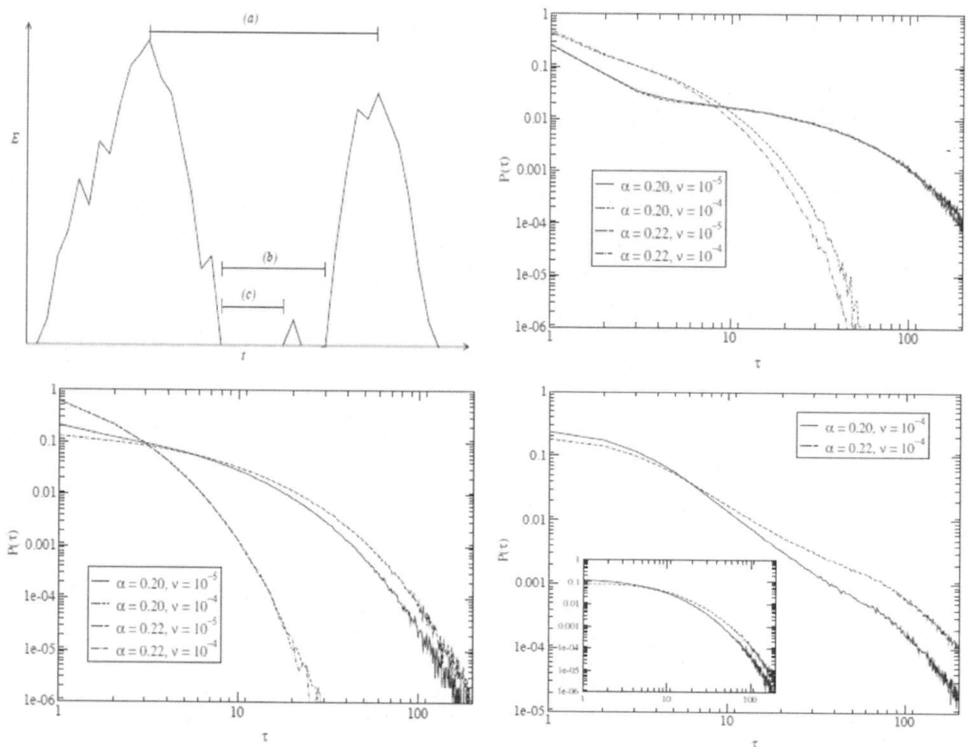

Fig. 5.6 Waiting-time distributions for events defined by (a) peak-to-peak (top right), (b) finite threshold (bottom right), and (c) no threshold (bottom left), numerically simulated for the Olami–Feder–Christensen model. In each case there are two simulations with weak and strong driving (Hamon et al. 2002).

longer waiting times occur, which produces a powerlaw-like fat tail in the waiting-time distribution (Fig. 5.6 right frames).

The ambiguity between powerlaw-like or exponential-like tails was also noticed when the Poissonian nature of solar flare statistics was tested (Wheatland et al. 1998; Lepreti et al. 2001; Wheatland and Litvinenko 2002), where the outcome depends on the definition of events and waiting times.

5.4 Waiting-Time Statistics in Geophysics

We will see that earthquakes cannot be represented by a simple Poisson random process of independent and uncorrelated events, at least not the aftershocks, which causes some complications for the waiting-time distribution. We introduced the powerlaw-like frequency distribution of earthquakes in Fig. 1.7 and Section 1.5, which can be described by (i) the Gutenberg–Richter law, i.e., $N(E) \propto E^{-5/3} \ldots E^{-2}$ (with E the size or magnitude and $N(E)$ the size distribution). There are other scaling laws known for earthquakes, such as: (ii) *Omori's law* (Omori 1895) for short-range temporal correlations between earthquakes, which is the scaling $N(\Delta t) \propto (\Delta t)^{-1}$ of the frequency $N(\Delta t)$ of aftershocks as a function of time Δt since the main shock; (iii) the *productivity law* $N(E) \propto E^{2/3}$, giving the number of earthquakes triggered by an event of energy E; (iv) the powerlaw distribution $N(L) \propto L^{-2}$ of fault lengths; (v) the fractal and multi-fractal structure of fault network and (vi) of earthquake epicenters (Saichev and Sornette 2006). There have been several attempts to unify these relationships into a single scaling law, which all play a role in the formulation of a waiting-time distribution.

Unified scaling laws for the distribution of waiting times of earthquakes were derived, e.g., by Bak et al. (2002), or Saichev and Sornette (2006). Bak et al. (2002) used a total of 335,076 earthquakes that occurred in California during 1984–2000 and defined a waiting time Δt as the time interval between the beginning of two successive earthquakes. They measured the probability distribution $P_{S,L}$ of waiting times Δt between earthquakes occurring within a range L whose magnitude was greater than $m = \log(S)$. Waiting times measured between all earthquakes have been found to have different distributions than those measured between local earthquakes only (e.g., Bak 1996, p.173; Ito 1995). In the study of Bak et al. (2002), the magnitudes m were found to follow the Gutenberg–Richter law $N(M > m) \propto m^{-b}$ with $b = 0.95$. Such probability distributions $P_{S,L}(\Delta t)$ of waiting times Δt are shown in Fig. 5.7 (left) for different cutoffs $m =^{10} \log(S)$ and sizes L. The waiting-time distributions of the different data sets coincide for small events, but diverge for large events. The mean powerlaw slope in the middle part fits $N(\Delta t) \propto \Delta t^{-1}$, while the fall-off at large waiting times approaches $N(\Delta t) \propto (\Delta t)^{-2}$, which is not as steep as we expect for a nonstationary Poisson process, i.e., $N(\Delta t) \propto (\Delta t)^{-3}$ (see Fig. 5.3).

A unified scaling law was achieved by modifying the waiting time Δt with a function of S and L in Bak et al. (2002). Choosing the x-axis as $x = \Delta t S^{-b} L^{d_f}$ and the y-axis as $y = \Delta t^p S^{-b} L^{d_f}$, the data could be collapsed to a single well-defined function $y = f(x)$,

$$(\Delta t)^p S^{-b} L^{d_f} = f(\Delta t S^{-b} L^{d_f}) \ , \qquad (5.4.1)$$

with the best-fit parameters $p = 1$ (which is Omori's law for aftershocks; Omori 1895), $b \approx 1$ (which is the value of the Gutenberg–Richter law), and $d_f \approx 1.2$ (which describes the 2-D fractal dimension of the location of epicenters projected onto the Earth's surface). The collapsed waiting-time distributions are shown in Fig. 5.7 (right), which shows two regimes separated by a kink. The interpretation of this dual behavior is that two successive earthquakes will be either correlated (for small values x to the left of the kink), or uncorrelated (for large values of x to the right to the kink). The parameter x is a combination of magnitude S, size L, and waiting time Δt, but equally applies to main shocks and aftershocks, constituting a unified scaling law for waiting times from tens of seconds to tens of years.

The same scaling (Eq. 5.4.1) was also fitted to other earthquake datasets and a third regime was found with a different powerlaw slope for very small values of x that did not follow the "universal" scaling law of Eq. (5.4.1) (Davidsen and Goltz 2004). Also Carbone et al. (2005) investigated different data sets from 4 different regions in Italy and found that the "universal" scaling law claimed by Bak et al. (2002) did not hold for all data sets and suggested a generalized function for Omori's law (Omori 1895), implying that the general earthquake process is non-Poissonian.

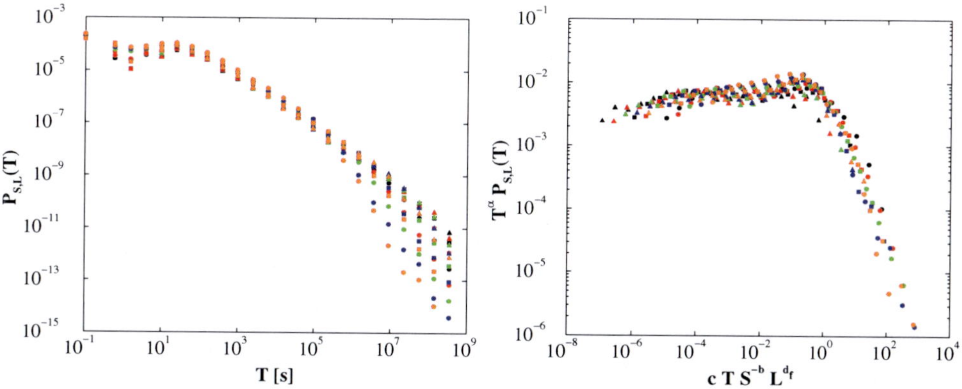

Fig. 5.7 *Left:* Waiting-time distributions $P_{S,L}(\Delta t)$ of earthquakes occurring in California during 1984–2000 for different data subsets with magnitude $m =^{10} \log(S) = 2,3,4$ within an area of length $L = 0.25°, 0.5°, 1°, 2°, 4°$ in a longitude–latitude grid covering California. (Note that the waiting time Δt is rendered with the symbol **T** in the original figure.) *Right:* The same data are plotted as parameter $y = (\Delta t)^p P_{S,L}(\Delta t)$ as a function of the variable $x = \Delta t S^{-b} L^{d_f}$, with the coefficients $p = 1$, $b = 1$, and $d_f = 1.2$. The function is constant for $x < 1$, corresponding to the correlated Omori law regime, while it is decaying for large earthquakes at $x > 1$, associated with the uncorrelated regime of earthquakes. Note that the datapoints are collapsed to a single function $y(x)$ for this "universal scaling law". Reprinted from Bak et al. (2002) with permission; Copyright by American Physical Society.

Along the same lines, Saichev and Sornette (2006) derived a "universal" distribution of waiting times (called *inter-earthquake* times therein) by generating a statistical probability function under the assumption of statistical stationarity, i.e., (i) the branching ratio (or average number of earthquakes/aftershocks per earthquake) be less than unity, and (ii) the average rate of the Poissonian distribution of spontaneous events be non-zero. This model is also based on the strategy that the impact of aftershocks can be neglected for events that occur outside some space-time window, while events in a given space domain are only considered if they were triggered by a source from the same space domain. The unified scaling does not distinguish between foreshocks, main shocks, and aftershocks, similar to the model of Bak et al. (2002).

Alternative modeling of waiting-time distributions for earthquakes include, for instance, avalanches triggered by externally-imposed (non-SOC) coherent noise (Newman and Sneppen 1996), or Lévy flights with anomalous diffusion in the subdiffusive regime (Sotolongo-Costa et al. 2000).

5.5 Waiting-Time Statistics in Magnetospheric Physics

Waiting times have been studied in magnetospheric physics for auroral emission, geomagnetic activity, magnetospheric substorms, and solar wind, to answer questions whether Poisson random processes, SOC processes, intermittent turbulence, finite system-size effects, or clusterization effects are the key players.

Regarding *auroral emission*, we discussed the SOC model of Chapman et al. (1998) in Section 2.5.1, which essentially represents a modified BTW sandpile model with the additional effect of a finite system size. The finite system size effect, which was observed for auroral blobs (Fig. 1.10), produces an excess of large events with a characteristic length scale L_0 corresponding to the maximum size of the system (Fig. 2.13), but also produces a waiting-time distribution with a well-defined mean (see Fig. 3 in Chapman et al. 1998), probably corresponding to the average propagation time of the phenomenon through the system. In addition, simulations with strong driving (i.e., with a higher loading rate) showed destruction of smaller events and hence a higher probability for larger events (Chapman et al. 2001), which produces a cutoff in the waiting-time distribution at the lower end as well. Thus, Chapman et al. (1998, 2001) conclude that the waiting-time distribution does not reflect SOC behavior, but is controlled by the rate of energy inflow and the scale size of the system.

A waiting-time distribution of the *auroral electron jet (AE) index*, a standard indicator of geomagnetic activity, sampled by 12 stations in the polar region of the Earth's northern hemisphere, was analyzed for the months of June 1988 and February 1981 by Lepreti et al. (2004), with a sampling time of 1 min (Fig. 5.8 top). The waiting-time distribution

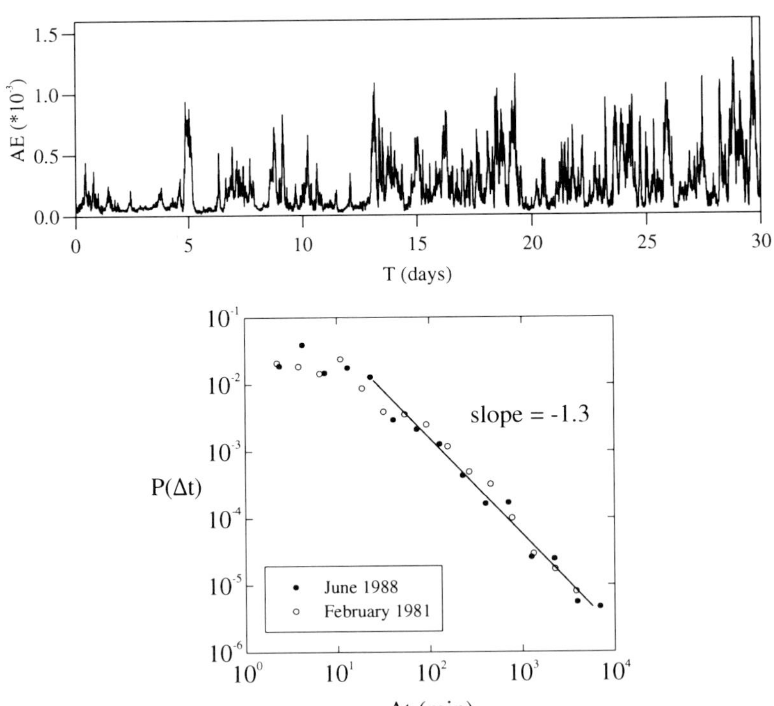

Fig. 5.8 *Top:* The AE index measured during the month June 1988. *Bottom:* The waiting-time distribution between successive AE index bursts for June 1988 and February 1981. The solid line represents a powerlaw fit with an exponent of $p = 1.3$ (Lepreti et al. 2004). (Reprinted with permission of Elsevier)

exhibits an approximate powerlaw with a slope of $p = 1.31 \pm 0.07$ for June 1988, and $p = 1.27 \pm 0.08$ for February 1981. The powerlaw shape of waiting times over 3 orders of magnitude is clearly not consistent with an exponential Poisson distribution (Eq. 4.4.3), as expected for randomly generated avalanches in the BTW SOC model. Lepreti et al. (2004) conclude that the deviation from the local Poisson hypothesis (where subsequent bursts are independent) implies a tendency of clusterization into larger bursts. Numerical simulations of energy dissipation in a turbulent MHD shell model is able to reproduce a powerlaw distribution of waiting times (although with different powerlaw slopes), and thus an interpretation of energy dissipation in auroral electron jet events by turbulence is more consistent with the data (Boffetta et al. 1999; Lepreti et al. 2004). Interestingly, although a nonstationary Poissonian process can also produce a powerlaw-like waiting-time distributions (Fig. 5.3), the expected powerlaw index of $p \approx 3.0$ does not match what is observed in AE events ($p \approx 1.3$).

Substorms in the Earth's magnetotail evolve through three phases: (i) a growth phase with gradual accumulation of energy from the solar wind, (ii) an expansion phase where the magnetotail passes an instability threshold (substorm onset), and (iii) a recovery phase with energy dissipation through the magnetosphere (Fig. 9.10). The waiting-time distribution has been measured by Borovsky et al. (1993) for a one-year period starting on October 1982. From 1,290 substorm events they found a waiting-time distribution with a peak around $\approx$3 hrs, a mean of $\mu = 5.1$ hrs, and a standard deviation of $\sigma = 4.7$ hrs. Thus, the distribution is not a scale-free powerlaw distribution as expected for SOC phenomena, but rather a peaked quasi-periodic distribution. Freeman and Morley (2004) were able to fit this distribution with the following simple 3-rule model: (i) The substorm is driven by power from the solar wind; (ii) a critical energy threshold exists for the magnetotail to become unstable (e.g., magnetic reconnection is inhibited until the current sheet is stretched sufficiently thin for electron inertial or gyrokinetic effects to become important in the generalized Ohm's law); and (iii) once the magnetotail is sufficiently stressed energy is released and the magnetotail moves into a lower energy state below the threshold. With this simple model Freeman and Morley (2004) could reproduce the observed waiting-time distribution, which implies that the variability of substorm waiting times is essentially attributable to the variability of the solar wind power input. Since the driver is an external force (i.e., the solar wind), this model corresponds to the concept of *forced criticality* (Chang 1992) rather than that of *self-organized criticality* by Bak et al. (1987).

Waiting-time distributions were also measured from fluctuations of ionospheric plasma velocities, using data from the *SuperDARN* radar network, where powerlaw distributions with slopes of $p \approx 1.8$–2.5 were found (Bristow 2008).

5.6 Waiting-Time Statistics in Solar Physics

The majority of solar flares is believed to be triggered independently, and thus should have a waiting-time distribution consistent with a Poissonian random process, but there is a subset of so-called *sympathetic flares* which have a causal connection and thus are not independent (e.g., Simnett 1974; Gergely and Erickson 1975; Fritzova-Svestkova et al. 1976; Pearce and Harrison 1990; Bumba and Klvana 1993; Biesecker and Thompson

2000; Moon et al. 2002), similar to aftershocks of large earthquakes. In addition, the mean flare rate varies by a substantial factor during the solar cycle, which constitutes a nonstationary Poisson process, for which we expect a powerlaw-like waiting-time distribution with a slope of $p \approx 3$ (Fig. 5.3). Further deviations from Poissonian waiting-time distributions may arise depending on the definition and detection threshold of flare events (Section 5.3). There are a number of solar flare studies that deal with these conclusions or probe alternative interpretations, which we describe in the following.

5.6.1 Solar Flare Hard X-Rays

One of the first waiting-time distributions of solar flares was published by Pearce et al. (1993), using 8,319 hard X-ray flare events detected with HXRBS/SMM, finding a powerlaw distribution $N(\Delta t) \propto \Delta t^{-0.75\pm0.1}$ in the range of $\Delta t = 1$–100 min, which could not be explained (Fig. 5.11, top left panel). A similar dataset observed with the more sensitive BATSE/CGRO was analyzed by Biesecker (1994), which showed no simple Poissonian distribution either, in the time interval range of $\Delta t = 2$–400 min, and an upper limit of $\lesssim 25\%$ was inferred for sympathetic flaring. For a dataset of 182 hard X-ray flares detected by WATCH/Granat (Fig. 5.11, left middle panel), a waiting-time distribution with a powerlaw slope of $p = 0.78 \pm 0.13$ was found in the range of $\Delta t =$ 10–300 min, with an exponential rollover at the upper end (Crosby 1996), which is similar to the HXRBS/SMM result. A dataset of 6,919 hard X-ray bursts at energies ≥ 30 keV detected by ICE/ISEE-3 (Fig. 5.11, bottom left) was sampled by Wheatland et al. (1998), showing a double-hump distribution of waiting times (similar to the second case in Fig. 5.3), which can be interpreted as an overabundance of short waiting times ($\Delta t \approx 10$ s to 10 min) compared with a nonstationary Poisson process fitted to the longer time intervals of $\Delta t \approx 10$–1,000 min (Wheatland et al. 1998; Wheatland and Eddey 1998). Modeling of the waiting-time distribution in terms of a nonstationary Poisson process with Bayesian blocks (Section 5.2) was attempted, but no suitable fit was found. The overabundance of short waiting times was interpreted as clustering of multiple hard X-ray bursts per flare event (Wheatland et al. 1998).

In Fig. 5.9 we show waiting-time statistics of 11,594 hard X-ray flare events observed with the *Ramaty High Energy Spectroscopic Solar Imager (RHESSI)* during 2002–2008. The daily flare rate shows a high degree of fluctuations, even during the solar minimum phase. The frequency distribution of waiting times can be fitted with a powerlaw function, i.e., $N(\Delta t) \propto \Delta t^{-2.0}$ in the time interval range of $\Delta t \approx 2$–1,000 hrs (Fig. 5.9 bottom right), so the longer waiting times $\Delta t \gtrsim 2$ hrs are consistent with a nonstationary Poisson random process in a SOC system.

At shorter time intervals there is an excess of events at waiting times near the orbital period of $\Delta t \lesssim t_{orbit} \approx 1.6$ hrs (Fig. 5.9, bottom left), which is caused by instrumental effects. A first reason is that events that start at spacecraft night cannot be detected until the spacecraft comes out of the night part of the orbit, which causes a clustering that systematically increases from one half to one full orbital period. A second reason is that large events that extend over more than one full orbital period are double-counted in each subsequent orbit. Additional data gaps result from irregular switch on/off periods during some parts of

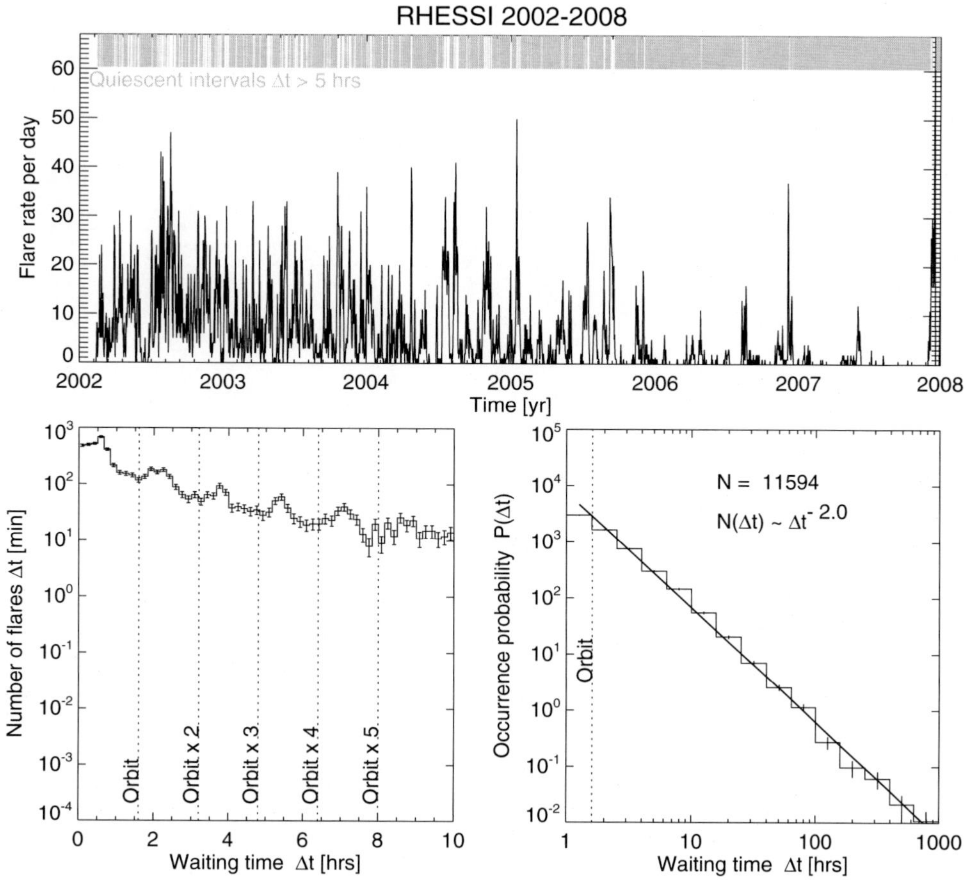

Fig. 5.9 *Top:* Flare rate per day observed with RHESSI during 2002–2008, containing a total of $\approx 12{,}000$ flare events. Quiescent time intervals with $\Delta t > 5$ hrs are marked in the form of a "bar code" at the top of the panel. *Bottom left:* The frequency distribution of waiting times $N(\Delta t)$ is shown for short time intervals $\Delta t \le 10$ hrs, which shows peaks at subharmonics of the orbital period of ≈ 1.6 hrs. *Bottom right:* The longer waiting-time intervals $\Delta t \approx 2$–24 hrs can be fitted with a powerlaw function with a slope of $p = 2.0$ (Aschwanden and McTiernan 2010).

the orbit, such as when the spacecraft passes through the *South Atlantic Anomaly (SAA)*. Such instrumental biases have been modeled with Monte-Carlo simulations in previous waiting-time studies (e.g., Biesecker 1994).

The effect of flux thresholds in the event definition on the distribution of waiting times of this particular dataset was investigated by Aschwanden and McTiernan (2010), an issue that has been raised in previous studies (e.g., Buchlin et al. 2005; Hamon et al. 2002). Hamon et al. (2002) finds for the Olami–Feder–Christensen model (Olami et al. 1992), which is a cellular automaton model for systems with self-organized criticality, that event definitions without any threshold lead to stretched exponential waiting-time distributions, while threshold-selected events produce an excess of longer waiting

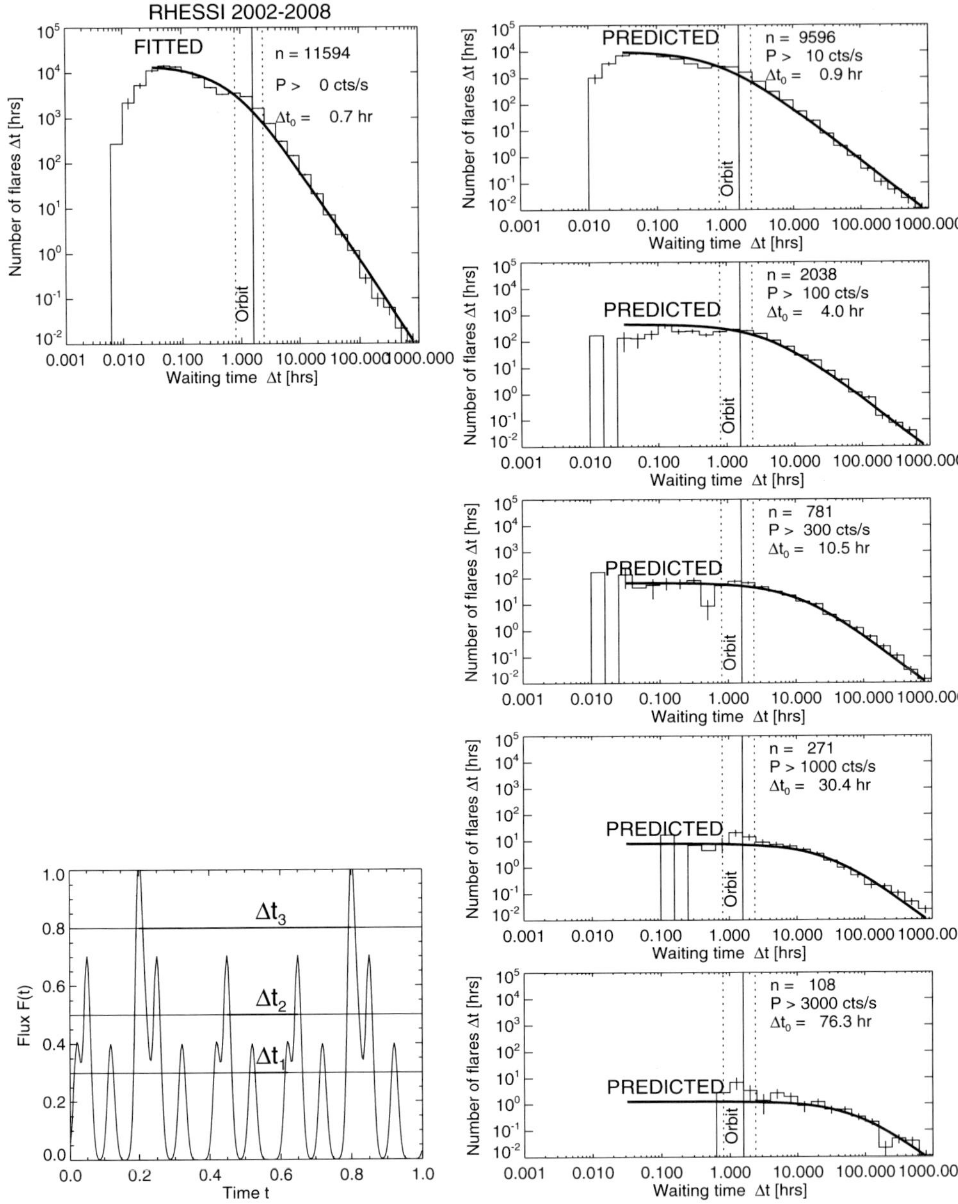

Fig. 5.10 *Top left:* Waiting-time distribution of solar flares function, representing a stationary Poisson process (thick solid curve), while the nonstationary Poisson process that has a function with a powerlaw tail of $p = 2$ is indicated with a thick curve. *Right:* Waiting-time distributions for five different subsets of the data selected by thresholds of $P \geq 10, 100, 300, 1{,}000, 3{,}000$ cts s^{-1}. The same model functions are predicted (using Eq. 5.6.2) for the threshold-selected subsets based on the number of detected events above the selection threshold. All subsets are consistent with the same nonstationary Poisson process. *Bottom left:* Schematic showing the relationship between the increase of waiting times $(\Delta t_1, \Delta t_2, \Delta t_3)$ as a function of progressively increasing thresholds (P_1, P_2, P_3) (adapted from Aschwanden and McTiernan 2010).

times (Fig. 5.6). Investigating this problem for RHESSI data, various flux thresholds, e.g., $P = 10, 100, 300, 1{,}000, 3{,}000$ cts s^{-1}, have been applied to the complete RHESSI flare catalog, and the waiting times have been resampled for events above these flux thresholds. The corresponding six waiting-time distributions are shown in Fig. 5.10. The waiting-time distributions clearly show an increasing excess of longer waiting times with progressively higher thresholds, which is expected due to the filtering out of shorter time intervals between flares with weaker fluxes below the threshold (Fig. 5.10, bottom left). Based on the reduction in the number n of events as a function of the flux threshold, we can make a prediction for how the mean waiting-time scales with increasing threshold, which is a reciprocal relationship, since the total duration T of all waiting times is constant,

$$T = \sum_i^n \Delta t_i = n \langle \Delta t \rangle = n_T \langle \Delta t_T \rangle \ . \tag{5.6.1}$$

Thus, from the number of events n_T detected above a selected threshold we can predict the mean waiting time,

$$\langle \Delta t_T \rangle = \frac{n}{n_T} \langle \Delta t \rangle \ . \tag{5.6.2}$$

Using the full set of data with $n = 11{,}594$ events and a mean waiting time of $\langle \Delta t \rangle = 0.71$ hrs (Fig. 5.10, top left), we can predict the distributions and average waiting times for the thresholded subsets, based on their detected numbers n_T using Eq. (5.6.2): $\langle \Delta t_T \rangle = 0.9, 4.0, 10.5, 30.4, 76.3$ hrs. We fit our theoretical model of the waiting-time distribution (Eq. 5.2.18) of a nonstationary Poisson process and predict the distributions for the threshold datasets, using the scaling of the predicted average waiting times $\rangle \Delta t_T \langle$. The predicted distributions (thick curves in Fig. 5.10) match the observed distributions of thresholded waiting times (histograms with error bars in Fig. 5.10) quite accurately, which demonstrates how the waiting-time distribution changes in a predictable way when flux thresholds are used in the event selection.

A compilation of waiting-time distributions observed for solar flares in hard X-rays is shown in Fig. 5.11, including datasets from ISEE-3/ICE, HXRBS/SMM, WATCH/Granat, BATSE/CGRO, and RHESSI. It demonstrates why earlier studies within a limited range of waiting times ($\lesssim 1.8$ decades) were characterized by a single powerlaw with a slope of $p \approx 0.8$ (Pearce et al. 1993; Crosby 1996), while datasets covering a larger range of waiting times over 4–6 orders of magnitude (ISEE-3, HXRBS, BATSE, RHESSI) can all be fitted by the same waiting-time distribution with a powerlaw function with slope $p \approx 2$ at the upper end (Eq. 5.2.18), as derived from a nonstationary Poisson process in the limit of highly intermittent flaring (Aschwanden and McTiernan 2010). This result of a highly intermittent flare rate suggests that the Sun has dual states of *flare-active* and *quiescent* periods, similar to accretion disks around black holes (with hard and soft states, see Section 4.7; Mineshige et al. 1994a), or earthquakes with different levels of activity between main shocks and during aftershocks (Omori's law; Omori 1895). For solar flare hard X-rays, we find an average breaking point at $\Delta t_0 = 0.80 \pm 0.14$ hrs (averaged from HXRBS/SMM, BATSE/CGRO, and RHESSI), above which the waiting-time distribution can be characterized with a powerlaw function. The flare-active state appears very intermittent throughout the entire solar cycle (see time profile in Fig. 5.9), since the daily and hourly fluctuations

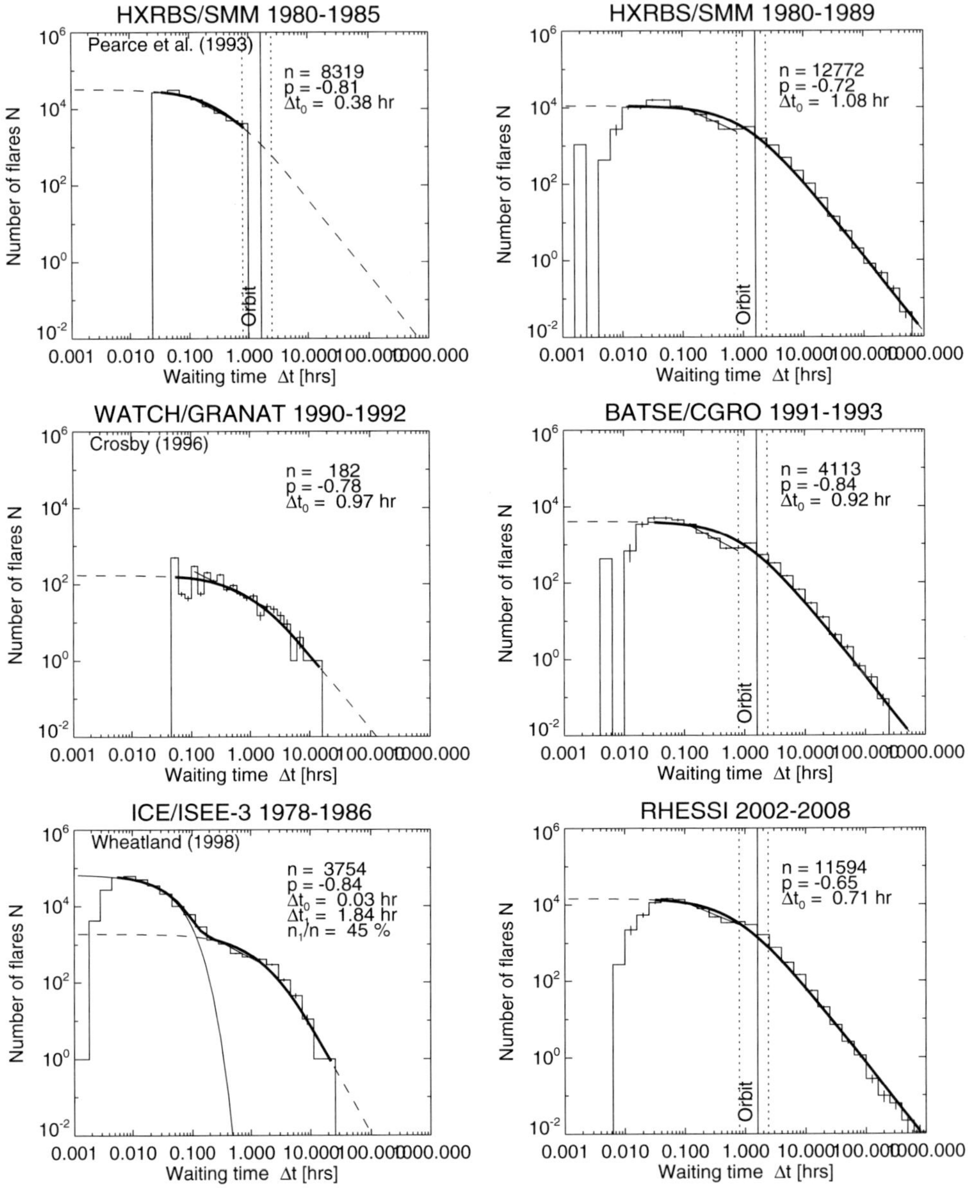

Fig. 5.11 Waiting-time distributions of six different datasets: HXRBS/SMM (top left and right), WATCH/Granat (middle left), ISEE-3/ICE (bottom left), BATSE/CGRO (middle right), and RHESSI (bottom right). The distribution of the observed waiting times are shown with histograms, fits of nonstationary Poisson processes with dashed curves (with a powerlaw tail with slope of $p = 2$), and the best fit in the fitted range with thick solid curves. Powerlaw fits in the range of $\Delta t \approx 0.1$–2.0 hrs as fitted in the original publications (Pearce et al. 1993; Crosby 1996) are also shown (straight line and slope p). The excess of events with waiting times near the orbital period ($t_{orbit} \approx 1.6$ hrs) is an artificial effect and is not included in the model fits (Aschwanden and McTiernan 2010).

are at any time much stronger than the long-term trends of the solar cycle. For the dataset of ISEE-3/ICE, an overabundance of short waiting times with a mean of $\Delta t = 0.03$ hrs (2 min) was found, which seems to be associated with the detection of clusters of multiple hard X-ray bursts per flare (Wheatland et al. 1998).

Thus, the main observational result of these hard X-ray studies is that the waiting-time distribution of solar flares is consistent with a powerlaw function for waiting times $\Delta t \gtrsim 1$ hr, consistent with a nonstationary Poissonian random process. The fact that the driver process for solar flares is nonstationary does not contradict with models of self-organized criticality (Bak et al. 1987, 1988; Lu and Hamilton 1991; Charbonneau et al. 2001), except that the input rate is highly variable. Alternative interpretations, such as intermittent turbulence (Boffetta et al. 1999; Lepreti et al. 2001), are discussed in Chapter 10.

5.6.2 Solar Flare Soft X-Rays

Further statistics of flare waiting times was gathered in soft X-rays, using the GOES database during 1975–1999 in the 1-8 Å wavelength range, yielding a waiting-time distribution that has an average powerlaw slope of $p = 2.4 \pm 0.1$ (Boffetta et al. 1999; Lepreti et al. 2001) or $p = 2.16 \pm 0.05$ (Wheatland 2000a) in the time interval range of $\Delta t \approx 6$–70 hrs (Fig. 5.12 top), with a flat rollover at shorter time intervals of $\Delta t \approx 0.1$–6 hrs, where the previous hard X-ray waiting-time distributions were measured. However, although these three studies present the same observational result, they offer three completely different interpretations. Boffetta et al. (1999) argue that the lack of a simple Poissonian distribution rules out SOC behavior and is more consistent with the powerlaw distribution obtained with a shell model of MHD turbulence, while Lepreti et al. (2001) fit a Lévy function (over a range of 1.5 orders of magnitude) and suggest that the underlying flare process is statistically self-similar but has some "memory", while Wheatland (2000a) is able to model the data with a nonstationary Poisson process, which is consistent with SOC behavior in a system with a variable driver (i.e., the solar activity cycle and its fluctuations). Wheatland (2000a) decomposed the 20-year time series into Bayesian blocks (Fig. 5.12, middle), and found an exponentially decreasing distribution for the fraction of time $f(\lambda)$ of a given flaring rate λ (which is a constant per Bayesian block, see also Section 5.2),

$$f(\lambda) = \frac{1}{\lambda_0} \exp\left(-\frac{\lambda}{\lambda_0}\right) , \qquad (5.6.3)$$

with $\lambda_0 \approx 0.15$ hr^{-1} (Fig. 5.12 bottom). The fit of this nonstationary (or piecewise-constant) Poisson process to the waiting-time distribution is shown in Fig. 5.12 (top), which is able to reproduce the observed waiting-time distribution over a large range (of 4 orders of magnitude). The time-varying flare rate applies also to individual active regions (Wheatland 2001). More detailed investigations revealed that the flare rate $f(\lambda)$ varies also during the solar minimum and maximum, which introduces a time variability into the powerlaw index p of waiting times,

$$N(\Delta t) \propto t^{p} \propto t^{-(3+\delta)} , \qquad (5.6.4)$$

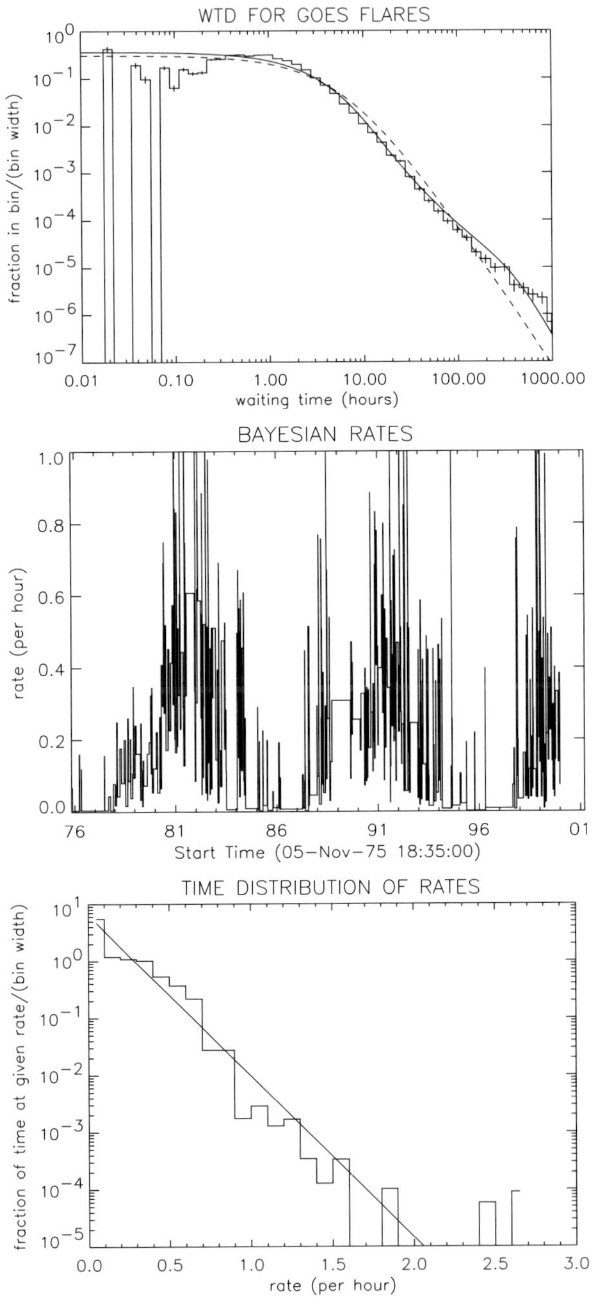

Fig. 5.12 *Top:* Waiting-time distribution for GOES 1-8 Å flares observed during 1975–1999 (histogram) and fit of nonstationary Poisson process with exponentially decreasing flare rate distribution (dashed curve). *Middle:* Bayesian blocks decomposition of the rate of occurrence of GOES flares. *Bottom:* Distribution of flaring rates, based on the Bayesian rate estimates (Wheatland 2000a; reproduced by permission of the AAS).

varying from $\delta = -0.1 \pm 0.5$ ($p = 2.95 \pm 0.5$) during the solar maximum to $\delta = -1.7 \pm 0.2$ ($p = 1.3 \pm 0.2$) during the solar minimum, or an overall average of $\delta = -0.9 \pm 0.1$ ($p = 2.2 \pm 0.1$) during the entire solar cycle (Wheatland and Litvinenko 2002). The variability of the mean flare rate λ was correlated with the sunspot number, which represents the input of subphotospheric magnetic energy into the corona, and a cross-correlation coefficient of CCC ≈ 0.8 was found for a time lag of ≈ 9 months, which seems to reflect the hysteresis of the coronal response (Wheatland and Litvinenko 2001). The Bayesian method of flare rate estimation in nonstationary Poisson processes can then be used for probabilistic forecast of solar flares (Wheatland 2004) or tests of sympathetic flaring. Sympathetic flaring was investigated in Wheatland (2006) and Wheatland and Craig (2006).

A follow-up study to Wheatland (2000a) confirmed the consistency of a nonstationary Poisson process with the data and found, in addition, stationary Poisson processes for flares in individual active regions (Moon et al. 2001). Alternatively, a subset of the data (2 years during the solar maximum) of the same waiting-time distribution was also modeled with the statistical method of *diffusion entropy*, which rests on the evaluation of the entropy of the diffusion process generated by the time series, which also could suitably fit the data and found a more accurate value for the powerlaw slope, i.e., $p = 2.14 \pm 0.01$ (Grigolini et al. 2002). At this point it is not clear whether the result of a nonstationary Poisson process can simply be interpreted as a threshold effect as for hard X-ray flares (Fig. 5.10), for which a theoretical value of $p = 3$ is expected (Fig. 5.3). Future studies are needed that investigate systematic differences and biases of soft X-ray and hard X-ray flare catalogs, which needs to be checked event by event.

Further theoretical modeling of waiting times of solar flares includes 1-D MHD models (Galtier 2001), the master (probability balance) equation (Wheatland and Glukhov 1998; Wheatland 2008, 2009), cellular automaton simulations with a variable driver with random walk (Norman et al. 2001) or deterministically-driven (energy-loading) driver (Charbonneau et al. 2007), or threshold-dependent statistics of inter-occurrence times (Baiesi et al. 2006; Paczuski et al. 2005). If the inter-occurrence (or waiting) time is rescaled by the rate of the events (i.e., by the flaring rate $\lambda(t)$), a universal distribution is found for the waiting times that is independent of the threshold (Baiesi et al. 2006).

Does the waiting time give us some information about the energy build-up in solar flares? Early studies suspected that the waiting time is the longer the more energy is built up, which predicts a correlation between the waiting time and the energy of the flare (Rosner and Vaiana 1978). However, several observational studies have clearly shown that no such correlation exists (e.g., Lu 1995b; Crosby 1996; Wheatland 2000b; Georgoulis et al. 2001), not even between subsequent flares of the same active region (Crosby 1996; Wheatland 2000b). This null-result is not surprising in the concept of SOC models. The original SOC model of BTW assumes that avalanches occur randomly in time and space without any correlation, and thus a waiting-time interval between two subsequent avalanches refers to two different independent locations (except for sympathetic flares), and thus bears no information on the amount of energy that is released in each spatially separated avalanche. During a SOC avalanche, also there is only a small amount of the available free energy depleted, which makes the amount of depleted energy even less correlated with the waiting time.

5.6.3 Coronal Mass Ejections

Coronal mass ejections (CMEs) are launched as a consequence of a magnetic instability in the solar corona, in most cases accompanied by a solar flare, which most likely is driven by a magnetic reconnection process. The two phenomena are therefore strongly connected and a similar statistics of waiting times is expected for both. Wheatland (2003) examined 4,645 CME events observed with SOHO/LASCO during 1996–2001, as well as soft X-ray flares observed by GOES during the same period. He found similar waiting-time distributions for both phenomena, as shown in Fig. 5.13. The waiting-time distribution is flat for short waiting times and turns into a powerlaw-like distribution for long waiting times,

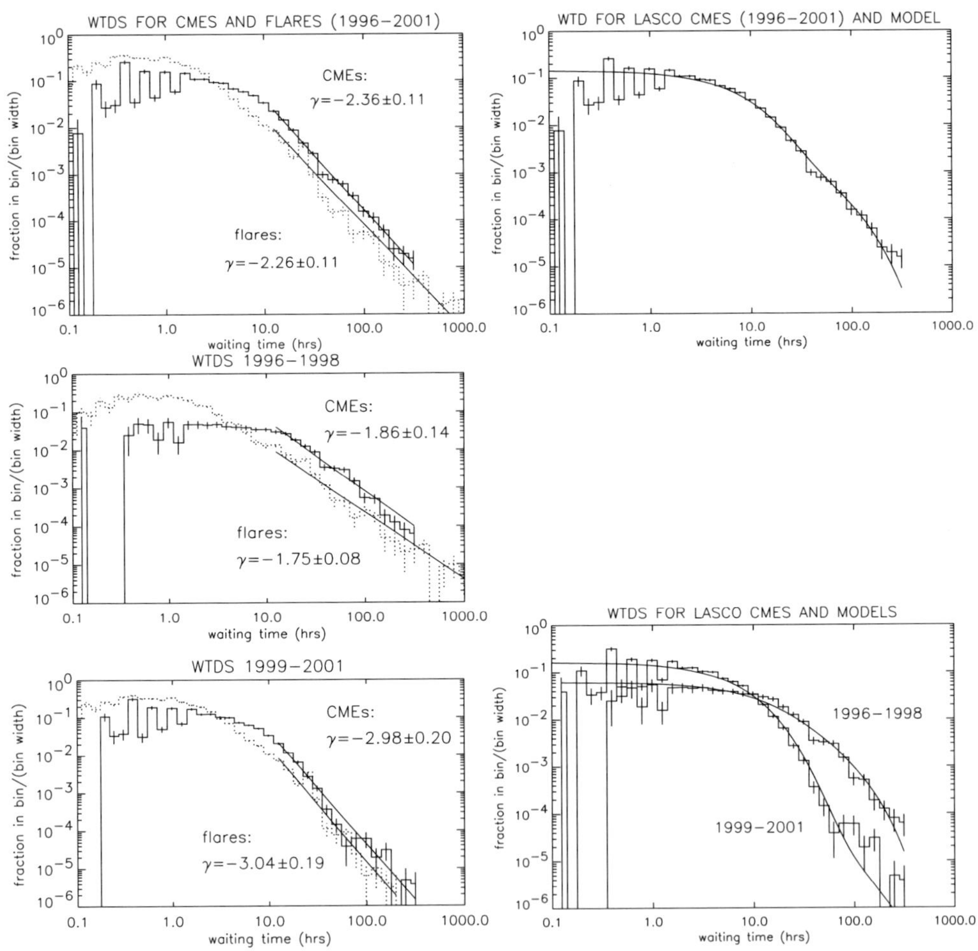

Fig. 5.13 *Left:* Waiting-time distributions of CMEs observed with SOHO/LASCO and flares observed with GOES, for the whole time- interval 1996–2001 (top), and the two epochs near solar minimum 1996-1998 (middle) and near solar maximum 1999–2001 (bottom), with fitted powerlaw tail. *Right:* The same three distributions fitted with a nonstationary Poisson process model with a time-varying flare rate (Wheatland 2003).

with slopes of $p = 2.36 \pm 0.11$ (all LASCO CMEs), $p = 1.86 \pm 0.14$ (during 1996–1998 near solar minimum), and $p = 2.98 \pm 0.20$ (during 1999–2001 near solar maximum), compared with $p = 2.26 \pm 0.11$ (all GOES flares), $p = 1.75 \pm 0.08$ (during 1996–1998), and $p = 3.04 \pm 0.19$ (during 1999–2001) (Fig. 5.13, left). Wheatland (2003) subdivided the time series into Bayesian blocks with constant rates and was able to fit the distributions with a nonstationary Poisson process for both flares and CMEs (Fig. 5.13, right panel). Thus, the fits are consistent with theoretical models of nonstationary Poisson processes, for which a powerlaw tail results with a maximum slope of $p \lesssim 3$ (Fig. 5.3). It is also satisfactory to see an identical variation of the powerlaw slope for CMEs and flares during the solar cycle, which implies a synchronous variation of the flaring rate $\lambda(t)$, because this supports the notion that CMEs and flares are driven by the same magnetic instability, which was also concluded by Yeh et al. (2005) with similar data. The same data were also analyzed by Moon et al. (2003) and fitted with the same model of a nonstationary Poisson process developed by Wheatland (2000a), without finding an excess of short waiting times that could possibly indicate sympathetic CMEs.

5.6.4 Solar Radio Bursts

The waiting times of solar radio bursts observed at 0.2–10 MHz with STEREO/ WAVES have been investigated during a type III storm that lasted from 2006 Nov 11 to 17 (Eastwood et al. 2010). Such type III storms are associated with an individual active region in the solar corona, often associated with a radio type I storm, but they are observed further out in the heliosphere at a distance of a few solar radii. The waiting-time distribution of this observation of some 800 radio bursts is shown in Fig. 5.14 (right frame). The distribution was modeled with a time-dependent Poisson process (Eq. 5.2.6) and a satisfactory fit was found. During the main phase of Nov 12–17, the average burst rate is almost constant and probably could be modeled with a stationary Poisson process. The Poissonian nature of these radio bursts implies that subsequent events, believed to be generated by a nonlinear conversion of Langmuir waves excited at the local plasma frequency by nonthermal electron beams, are stochastically produced and do not interact with each other, like subsequent avalanches in the sandpile SOC model.

5.6.5 Solar Wind

The solar wind is believed to be subjected to MHD turbulence, which produces intermittent cascading of spatial structures from larger to smaller scales (e.g., Veltri 1999; Podesta et al. 2006a,b, 2007). The phenomenology of nonlinear cascades in the solar wind can be characterized by Kraichnan or Kolmogorov power spectra (Horbury and Balogh 1997). The identification of the most intermittent structures includes shock waves, small random currents, current cores, and one-dimensional current sheets.

The waiting-time distribution (or probability density function) of energy density fluctuations ($\propto B^2$) in the solar wind was found to have powerlaw tail during solar maximum (Freeman et al. 2000a), similar to the Lévy flight model (Hnat et al. 2007), which was also applied to waiting-time distributions of solar energetic particle (SEP) events (Gabriel and

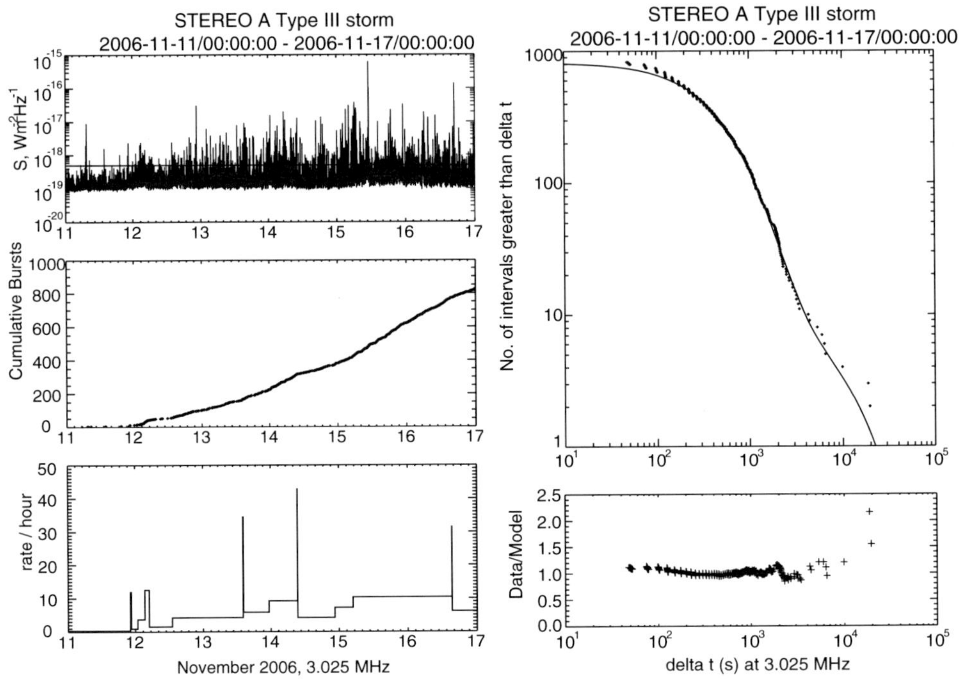

Fig. 5.14 *Top left:* Time series of radio bursts at 3.025 MHz. *Middle left:* Cumulative number of radio bursts. *Bottom left:* Bayesian blocks decomposition. *Top right:* Cumulative waiting-time distribution of radio bursts at 3.025 MHz, fitted with a piecewise constant Poisson process. *Bottom right:* Ratio of data to model distribution. (Eastwood et al. 2010; reproduced by permission of the AAS).

Patrick 2003). Freeman et al. (2000a) measure the waiting-time distribution from WIND spacecraft data in the range of $\Delta t \approx 0.5 \times 10^2$–$10^5$ s ($\approx$ 0.1–30 hrs) and find an approximate fit with a powerlaw function with a slope of $p = 1.67$ (Fig. 5.15). Comparisons of solar wind data from the *Advanced Composition Explorer (ACE)* and MHD turbulence simulations exhibited a good agreement in the waiting-time statistics of magnetic field increments, which was interpreted as evidence of the solar wind magnetic structures emerging from MHD turbulence (Greco et al. 2009a). However, the distribution of waiting times was found to be closer to a powerlaw for spatial separation scales smaller than the correlation scale, while they are closer to an exponential distribution for events separated by more than a correlation scale, which raises the question whether the phenomenon of clusterization occurs (Greco et al. 2009b).

We can ask the question whether the prevailing observation of a powerlaw distribution of inter-burst intervals (waiting times) in the solar wind is sufficient proof for the operation of MHD turbulence. While it is certainly consistent with some theoretical simulations, SOC behavior cannot be ruled out because a nonstationary Poisson process produces also powerlaw-like waiting-time distributions. However, the expected powerlaw slope is typically around $p \lesssim 3$ for this model (Fig. 5.3), which is somewhat steeper than the index observed in the solar wind ($p \approx 1.7$).

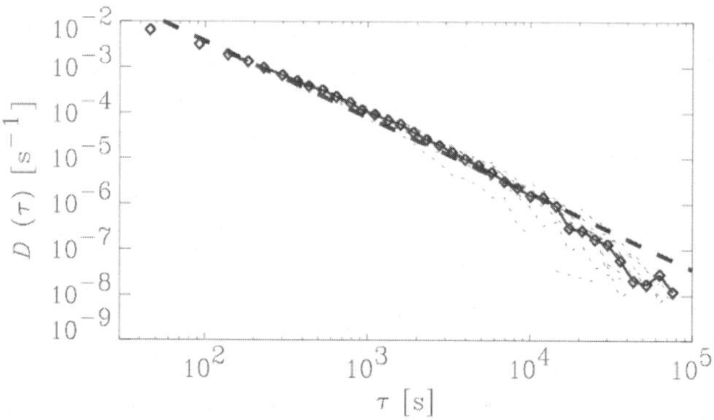

Fig. 5.15 Waiting-time distribution of inter-burst intervals between solar wind fluctuations ($\propto B^2$) measured from WIND spacecraft data. The powerlaw fit (dashed line) has a slope of $p = 1.67$. Reprinted from Freeman et al. (2000a) with permission; Copyright by American Physical Society.

5.7 Waiting-Time Statistics in Astrophysics

5.7.1 Flare Stars

Stars with high magnetic activity reveal in soft X-rays and EUV a high level of flux variability on similar timescales as flares from the Sun (Butler et al. 1986; Ambruster et al. 1987; Collura et al. 1988; Kashyap and Drake 1999; Audard et al. 1999, 2000; Osten and Brown 1999). Large fluctuations have been identified as flare-like phenomena similar to the Sun, based on spectral soft X-ray temperature measurements and simultaneous non-thermal emission detected in radio wavelengths, and thus are believed to be the analogs to large solar flares. The *quiescent emission* was initially interpreted as quasi-stationary thermal emission of hot plasma from the stellar corona, but later it was interpreted as a superposition of small fluctuations that correspond to solar microflares and nanoflares, which potentially could play a role in the coronal heating (Audard et al. 2003; Güdel et al. 2002; Güdel 2004; Arzner and Güdel 2004). An example of such quiescent emission from the flare star AD Leo is shown in Fig. 5.16, recorded over 5 weeks. The question arises whether these small fluctuations represent SOC phenomena, such as solar flares.

The time profile of the AD Leo observation shown in Fig. 5.16 shows a very low flux of less than one photon count per second in the average, and thus is integrated over time bins as long as one spacecraft orbit (1.57 hrs). The original data input thus consists of time-tagged single photons, each one recorded at a different time t_i. In order to extract the maximum amount of information from such low count rates we have to analyze the statistics of *photon waiting times*, the time intervals $\Delta t_i = t_i - t_{i-1}$ between two subsequent photon arrivals, rather than the time intervals between two subsequent flares in the case of solar data. An example of such an analysis is presented in Arzner and Güdel (2004), which we briefly summarize here. One method consists of the assumption of a generic time profile for each flare or subflare, such as a sharp rise with an exponential decay, corresponding to the cooling time of the flare. Such a generic time profile is shown in

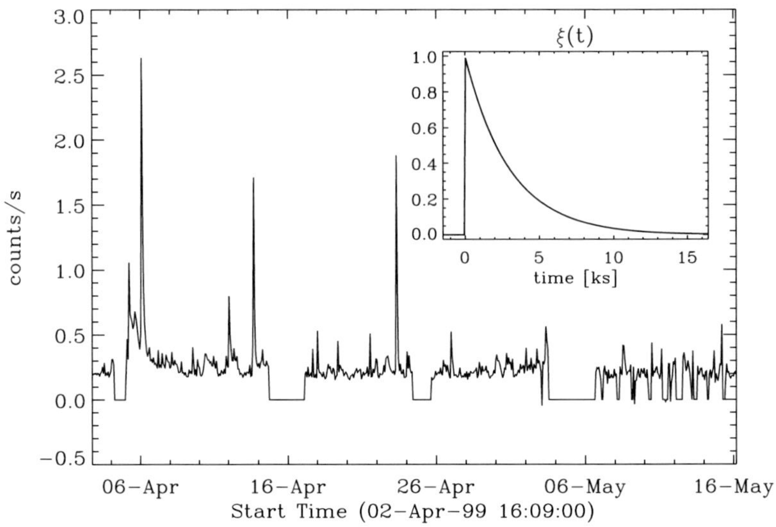

Fig. 5.16 EUVE/DS light curve from flare star AD Leo obtained between 1999 April 2 and May 16, integrated over one spacecraft orbit (5,663 s). The light curve of a model flare with an exponential decay time of 3,000 s is shown in the insert (Arzner and Güdel 2004; reproduced by permission of the AAS).

the insert in Fig. 5.16. A second model parameter is the distribution of peak fluxes, for which a powerlaw distribution (extending to lower energies) can be assumed for SOC models (Fig. 5.17, top right). A numerical simulation can then mimic a time profile by superposing the generic time profiles (Fig. 5.17, top left), randomly distributed in time, with amplitudes drawn from the peak flux distribution. The same Monte-Carlo simulation will then also provide the distribution of photon waiting times (Fig. 5.17, bottom right) and the histogram of binned counts (Fig. 5.17, bottom left). Special provisions have to be made for treatment of the flare-unrelated background subtraction. If the theoretical and observed histograms of binned counts agree, we can conclude that the time profile is consistent with a Poissonian random distribution of stellar flares and microflares. With this method, a powerlaw slope of $\alpha = 2.3 \pm 0.1$ could be constrained for the distribution of flare peak fluxes for the time profile observed from AD Leo (Arzner and Güdel 2004). A powerlaw index greater than two implies that the integral of the peak flux distribution $N(P)$ diverges for the smallest events (Section 7.1.5), which thus could dominate the heating of the stellar atmosphere. In addition, the match of the model with the data implies also that flaring on this star is consistent with a stationary Poisson process. It would be interesting to perform the same exercise for 5 weeks' worth of solar data in the same wavelength range, for which we found a nonstationary Poisson process (Aschwanden and McTiernan 2010). Flare stars such as AD Leo have clearly a higher flare rate than the Sun at the same flux threshold level (Audard et al. 2000).

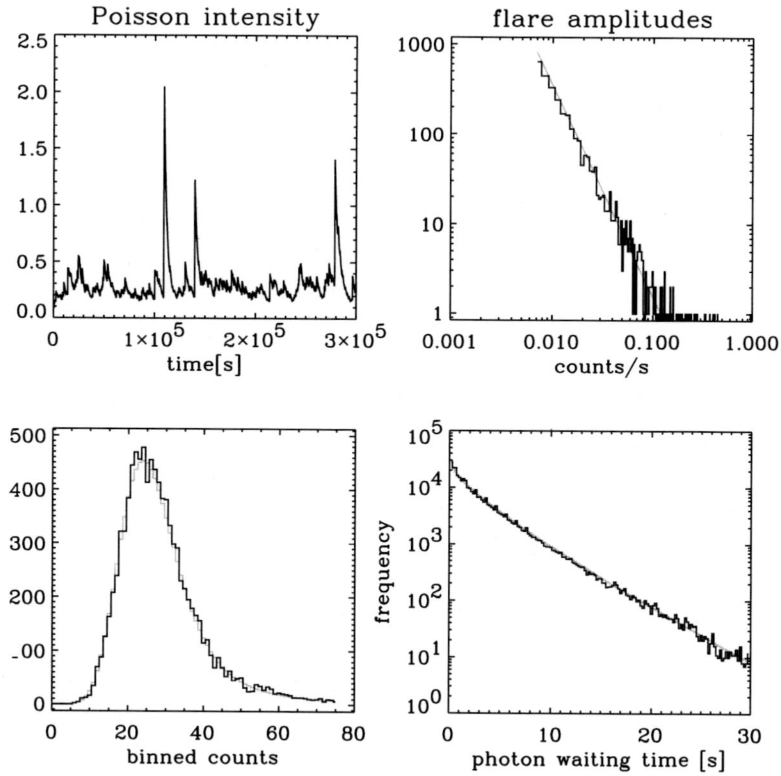

Fig. 5.17 Simulation of an intensity time profile (top left), a powerlaw distribution of amplitudes (top right), histograms of binned counts (bottom left), and photon waiting times (bottom right) in the context of data analysis of flaring events for the flare star AD Leo. The model flare used in this simulation has an exponential decay time of 3,000 s, as shown in the insert of Fig. 5.16 (Arzner and Güdel 2004; reproduced by permission of the AAS).

5.7.2 Black Hole Accretion Disks

Cygnus X-1 is the longest-known black-holes candidate, exhibiting a highly variable flickering in X-rays on time scales of some 10 ms, which corresponds to a light travel time of $\approx 3{,}000$ km, rendering it a very compact object. We discussed $1/f$ power spectra of flicker noise in Section 4.7 and the shot-noise model in Section 4.8. In the shot-noise model proposed by Terrell (1972), the time profile is assumed to consist of identical pulses (or shots), occurring randomly in time, so that the waiting-time distribution is expected to follow a Poisson distribution. The randomness of the pulses was investigated with power spectra (Terrell 1972) as well as with waiting-time distributions (Negoro et al. 1995). Similar tests with the timing of gamma-ray bursts showed evidence for chain reactions in a near-critical regime (Stern and Svensson 1996).

An observed waiting-time distribution obtained from Ginga observations of Cygnus X-1 in its low (hard) state over 4.5 hrs in the energy range of 1.2–58.4 keV (Negoro et

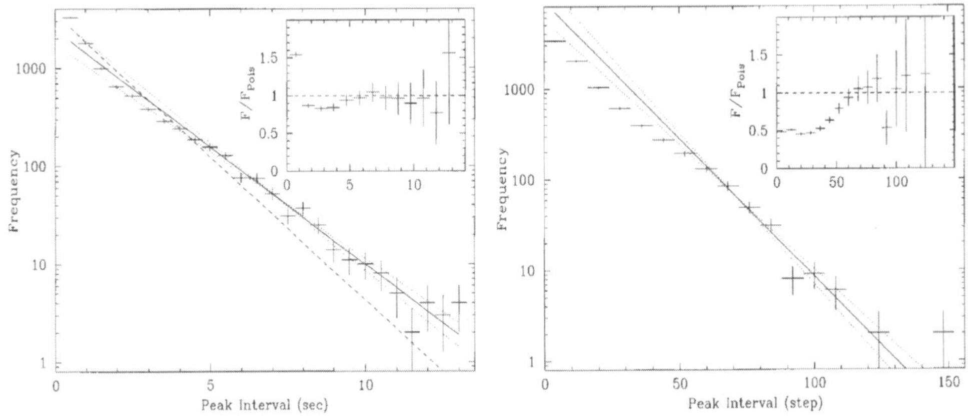

Fig. 5.18 *Left:* Observational waiting-time distribution for shots with peak intensities of $P \geq 1.5\langle P\rangle$, measured from the stellar black-hole candidate Cygnus X-1 with the *Ginga* spacecraft. The reference Poisson distribution and 90% confidence regions are indicated by the solid and dotted lines. The insert shows the ratio of the observed to the reference distributions. *Right:* Waiting-time distribution simulated with a theoretical SOC model. The photon fluctuations are not included in the error bars (Negoro et al. 1995; reproduced by permission of the AAS).

al. 1995) is shown in Fig. 5.18 (left). Shot events were selected with a flux threshold of $P \geq 1.5\langle P\rangle$, where $\langle P\rangle$ is the local mean number of counts. The distribution shown in Fig. 5.18 (left) contains 9,016 events. The distribution is approximately consistent with a Poisson random distribution, indicated with a solid line in Fig. 5.18 (top left), including the 90% confidence regions (dashed lines), but a detailed inspection reveals significant deviations at waiting times of $\Delta t \lesssim 5$ s (see insert in Fig. 5.18, left). This small, but significant deviation was also confirmed using different thresholds of $P \geq 2.0\langle P\rangle$ and $P \geq 2.35\langle P\rangle$. Thus the data show an underabundance of short waiting times by comparison with the constant-rate Poisson model.

A theoretical model in terms of a cellular automaton applied to accretion disks around black holes was developed by Mineshige et al. (1994a), which we described in Section 2.7.1. The inner portion of a black hole accretion disk is assumed to be composed of numerous small reservoirs. If a critical mass density is reached in a reservoir, an unknown instability occurs and the accumulated material drifts inward as an avalanche, emitting X-rays by bremsstrahlung during this process. Once a major avalanche occurs, the local mass reservoir becomes depleted and some time is required to replenish the hole. Negoro et al. (1995) performed Monte-Carlo simulations based on this SOC model, constrained by the observed exponential pulse flux distribution $N(P)$. The resulting simulated waiting-time distribution is shown in Fig. 5.18 (right panel), which indeed reproduces an underabundance of shorter waiting times, or an overabundance of longer waiting times, with respect to a Poissonian distribution, similar to that observed in the data (Fig. 5.18, left panel). So, the observed suppression of the shot appearance after each big event suggests the existence of reservoirs in black hole accretion disks. The irregular timing and peak fluxes of individual shots suggest many different reservoirs with different capacities. Comparing with an ideal SOC model where the peak flux distribution is a powerlaw, in contrast to the observed exponential distribution from Cygnus X-1, one might think that some additional processes might be in play, such as gradual diffusion of mass, even when the critical condition is not satisfied (Mineshige et al. 1994b). The underlying critical condition of the SOC state is envisioned to be of magnetic origin, such as a magnetic reconnection process.

5.8 Summary

The basic definition of a random process is the Poissonian distribution of waiting times between subsequent events, which can be approximated by an exponential probability distribution (Section 5.1). The distribution of waiting times is an exponential function for stationary Poisson processes only, where the mean event rate is time-independent, while the generalized concept of Bayesian statistics is needed (with piecewise constant event rates during individual Bayesian blocks) for nonstationary Poisson processes (Section 5.2). We showed mathematically that several nonstationary Poisson processes with a gradually varying event rate lead to a powerlaw-like distribution of waiting times with a slope of $p = 3$ in the tail, while highly intermittent flare rates produce a powerlaw slope as flat as $p = 2$ (Section 5.2). Thus, waiting-time distributions can exhibit both exponential as well as powerlaw-like distribution functions, depending on whether they have a stationary or a (time-varying) nonstationary mean event rate. The measurement of waiting-time distributions is very sensitive to the event definition and the flux thresholds used in the event detection (Section 5.3). For one dataset we demonstrated that event statistics without thresholding is consistent with a stationary Poisson process, while significant flux thresholds lead to a predictable increase of the mean waiting time and to powerlaw-like tails in the waiting time distribution (Figure 5.10). Generally, the observed waiting-time distributions rarely show a simple exponential function as expected for an ideal SOC system driven at a constant mean rate. Waiting-time distributions observed in earthquakes exhibit different distributions for main shocks and aftershocks (Section 5.4). In magnetospheric physics (Section 5.5), waiting-time distributions of auroral emission exhibit finite-size effects, auroral electron jet (AE) indices exhibit flat powerlaws, and substorms in the Earth's magnetotail exhibit forced criticality by the external solar wind rather than classical SOC behavior. Waiting-time distributions in solar physics (Section 5.6) are consistent with a nonstationary Poisson process for flare events observed in hard X-rays (which show flare-active and quiescent states similar to black holes), for flare events in soft X-rays, for coronal mass ejections, and for radio bursts. Flat powerlaw waiting-time distributions are observed for the solar wind. In astrophysics (Section 5.7), waiting times of photon counts are modeled for flare stars (such as AD Leo), and waiting-time distributions of pulses from accretion disks around black holes (such as Cygnus X-1) are found to reveal deviations from a simple stationary Poisson process. In summary, the complexity of the observed waiting-time distributions is not necessarily inconsistent with SOC models, but requires modifications of ideal SOC models in terms of system size effects, clusterization, and time-varying drivers.

5.9 Problems

Problem 5.1: Calculate the theoretical waiting-time distribution for a Poisson process that has another time-varying rate than that shown in Fig. 5.3, for instance a sinusoidal variation, using either Eq. (5.2.6) or Eq. (5.2.7). Verify the analytical solution with numerical Monte-Carlo simulations as shown in Fig. 5.3 (right panels).

Problem 5.2: Can you find a time-varying rate $\lambda(t)$ that produces waiting-time distributions with a powerlaw tail that has a different value than $p = 3$ (Eq. 5.2.12), such as $p = 2$ or $p = 1$? Could such a model explain the observations of the AE index (Fig. 5.8) or solar wind (Fig. 5.15)?

Problem 5.3: Simulate a time series with a random distribution of pulses with a constant width and a powerlaw distribution of peak fluxes. Sample the waiting times of this time profile for thresholds from 10% to 50% of the peak flux. How does the functional form of the waiting-time distribution change for the various thresholds? Can you predict the thresholded waiting-time distributions analytically with a relationship like Eq. (5.6.2)?

Problem 5.4: Simulate a solar flare catalog for a hypothetical spacecraft that has a 90-minute orbit, assuming a stationary flare rate of $\langle \Delta t \rangle = 1$ hr. How do the orbital gaps of the spacecraft night affect the waiting-time distribution, compared with a spacecraft that sees the Sun uninterruptedly?

Problem 5.5: Verify the distribution of binned counts and photon waiting times shown for a flare star in Fig. 5.17 with a Monte-Carlo simulation, assuming the same powerlaw distribution of flare amplitudes. What is the mean flaring rate for this case?

6. Event Detection Methods

The only reason for time is so that everything doesn't happen at once.

Albert Einstein

Everything happens to everybody sooner or later if there is time enough.

George Bernard Shaw

The phenomenon of self-organized criticality (SOC) can only be identified and validated by event statistics, which requires measurements of relevant observables, such as time scales and size scales of avalanches. The most defining predictions of the ideal SOC model are the scale-free powerlaw distributions of time and size scales, as well as the Poissonian randomness of waiting times. In this Chapter we focus on the methods of measuring time and size scale distributions, mostly based on events detected in astrophysical observations. Apart from particle in-situ measurements in the heliosphere, astrophysical observations generally provide light curves in some wavelength as input for event statistics. In a time series, an event can then be defined by a start time, a peak time, and an end time, which yields a convenient definition for the duration, peak flux, and integrated flux of an event, to be used for statistics of SOC avalanches in terms of durations, peak energies, and total energies. However, the devil is always in the detail. There is no sure-fire method of measuring a unique duration and a peak flux of an event from a light curve. There are numerous diabolical effects such as the separation of events from a fluctuating background, flux threshold biases, confusion from near-simultaneous events, the ambiguity of an event definition in multiple-peak light curves, instrumental irregularities and data gaps, to name just a few. The resulting distributions of measured values are often not robust, unless every event can be uniquely defined, but rather subject to the event definitions and detection algorithms. Thus, the methods used to detect events and to measure their parameters have a profound impact on the results whether we find powerlaw distributions or different functional forms, being our arbiters of SOC phenomena. The best way to investigate various measurement biases is always to conduct a numerical simulation of data and to test a detection algorithm with such well-defined data. The output of the resulting event statistics, in the form of parameter distributions and correlations, can then be cross-compared with those of the input parameters and measurement biases can be quantified. In this Chapter we will start with the simulation of a time series that contains SOC events according to our standard model described in Section 3.1 and will scrutinize the performance and biases of various detection methods that have been previously used for SOC event statistics.

6.1 Test Data for Event Detection

In order to test various structure detection algorithms it is useful to simulate first a realistic model of a time series $f(t)$. Since we are mostly interested in SOC phenomena, we use our standard analytical model described in Section 3.1, which consists of avalanche events that have an exponential growth during the rise time and a linear decay after the peak. The canonical time profile of such a pulse is shown in Fig. 3.2. The pulses have random waiting times and random rise times.

We simulate a time series containing $n = 1{,}000$ pulses with an average waiting time of $\Delta t_0 = 5$ and a time resolution of $dt = 0.1$, so the time series contains $n\Delta t_0/dt = 50{,}000$ data points. The start times t_i are randomly distributed in the time interval $0 < t_i < t_{max} = n\Delta t_0 = 5{,}000$, produced by a numerical random generator. We verify that the distribution of waiting times Δt_i follows the prescribed random (exponential) distribution $N(\Delta t)$ (Eq. 5.1.1), which is shown in Fig. 6.1 (bottom row, left). The rise times have a mean of $\langle \tau \rangle = t_S = 1$ and are numerically generated with the function (Eq. 7.1.30),

$$\tau_i = -t_S \log(1 - \rho_i) \ , \quad \rho_i = [0, 1] \ , \tag{6.1.1}$$

where ρ_i is a uniformly distributed random number in the interval $[0, 1]$, which is drawn from a numerical random generator. We verify that the distribution of rise times τ_i follow the prescribed random (exponential) distribution $N(\tau)$ (Eq. 3.1.4), which is shown in Fig. 6.1 (third row, left).

Next we simulate the peak energies P_i according to Eq. (3.1.3),

$$P_i = W_0 \left[\exp\left(\frac{\tau_i}{\tau_G} \right) - 1 \right] \ , \tag{6.1.2}$$

using the constants $W_0 = 1$ and $\tau_G = 1$. The resulting distribution of peak energies P_i has a powerlaw slope of $\alpha_P = 1.89$ (Fig. 6.1, third row right). The theoretical distribution of peak energies P_i is expected to be a powerlaw distribution with a slope of $\alpha_P = (1 + \tau_G/t_S) = 2.0$ (Eq. 3.1.28). This is compatible with the simulation, since we found that different random number representations typically change the powerlaw slope of the distributions in the order of about $\pm 10\%$.

Next we can simulate the decay times D_i of the pulses, which depend linearly on the peak energy P_i (Eq. 3.1.13),

$$D_i = t_D \left(\frac{P_i}{W_0} \right) \ , \tag{6.1.3}$$

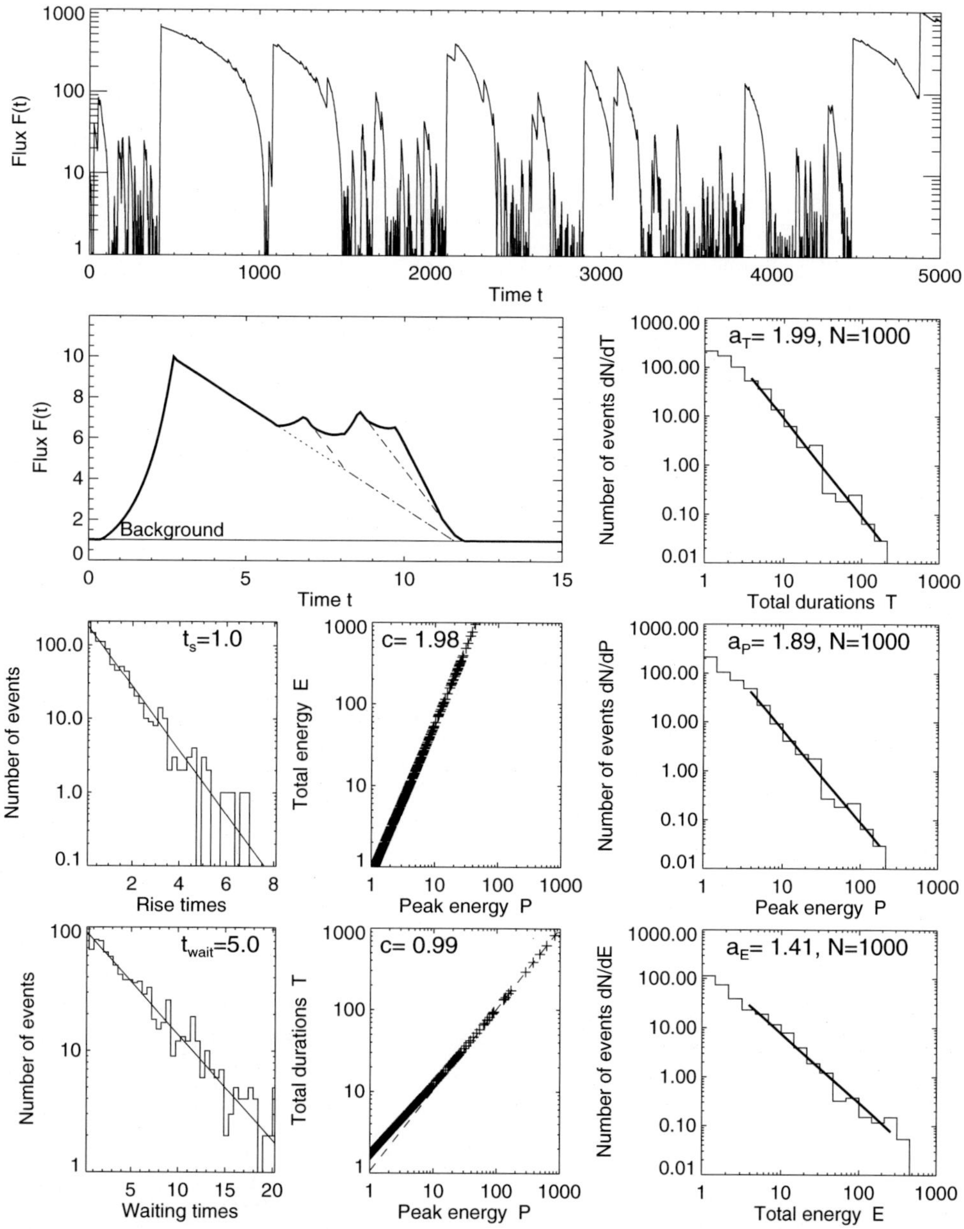

Fig. 6.1 Simulation of a time series (top panel) consisting of pulses with an exponential rise and a linear decay as defined in our analytical exponential-growth SOC model (Section 3.1), with the number of events $n = 1{,}000$, average waiting time $\Delta t_0 = 5$, time resolution $dt = 0.1$, growth time $\tau_G = 1$, average saturation time $t_S = 1$, and decay time $\tau_D = 1$. The different panels show the first pulses enlarged (second row left), the distributions or rise times and waiting times (left column), correlations between E, T, and P (middle column), and the frequency distributions $N(T)$, $N(P)$, and $N(E)$ with fitted powerlaws (right column).

using the constant $t_D = 1$. Adding the rise time τ_i and the decay time D_i together, we obtain the total pulse duration T_i (Eq. 3.1.14),

$$T_i = \tau_i + D_i \ . \tag{6.1.4}$$

The resulting distribution of total durations T_i has a powerlaw slope of $\alpha_T = 1.99$ (Fig. 6.1, second row right). This value is compatible with the theoretical expectation $\alpha_T = \alpha_P = (1 + \tau_G/t_S) = 2.0$ (Eq. 3.1.28).

Now we can calculate the total energy E_i (with the background level W_0 subtracted), according to Eq. (3.1.20),

$$E_i = P_i t_G - W_0 \tau_i + \frac{1}{2} P_i D_i \ . \tag{6.1.5}$$

The resulting distribution of total energies E_i has a powerlaw slope of $\alpha_P = 1.41$ (Fig. 6.1, bottom row right). Theoretically, we expect the relation $\alpha_E = (\alpha_P + 1)/2 = (2+1)/2 = 1.50$, which indeed is compatible. For our particular random representation shown in Fig. 6.1 we actually expect $\alpha_E = (\alpha_P + 1)/2 = (1.89 + 1)/2 = 1.44$, which is even closer to the measured value.

We can also plot the correlations between E and P and find a powerlaw fit of $E \propto P^{1.98}$ (Fig. 6.1, third row middle), which agrees with the theoretical limit of $E \propto P^2$ for large values (Eq. 3.1.28). Plotting the correlation between T and P we obtain $T \propto P^{0.99}$ (Fig. 6.1, bottom row middle), which agrees with the theoretical limit of $T \propto P^1$ for large values (Eq. 3.1.28).

Fig. 6.1 shows such a simulated flux time profile $f(t)$, with the full time series shown (Fig. 6.1, top), as well as an enlargement of the first pulses (Fig. 6.1, second row left). Thus we have a simulation of frequency distributions with well-defined values which can be cross-compared with those obtained from different pulse detection algorithms, as we will carry out in the following sections.

When we refer to the peak energy P and total energy E we follow the nomenclature of the SOC terminology, but actually mean the peak flux P and total (time-integrated) flux or fluence E when we detect events from the photon flux of an astrophysical time series. Strictly speaking, the energy rate is then defined in terms of the radiated energy $dE/dt = (n_{ph}/dt)h\nu$, where (n_{ph}/dt) is the number of photons per seconds. We will discuss other definitions of energies for SOC avalanches in Chapter 9 on physical SOC models.

6.2 Threshold-Based Event Detection

The probably most common and conceptually simplest method of defining events and measuring their duration in astrophysical time series is based on flux thresholds. The essential assumption is that background fluctuations have fluxes below the threshold, while events of interest exceed the flux threshold. This method of event detection is straightforward and pretty safe for large events and high thresholds, but systematically degrades when we detect weaker events, as desirable for SOC studies spanning over a large logarithmic range of time or size scales. In order to obtain some insight how methods of threshold-based event detection affect the statistical distribution of SOC parameters, we apply this method to the test data shown in Fig. 6.1.

We detect an event simply by identifying the start and end times when the flux profile $f(t)$ crosses the threshold F_{th} at those times according to,

$$\begin{array}{l} f(t) < F_{th} \text{ for } t < t_{start} \\ f(t) \geq F_{th} \text{ for } t_{start} < t < t_{end} \\ f(t) < F_{th} \text{ for } t > t_{end} \end{array} \quad (6.2.1)$$

The total time duration T is then defined by the time difference, the peak energy P by the maximum flux during this time interval, and the total energy E by the integral of the flux above the threshold,

$$\begin{array}{ll} T_i & = t_{end} - t_{start} \\ P_i & = max[f(t_{start}), ..., f(t_{end})] - W_0 \\ E_i & = \sum_{i=i_{start}}^{i_{end}} [f(t_i) - W_0] dt \end{array} \quad (6.2.2)$$

The frequency distributions and parameter correlations obtained with this detection method is shown in Fig. 6.2, which can be compared with the input parameters of the simulated time series shown in Fig. 6.1. The biggest difference is that we detect only 172 pulses out of the 1,000 simulated pulses, where we lose progressively more pulses towards weaker fluxes, because they either are below the threshold, or occur near-simultaneously during larger pulses. This progressive insensitivity towards weaker events leads to a peak flux-dependent under-abundance of shorter pulse durations, and thus to an underestimate of the powerlaw slope, i.e., $\alpha_T = 1.48$ for a threshold of $F_{th} = 1.5W_0$ (Fig. 6.2, second row right), compared with $\alpha_T = 1.99$ of the input data. Similarly, also the powerlaw slope of the peak energy distribution is underestimated, i.e., $\alpha_P = 1.50$ (Fig. 6.2, third row right), compared with $\alpha = 1.89$ of the input data. Also the powerlaw slope of the total energy distribution is underestimated, i.e., $\alpha_P = 1.28$ (Fig. 6.2, bottom row right), compared with $\alpha = 1.41$ of the input data. The correlations between the parameters are less severely affected, because the threshold effects cancel out to some degree, so we find $E \propto P^{1.88}$ (instead of $\propto P^2$) and $T \propto P^{0.93}$ (instead of $\propto P^1$) (Fig. 6.2, left).

This example clearly demonstrates that a flux threshold leads to a significant underestimate of the powerlaw slopes, even for noise-free data (Fig. 6.2). The bias mostly occurs for pulse durations that are shorter than the average waiting time. Only for time series containing well-separated events this bias vanishes in the limit of infinitely small thresholds.

Light curves from astrophysical objects often have a low signal-to-noise ratio, in which case the rate of arriving photons at our detector exhibits fluctuations of comparable magnitude as the weakest SOC events we try to detect. The observed photon flux $F_{obs}(t)$ follows Poisson statistics,

$$f_{obs}(t_i) \approx f(t_i) + \rho(t_i)\sqrt{f(t_i)} \ , \quad (6.2.3)$$

where $\rho(t_i)$ is a random number with a mean of $\mu = \langle \rho(t_i) \rangle = 0$ and a standard deviation of $\sigma = 1$ in the Gaussian approximation (Section 4.1). The addition of this photon noise to the simulated time profile shows clearly a noisier background, while the signal-to-noise ratio is better at the peak of large pulses (Fig. 6.3, top). If we detect temporal structures with a threshold of $F_{th} = 4.0W_0$, we detect a total of $N = 1{,}015$ structures (Fig. 6.3), compared

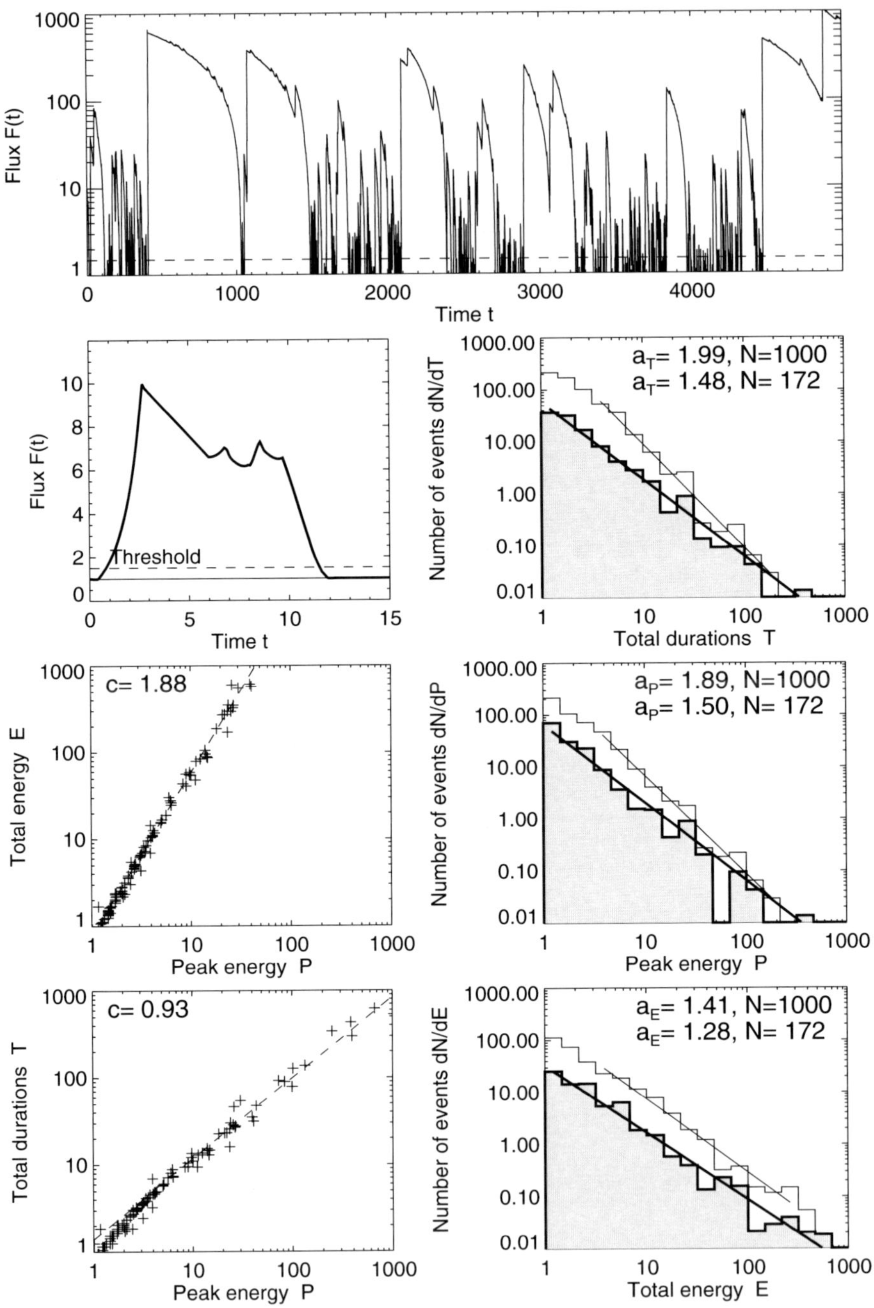

Fig. 6.2 The same time series as shown in Fig. 6.1 (top panel), with frequency distributions of event parameters (right) and correlations between event parameters (left), for events detected above a flux threshold of $F_{th} \geq 1.5W_0$. The histograms of detected structures are shown in gray, while the histograms of the input data (Fig. 6.1) are shown in white.

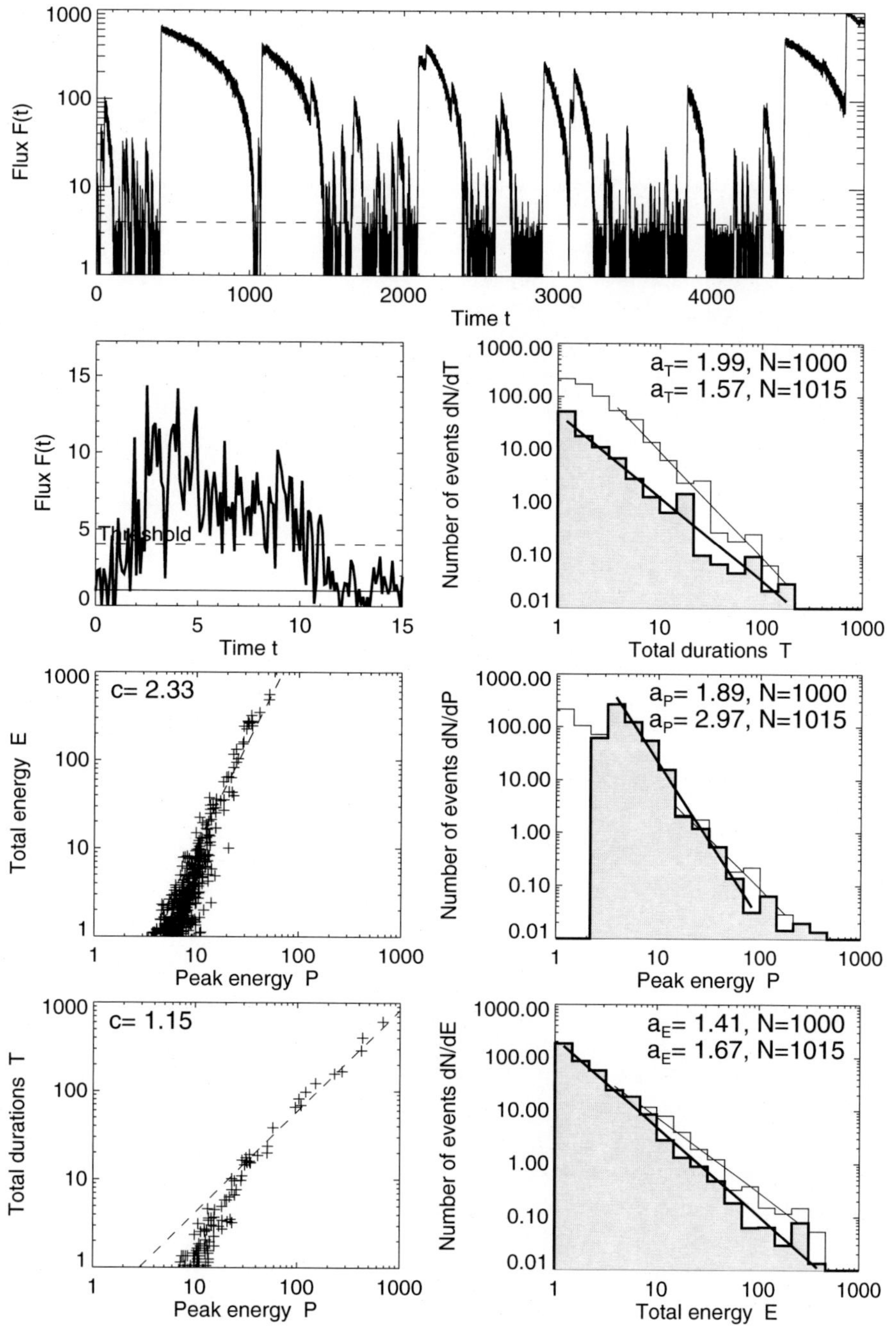

Fig. 6.3 The same time series as shown in Fig. 6.1 but with added Poissonian photon noise (top panel). The frequency distributions of event parameters (right) and correlations between the event parameters are shown (left) for events detected above a flux threshold of $F_{th} \geq 4.0 W_0$.

with $N = 172$ in the noise-free data (Fig. 6.2), out of the simulated $N = 1{,}000$ pulses. The threshold of $F_{th} = 4.0W_0$ corresponds to a 3-sigma significance level (for a Gaussian distribution), which has a confidence level of 99.73%, so we expect about 0.27% fluctuations beyond the threshold level (i.e., positive and negative), or half of it (0.135%) above the positive threshold, which amounts to excess fluctuations in 68 time bins for our time series of $n = 50{,}000$ data points, which explains the observed excess of 15 events. If we compare the observed frequency distribution of time scales between the input parameters (Fig. 6.1) and the noisy data (Fig. 6.3), we can see that we detect an over abundance of events with small peak energies, leading to a much steeper slope for peak fluxes ($\alpha_P = 2.97$ instead of $\alpha_P = 1.89$), a slightly steeper slope for energies ($\alpha_E = 1.67$ instead of $\alpha_E = 1.41$), and a flatter slope for durations ($\alpha_T = 1.57$ instead of $\alpha_T = 1.99$). Consequently, event detection with noisy data lead to substantially modified frequency distributions, hence, we have to apply suitable procedures to suppress the photon noise. This is more of a problem for photon-starved astrophysical time series, but much less so for the generally photon-rich solar data.

An efficient and robust technique to get rid of photon noise is the smoothing of a time series with a boxcar function, which is defined as the replacement of the flux value $f_i^{sm} = f^{sm}(t_i)$ at each time point with the average within a "boxcar" centered around the time point t_i,

$$f_i^{sm} = f^{sm}(t_i) = \sum_{j=i-n_{sm}/2}^{i+n_{sm}/2} f(t_j) \ . \tag{6.2.4}$$

We demonstrate this technique by applying a boxcar smoothing with a length of $n_{sm} = 2t_S/dt = 20$ datapoints to the noisy time series of Fig. 6.3, which is shown in Fig. 6.4. We notice that the data noise is mostly gone; an individual pulse shape is again recognizable with a well-defined rise and decay time (Fig. 6.4, second row left), and the resulting frequency distributions have similar powerlaw slope as the noise-free data (Fig. 6.2), i.e., for durations $\alpha_T = 1.42$ versus $\alpha_T = 1.48$, for peak energies $\alpha_P = 1.43$ versus $\alpha_T = 1.50$, for energies $\alpha_E = 1.23$ versus $\alpha_T = 1.28$. What is even more important, we detect a similar number ($N = 211$) of structures in the smoothed data for a threshold of $F_{th} = 1.5$ as in the noise-free data with the same threshold ($N = 172$), so we get rid of most false structures produced by the photon noise.

The consequences of time-overlapping pulses, photon noise, thresholding, and photon noise on the measurement of time scale distributions, as we exemplified here, apply specifically to pulses that have a powerlaw distribution in their amplitude and duration, as expected for the classical SOC model. Statistics of pulses with other amplitude and duration distributions may exhibit a different behavior. The analysis of astrophysical time series consisting of pulses with random (exponential) distributions in amplitude and durations has been studied in Scargle (1981), with a model called the *moving average model* therein, in contrast to the *autoregressive model* (Scargle 1981), where pulses are clustered in time and have a memory over their recent past. The effects of the event definition and threshold (see Fig. 5.4, middle column) on the measured time scale distribution was also quantitatively studied in a particular time series constructed from an MHD turbulence model (Buchlin et al. 2005), and the main drawbacks were found to be similar as demonstrated here: (i) the loss of small events that produces a cutoff for short time scales, (ii)

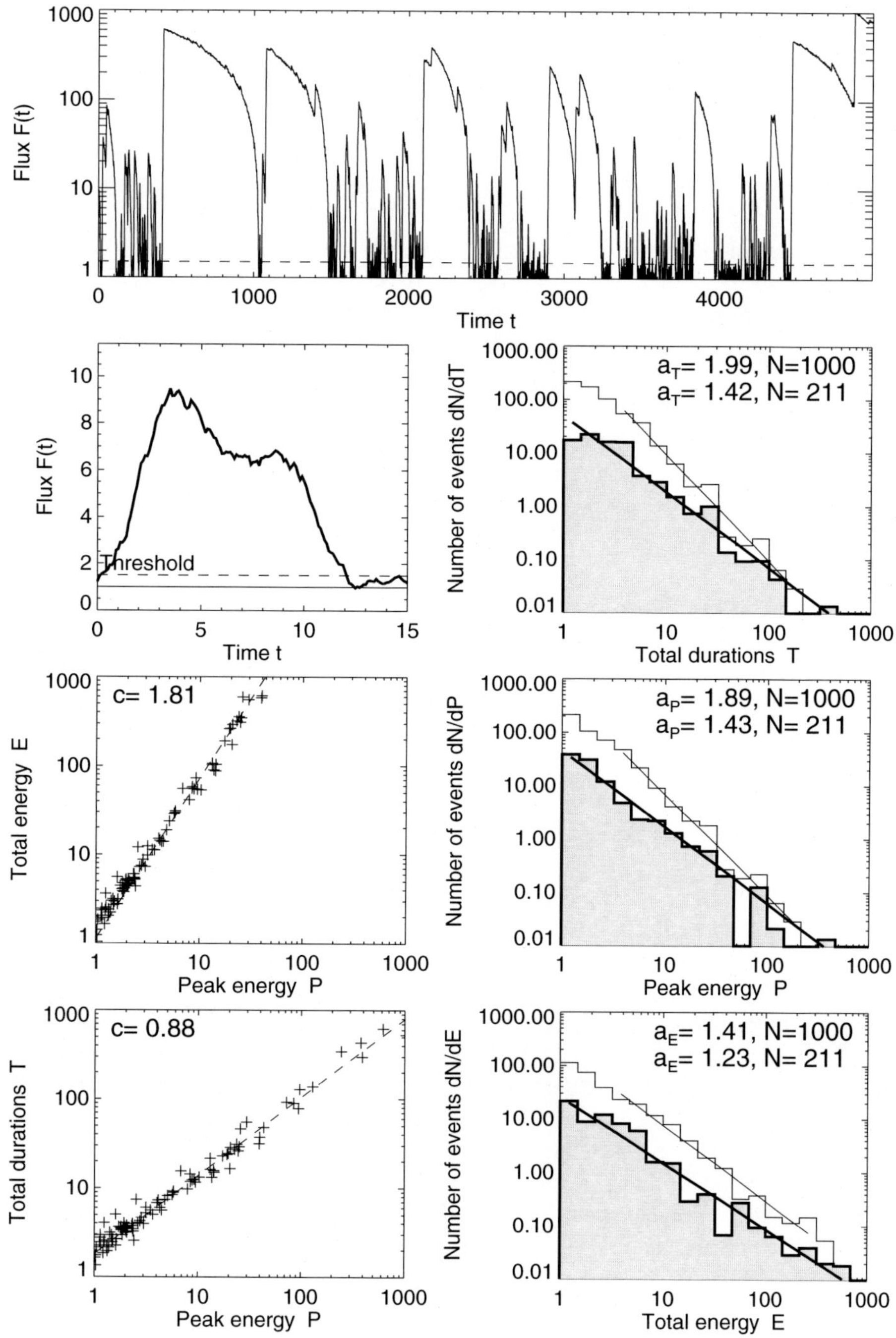

Fig. 6.4 The same simulation of a time series (top panel), correlations (left), and frequency distributions (right) as shown in Fig. 6.3, but a smoothing with a boxcar length of $n_{sm} = 2t_S/dt = 20$ datapoints has been applied.

the inability to separate closely-spaced events, and (iii) the adjustment of thresholds in nonstationary time series (with varying mean rates), which requires Bayesian statistics. Multi-level detection of pulses using the flux levels in the preceding and following time intervals has been applied in the automated detection of gamma ray bursts (Quilligan et al. 2002).

6.3 Highpass-Filtered Event Detection

In order to overcome the non-detection of weaker pulses that occur simultaneously during longer pulses, a fundamental difficulty with threshold-based detection methods (Section 6.2), it is sometimes useful to use a highpass filter, which filters out the slowly-varying time components, so that small pulses on top of a longer pulses can be detected. We demonstrate this method in Fig. 6.5, where we use the same smoothed time series as shown in Fig. 6.4, but subtract a moving-average time profile that is smoothed with a 20 times longer boxcar, $n_{sm2} = n_{sm1} \times 20$. The method of subtracting a smoothed time profile from the original data is also called *unsharp masking*, and is defined as,

$$f_i^{HP} = f^{HP}(t = t_i) = f_i - \sum_{j=i-n_{sm2}/2}^{i+n_{sm2}/2} f(t_j) \ . \tag{6.3.1}$$

Since we already smoothed the original data with a smoothing boxcar n_{sm1}, which represents a lowpass filter, we have actually a *bipass filter*,

$$f_i^{BF} = f^{BF}(t = t_i) = \sum_{j=i-n_{sm1}/2}^{i+n_{sm1}/2} f(t_j) - \sum_{j=i-n_{sm2}/2}^{i+n_{sm2}/2} f(t_j) \ , \tag{6.3.2}$$

where the lowpass filter constant has to be longer than the highpass filter constant, i.e., $n_{sm2} > n_{sm1}$. A bipass filter is essentially sensitive to time structures in the time range of approximately $n_{sm1} \times dt < T < n_{sm2} \times dt$. In our example shown in Fig. 6.5 we expect a lower cutoff of $T_1 = n_{sm1} * dt/2 = 1.1$ and an upper cutoff of $T_2 = n_{sm2} * dt/2 = 20$. As we can see from the number of detected time structures, the highpass filter method yields a more complete sample, i.e., $N = 939$ for a threshold of $F_{th} = 0.01\ W_0$, compared with $N = 172$ in the noise-free data (Fig. 6.2). The resulting frequency distributions of the bipass-filtered structures are remarkably robust in retrieving the powerlaw slopes of fluxes ($\alpha_P = 1.74$ versus $\alpha_P = 1.89$) and total energies ($\alpha_E = 1.27$ versus $\alpha_P = 1.41$), and durations ($\alpha_T = 1.87$ versus $\alpha_P = 1.99$), but impose an upper cutoff at the filter time scale ($T \lesssim n_{sm2}\ dt = 40$) (dashed line in Fig. 6.5, second row right). Thus, except for the filter cutoff of time scales, this method yields robust powerlaw slopes and has a detection efficiency of $\approx 94\%$ for the example analyzed here. However, the parameter correlations are significantly distorted, i.e., $E \propto P^{1.44}$ (instead of P^2), and $T \propto P^{0.84}$ (instead of P^1) (Fig. 6.5, left). Apparently, the distortion of the parameter correlations cancel out to a large extent in the frequency distributions.

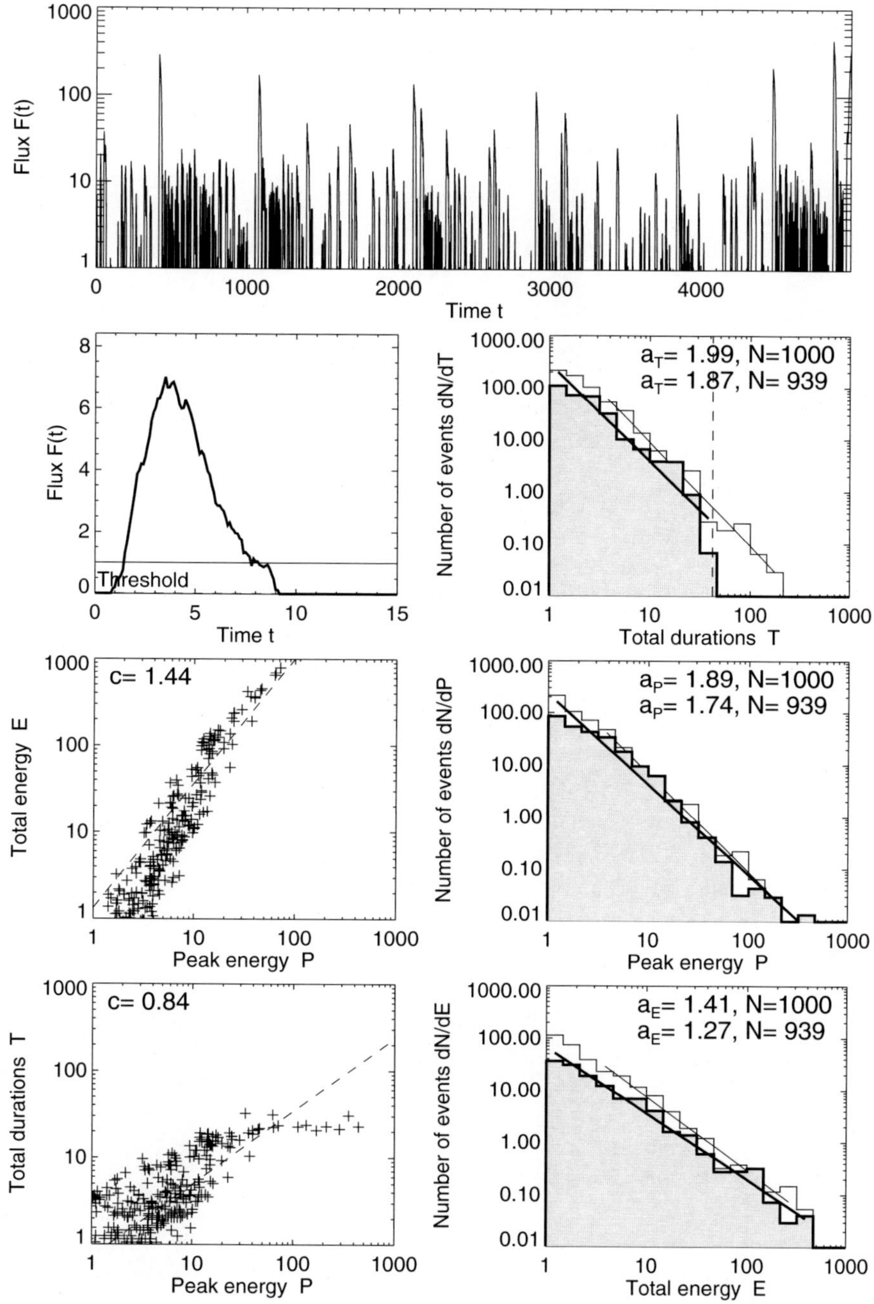

Fig. 6.5 The same simulation of a time series (top panel), correlations (left), and frequency distributions as shown in Fig. 6.3, except for application of a lowpass filter (smoothing) with a boxcar of $n_{sm1} = 2t_S/dt = 20$ datapoints and a highpass filter with a boxcar introduces a cutoff at $T \lesssim 40$ in the time duration histogram (marked with a dashed line in second row right).

Taking advantage of these properties, we could conceive an improved method by combining multiple bipass filters with adjacent but not overlapping time scale ranges, in order to obtain a complete sampling in each filter. Extending this method into the continuum limit we arrive at the so-called *multi-scale* methods, which include the wavelet method (Section 6.7) or principal component analysis (Section 6.8).

6.4 Peak-Based Event Detection

A method of pulse detection that is independent of flux thresholds is the detection of pulse peaks, which of course strongly depends on the pulse shape. In the case of noise-free data (Fig. 6.1), the detection efficiency could approach 100%, because every pulse has a single peak that can be separated temporally, except for near-simultaneous pulses within $\lesssim 2dt$, where dt is the time resolution of the data. In noisy data, however (Fig. 6.3), every pulse has multiple peaks, which even persist in smoothed (Fig. 6.4) and bipass-filtered data (Fig. 6.5). Effects of the event definition by peak times (Fig. 5.4, left) on the frequency distribution of pulse durations are studied in Buchlin et al. (2005) for a particular dataset of MHD turbulence. The problem is mostly to discriminate between peaks of significant pulses and noise peaks. As the enlarged pulse in Figs. 6.3 to 6.5 show, a single pulse can have a multitude of noise peaks. Generally, a decomposition of a multi-peak structure with n local peaks into a denoised structure is ambiguous, because there are $n(n-1)/2$ possible combinations to form subgroups, which can have a significant combined flux about the local background. Thus, a denoising method has first to be applied to the time series, such as smoothing (Fig. 6.4) or a Fourier lowpass filter (Fig. 6.6), before unique peaks can be attributed to individual pulse structures. A peak-based event detection is the less problematic the better the signal-to-noise ratio of the data is.

Once the peaks of significant structures have been identified with a unique peak time, we still have to find the start and end times in order to obtain the total duration. The start time can usually be found by tracking the next significant local minimum (valley) before the peak time. The end time can be estimated the same way by the next local minimum after the peak time, but since our pulses naturally have a longer decay time than rise time, there may be multiple peaks that occur during the decay time. In order to bypass those secondary structures, one would have to require the following minimum to be as low in flux as the flux minimum at the start time.

6.5 Fourier-Filtered Event Detection

A common method of denoising a time series is the application of a Fourier lowpass filter, which we demonstrate in Fig. 6.6. Strictly speaking, this is a special method of the more general category of threshold-based event detection methods (Section 6.2). We use a *Fast Fourier Transform (FFT)* to produce a power spectrum, apply a cutoff to frequencies, $\nu < \nu_{cutoff} = 1/t_{filter}$, or time scales of $T \gtrsim t_{filter}$, and apply the inverse Fourier transform to return a smoothed time profile, so it is a three-step process,

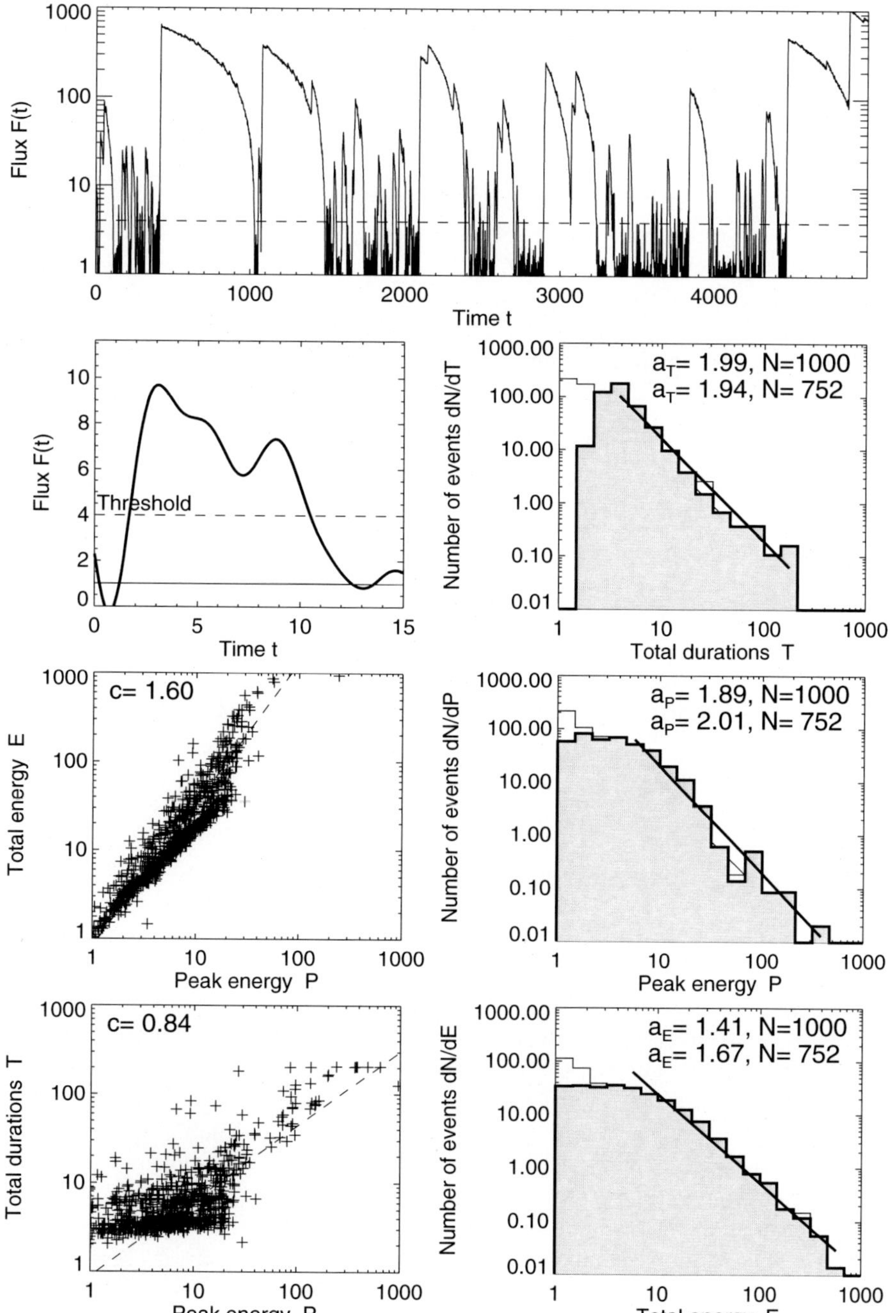

Fig. 6.6 The same simulation of a time series (top panel), correlations (left), and frequency distributions (right) as shown in Fig. 6.1, except for application of a Fourier lowpass filter with a filter passband for time scales $t \le t_{filter} = 3$ and threshold $F_{th} = 4$.

$$
\begin{aligned}
f(t) &\mapsto P(\nu) = FFT[f(t)] \ , \\
P(\nu) &\mapsto P(\nu \geq \nu_{cutoff}) \ , \\
P(\nu \geq \nu_{cutoff}) &\mapsto f_{lowpass}(t) = FFT^{-1}[P(\nu \geq \nu_{cutoff})] \ .
\end{aligned}
\tag{6.5.1}
$$

Once we obtain the Fourier lowpass-filtered time profile, either a threshold-based (Section 6.2) or a peak-based (Section 6.4) method can be applied to produce statistics of peak fluxes and time durations.

In Fig. 6.6 we demonstrate a peak-based detection method applied to a denoised time profile using a Fourier filter with a cutoff of $t_{filter} = 3$. In addition we require a threshold of $F_{th} = 4$. The detection of a structure requires a local peak above this threshold. The start time is found at the next local minimum before the peak, while the end time is found at the next local minimum after the maximum that has a lower flux than at the start time. With this detection scheme we detect 752 (75%) structures in the example shown in Fig. 6.6. As the enlarged time profile in Fig. 6.6 shows, there are two local peaks during the fist time segment of $t = [0, 15]$, which can be compared with the input of four structures that appear clustered (Fig. 6.1). Thus, the Fourier filter has some capability to discriminate near-simultaneous structures, but the number of discriminated structures depends on the filter cutoff. The resulting frequency distributions are relatively robust for this cutoff filter and flux threshold, i.e., for the duration $\alpha_T = 1.94$ versus $\alpha_T = 1.99$, for the peak energy $\alpha_P = 2.01$ versus $\alpha_P = 1.89$, and the total energy $\alpha_E = 1.67$ versus $\alpha_T = 1.41$ (Fig. 6.6). Thus the application of a Fourier lowpass filter is a quite robust technique in retrieving the correct slope of the frequency distributions, has a high detection efficiency ($\approx 75\%$), but has the disadvantage of missing time structures longer than the time scale cutoff ($T \gtrsim t_{filter}$). The results, of course, depend very much on the Fourier filter cutoff ν_{filter} and flux threshold F_{th}.

6.6 Time Scale Statistics from Power Spectra

A Fourier spectrum of a time series is primarily used to detect hidden periodicities, one famous example in astrophysics being the periodic signals from pulsars. Alternatively, a power spectrum can also serve to characterize the occurrence of time structures, such as the $1/f$ noise spectrum, which corresponds to a Poissonian (random) distribution of time scales (Section 4.6). In Section 4.8.2 we derived the specific function of power spectra $P(\nu)$ for randomly distributed pulses with a fixed or mean duration $\langle T \rangle$ according to the shot noise model, as well as for a powerlaw distribution $N(T)$ of pulse durations (Section 4.8.4). Therefore, we can invert the distribution of time scales $N(T)$ analytically from the power spectra $P(\nu)$ of a time series for those special cases.

We demonstrate the inversion of the time scale distribution $N(T)$ from the power spectrum $P(\nu)$ with the example simulated in Fig. 6.3. In Fig. 6.7 we show the Fourier spectrum of the noisy time series $f(t)$ simulated in Fig. 6.3, computed with a standard FFT. We fit a powerlaw spectrum and obtain a slope of $p = 1.03$ for the spectrum $P(\nu) \propto \nu^{-p}$. The time series was simulated with a growth time $\tau_G = 1$ and a mean saturation time $t_S = 1$, which yields a peak energy distribution with a powerlaw slope of $\alpha_P = (1 + \tau_G/t_S) = 2$. The distribution of total durations has a powerlaw slope of $\alpha_T = \alpha_P = 2$, while the distribution

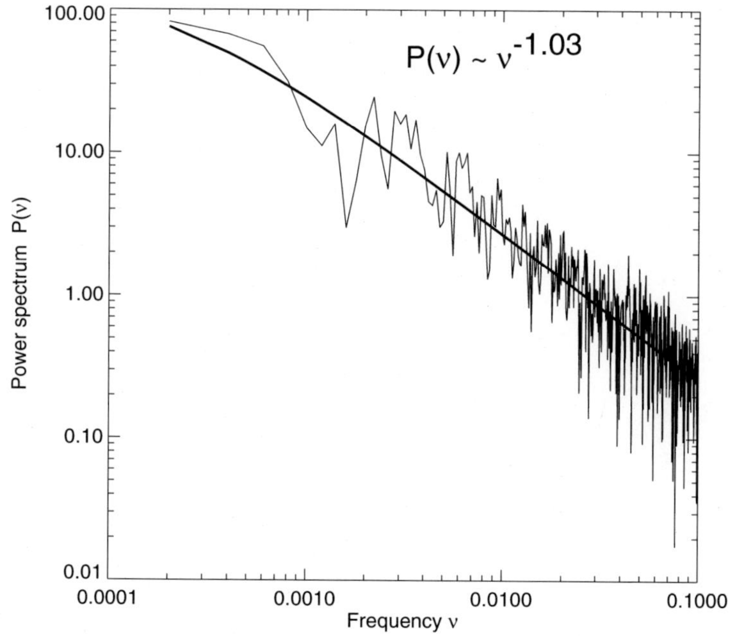

Fig. 6.7 Fourier power spectrum $F(\nu)$ of time series $f(t)$ shown in Fig. 6.3 (top panel). The spectrum is fitted with a powerlaw function, $P(\nu) \propto \nu^{-p}$ with a slope of $p \approx 1.03$.

of energies is $\alpha_E = (\alpha_P + 1)/2 = 1.5$ (Eq. 3.1.28). From the relation Eq. (4.8.23), i.e., $-\alpha_E(1+\gamma) + \gamma = -\alpha_T$ we can constrain the correlation coefficient γ between energies and time durations, $E \propto T^{(1+\gamma)}$,

$$\gamma = \frac{\alpha_T - \alpha_E}{\alpha_E - 1} , \tag{6.6.1}$$

which yields $\gamma = (2-1.5)/(1.5-1) = 1$. The corresponding powerlaw index of the power spectrum, $P(\nu) \propto \nu^{-p}$, is then $p = (2-\alpha_E)(1+\gamma) = (2-1.5)(1+1) = 1.0$ according to Eq. (4.8.27), which agrees with our measurement of $p \approx 1.03$ obtained in Fig. 6.7. The inversion of the time scale distribution requires the knowledge of both the power spectrum with slope p and the correlation coefficient γ, as we can infer from Eqs. (4.8.22) and (4.8.27),

$$\alpha_T = \alpha_E(1+\gamma) - \gamma = 2(1+\gamma) - p - \gamma \tag{6.6.2}$$

so the correlation $E \propto T^{(1+\gamma)}$ between the total energies E and total durations T has to be measured too, at least for the larger pulses.

A power spectrum from a solar light curve observed with the GOES 6 satellite over a total of 32 months during the years 1991–1994 was measured by Ueno et al. (1997), who finds three different spectral components, which can be characterized with the following powerlaw slopes (Fig. 6.8): $p = 1.50 \pm 0.02$ in the frequency range of $\nu \geq 10^{-3.8}$ Hz ($\lesssim$1.8 hrs), $p = 0.95 \pm 0.03$ in the frequency range of $10^{-4.7} < 10^{-3.8}$ Hz (1.8–14 hrs), and $p = 0.45 \pm 0.08$ in the frequency range of $\nu \leq 10^{-4.7}$ Hz ($\gtrsim$14 hrs). If we translate

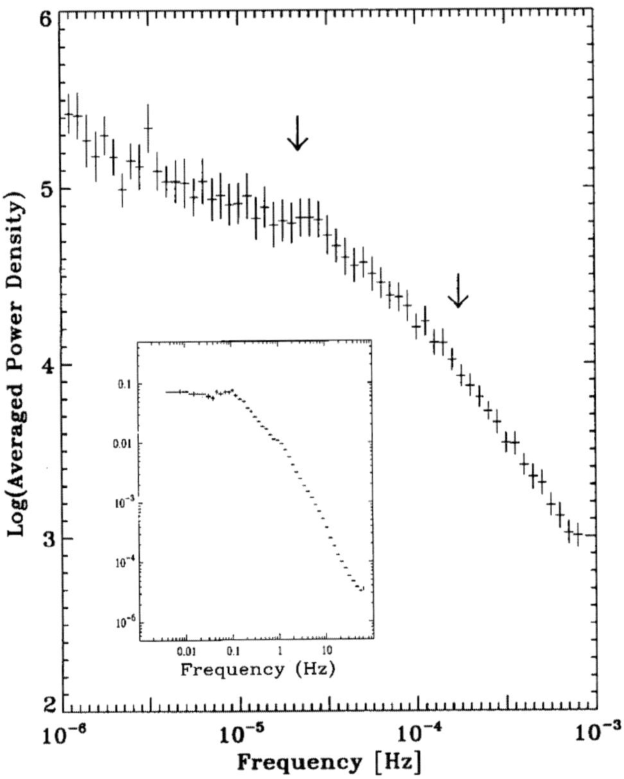

Fig. 6.8 Power spectrum from a solar soft X-ray time profile observed by GOES during 1991-1994, showing three segments with different powerlaw slopes of $p \approx 0.45, 0.95$, and 1.50 (separated by arrows). The insert shows a power spectrum from Cygnus X-1 obtained by Negoro (1992) which has a similar spectrum (Ueno et al. 1997; reproduced by permission of the AAS).

these powerlaw slopes of the power spectrum p into the powerlaw slopes α_T of a time scale distribution, using $\gamma = 1$ as above, we would obtain values of $\alpha_T = 1.55, 2.05$, and 2.55 in order of increasing time scales, which appears to be similar to those found from numerical SOC simulations of cellular automatons by Lu and Hamilton (1991), where a mean slope of $p = 2.17$ was found, being somewhat flatter at smaller time scales and somewhat steeper at longer time scales (Eq. 2.6.15). Interestingly, this three-part power spectrum for the Sun resembles also a similar power spectrum observed from the black hole candidate Cygnus X-1 (Negoro et al. 1992; Ueno et al. 1997); see insert in Fig. 6.8.

Other applications of Fourier power spectra analysis to astrophysical time series address the restoration and enhancement of astronomical data (Brault and White 1971), unevenly spaced data (Scargle 1982, 1989), and uncertainties in the Fourier power spectrum due to noise (Hoyng 1976).

6.7 Wavelet-Based Time Scale Statistics

When analyzing non-periodic time structures, to be expected for randomly occurring SOC events, a Fourier decomposition of a time profile is not a natural tool, since the harmonic modes used in the expansions are themselves periodic, while the time profile is non-periodic. A better approach is to use other *spectral methods*, such as the *Windowed Fourier Transform*, a *wavelet-based method*, or a *multi-resolution method*. A windowed Fourier transform chops up a time series $f(t)$ into a sequence of windows and yields a Fourier spectrum $P(\nu, t_i)$ for every window t_i as a function of time. The wavelet transform can be considered as a generalization of the windowed Fourier transform, which also yields a gliding power spectrum as a function of time, but uses a better adapted functional decomposition of pulses in a time series, using a so-called *mother wavelet* function, rather than the harmonic sinusoidal functions used in the windowed Fourier transform. There is extensive literature on wavelet methods in general (e.g., Mallat 1989; Daubechies 1992; Meyer and Ryan 1993; Kaiser 1994, Chan 1995), which we do not review here.

Wavelet methods in the analysis of astrophysical time series have been introduced by Scargle (1993), and applied, e.g., to study geomagnetic time series (Kovacs et al. 2001; Vieira et al. 2003), solar helioseismology (Fröhlich et al. 1997), solar diameter variations (Vigoroux and Delache 1993), solar cycle variability (Watari 1995, 1996a; Polygiannakis et al. 2003), solar irradiance (Willson and Mordvinov 1999), solar chromospheric oscillations (Bocchialini and Baudin 1995), sunspot oscillations (Jess et al. 2007), solar hard X-ray flares (Aschwanden et al. 1998a; McAteer et al. 2007), solar radio bursts (Schwarz et al. 1998), stellar chromospheric oscillations (Frick et al. 1997), quasi-periodic oscillations in accretion disks (Scargle et al. 1993), or gamma-ray bursts (Young et al. 1995), just to name a few that deal with wavelet-based analysis of time series. In addition, wavelet-based analysis has also been applied in the spatial domain, especially in solar imaging data.

For statistics of SOC parameters, in particular for durations T of pulses, we are interested whether the output of standard wavelet algorithms, the time-dependent Fourier power spectral density $P(\nu, t_i)$, consisting of *scalegrams* $S(T)$ for each time interval $[t_i, t_{i+1}]$,

$$S(T) = \langle |P(\nu[T], t)|^2 \rangle \ , \qquad \text{for } t_i < t < t_{i+1} \tag{6.7.1}$$

where $\nu[T] = 1/t$, can be transformed into a distribution $N(T)$ of time scales T. Such a transformation has been developed in Aschwanden et al. (1998a). The procedure is sketched in Fig. 6.9 and examples are simulated in Fig. 6.10. Essentially, a scalegram $S(T)$ can be considered as a convolution of a distribution function $N(T)$ of time scales T with a kernel function $p(T)$. The distribution $N(T)$ of time scales can then be obtained from the inversion of a scalegram $S(T)$ using the kernel function $p(T)$ that corresponds to a particular *mother wavelet* function. The schematic in Fig. 6.9 shows a rectangular distribution $N(T)$ of time scales and how the convolution of a (double-powerlaw) kernel function $p(T)$ produces the scalegram $S(T)$. The numerical simulations in Fig. 6.10 show artificial time profiles $f(t)$, the wavelet scalegrams $S(T)$, and the inverted time scale distributions $N(T)$ (histograms in Fig. 6.10, right), compared with the theoretical distributions $N(T)$ (delta-functions and Gaussians) that have been used as input in the generation of the

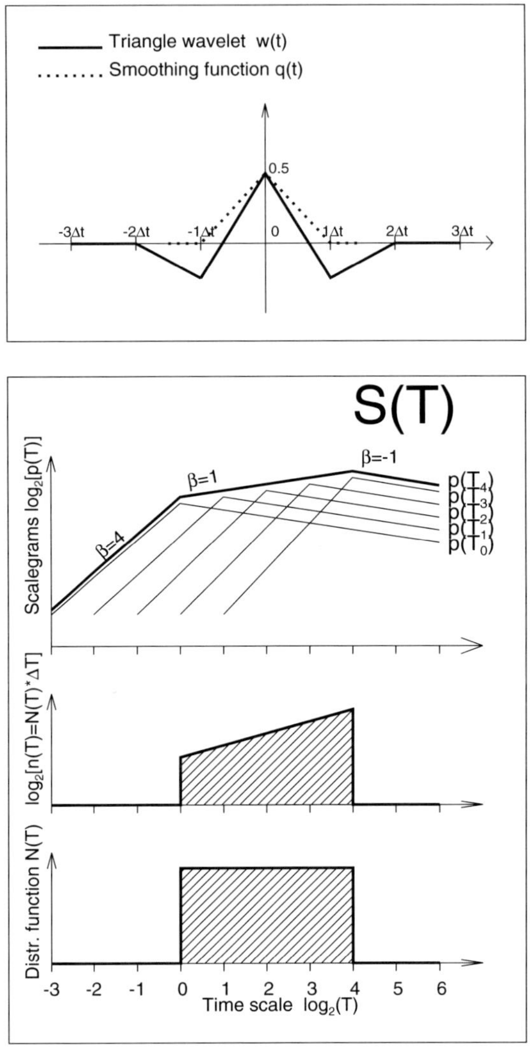

Fig. 6.9 *Top:* Triangle mother wavelet function $w(t)$ (thick line) and smoothing function $q(t)$ (dotted). *Bottom:* Schematic illustration of the convolution of a standard distribution function $N(T)$ (bottom) of time scales with kernel functions $p(T_i)$ that sum up to a scalegram $S(T)$ (Aschwanden et al. 1998a).

time profiles $f(t)$. These examples demonstrate that the wavelet-based inversion method can retrieve the original time scales.

A practical example of a wavelet scalegram of an observed time profile observed in a solar flare is shown in Fig. 6.11, along with the inverted time scale distributions $N(T)$ in four different time intervals. This method has been applied to the time profiles of 46 solar flare events and exponential distributions of time scales were found, in contrast to powerlaw-like distributions expected for SOC models. In the study of McAteer et al. (2007), the

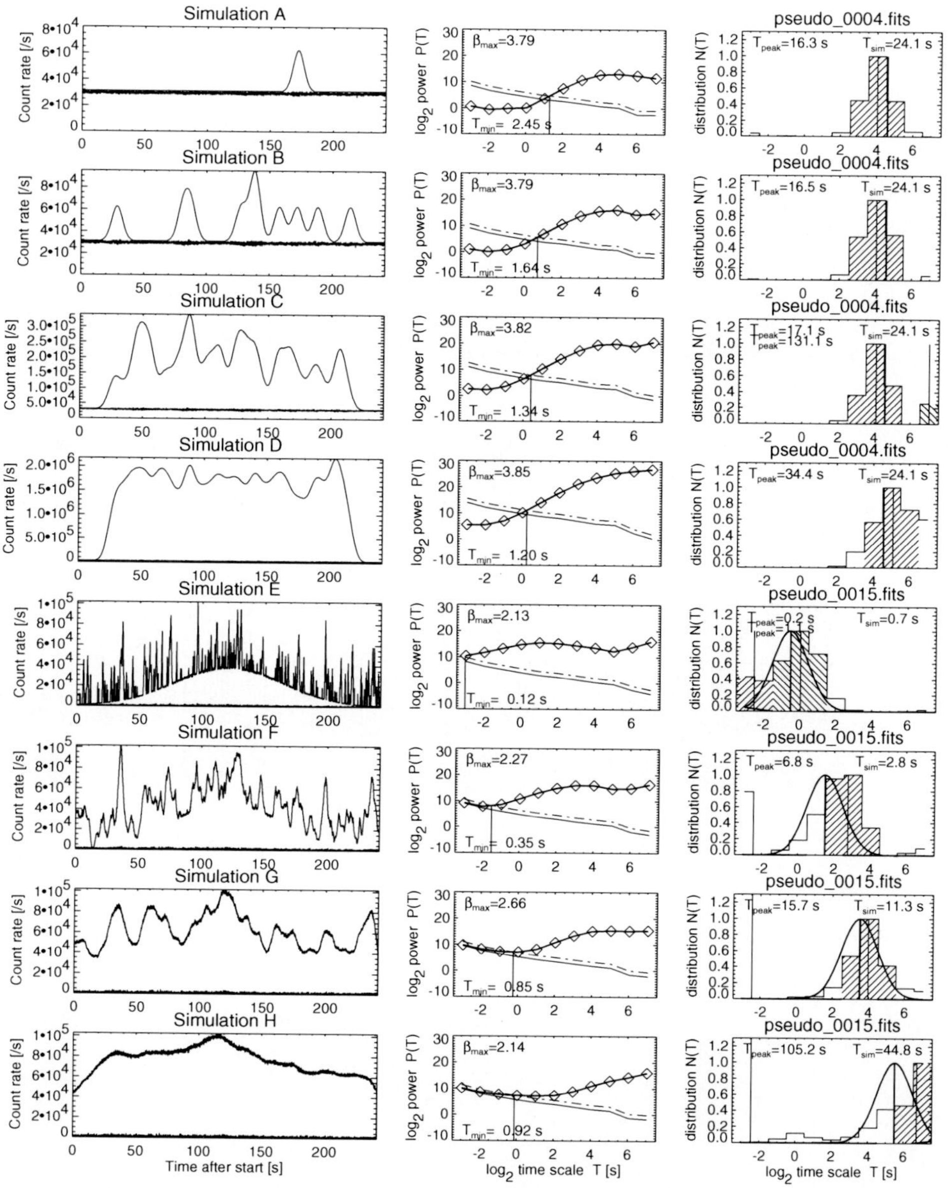

Fig. 6.10 Eight numerical simulations (A–H) of time profiles $f(t)$ (left panels), the computed wavelet scalegrams $S(T)$ (middle panels), and inverted time scale distribution functions $N(T)$ (right panels). The time profiles contain also the noise templates (left panels). The scalegrams (diamonds in middle panels) contain also the noise scalegrams (thin solid line) with the 3σ-limit (dashed line). The slope β_{max} is measured at the steepest part of the scalegrams. The inverted time scale distribution functions (histograms in right panels) are compared with the theoretical distribution functions (thick curve) used in the simulation of $f(t)$, with mean time scales T_{sim}, and are compared with the inverted peak times T_{peak} (weighted over hatched part of histogram) (Aschwanden et al. 1998a).

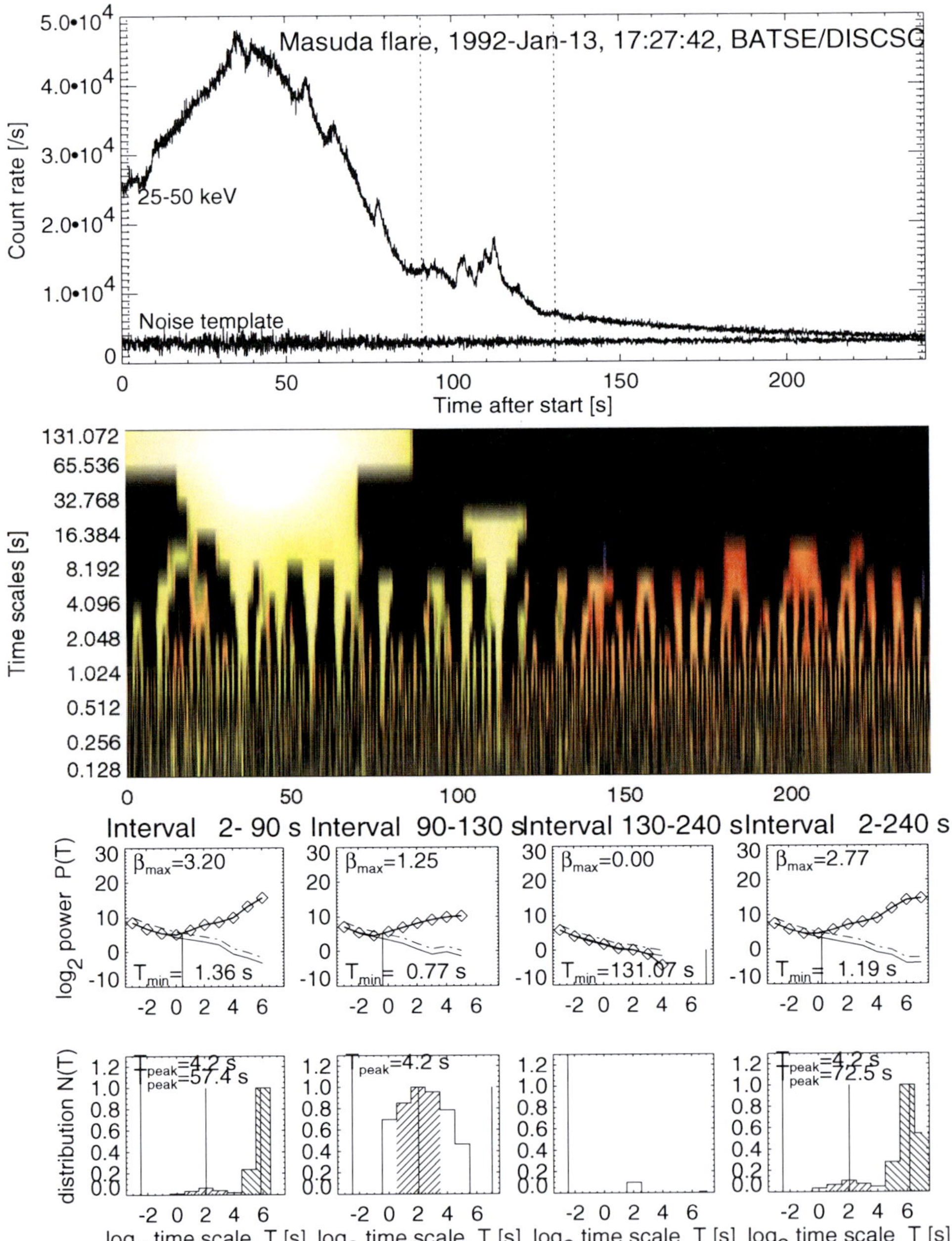

Fig. 6.11 Time profile (top), scalogram $P(T,t)$ (middle panel with grayscale), time-averaged scalegrams $S(T)$ (third row), and inverted time scale distribution functions $N(T)$ (fourth row) for the Masuda flare, 92-Jan-13, 17:27:42 UT, observed with *BATSE/CGRO* (Aschwanden et al. 1998a).

wavelet-based analysis of a solar flare revealed a Hölder exponent that indicates a high degree of memory between subsequent hard X-ray peaks, which is also in contrast to the supposed independent events in a SOC process. Wavelet-based statistics of time scales has not been exploited to the full extent yet, but appears to be a very promising method for obtaining statistics of temporal structures from convolved time series, which contain near-simultaneous events with differing time scales.

6.8 Principal Component Analysis

While a Fourier analysis decomposes a time series into harmonic sinusoidal components, and a wavelet analysis decomposes into stretched and shifted mother wavelet functions (e.g., a Mexican hat function), there is an even better adjusted decomposition method that attempts to find a minimum number of best-fit components, which is called *Principal Component Analysis (PCA)*, *Independent Component Analysis (ICA)*, *Proper Orthogonal Decomposition (POD)*, *Complex Empirical Orthogonal Function (CEOF)* analysis, *Hotelling transform*, or *Karhunen–Loève transform (KLT)*. The mathematical procedure transforms a number of possibly correlated variables into a smaller number of uncorrelated (independent, or orthogonal) variables, called principal components, and involves the calculation of eigenvalues of a data covariance matrix, or a singular value decomposition of a data matrix. The PCA method can be used for automated detection of spatial or temporal features.

In astrophysical time series (or image time series), the PCA method (or a PCA-like extension) has been applied, e.g., to solar EUV data to detect propagating waves (Terradas et al. 2004), to solar cycle synoptic data to characterize the "butterfly diagram" over $\approx$25–50 years (Lawrence et al. 2005; Vecchio et al. 2005a), to solar magnetogram data to identify low-frequency oscillations in photospheric motion (Vecchio et al. 2005b), to interplanetary magnetic field polarity data (Cadavid et al. 2008), or to $\approx$17,000 light curves of variable stars such as RR Lyrae's, Cepheids, and Mira variables (Deb and Singh 2009).

An example of a PCA analysis is shown in Fig. 6.12 for a time series of solar EUV data (Terradas et al. 2004). A decomposition into oscillatory components with independent periods is attempted, called *empirical mode decomposition (EMD)*, which obtains from the analyzed signal six different components with different periods and amplitudes. The sum of the six decomposed components represent the original data within the data noise. The EMD method produces a decomposition into frequency band-limited components by using information from the signal itself instead of prescribing basis functions with fixed frequency, such as in Fourier or wavelet methods. The decomposition is not unique, but attempts to represent a time series with a minimum number of time scale ranges. Another example is shown in Fig. 6.13 for a time series from a variable star (Deb and Singh 2009). The full quasi-periodic time series was decomposed with both Fourier and PCA decomposition techniques. Fig. 6.13 shows the reconstruction of a fundamental mode (FU) Cepheid light curve using the first 1, 3, 7, and 10 principal components. It was found that 10 principal components contain nearly 90% of the variance in the data.

These examples demonstrate how an arbitrary time profile can be decomposed into a relatively small number of noise-free time time profiles $f_T(t)$, each one having a characteristic time scale T within a prescribed bandwidth Δt. From these time profiles $f_T(t)$,

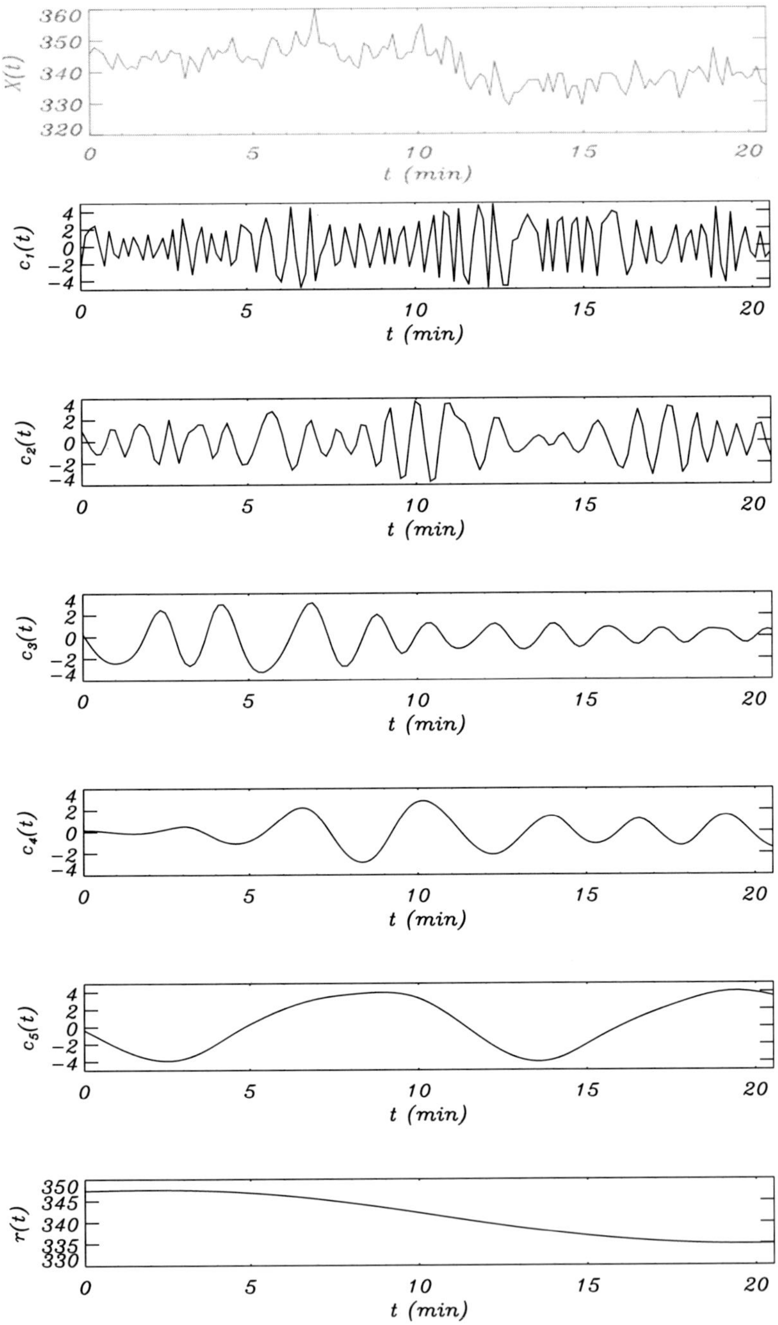

Fig. 6.12 A time series of EUV flux observed in a loop in the solar corona (top), decomposed into six principal components with increasing time scales, according to an empirical mode decomposition (EMD) method. Each component is not strictly periodic (as a Fourier mode), but has its own characteristic time scale within a small tolerance range, $T \pm \Delta T$ (Terradas et al. 2004; reproduced by permission of the AAS).

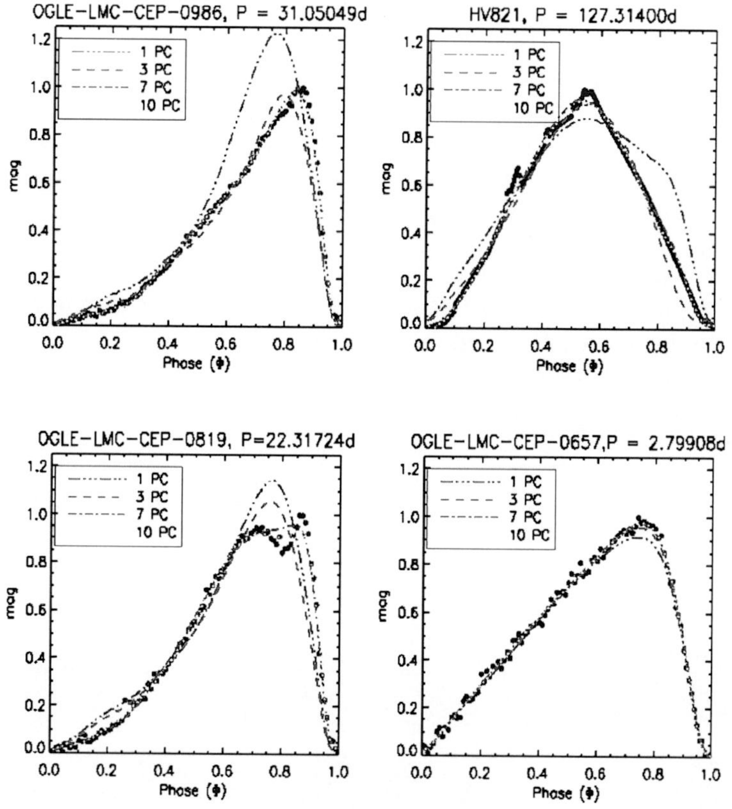

Fig. 6.13 Reconstruction of four FU Cepheid light curves using the first 1, 3, 7, and 10 principal components (Deb and Singh 2009).

the peak amplitudes P_T, and total time-integrated fluxes E_T of pulse structures with a time scale T can be sampled, using either a threshold-based or peak-based event detection method. This would allow us, after proper normalization, to obtain the frequency distributions of time scales $N(T)$, peak energies $N(P)$, and total energies $N(E)$. Therefore, the PCA method appears to be a useful and efficient method for SOC statistics, probably better adapting to unknown pulse shapes than Fourier-based and wavelet-based methods.

6.9 Image-Based Event Detection

In the previous sections we discussed SOC event statistics obtained from one-dimensional (1-D) time series data $f(t)$, as we usually obtain from astrophysical observations, but statistics of SOC events has also been inferred from time sequences of images $f(x,y;t)$, such as from magnetospheric or solar imaging data. Although the automated processing of three-dimensional (3-D) data $f(x,y;t)$ is more complex, the chief advantage for SOC statistics is the spatial separation of near-simultaneous events, which can conveniently be discriminated in the space domain, while they coincide in the time domain. This is particularly important for SOC statistics because spatial correlations can introduce time clustering and deviation from Poissonian random statistics (e.g., aftershocks of earthquakes, or sympathetic solar flares), in contrast to spatially independent events that are expected to obey a true random behavior of waiting times. Event or feature detection in imaging data became a growing industry and for general introductions into digital image processing we refer to the textbooks of Gonzales and Woods (2008), Jain (1989), Castleman (1996), Jähne (2005), Woods (2006) and Mallat (2008), and more specifically for astrophysical data see Starck et al. (1998) and Starck and Murtagh (2002), or for image processing techniques and feature recognition in solar physics see Aschwanden (2009). Here we will discuss only a few examples that are most relevant for SOC statistics.

A threshold-based detection method of temporal features essentially involves a criterion $f(x,y;t) \geq F_{th}$ in 3-D space, but additionally requires the automated detection of spatially coherent and contiguous features. A common procedure is to perform an image segmentation in an image $f(x,y;t=t_i)$ that detects spatially coherent shapes above a prescribed threshold, with subsequent identification of co-spatial structures in the preceding or following images $f(x,y;t=t_{i-1},t_{n+1})$ that have at least one pixel above the prescribed threshold in common. An example is given in Fig. 6.14, which shows a result of detected

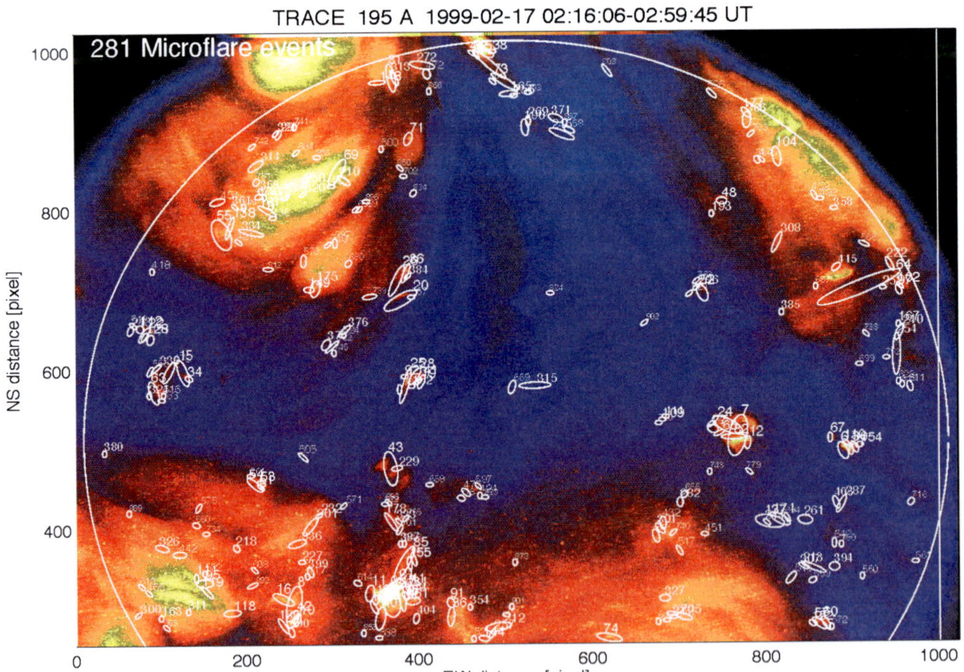

Fig. 6.14 This synthesized TRACE 195 Å image is summed from 22 images recorded during 1999-Feb-17 02:16:06–02:59:45 UT. The circle encompasses the analyzed field-of-view with a diameter of ≈ 8 arcmin. The numbered ellipses mark 281 flare-like events that fulfill the flare definition criterion, out of a total of 901 EUV brightening events. The geometric size and orientation of the ellipses is on scale, encompassing the simultaneously-varying pixels of a flare event (Aschwanden et al. 2000b).

solar nanoflares in a solar EUV data cube (Aschwanden et al. 2000a,b). Since SOC phenomena are dynamic events, the threshold criterion for an event detection is in this case not simply a flux threshold, but rather a variability threshold Δf_{th}, which can be defined in terms of a flux change that exceeds the level of random fluctuations,

$$f(x,y;t_{i+1}) - f(x,y;t_i) \geq \Delta f_{th} = 3\sigma_f \ , \tag{6.9.1}$$

where σ_f is the standard deviation of the photon Poisson noise in a time bin corresponding to the exposure time of the image. Examples of such variability maps are shown in Fig. 6.15, where it can be seen that those pixels with high fluxes (indicated with flux contours in Fig. 6.15) are not necessarily identical with those of significant flux variabilities from one to the next time frame. To ensure a proper tracking of a coherent event in time, only pixels that exhibit a co-spatial variability in the previous and/or subsequent time frame are considered as part of the same coherent event, or SOC avalanche (marked with diamonds in Fig. 6.15), while other pixels with significant variability occurring in one single time frame only are considered as event-unrelated (instrumental or unresolved) brightness fluctuations. The time evolution of such automatically traced features observed in two different wavelengths (Fig. 6.16) exhibits the typical fast rise and exponential decay of a solar flare event. The multi-wavelength coverage of these events moreover ensures the self-consistent physical evolution of an elementary solar flare process, which consists of a rapid impulsive heating phase with subsequent plasma cooling by thermal conduction and radiative cooling. It exhibits the typical exponential decay, which appears delayed in the wavelength with the cooler temperature. Proper definition of events are extremely important in image-based feature detection methods, because multi-dimensional data are more prone to erroneous event detections of unrelated other variabilities contained in the data than 1-D time series.

The numerical event detection code used for the examples shown in Fig. 6.14–6.16 was especially designed to detect solar microflares and nanoflares, which represent the faintest counterparts of solar flares, and thus are important to extend the dynamic range of frequency distributions of flare energies over nine orders of magnitude. Similar codes were also developed by Krucker and Benz (1998) and Parnell and Jupp (2000), which spurred controversial results on the powerlaw slopes in the nanoflare regime. A number of issues were considered that contribute to the initially discrepant results of powerlaw slopes, such as event definition, selection, and discrimination, sample completeness, observing cadence and exposure times, pattern recognition algorithms, threshold criteria, instrumental noise, wavelength coverage, fractal geometry, but also physical modeling issues of energy, temperature, electron density, line-of-sight integration, and fractal volume (e.g., Aschwanden and Parnell 2002; Benz and Krucker 2002). The issue of the correct powerlaw slope of the frequency distribution of nanoflare energies was further aggravated by the fact that the initially discrepant results scattered on both sides of the critical value (with a slope of $\alpha_E = 2$) that decides whether the energy of nanoflares is more important for coronal heating (if $\alpha_E > 2$). We will come back to this issue when we discuss physical energy models of SOC events in Section 9.3.

A similar task is the automated detection of solar *bright points*, which are small, bipolar magnetic fields in the photosphere and can be detected best in EUV. These events

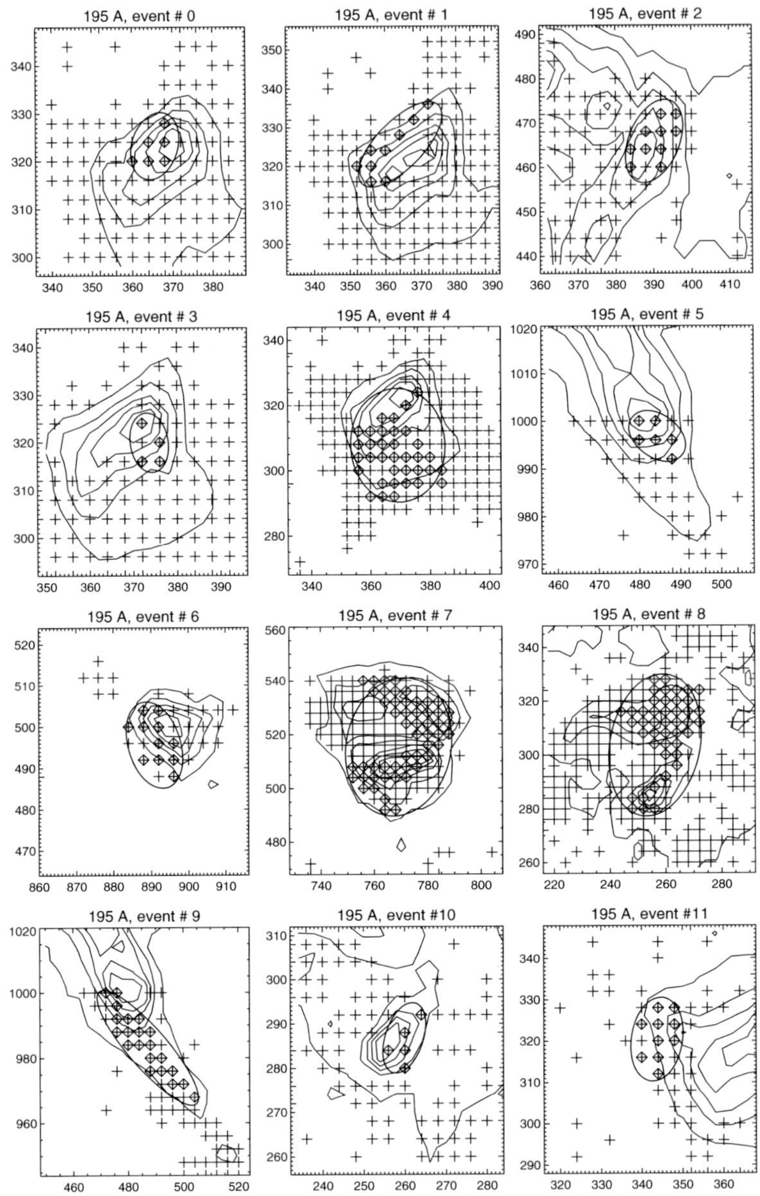

Fig. 6.15 The spatial clustering of the pattern recognition code is illustrated for the 12 largest events on 99-Feb-17, 02:15–03:00 UT. The contours outline local EUV intensity maps around the detected structures. The crosses mark the positions of macropixels with significant variability ($N_\sigma > 3$). The spatio-temporal pattern algorithm starts at the pixel with the largest variability, which is located at the center of each field of view, and clusters nearest neighbors if they fulfill the time coincidence criterion ($t_{peak} \pm 1\Delta t$). These macropixels that fulfill the time coincidence criterion define an event, marked with diamonds, and encircled with an ellipse. Each macropixel that is part of an event, is excluded in subsequent events. Note that events 0,1,3,11 belong to the same active region, where the four near-cospatial zones have peaks at different times and thus make up four different events (Aschwanden et al. 2000a).

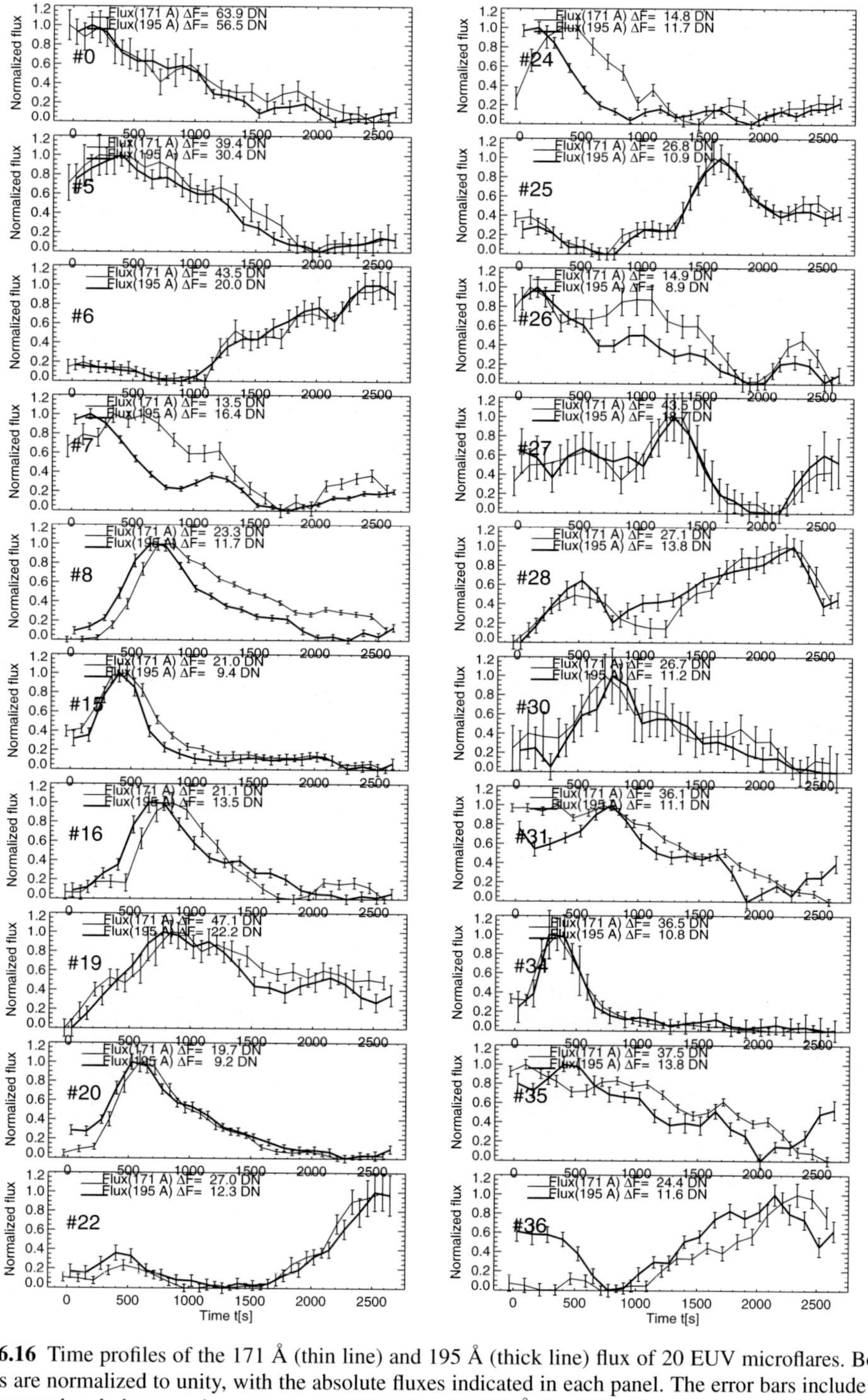

Fig. 6.16 Time profiles of the 171 Å (thin line) and 195 Å (thick line) flux of 20 EUV microflares. Both fluxes are normalized to unity, with the absolute fluxes indicated in each panel. The error bars include all instrumental and photon noise components. Note that the 171 Å flux is highly correlated with the 195 Å flux, but generally delayed, as expected for a plasma cooling process (Aschwanden et al. 2000b).

have been detected with a highpass-filter method with proper noise threshold estimates from a long-time data series over 9 years to the extent of an unheard number of 1.3×10^8 events (McIntosh and Gurman 2005). Such large statistics is extremely useful for SOC statistics, regarding the accurate functional form of the frequency distributions and their time-dependent changes.

Automated detection of *coronal mass ejection (CME)* events in coronagraph images represent a specially challenging task because of their highly transient nature and complex and inhomogeneous spatial morphology. A CME rapidly expands in 3-D space, the observed brightness becomes quickly diluted, and the morphology evolves from an initial fan pattern to a turbulent spherical shape, possibly containing multiple shock fronts with accelerating and decelerating speeds. Thus, typical SOC parameters like a peak flux P, total flux E, and duration T are difficult to define for such dramatically changing morphological structures. Even waiting times Δt of CMEs (Fig. 5.13) are problematic to measure, because multiple CMEs interfere with each other in an observed field-of-view. Frequency distributions of CMEs have only been sampled for their (angular) sizes, which were found to exhibit invariant powerlaw slopes during a solar cycle (Robbrecht et al. 2009). Thus, typical CME observables entail an angular width, an apparent latitude, and apparent velocities, which are not straightforward to translate into SOC parameters. These parameters would correspond to a geometric aspect ratio, location, and velocity of sandpile avalanches. Nevertheless, automated detection algorithms for CME events have been developed by using a threshold-segmentation technique of radial off-limb images (Olmedo et al. 2008), a wavelet-based multi-scale edge detection technique (Young and Gallagher 2008), or a Hough transform with a morphological opening operator (Robbrecht and Berghmans 2004). An example of a CME detection by Young and Gallagher (2008) is shown in Fig. 6.17. The optical brightness of CMEs is usually so weak that they can only be detected in running time-difference or in polarized brightness images. CME-related phenomena are so-called *EIT waves*, which according to one model propagate concentrically to the CME over the solar surface and can be traced by means of flux threshold-based detection of spherically propagating ring patterns (Podladchikova and Berghmans 2005).

Future automated detection algorithms of spatio-temporal patterns are expected to involve more artificial-intelligence algorithms or neural-network-learning techniques, which can adjust to the unknown or unquantified morphological shapes progressively with the increasing number of detected events.

6.10 Summary

Feature and event detection methods represent the input for SOC statistics and thus it is extremely important to simulate and understand their statistical biases on the resulting frequency distributions of SOC events. We simulated a time series that is particularly designed for typical SOC events, characterized by powerlaw distributions of amplitudes and durations, as well as by Poisson statistics of waiting times (Section 6.1). Armed with such test data, we simulate the detection biases for threshold-based event detection (Section 6.2), for both noise-free data and data affected by heavy photon noise. We test how smoothing of a time series, which is one option to suppress the data noise, affects the

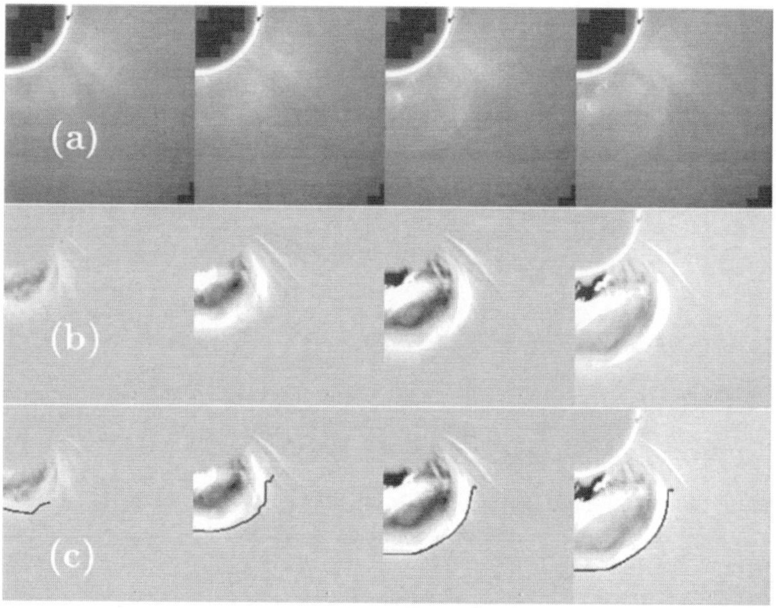

Fig. 6.17 Illustration of a CME edge detection in subsequent images. (a) The original LASCO C2 images, (b) running difference images of the LASCO C2 images, and (c) application of the multiscale edge detection algorithm to the sequence of the original images. The edges are the black lines displayed over the running difference images. The CME erupted on 18 April 2000, the times for the frames are (from left to right) 16:06 UT, 16:30 UT, 16:54 UT, and 17:06 UT (Young and Gallagher 2008).

resulting frequency distribution. To overcome the main disadvantage of threshold-based event detection, namely the loss of weak events in the presence of large events, we test a highpass-filtered and bipass-filtered detection method (Section 6.3). Other alternatives are peak-based detection methods (Section 6.4) and Fourier-filtered time series (Section 6.5). We demonstrate that Fourier power spectra $P(\nu)$ (Section 6.6) or wavelet-based methods (Section 6.7) can be used to retrieve the frequency distribution of time scales $N(T)$. A related method is the principal component analysis (Section 6.8), which has not been much used for SOC statistics yet. Finally, given the availability of imaging data in magnetospheric and solar physics, image-based spatio-temporal detection methods are appropriate for SOC statistics, which have the chief advantage of spatial discrimination of cotemporaneous events. We illustrate such spatio-temporal feature recognition techniques for the detection of solar nanoflares and coronal mass ejections. The simulation and testing of any automated event detection technique cannot be taken seriously enough, because systematic errors and biases occur most dramatically for the weakest events, which populate the largest fraction of the logarithmic scale range covered in SOC statistics.

6.11 Problems

Problem 6.1: Simulate the time series described in Section 6.1 for longer mean waiting time (e.g., $\Delta t_0 = 10, 50, 100$) and find the scaling how the number of time-overlapping events reduces with increasing waiting time.

Problem 6.2: Using the test time series simulated in Problem 6.1, develop a simple peak-based event detection algorithm and test whether you can retrieve the frequency distribution of the input parameters for longer waiting times.

Problem 6.3: With a threshold-based event detection algorithm determine how the number of detected events scales with the threshold.

Problem 6.4: Calculate the powerlaw slopes of the frequency distribution of times for the three spectral segments shown in the power spectrum of Cygnus X-1 (Fig. 6.8), using Eq. (6.6.2) and assuming correlations of $E \propto T$ and $E \propto T^2$.

Problem 6.5: Discuss the pro's and con's of Fourier-based, Windowed Fourier transform, wavelet-based, and principal component analysis for periodic, quasi-periodic, and non-periodic time series.

Problem 6.6: Discuss and simulate two different strategies for an image-based event detection method: (1) Detect temporal structures in $n_x \times n_y$ time series $f(t)$ for every image pixel first and then identify co-spatial patterns; or (2) detect spatial structures in each image plane $f(x,y)$ first and then track co-spatial structures in time.

7. Occurrence Frequency Distributions

Probability is expectation founded upon partial knowledge. A perfect acquaintance with all the circumstances affecting the occurrence of an event would change expectation into certainty, and leave neither room nor demand for a theory of probabilities.

George Boole (1815–1864)

Nevertheless, as is a frequency occurrence in science, a general hypothesis was constructed from a specific instances of a phenomenon.

Sidney Altman (born 1939)

It is customary in the statistics of nonlinear processes to histogram the logarithmic number of events versus a logarithmic size scale, which is called a *log N − log S diagram*, *size distribution*, *occurrence frequency distribution*, or simply *frequency distribution*. In such log-log representations, the difference between (i) a Poissonian random process, which can be characterized by an exponential distribution function that drops off sharply above an e-folding size scale, and (ii) nonlinear processes governed by self-organized criticality, which ideally produce a scale-free powerlaw distribution function, appears most striking. Frequency distributions thus have become the arbiters of SOC versus non-SOC processes, starting from the famous magnitude diagram of earthquakes discovered by Beno Gutenberg and Charles Francis Richter in 1954 (i.e., the Gutenberg–Richer law). Frequency distributions of SOC phenomena obtained from astrophysical data were first identified in solar flare data by Ed Lu and Russell Hamilton (1991), based on log-log histograms published earlier without an interpretation in terms of SOC (e.g., Dennis 1985). Frequency distributions can be plotted for any conceivable parameter, preferably a (model-independent) observable, which does not require an arbitrary choice of a physical model. For earthquake statistics, the most commonly used parameter is the magnitude, measured by well-calibrated seismometers. In astrophysical data, where the observable is typically a time series of flux intensity in some given wavelength range, obvious parameters used for frequency distributions are the peak flux P, the total (time-integrated) flux or fluence E, and the total duration T of an event. While such observables can be unambiguously measured from well-calibrated detectors, the ultimate desire is to obtain frequency distributions of physical parameters, such as the thermal energy E_{th}, the nonthermal energy E_{nth}, the magnetic energy E_B, the kinetic energy E_{kin}, or the potential energy E_{pot}, which of course are all model-dependent, to be discussed in Chapter 9. In the following section we review the occurrence frequency distributions of (preferably) observables measured from astrophysical events that are (hypothetically) associated with SOC processes and we will compare the observations with analytical SOC models (Chapter 3) in order to evaluate their consistency with SOC theory.

7.1 Basics of Frequency Distribution Functions

The data input for an occurrence frequency distribution is usually a list or a catalog of events, characterized by some size parameter x_i for $i = 1, ..., n$ events, regardless whether the list was generated by visual inspection or by an automated computer algorithm (Chapter 6). How do we construct a log-log histogram from an event catalog? There are essentially two ways, either a logarithmically binned histogram if large statistics is available, or a rank-order plot if the size of the statistical sample is rather small.

7.1.1 Differential Frequency Distributions

If we have large statistics (at least $n \gtrsim 10^2, ..., 10^3$), we can first establish a logarithmic binning axis bound between the minimum and maximum value, $x_{min} \le x_i \le x_{max}$, with n_j bins,

$$x_j^{bin} = x_{min} \left(\frac{x_{max}}{x_{min}} \right)^{(j-1)/(n_j-1)} , \qquad j = 1, ..., n_j \tag{7.1.1}$$

which is equidistant on a logarithmic scale, but has variable intervals on a linear scale,

$$\Delta x_j^{bin} = x_{j+1}^{bin} - x_j^{bin} = x_j^{bin} \left[\left(\frac{x_{max}}{x_{min}} \right)^{1/(n_j-1)} - 1 \right] . \tag{7.1.2}$$

In a next step we can count the number of events N_j^{bin} that fall in each bin with interval $x_j^{bin} \le x_i \le x_{j+1}^{bin}$. We have to be aware that this number N_j^{bin} depends on the particular bin size Δx_j^{bin} we have chosen. In order to obtain the functional form of the frequency distribution, which should be independent of the binning, we have to divide the number of events by the bin size,

$$N_j = N(x_j) = \frac{N_j^{bin}}{\Delta x_j^{bin}} , \tag{7.1.3}$$

and can plot the frequency distribution with N_j on the y-axis versus the size x_j on the x-axis. This representation normalizes the distribution to the total number of events n, which we should obtain by integration over the x-axis, or summing over all (non-equidistant) bins on the x-axis,

$$\int_0^{\infty} N(x)\,dx = \sum_{j=0}^{n_j} N_j\,dx_j = \sum_{j=0}^{n_j} N(x_j)\Delta x_j^{bin} = \sum_{j=0}^{n_j} N_j^{bin} = n\,. \tag{7.1.4}$$

Instead of expressing the number of occurrences by the actually observed numbers N_j, it is also customary to use a *probability distribution function* $P(x)$, which is simply the number of events in each bin normalized by the total number of events n,

$$P(x) = \frac{N(x)}{n} \tag{7.1.5}$$

which has the total integral normalized to unity,

$$\int_{x_{min}}^{x_{max}} P(x)\,dx = \int_{x_{min}}^{x_{max}} \frac{N(x)}{n}\,dx = 1\,. \tag{7.1.6}$$

Both representations, $N(x)$ or $P(x)$, are called a *differential frequency distribution*, because they express the number of events in a "differential" bin dx.

7.1.2 Cumulative Frequency Distributions

An integrated *differential frequency distribution* $N(x)dx$ is called a *cumulative frequency distribution* $N^{cum}(>x)$, which expresses in each bin the sum of all events that are larger than the size parameter of the bin x,

$$N^{cum}(>x) = \int_x^{x_{max}} N(x)dx\,, \tag{7.1.7}$$

which we denote by $N^{cum}(>x)$, in contrast to the differential distribution $N(x)$. The cumulative frequency distribution contains more statistics in the rarer bins at larger sizes, and thus appears smoother at the upper end than differential distribution functions. However, the values in each bin are statistically not independent, but always contain information from all other bins on the right-hand side. The particular functional shape at the upper cutoff can dominate the entire distribution function.

If the differential frequency distribution is a powerlaw function with slope α, the cumulative frequency distribution is expected to have a flatter powerlaw slope by one,

$$\begin{aligned} N(x) &\propto x^{-\alpha} \\ N^{cum}(>x) &\propto x^{-\beta} \lesssim x^{-(\alpha-1)} \end{aligned}\,. \tag{7.1.8}$$

For instance, both the differential and cumulative frequency distributions for earthquakes are shown in the same plot (Fig. 1.7), with slopes of $\alpha = 2$ and $\beta = 1$. However, the powerlaw relationship $\beta = \alpha - 1$ is only true when the differential frequency distribution extends to infinite, which is never the case. In reality, there is always a largest event at x_{max},

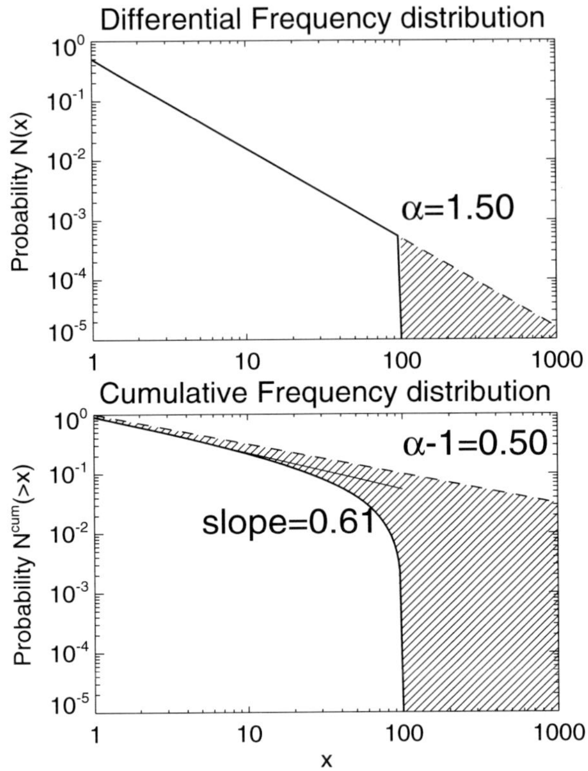

Fig. 7.1 A differential frequency distribution function $N(x)$ is shown with a powerlaw slope of $\alpha = 1.5$ and cutoff at $x_{max} = 100$ (top). The corresponding cumulative frequency distribution function $N^{cum}(>x)$ is fitted in the lower half (logarithmic) range $x = [1, ..., 10]$, which gives a powerlaw slope of $\beta = 0.61$, which is steeper than expected for an ideal powerlaw distribution without upper cutoff x_{max}, i.e., $\beta = \alpha - 1 = 0.5$.

which causes a sharp cutoff in the differential frequency distribution $N(x)$, but a gradual steepening in the cumulative frequency distribution, because of the missing contributions from $x > x_{max}$. This detail is quite important, because it leads to a significant over-estimate of the powerlaw slope when the rule $\alpha = \beta + 1$ is applied. We demonstrate this in the following. We define a powerlaw distribution function with a sharp cutoff at x_{max},

$$N(x) = (\alpha - 1)x^{-\alpha} , \quad x \le x_{max} . \tag{7.1.9}$$

The cumulative frequency distribution function can then be calculated by integrating over the range from x to x_{max},

$$N^{cum}(>x) = n \frac{\int_x^{x_{max}} N(x')\, dx'}{\int_{x_{min}}^{x_{max}} N(x')\, dx'} = n \frac{\int_x^{x_{max}} x'^{-\alpha}\, dx'}{\int_{x_{min}}^{x_{max}} x'^{-\alpha}\, dx'} = n \frac{(x^{1-\alpha} - x_{max}^{1-\alpha})}{(x_{min}^{1-\alpha} - x_{max}^{1-\alpha})} . \tag{7.1.10}$$

We see that the second term in the integral, $x_{max}^{1-\alpha}$ steepens the slope and lets the cumulative distribution drop to zero when x approaches x_{max}. We plot the two distribution functions in Fig. 7.1 for $\alpha = 1.5$. The powerlaw slope of the cumulative distribution function is expected to be $\beta = \alpha - 1 = 1.5 - 1 = 0.5$ at the lower end $x \ll x_{max}$, but becomes systematically steeper near the upper cutoff. If we were to fit a powerlaw over the powerlaw-like range of x, say in the range $x_{min} \le x \le x_{max}/10$, we would measure a slope of $\beta = 0.61$ (Fig. 7.1). This steepening effect on the slope due to the presence of an upper cutoff does not occur in the differential distribution, so it is important to take this effect into consideration when dealing with cumulative frequency distribution functions, e.g., see Fig. 1.7 for earthquakes or Fig. 1.15 for stellar flares.

How can this upper cutoff effect be taken properly into account? The best way is to fit the exact analytical function of the cumulative frequency distribution function, which is $N^{cum}(>x) \propto (x^{1-\alpha} - x_{max}^{1-\alpha})$ (Eq. 7.1.10), rather than the powerlaw approximation $N^{cum}(>x) \propto x^{1-\alpha}$. Alternatively, if the original data are not available, but only a powerlaw fit to the cumulative distribution is known (e.g., from literature),

$$N^{cum}(>x) = n\left(\frac{x}{x_{min}}\right)^{-\beta} , \qquad (7.1.11)$$

we can calculate the relationship between the cumulative powerlaw slope β and the differential powerlaw slope α. A practical way is to assume that the cumulative powerlaw slope β gives a good fit of the parameter x in the lower half (logarithmic) range $[x_{min}, x_{max}]$ (see Fig. 7.1), which we define with the fractions $[q_1, q_2]$ with $q_1 = x_{min}/x_{max}$ and $q_2 = q_1^{1/2}$. For instance, for $x_{min} = 1$ and $x_{max} = 100$ (Fig. 7.1), the lower logarithmic half has the fractions $q_1 = 0.01$ and $q_2 = 0.1$. From the cumulative powerlaw fit we have the following occurrence ratio between these two points (using $q_1 = q_2^2$),

$$\frac{N^{cum}(>x_2)}{N^{cum}(>x_1)} = \left(\frac{x_2}{x_1}\right)^{-\beta} = \left(\frac{q_2}{q_1}\right)^{-\beta} = q_2^{\beta} . \qquad (7.1.12)$$

On the other side, from Eq. (7.1.10) we have, using $q_1 = q_2^2$ and applying $(x^2 - 1) = (x-1)(x+1)$ for $x = q_2^{1-\alpha}$,

$$\frac{N^{cum}(>x_2)}{N^{cum}(>x_1)} = \left(\frac{q_2^{1-\alpha} - 1}{q_1^{1-\alpha} - 1}\right) = \left(\frac{q_2^{1-\alpha} - 1}{q_2^{2(1-\alpha)} - 1}\right) = \frac{1}{q_2^{(1-\alpha)} + 1} . \qquad (7.1.13)$$

Setting these two expressions (Eqs. 7.1.12 and 7.1.13) equal, we obtain the following relationship for α as a function of β,

$$\alpha = 1 - \frac{\log[q_2^{-\beta} - 1]}{\log(q_2)} . \qquad (7.1.14)$$

or vice versa, the relationship for β as a function of α,

$$\beta = -\frac{\log [q_2^{(1-\alpha)} + 1]}{\log(q_2)} . \quad (7.1.15)$$

We see that both expressions yield the approximation $\alpha \approx 1+\beta$, if $q_2^{-\beta} \gg 1$ or $q_2^{1-\alpha} \gg 1$, which comes down to the condition of large logarithmic ranges, i.e. $x_{max} \gg x_{min}$. For the case shown in Fig. 7.1 with $q_2 = 0.1$ and $\alpha = 1.5$, we obtain $\beta = \log(1+0.1^{-0.5}) = 0.62$, which is significantly different from the approximation $\beta \approx \alpha - 1 = 0.5$. The difference is even larger for smaller logarithmic ranges, say for one decade ($x_{min}/x_{max} = 0.1$), as it is the case for small samples, such as statistics of stellar flares (Fig. 1.15).

7.1.3 Rank-Order Plots

If the statistical sample is rather small, in the sense that no reasonable binning of a histogram can be done, either because we do not have multiple events per bin or because the number of bins is too small to represent a distribution function, we can create a rank-order plot. A rank-order plot is essentially an optimum adjustment to minimum statistics that gives every single event a single bin. From an event list of a parameter $x_i, i = 1,...,n$, which is generally not sorted, we have first to generate a rank-ordered list by ordering the events according to increasing size,

$$x_1 \le x_2 \le ... \le x_j \le ... \le x_n , \quad j = 1,...,n . \quad (7.1.16)$$

The bins are generally not equidistant, neither on a linear nor logarithmic scale, defined by the difference between subsequent values of the ordered x_j,

$$\Delta x_j^{bin} = x_{j+1}^{bin} - x_j^{bin} . \quad (7.1.17)$$

In a rank-ordered sequence of n events, the probability for the largest value is $1/n$, for events that are larger than the second-largest event it is $2/n$, and so forth, while events larger than the smallest event occur in this event list with a probability of unity. Thus, the cumulative frequency distribution is simply the reversed rank order,

$$N^{cum}(>x_j) = (n+1-j) , \qquad j = 1,...,n , \quad (7.1.18)$$

and the distribution varies from $N^{cum}(>x_1) = n$ for $j = 1$ to $N^{cum}(>x_n) = 1$ for $j = n$. We can plot a cumulative frequency distribution with $N^{cum}(>x_j)$ on the y-axis versus the size x_j on the x-axis. The distribution is normalized to the number of events n,

$$\int_{x_1}^{x_n} N(x)\, dx = N^{cum}(>x_1) = n . \quad (7.1.19)$$

The differential frequency distribution function $N(x)$ could in principle be computed from the derivative of the cumulative distribution, but there is usually considerable noise between subsequent events in a rank order, so that smoothing is recommended before differentiation.

We show two examples of rank-ordered plots in Fig. 7.2. The first example is based on a differential frequency distribution of time scales that correspond to an exponential function with time scale $\tau = 1$,

$$N(t) = \frac{1}{\tau} \exp\left(-\frac{t}{\tau}\right) , \tag{7.1.20}$$

Using a random generator we are producing $n = 100$ values of time scales $t_i, i = 1, ..., n$ that correspond to this differential distribution according to the method described in the following Section 7.1.4, which we plot in a rank-ordered diagram as shown in Fig. 7.2 (left; diamonds). In order to prove that this rank-order plot corresponds to the cumulative distribution, we calculate the distribution analytically by integrating Eq. (7.1.20),

$$N^{cum}(> t) = \int_t^{\infty} N(t')\, dt' = \int_t^{\infty} \frac{1}{\tau} \exp\left(-\frac{t'}{\tau}\right) dt' = \exp\left(-\frac{t}{\tau}\right) . \tag{7.1.21}$$

which agrees (Fig. 7.2 left, thick solid curve) with the rank-ordered values.

The second example is based on a differential frequency distribution of energies that have a powerlaw function with slope of $\alpha = 1.5$,

$$N(E) = (\alpha - 1) E^{-\alpha} . \tag{7.1.22}$$

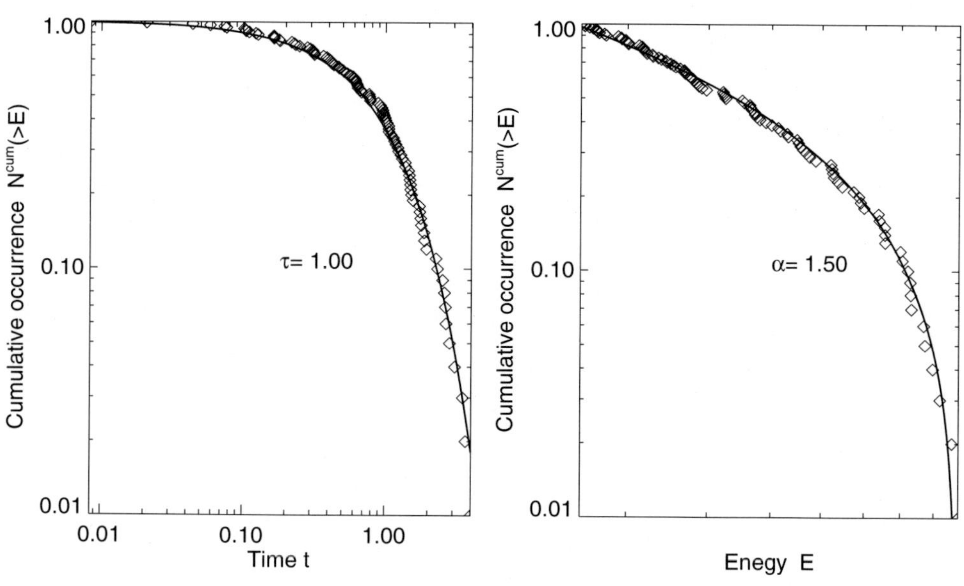

Fig. 7.2 Numerically generated rank-order plots (diamonds) and theoretical cumulative frequency distribution functions (thick solid curves) for an exponential function with $\tau = 1$ (left) and a powerlaw function with $\alpha = 1.5$ (right).

Again, using a random generator we are producing $n = 100$ values of energy $E_i, i = 1, ..., n$ that correspond to this differential distribution according to the method described in the following Section 7.1.4, which we plot in a rank-order diagram as shown in Fig. 7.2 (right; diamonds). In order to prove that this rank-order plot corresponds to the cumulative distribution, we calculate the distribution analytically by integrating Eq. (7.1.22),

$$N^{cum}(>E) = \int_E^{E_{max}} (\alpha - 1)\varepsilon^{-\alpha} \, d\varepsilon = E^{(1-\alpha)} - E_{max}^{1-\alpha} \, . \tag{7.1.23}$$

which agrees (Fig. 7.2 right, thick solid curve) with the rank-ordered values.

An observational example of cumulative frequency distributions based on a rank-order plot is shown in Fig. 1.15 for stellar flares, where the statistics literally does not include more than about a dozen events per star (Audard et al. 2000).

Sometimes it is also convenient to plot the size versus the rank, such as the ranking of cities by population size shown in Fig. 1.4. This is essentially the rank-order plot defined in Eq. (7.1.18), but with exchanged x- and y-axis. The rank order on the x-axis is the independent variable N_j, while the y-axis is the dependent variable $x_j = x(N_j)$. Since the axes are exchanged, a powerlaw function would have approximately the reciprocal value for the slope,

$$\begin{aligned} N(x) &\propto x^{-\alpha} \\ x(N) &\approx N^{-1/\alpha} \end{aligned} \tag{7.1.24}$$

This type of rank-order plot with size versus rank was originally used by Zipf (1949) for statistics of word usage (Section 1.3), and thus is also called *Zipf plot*.

7.1.4 Numerical Generation of Frequency Distributions

For numerical simulations of frequency distributions, for instance Monte-Carlo simulations of SOC models, we need to create randomly distributed values x_i that have a particular prescribed function of their frequency distribution, such as an exponential function for waiting times, or a powerlaw function for energies. Let us prescribe the form of the frequency distribution with a probability function $p(x)$ in the interval $[x, x+dx]$, which has the normalization,

$$\int_0^\infty p(x) \, dx = 1 \, . \tag{7.1.25}$$

The total probability $\rho(x)$ to have a value in the range of $[0, x]$ is then the integral,

$$\rho(x) = \int_0^x p(x') \, dx' \, . \tag{7.1.26}$$

If the analytical function $\rho(x)$ can be inverted, say with the analytical inverse function ρ^{-1}, so that

$$x = \rho^{-1}(\rho) = \rho^{-1}(\rho[x]) \, , \tag{7.1.27}$$

we have a transformation that allows us to generate values x_i from a distribution of probability values ρ_i. There are many numerical random generator algorithms available that produce a random number ρ_i in a unity range of $[0,1]$, which can then be used to generate values x_i with the mapping transform $x_i = \rho^{-1}(\rho_i)$. The frequency distribution of these values x_i will then fulfill the prescribed function $p(x)$.

As an example we demonstrate the numerical generation of a sample of time scales t that has a frequency distribution function following an exponential function,

$$p(t) = \frac{1}{\tau} \exp\left(-\frac{t}{\tau}\right) , \tag{7.1.28}$$

which fulfills the normalization $\int_0^1 p(t)\,dt = 1$. The total probability $\rho(t)$ to have a value in the range $[0,t]$ is then the integral function of $p(t)$,

$$\rho(t) = \int_0^t p(t')\,dt' = \int_0^t \frac{1}{\tau} \exp\left(-\frac{t'}{\tau}\right) dt' = \left[1 - \exp\left(-\frac{t}{\tau}\right)\right] . \tag{7.1.29}$$

The inverse function $t(\rho)$ of $\rho(t)$ is

$$t(\rho) = -\tau \ln(1-\rho) . \tag{7.1.30}$$

In Fig. 7.3 (left) we use a random generator that produces 10,000 values ρ_i, uniformly distributed in the range of $[0,1]$, and use the transform Eq. (7.1.30) to generate values

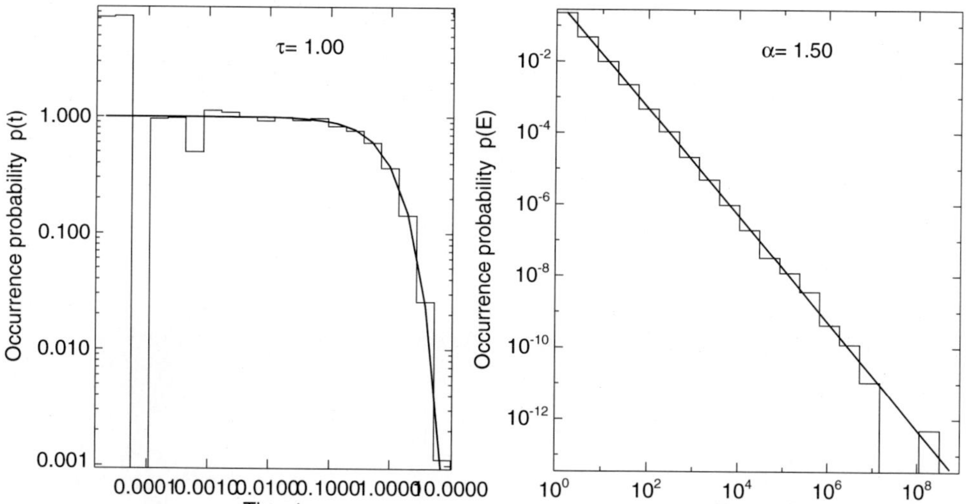

Fig. 7.3 *Left:* An exponential frequency distribution is numerically generated, based on $n = 10{,}000$ uniformly distributed values ρ_i in the range $[0,...,1]$ and times $t_i = -\tau \ln(1-\rho_i)$ (Eq. 7.1.30) (histogram), leading to the occurrence probability function $p(t)$ as defined in Eq. (7.1.28) (thick curve). *Right:* A powerlaw frequency distribution is numerically generated, based on $n = 10{,}000$ values ρ_i uniformly distributed in the range $[0,...,1]$, with energies $E_i = (1+\rho_i)^{1/(1-\alpha)}$ with $\alpha = 1.5$ (Eq. 7.1.33) (histogram), leading to the occurrence probability function $p(E)$ as defined in Eq. (7.1.31) (thick curve).

$t_i = -\tau \ln(1-\rho_i)$ with $\tau = 1$ and sample the frequency distribution of the 10,000 values t_i, which follows the prescribed exponential function $p(t)$ defined in Eq. (7.1.28).

As a second example we prescribe a powerlaw function $p(E)$ for the frequency distribution,

$$p(E) = (\alpha - 1)E^{-\alpha} \ , \tag{7.1.31}$$

which fulfills the normalization $\int_1^\infty p(E)\ dE = 1$. The total probability $\rho(E)$ in the range $[0, E]$ is then the integral function of $p(E)$ (Eq. 7.1.31),

$$\rho(E) = \int_0^E p(\varepsilon)\ d\varepsilon = \int_0^E (\alpha - 1)\varepsilon^{-\alpha}\ d\varepsilon = \left[1 - E^{(1-\alpha)}\right] \ . \tag{7.1.32}$$

The inverse function $E(\rho)$ of $\rho(E)$ (Eq. 7.1.32) is

$$E(\rho) = [1-\rho]^{1/(1-\alpha)} \ . \tag{7.1.33}$$

In Fig. 7.3 (right) we use a random generator that produces 10,000 values ρ_i uniformly distributed in the range of $[0, 1]$, choose a powerlaw index of $\alpha = 1.5$, and use the transform Eq. (7.1.33) to generate values $E_i = [1-\rho_i]^{-2}$ and sample the frequency distribution of the 10,000 values E_i, which follows the prescribed powerlaw function $p(E) = 0.5E^{-1.5}$ as defined in Eq. (7.1.31).

7.1.5 Integrals of Powerlaw Distributions

For normalization purposes or when the total number n of events needs to be evaluated from a powerlaw distribution $N(x) = (\alpha - 1)x^{-\alpha}$, we have to integrate over the valid range bound by $x_{min} < x < x_{max}$,

$$n = \int_{x_{min}}^{x_{max}} N(x)\ dx = \int_{x_{min}}^{x_{max}} (\alpha - 1)x^{-\alpha}\ dx = x_{min}^{1-\alpha} - x_{max}^{1-\alpha} \ , \tag{7.1.34}$$

which is defined for $\alpha \neq 1$. Generally both boundaries contribute significantly to the total number, unless the powerlaw distribution extends over a very large range, say more than three orders of magnitude. For such large ranges, the following approximations can be used,

$$n = \begin{cases} \approx x_{min}^{1-\alpha} \text{ for } (x_{max} \gg x_{min}) \text{ and } (\alpha > 1) \\ \approx x_{max}^{1-\alpha} \text{ for } (x_{max} \gg x_{min}) \text{ and } (\alpha < 1) \end{cases} \ . \tag{7.1.35}$$

The total integral (or first moment) of a powerlaw distribution function, for instance the total energy of an occurrence frequency distribution of energies, can be obtained by convolving the variable x with the powerlaw distribution $N(x) = (\alpha - 1)x^{-\alpha}$ over the valid range $x_{min} \leq x \leq x_{max}$,

$$x_{tot} = \int_{x_{min}}^{x_{max}} x\,N(x)\ dx = \int_{x_{min}}^{x_{max}} (\alpha - 1)x^{1-\alpha}\ dx = \left(\frac{\alpha - 1}{2-\alpha}\right) \left[x_{max}^{2-\alpha} - x_{min}^{2-\alpha}\right] \ , \tag{7.1.36}$$

which is only defined for $\alpha \neq 2$. Again, generally both boundaries contribute significantly to the total number, unless the powerlaw distribution extends over a very large range, in which case the following approximations can be used,

$$x_{tot} \approx \left(\frac{\alpha-1}{2-\alpha}\right) \begin{cases} x_{min}^{2-\alpha} \text{ for } (x_{max} \gg x_{min}) \text{ and } (\alpha > 2) \\ x_{max}^{2-\alpha} \text{ for } (x_{max} \gg x_{min}) \text{ and } (\alpha < 2) \end{cases} . \tag{7.1.37}$$

The critical value is $\alpha = 2$, which decides whether the integral diverges at the lower bound (if $\alpha > 2$) or upper bound (if $\alpha < 2$). A far-reaching application of this integral is the total energy contained in the distribution of solar or stellar flares, which is also responsible for heating of the solar or stellar corona, and could be dominated by nanoflares if $\alpha > 2$ applies over a large energy range, as pointed out by Hudson (1991).

7.1.6 Powerlaw Scaling Laws and Correlations

We consider the case where two parameters x and y are correlated with each other by a powerlaw function with the power coefficient β,

$$y \propto x^{\beta} , \tag{7.1.38}$$

where x and y could be the peak energy P, the total energy E, duration T, or any other SOC parameter. Since every SOC parameter has a powerlaw-like distribution function in our SOC standard model (Section 3.1),

$$\begin{aligned} N(x) &\propto x^{-\alpha_x} \\ N(y) &\propto y^{-\alpha_y} \end{aligned} , \tag{7.1.39}$$

it is useful to calculate the relationship between the power indices α_x, α_y, and β. The general way to substitute a variable $y(x)$ in a frequency distribution $N(x)$ is,

$$N(y)\, dy = N[x(y)] \left| \frac{dx(y)}{dy} \right| dy , \tag{7.1.40}$$

which yields, for the function $y(x) \propto x^{\beta}$ defined in Eq. (7.1.38), using the inverse function $x(y) \propto y^{1/\beta}$ and derivative $dx/dy \propto y^{(1/\beta-1)}$,

$$N(y)\, dy = y^{-\alpha_x/\beta+1/\beta-1}\, dy = y^{-\alpha_y}\, dy , \tag{7.1.41}$$

leading to the following relationship between the power indices,

$$\beta = \frac{(\alpha_x - 1)}{(\alpha_y - 1)} . \tag{7.1.42}$$

This is a useful relationship to compute (or verify) the power index β of two correlated parameters from their frequency distributions. For instance in Fig. 6.1, we have for $x = P$ and $y = E$ the frequency distributions with a powerlaw slope of $\alpha_x \approx 2.0$ and $\alpha_y \approx 1.5$, which predicts $\beta = (\alpha_x - 1)/(\alpha_y - 1) = (2.0 - 1)/(1.5 - 1) = 1/0.5 = 2$, which indeed corresponds to the power index $\beta \approx 2.0$ in the scatterplot of $E \propto P^2$ (Fig. 6.1, third row middle).

7.1.7 Accuracy of Powerlaw Fits

There are a number of effects that determine the accuracy of derived powerlaw indices in frequency distributions, $N(x) \propto x^{-\alpha}$, and underlying correlations, $y(x) \propto x^{\beta}$, such as formal errors of linear regression fits, the choice of dependent and independent variables, the statistical uncertainty of the number of events, a small logarithmic range, statistical weighting, histogram binning, cutoffs, truncations, deviations from powerlaws, sensitivity limits, etc. In addition, the sampling of events is also affected by data pre-processing, such as dead-time corrections, spike removals, or background subtractions. When error bars are given in literature, they usually refer to formal errors of a least-squares fit, but do not include systematic errors that result from numerous biases of the data set or model assumptions. The best way to obtain a realistic error is often to use a Monte-Carlo simulation of the data and measurement procedure based on a realistic model of the data. In order to give a typical assessment of various errors in our study of occurrence frequency distributions and parameter correlations we conduct some Monte-Carlo simulations of our standard SOC model (Section 3.1).

Following the numerical simulation procedure outlined in Section 7.1.4 we generate two sets of $n = 100$ variables, one for peak energies $P_i, i = 1, ..., n$ and one for total energies $E_i, i = 1, ..., n$, which both have powerlaw distribution functions, with powerlaw slopes of $\alpha_P = 2$ and $\alpha_E = 1.5$,

$$\begin{aligned} N(P) &= (\alpha_P - 1)P^{-\alpha_P} \\ N(E) &= (\alpha_E - 1)P^{-\alpha_E} \end{aligned} \quad . \tag{7.1.43}$$

They can be generated using two sets of random numbers ρ_i and ρ_j uniformly distributed in the range $[0, ..., 1]$, produced with a numerical random generator and the transform (Eq. 7.1.33),

$$\begin{aligned} P_i &= [1 - \rho_i]^{1/(1-\alpha_P)} \\ E_j &= [1 - \rho_j]^{1/(1-\alpha_E)} \end{aligned} \tag{7.1.44}$$

In order to ensure a parameter correlation we sort each set in increasing number, but add some random noise in both parameters. We mimic also an instrumental sensitivity limit by applying a flux threshold of $P \geq 0.5$, which causes a truncation error in P.

We plot the two sets of variables in form of a scatterplot E_j versus P_i in Fig. 7.4 (top left), which show a linear regression fit of $E \propto P^{1.78}$ instead of the theoretically expected relationship $E \propto P^2$. A linear regression fit with inverted axis yields $P \propto E^{0.52}$, which corresponds to $E \propto P^{(1/0.52)} = P^{1.93}$, so part of the difference results from the choice of the independent variable. There are other linear regression fits that treat both variables equally, such as the bisector method or the minimization of the orthogonal distance to the linear regression fit, which eliminate this bias. However, there is still an additional bias introduced by the flux threshold, which affects a truncation of data for P but not for E. This could be corrected by using a truncation limit that is orthogonal to the linear regression fit.

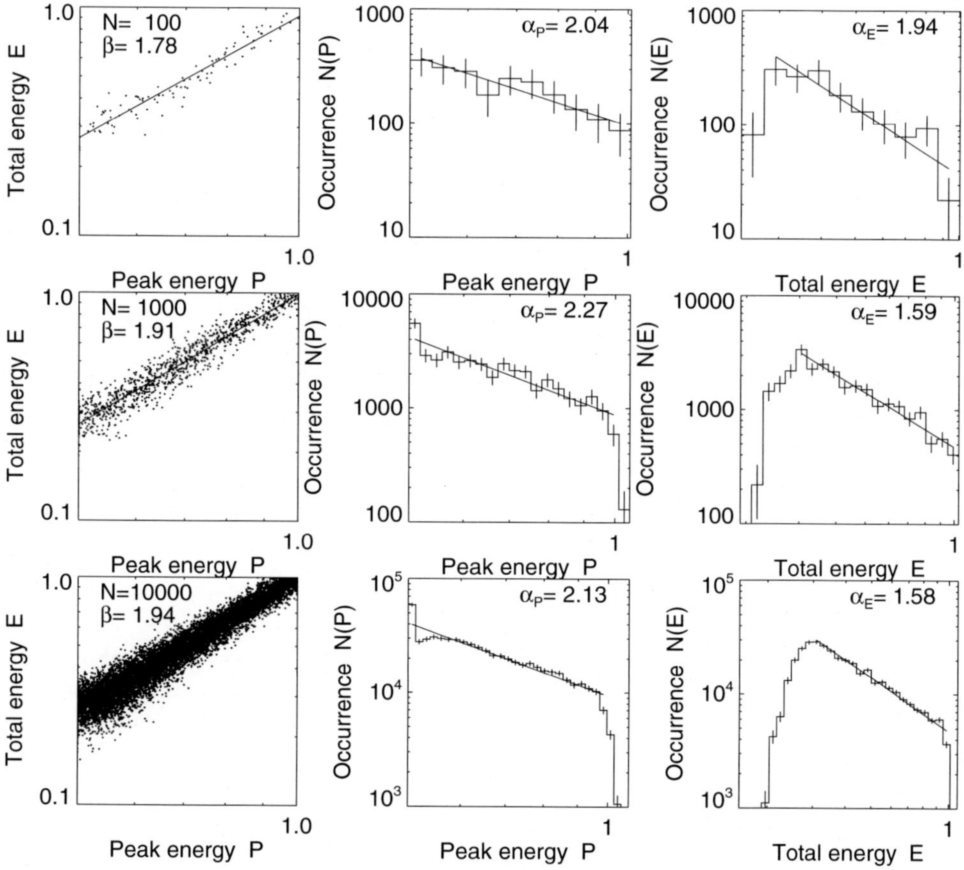

Fig. 7.4 Three Monte-Carlo simulations of peak energies P (middle column) and total energies E (right column) from random samples of prescribed powerlaw distributions $N(P) \propto P^{2.0}$ and $N(E) \propto E^{1.5}$, for sample sizes of $n = 100$ (top row), $n = 10^3$ (middle row), and $n = 10^4$ (bottom row). Note the truncation bias for a threshold at $P \geq 0.5$, which causes a lower rollover in the frequency distributions. The parameter correlations $E \propto P^{\beta}$ were fitted with a linear regression fit (left column) and the powerlaw slopes were fitted in the decreasing part on the right-hand side of the maximum of the distributions.

We bin the range of P and E each with 10 bins and determine the powerlaw slope with a linear regression for the bins right to the peak of the frequency distributions, in order to eliminate the rollover due to the sensitivity loss at low values, and find slopes of $\alpha_P = 2.04$ and $\alpha_E = 1.94$ (Fig. 7.4, top row), while we expect theoretically $\alpha_P = 2.0$ and $\alpha_E = 1.5$. Part of the discrepancy results from the small number statistics and the choice of bins in the linear regression fit, while part of the difference is caused by the generation of random numbers.

We repeat the same simulation for two larger sets of $n = 10^3$ (Fig. 7.4, middle row) and $n = 10^4$ events (Fig. 7.4, bottom row). We see that the correlation converges to a value of $\beta = 1.94$, which is still slightly different from the theoretical value of $\beta = 2.0$, either because of the truncation bias of the flux threshold or the random number generation. The powerlaw slopes of the distribution of peak energies P converges to $\alpha_P = 2.13$, and the slope of total energies E to $\alpha_E = 1.58$. The latter result is closer to the theoretical value of $\alpha_E = 1.5$, so increased statistics helps. The remaining difference in the powerlaw slope in the order of $\approx 5\%$ is caused by a combination of the truncation bias, the choice of fitted bins, and the random number generation. A bin-free powerlaw fitting procedure is described in Parnell and Jupp (2000), which may be preferable in some cases. A different option is also weighting of the bins by the number of events per bin. However, a best-fit in a log-log plot is generally achieved when equidistant bins on a logarithmic scale have equal weights, which is different from the weighting by number of events per bin. Whatever fitting strategy is considered best, we have always to keep in mind that the data often do not exactly correspond to a theoretical model, in which case a fitting parameter is in principle ill-defined. For instance, if the data obey an exponential distribution, a fit with a powerlaw model will yield a variable slope that starts from very flat at the lower end to very steep at the upper end of the distribution, so the powerlaw slope is ill-defined.

Let us also test the consistency between the linear regression fit of the correlation and the powerlaw fits of the frequency distributions. From the statistically largest sample we measure $\alpha_P = 2.13$ and $\alpha_E = 1.58$ (Fig. 7.4, bottom row), which provides a prediction of $\beta = (\alpha_P - 1)/(\alpha_E - 1) = (2.13 - 1)/(1.58 - 1) = 1.948$ that indeed agrees with the linear regression fit of the correlation plot, $\beta = 1.94$, with high accuracy ($\lesssim 0.4\%$), confirming the robustness of the fitting procedures used in this case. The discrepancy to the theoretical values of $\beta = 2$ seems to be caused by the particular random number representation used in the Monte-Carlo simulation for this case.

7.2 Frequency Distributions in Magnetospheric Physics

Let us now turn to observed occurrence frequency distributions of SOC events, starting from the magnetosphere. Some examples of frequency distributions of area sizes and dissipated power of magnetospheric substorm events are shown in Fig. 1.10, and lifetime distributions of substorm-related events are shown in Fig. 7.5, which are listed in Table 7.1.

Table 7.1 Frequency distributions measured in magnetospheric physics. References: 1, Lui et al. (2000); 2, Angelopoulos et al. (1999); 3, Takalo (1993); 4, Takalo et al. (1999a); 5, Freeman et al. (2000b); 6, Chapman and Watkins (2001), 7, Crosby et al. (2005).

Phenomenon	Parameter	Powerlaw slope α	Reference
Substorms (active)	area size	1.21 ± 0.08	1
	power	1.05 ± 0.08	1
Substorms (quiet)	area size	1.16 ± 0.03	1
	power	1.00 ± 0.02	1
Substorms	flow burst durations	1.59 ± 0.07	2
AE index	lifetimes	1.24	3,4
AU index	burst lifetimes	1.3	5,6
Outer radiation belt	electron counts	1.5–2.1	7

Frequency distributions of substorms have been obtained by measuring the projected area (in units of square kilometers) of auroral blobs with the POLAR spacecraft (Fig. 1.9), as well as by measuring the dissipated power (in units of watts) with the Ultraviolet Imager UVI (Lui et al. 2000). As a first-order approximation we can consider both the area or the dissipated power as a measure of the total energy E of substorm events. The frequency distributions (Fig. 1.10) show powerlaw distributions during quiet time intervals with slopes of $\alpha_E \approx 1.00$–1.16, and a similar slope of $\alpha_E \approx 1.05$–1.21 during substorm active time intervals, although there is in addition a Gaussian hump at the upper end of the distribution, which has been interpreted as a finite-size effect (Chapman et al. 1998), as simulated with a numerical model (Fig. 2.13; Section 2.5.1). Count rates of electrons accumulated by microsatellites during each crossing of the Earth's outer radiation belt revealed also powerlaw distributions, with slopes of $\alpha_E \approx 1.5$–2.1 (Crosby et al. 2005). It would be interesting to compare the same dataset $N(E)$ with the frequency distributions $N(P)$ of peak energies and $N(T)$ of burst durations, in order to test SOC models. However, we find related frequency distributions of burst durations measured from the AE index (Fig. 7.5 top; Takalo 1993) and AU index (Fig. 7.5 bottom; Freeman et al. 2000b), as well as from the durations of bursty bulk flow bursts in the magnetotail plasma sheet (Fig. 7.5 middle; Angelopoulos et al. 1999), which all exhibit powerlaw-like distributions in the range of $\alpha_T \approx 1.2$–1.6 (Table 7.1).

If we interpret these magnetospheric substorms as SOC events and apply our standard model of SOC avalanches with exponential growth and linear decay (Section 3.1), we expect the following relations for powerlaw slopes as summarized in Eq. (3.1.28),

$$\begin{aligned} \alpha_P &= 1 + \tau_G / t_S \\ \alpha_T &= \alpha_P \\ \alpha_E &= (\alpha_P + 1)/2 \end{aligned} \quad . \tag{7.2.1}$$

The fact that powerlaw slopes of $\alpha_T \approx 1.24$–1.6 are measured in substorms, would indicate that the ratio of the exponential growth rate τ_G to the mean saturation time t_S has a relatively low value of $(\tau_G/t_S) = \alpha_P - 1 = \alpha_T - 1 = 0.2$–$0.6$, which implies that the responsible instability saturates after $(t_S/\tau_G) = 1.6$–5.0 growth times. This implies relatively large amplification factors of $\exp(t_S/\tau_G) \approx 5$–$150$, which could be verified from the exponential growth during the rise time of substorm events. Such high amplification factors require coherent growth without a competing damping mechanism or collisional interactions, a characteristic that could constrain possible physical mechanisms responsible for geomagnetic substorms. Another prediction from our standard model is the powerlaw slope of the frequency distribution of energies, which based on $\alpha_T \approx 1.2$–1.6 would fall according to Eq. (7.2.1) in the range of $\alpha_E = (\alpha_P + 1)/2 = (\alpha_T + 1)/2 \approx 1.1$–$1.3$, which indeed is close to the values that Lui et al. (2000) observed during active time intervals of substorms, i.e., $\alpha_E \approx 1.21 \pm 0.08$ for the areas of substorms, and $\alpha_E \approx 1.05 \pm 0.08$ for the dissipated power, respectively.

In summary, we can conclude that the observed frequency distribution of magnetospheric substorms exhibit powerlaw-like functions for energy E and time duration T parameters, which are consistent with the statistics of an exponentially growing insta-

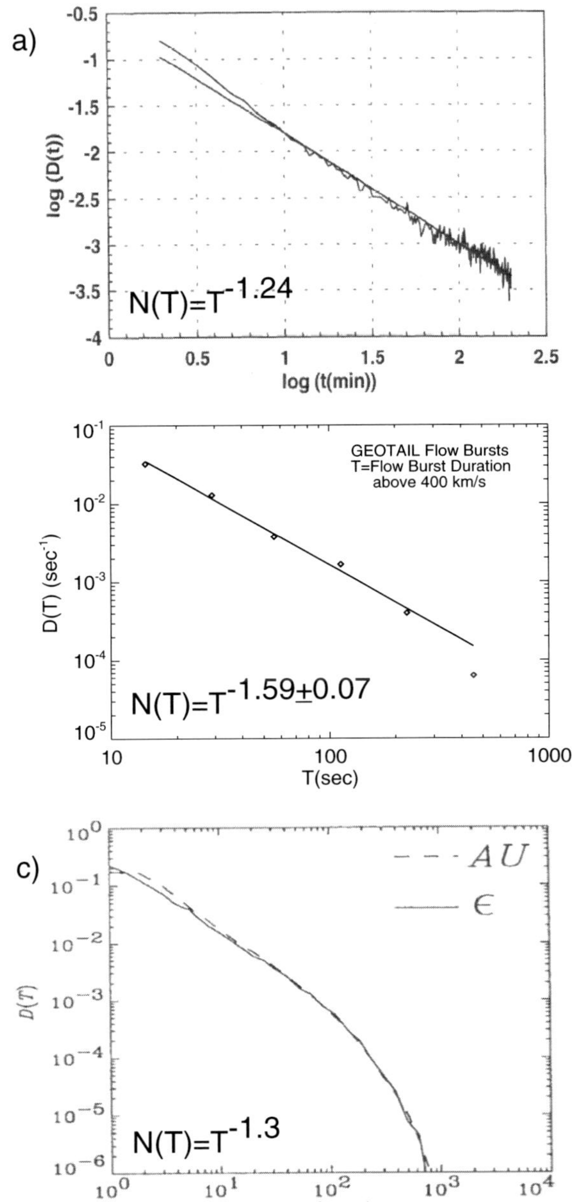

Fig. 7.5 *Top (a):* Lifetime distribution of magnetospheric disturbances as measured from the AE index of the years 1978–1985 (Takalo 1993; 1999a); *Middle (b):* Probability density distribution $N(T)$ of continuous flow magnitude durations T measured with the *GEOTAIL* satellite when it encountered the magnetotail plasma sheet during the period of Jan 1996–Oct 1998 (Angelopoulos et al. 1999); *Bottom (c):* Probability density function $N(T)$ of lifetimes of bursty bulk flow events in substorms measured of AU (Jan 1978–Jun 1988) and ε calculated from WIND SWE and MFI data for 1984–1987 (Freeman et al. 2000b; Chapman and Watkins 2001).

bility that has saturation times in the order of $t_S/\tau_G \approx 1.6$–5.0, amplification factors of $\exp(t_S/\tau_G) \approx 5$–$150$, leading to powerlaw slopes of $\alpha_T = \alpha_P \approx 1.2$–$1.6$ for durations, and $\alpha_E = 1.1$–1.3 for energies of substorm events.

7.3 Frequency Distributions in Solar Physics

Solar flares are probably the best-studied datasets regarding SOC statistics in astrophysics. Solar flares are catastrophic events in the solar corona, most likely caused by a magnetic instability that triggers a magnetic reconnection process, producing emission in almost all wavelengths, such as in gamma rays, hard X-rays, soft X-rays, extreme ultraviolet (EUV), Hα emission, radio wavelengths, and sometimes even in white light. Since the emission mechanisms are all different in each wavelength, such as nonthermal bremsstrahlung (in hard X-rays), thermal bremsstrahlung (in soft X-rays and EUV), gyrosynchrotron emission (in microwaves), plasma emission (in metric and decimetric waves), etc., we expect that the calculation of energies contained in each event strongly depends on the emission mechanism, and thus on the wavelength. It is therefore advisable to investigate the statistics of SOC events in each wavelength domain separately. The most unambiguous SOC parameters to report are the peak flux P, the total flux or fluence E, defined as the time-integrated flux over the entire event, and the total time duration T of the event. Conversions of fluxes and fluences into energy release rates and total energies require physical models, which will be discussed in Chapter 9.

7.3.1 Solar Flare Hard X-rays

Hard X-ray emission in solar flares mostly results from thick-target bremsstrahlung of nonthermal particles accelerated in the corona that precipitate into the dense chromosphere. Thus, the hard X-ray flux is the most direct measure of the energy release rate, and thus is expected to characterize the energy of SOC events in a most uncontaminated way, while emission in other wavelengths exhibit a more convolved evolution of secondary emission processes.

One of the earliest reports of a frequency distribution of solar hard X-ray flare fluxes was made by Datlowe et al. (1974), who published the cumulative frequency distribution of 123 flare events detected in the 20–30 keV energy range above a threshold of $\gtrsim 0.1$ photons (cm^{-2} s^{-1} keV^{-1}) with the OSO-7 spacecraft during 10 Oct 1971–6 June 1972, finding a powerlaw slope of $\beta_P \approx 0.8$. For compatibility we list only powerlaw slopes of differential frequency distributions in Table 7.2, and use the conversion $\alpha = \beta + 1$ if needed (but see bias described in Section 7.1.2). We list also the logarithmic ranges of the x-axis over which the powerlaw fit was obtained, e.g., ${}^{10}\log(P_{max}/P_{min}) = {}^{10}\log(30/0.3) \approx 2$ in the case of Datlowe et al. (1974), which is a good indicator of the accuracy of the powerlaw slope fit.

A sample of 25 microflares of smaller size were detected at 20 keV with a balloon-borne instrumentation of *University of California Berkeley (UCB)* during 141 minutes of

Table 7.2 Frequency distributions measured from solar flares in hard X-rays and gamma-rays. References: 1, Datlowe et al. (1974); 2, Lin et al. (1984); 3, Dennis (1985); 4, Schwartz et al. (1992); 5, Crosby et al. (1993); 6, Biesecker et al. (1993); 7, Biesecker et al. (1994); 8, Crosby (1996); 9, Lu et al. (1993); 10, Lee et al. (1993); 11, Bromund et al. (1995); 12, Perez-Enriquez and Miroshnichenko (1999); 13, Georgoulis et al. (2001); 14, Su et al. (2006); 15, Christe et al. (2008); 16, Lin et al. (2001); 17, Tranquille et al. (2009).

Powerlaw slope of peak flux α_P	Powerlaw slope of total fluence α_E	Powerlaw slope of durations α_T	log range	Instrument	Reference
1.8			2	OSO-7	1
2.0			1	UCB	2
1.8			4	HXRBS	3
1.73 ± 0.01			3.5	HXRBS	4
1.73 ± 0.01	1.53 ± 0.02	2.17 ± 0.05	3.5	HXRBS	5
1.61 ± 0.03			3.5	BATSE	4
1.75 ± 0.02			4	BATSE	6
1.68 ± 0.02			3.5	BATSE	7
1.59 ± 0.02		2.28 ± 0.08	3	WATCH	8
1.86	1.51	1.88	3	ISEE-3	9
1.75	1.62	2.73	2.5	ISEE-3	10
1.86 ± 0.01	1.74 ± 0.04	2.40 ± 0.04	3.5	ISEE-3	11
1.80 ± 0.01	1.39 ± 0.01		1	PHEBUS	12
1.59 ± 0.05	1.39 ± 0.02	1.09-1.15	2	WATCH	13
1.80 ± 0.02		3.6 [0.9]	3.5	RHESSI	14
1.58 ± 0.02	1.7 ± 0.1	2.2 ± 0.2	2	RHESSI	15
1.6			3	RHESSI	16
1.61 ± 0.04			2	ULYSSES	17

observations on 1980 June 27, yielding a powerlaw distribution with a slope of $\beta \approx 1$ (Lin et al. 1984).

A much larger amount of statistics was obtained with the *Hard X-Ray Burst Spectrometer (HXRBS)* onboard the *Solar Maximum Mission (SMM)* spacecraft, which recorded 6,775 flare events during the 1980–1985 period, exhibiting a powerlaw distribution of peak count rates with a slope of $\alpha_P = 1.8$ over four orders of magnitude (Dennis 1985), see Fig. 1.13.

A next mission with hard X-ray detector capabilities was the *Compton Gamma Ray Observatory (CGRO)*. Although it was designed to detect gamma-ray flashes from astrophysical objects, it detected also solar flares systematically during the period of 1991–2000. Using the *Burst And Source Transient Experiment (BATSE)*, statistics of flares with energies >25 keV was sampled and more detailed powerlaw distributions of peak fluxes were reported with values of $\alpha_P = 1.61 \pm 0.03$ (Schwartz et al. 1992), $\alpha_P = 1.75 \pm 0.02$ (Biesecker et al. 1993). and $\alpha_P = 1.68 \pm 0.02$ (Biesecker et al. 1994) for BATSE. Biesecker et al. (1994) noticed slight differences of the powerlaw slope during low activity ($\alpha_P = 1.71 \pm 0.04$) and high activity periods ($\alpha_P = 1.68 \pm 0.02$), which appear not to be significant.

A systematic study of flares observed with HXRBS over the entire mission duration of 1980–1989 was conducted by Crosby et al. (1993), measuring peak count rates P_{cts} (cts

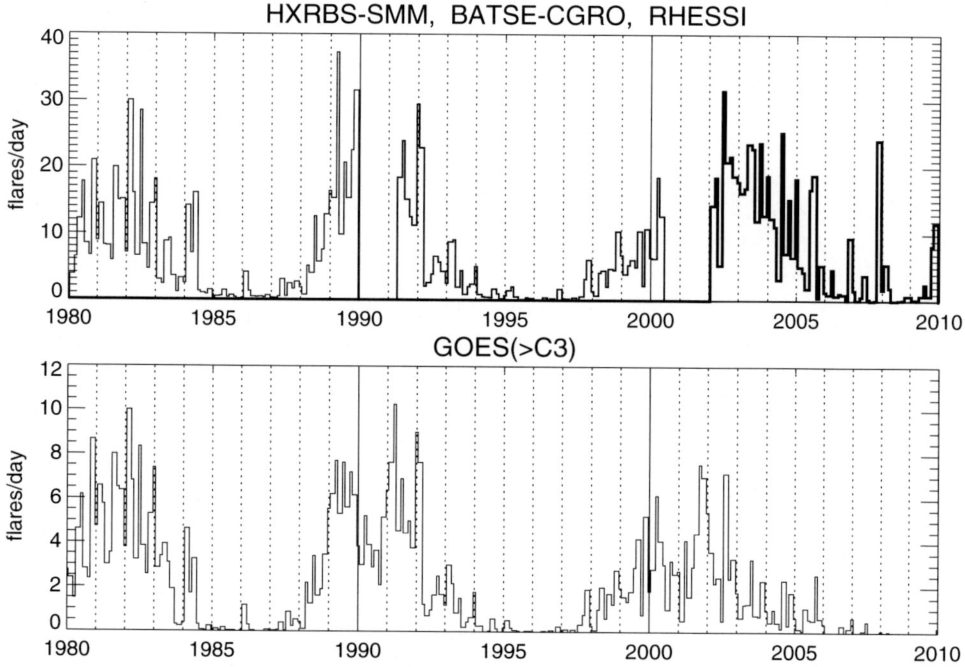

Fig. 7.6 *Top:* Monthly averages of solar flare rates observed during the last three solar cycles in hard X-rays with HXRBS/SMM (1980–1989), BATSE/CGRO (1991–2000), and RHESSI (2002–2010), corrected for the duty cycles of the instruments. *Bottom:* Monthly averages of the solar flare rate observed in soft X-rays with GOES, including events above the C3-class level.

s^{-1}), converted into photon fluxes P_{ph} (photons cm^{-2} s^{-1} keV^{-1}) at energies >25 keV, peak HXR spectrum-integrated fluxes P_X (photons cm^{-2} s^{-1}), peak electron fluxes P_e (ergs s^{-1}), flare durations T, and time-integrated total energies in electrons E_e (ergs), for four different time intervals of the solar cycle. The variability of the solar flare rate during the last three solar cycles can be seen in form of monthly averages in Fig. 7.6. In Table 7.2 we list the values for the time range of 1980–1982, which covers the solar maximum and has the largest statistics. The values of the powerlaw slopes change only by $\lesssim 2\%$ during the solar minimum. The multi-parameter statistics of P, E, and T allowed also to derive the following parameter correlations (see Section 7.1.6),

$$\begin{aligned} T &\propto P_{ph}^{0.43[0.41]} \\ P_X &\propto P_{ph}^{0.95[1.06]} \\ P_e &\propto P_{ph}^{1.02[0.93]} \\ E_e &\propto P_{ph}^{1.21[1.25]} \end{aligned} , \qquad (7.3.1)$$

where the powerlaw index is derived by linear regression between the parameters, as well as from the slopes of each frequency distribution with Eq. (7.1.42) (indicated in brackets

[...]). The peak parameters seem to be all close to proportional to each other, i.e., $P_X \propto P_e \propto P_{ph}$, so it does not matter much which one is used to characterize the peak energies of SOC avalanches. It is interesting to compare these correlation coefficients with our standard SOC model (Section 3.1), which predicts $\alpha_P = \alpha_T$ and $\alpha_E = (\alpha_P + 1)/2$, and the correlations $T \propto P^1$ and $E \propto P^2$. We have to investigate observations in other wavelengths and explore whether different definitions of event durations and energies can explain the discrepancy of observed correlations (Eq. 7.3.1) to the theoretical model.

From the *Wide Angle Telescope for Cosmic Hard X-Rays (WATCH)* onboard the Russian satellite *Granat*, a sample of 1,546 flare events was observed at energies of 10–30 keV or 14–40 keV during 1990–1992, yielding similar powerlaw slopes for peak count rates, $\alpha_P = 1.59 \pm 0.02$, and flare durations $\alpha_T = 2.28 \pm 0.08$ as reported before (Crosby 1996; Crosby et al. 1998). However, it was noted that the frequency distribution of flare durations exhibits a gradual rollover for short flare durations, approaching a slope of $\alpha_T \approx 1$, so it cannot be fitted with a single powerlaw distribution over the entire range of flare durations. From the *PHEBUS* instrument on *Granat*, which is sensitive to gamma-ray energies, Perez-Enriquez and Miroshnichenko (1999) analyzed 110 high-energy solar flares observed in the energy range of 100 keV–100 MeV and found the following powerlaw slopes: $\alpha_P = 1.80 \pm 0.01$ for (bremsstrahlung) hard X-ray fluxes at >100 keV, $\alpha_P = 1.38 \pm 0.01$ for photon energies at 0.075–124 MeV, $\alpha_P = 1.39 \pm 0.01$ for bremsstrahlung at 300–850 keV, $\alpha_E = 1.50 \pm 0.03$ for the 511 keV electron-positron annihilation line fluence, $\alpha_E = 1.39 \pm 0.02$ for the 2.223 MeV neutron capture line fluence, and $\alpha_E = 1.31 \pm 0.01$ for the 1–10 MeV gamma-ray line fluence. We have to be aware that this selection of high-energy (gamma-ray) flares is not representative for all hard X-ray flares, and thus has a biased distribution towards the largest events, which explains that most frequency distributions in gamma rays have a flatter slope than in hard X-rays. The flatter slope corresponds according to our standard model ($\alpha_P = 1 + t_S/\tau_G$) also to events with higher exponential growth factors $\exp(\tau/t_S)$, which is certainly expected for gamma ray-producing flare events.

Using data from a >25 keV hard X-ray detector onboard the ISEE-3/ICE spacecraft during 24 Aug 1978 and 11 Jul 1986, Lu et al. (1993) determined the frequency distributions of the peak luminosity P (erg s^{-1}), the energy E (erg), and flare duration T (s) and found that the measured distributions could be best fitted with a cellular automaton model that produced powerlaw slopes of $\alpha_P = 1.86$, $\alpha_E = 1.51$, and $\alpha_T = 1.88$. The fits of the distributions included an exponential rollover at the upper end, which explains that they inferred a less steep slope for durations than previously reported. Interestingly, these values agree much more closely with our standard model, which predicts for $\alpha_P = 1.86$ the slopes $\alpha_T = \alpha_P = 1.86$ and $\alpha_E = (\alpha_P + 1)/2 = 1.43$. This tells us that the rollovers at the lower and upper end of the distributions have to be included in the model fits in order to obtain proper powerlaw slopes. Lee et al. (1993) analyzed the same data and determined the correlations and frequency distribution powerlaw slopes with special care of truncation biases and obtained similar values for ISEE-3 ($\alpha_P = 1.75$, $\alpha_E = 1.62$, $\alpha_T = 2.73$) as Crosby et al. (1993) for HXRBS. A third study was done with the same data (Bromund et al. 1995), where the energy spectrum was also calculated to determine different energy parameters, similar to the study of Crosby et al. (1993), finding the following powerlaw slopes: $\alpha_P = 1.86, ..., 2.00$ for the peak photon flux P_{ph} (photons cm^{-2} s^{-1}),

$\alpha_P = 1.92,...,2.07$ for the peak electron power P_e (erg s^{-1}), $\alpha_E = 1.67,...,1.74$ for the total electron energy E_e (erg), and $\alpha_T = 2.40,...,2.94$ for the total duration T (s), where the range of powerlaw slopes results from the choice of the fitting range. The flare duration T was defined at a level of 1/e times the peak count rate. The following parameter correlations were found,

$$\begin{aligned} P &\propto T^{0.75[1.52]} \\ E &\propto T^{1.60[2.08]} \\ E &\propto P^{1.35[1.36]} \end{aligned} , \qquad (7.3.2)$$

where the powerlaw index is derived by orthogonal linear regression fits, as well as from the slopes of the frequency distributions Eq. (7.1.42) (indicated in brackets [...]). Interestingly, the first two correlations are consistent with our standard model, which predicts $P \propto T^1$ and $E \propto T^2$, within the uncertainty of the two methods.

From the latest solar mission with hard X-ray capabilities, the *Ramaty High-Energy Solar Spectroscopic Imager (RHESSI)* spacecraft, frequency distributions were determined in the 12–25 keV energy band from 2002–2005 (Su et al. 2006), finding powerlaw slopes of $\alpha_P = 1.80 \pm 0.02$ for the peak fluxes, and a broken powerlaw $\alpha_T = 0.9$–3.6 for the flare duration, similar to previous findings (e.g., Crosby et al. 1998). Christe et al. (2008) conducted a search of microflares and identified a total of $\approx$25,000 events observed with RHESSI during 2002–2007 and investigated the frequency distributions at lower energies, finding powerlaw slopes of $\alpha_P = 1.50 \pm 0.03$ for 3–6 keV peak count rates P_{ph} (cts s^{-1}), $\alpha_P = 1.51 \pm 0.03$ for 6–12 keV peak count rates, and $\alpha_P = 1.58 \pm 0.02$ for 12–25 keV peak count rates. Converting the peak count rates P into total energy fluxes by integrating their energy spectra, Christe et al. (2008) find an energy distribution with a powerlaw slope of $\alpha_E = 1.7 \pm 0.1$, with an average energy deposition rate of $\lesssim 10^{26}$ erg s^{-1}. It is interesting that this microflare statistics is fairly consistent with overall flare statistics, even if it represents only a subset in the lowest energy range.

Flare statistics was also gathered from the *Solar X-ray/Cosmic Gamma-Ray Burst Experiment (GRB)* onboard the *Ulysses* spacecraft (Tranquille et al. 2009), finding similar results for >25 keV events, i.e., a powerlaw slope of $\alpha_P = 1.61 \pm 0.04$ for the peak count rate, which steepens to $\alpha_P = 1.75 \pm 0.08$ if the largest events with pulse pile-up are excluded.

A specialized study investigated also how the frequency distribution of hard X-ray peak fluxes depends on the associated size of the active region and found evidence for an upper cutoff due to a finite size limit (Kucera et al. 1997). Besides flare events *per se*, one can also consider substructures in flares and conduct SOC statistics. If the substructures are self-similar to the overall structure, one would expect similar powerlaw slopes of the frequency distributions, which was indeed found to be the case for hard X-ray subpulses (Aschwanden et al. 1995), although subpulses from a single flare can have exponential distributions, while their superposition from many flares converges towards powerlaw distributions (Aschwanden et al. 1998b). These subpulses have typical time scales of $T_{sub} = 0.5$–1.5 s (Aschwanden et al. 1995) and $T_{sub} = 1.9 \pm 0.5$ s (Qiu and Wang 2006).

A summary plot of frequency distributions of the peak count rate P is shown for the three instruments HXRBS/SMM, BATSE/CGRO, and RHESSI in Fig. 7.7, yielding an average powerlaw slope of $\alpha_P = 1.75 \pm 0.05$. The corresponding frequency distributions

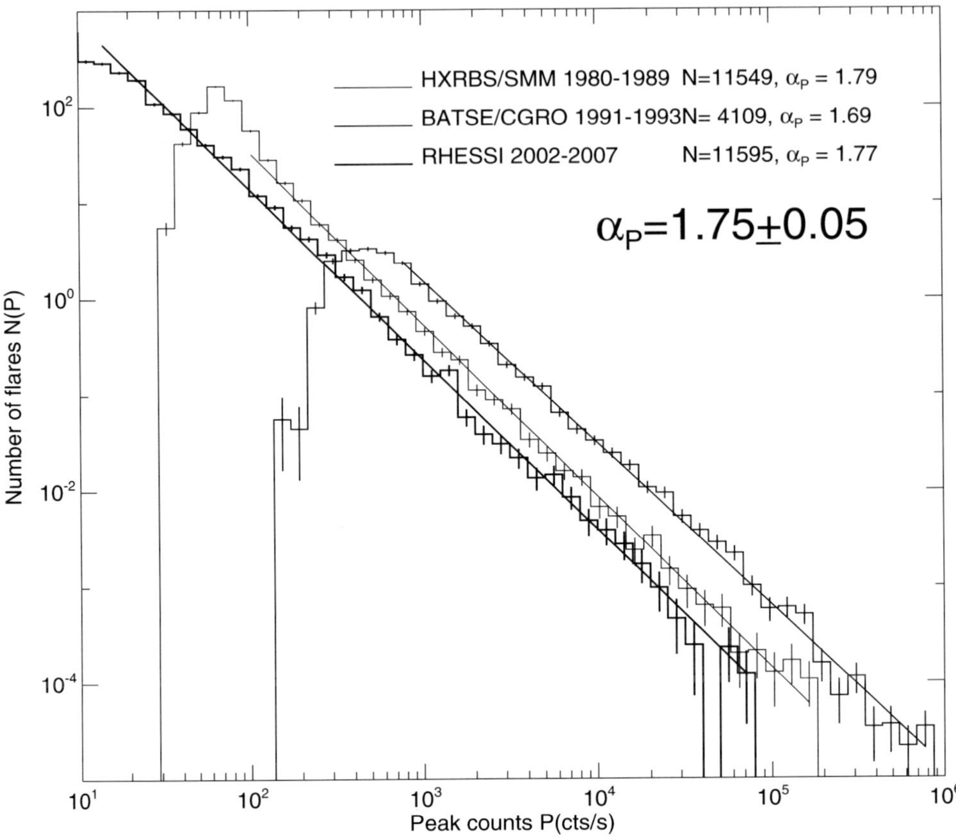

Fig. 7.7 Occurrence frequency distributions of hard X-ray peak count rates $P(cts/s)$ observed with HXRBS/SMM (1980–1989), BATSE (1991–1993), and RHESSI (2002–2007), with powerlaw fits. Note that BATSE/CGRO has larger detector areas, and thus records higher count rates. RHESSI flares were detected at energies of $\geq$12 keV, while HXRBS and BATSE flares were detected at energies of $\geq$25 keV. The average slope value is $\alpha_P = 1.75 \pm 0.05$.

of total counts or fluences are shown in Fig. 7.8, which have an average powerlaw slope of $\alpha_E = 1.61 \pm 0.04$. The distributions of flare durations are shown in Fig. 7.9, which exhibit an average of $\alpha_T = 2.08 \pm 0.10$, with a tendency toward a rollover at the low end. Thus, our best values are,

$$\begin{array}{ll} N(P) \propto P^{-\alpha_P} & \alpha_P = 1.75 \pm 0.05 \\ N(E) \propto E^{-\alpha_E} & \alpha_E = 1.61 \pm 0.04 \\ N(T) \propto T^{-\alpha_T} & \alpha_T = 2.08 \pm 0.10 \end{array} \tag{7.3.3}$$

It is interesting to note that these best observational values closely correspond to the numerically simulated values in the cellular automaton model of Lu and Hamilton (1991), see Eq. (2.6.15), i.e., $N(P) \propto P^{-1.67\pm0.04}$, $N(E) \propto E^{-1.53\pm0.02}$, and $N(T) \propto T^{-2.17\pm0.05}$. Using the averaged values of Eq. (7.3.3), we expect the following correlations between

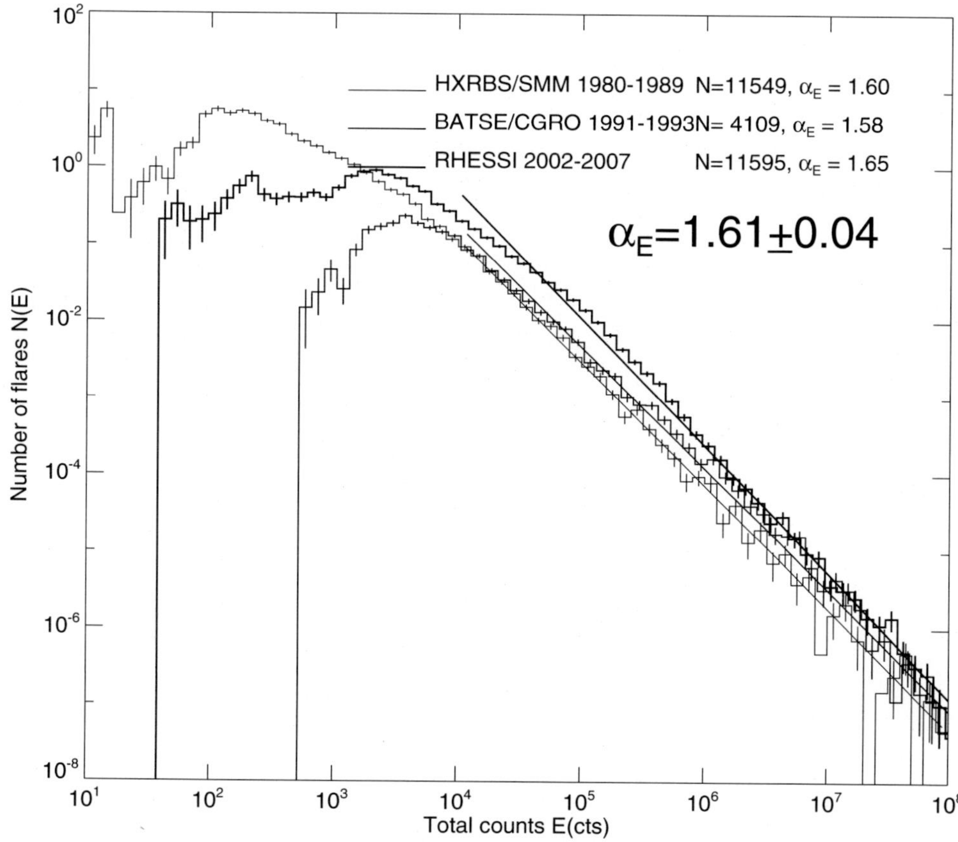

Fig. 7.8 Occurrence frequency distributions of hard X-ray total counts or fluence $E(cts)$ observed with HXRBS/SMM (1980–1989), BATSE (1991–1993), and RHESSI (2002–2007), with powerlaw fits. The average slope value is $\alpha_E = 1.61 \pm 0.04$.

these three parameters (using Eq. 7.1.42),

$$\begin{array}{ll} E \propto P^{\beta} & \beta = (\alpha_P - 1)/(\alpha_E - 1) = (1.75 - 1)/(1.61 - 1) = 1.23 \pm 0.09 \\ T \propto P^{\beta} & \beta = (\alpha_P - 1)/(\alpha_T - 1) = (1.75 - 1)/(2.08 - 1) = 0.70 \pm 0.07 \\ E \propto T^{\beta} & \beta = (\alpha_T - 1)/(\alpha_E - 1) = (2.08 - 1)/(1.61 - 1) = 1.77 \pm 0.16 \end{array} \quad (7.3.4)$$

In Fig. 7.10 we show the actual correlation plots between the parameters and determine linear regression fits, which give a comparable result, with $E \propto P^{1.26\pm0.04}$ and $T \propto P^{0.44\pm0.04}$. The latter correlation, of course, cannot be determined accurately from linear regression fits due to the large scatter in flare duration values. Thus, we have obtained representative values of the powerlaw slopes α and correlation coefficients β for solar flare hard X-ray parameters, averaged from three major missions over the last 30 years and three solar cycles, which can serve as reference for other wavelengths.

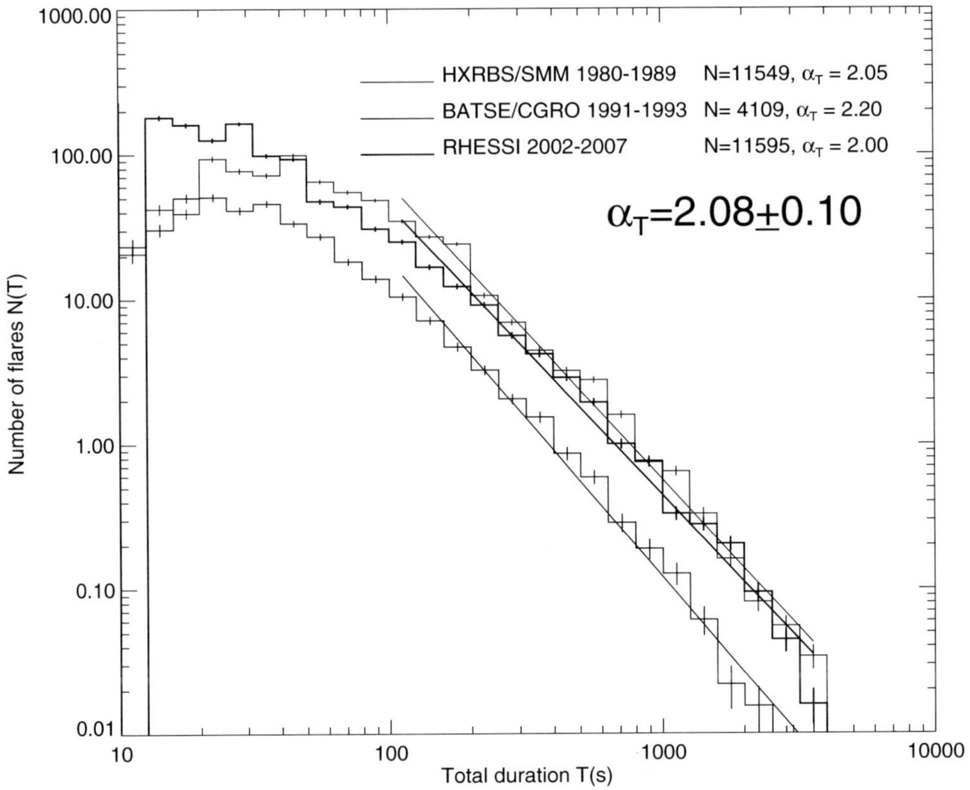

Fig. 7.9 Occurrence frequency distributions of hard X-ray flare durations $T(s)$ observed with HXRBS/SMM (1980-1989), BATSE (1991-1993), and RHESSI (2002-2007), with powerlaw fits. The flare durations for RHESSI were estimated from the time difference between the start and peak time, because RHESSI flare durations were determined at a lower energy of 12 keV (compared with 25 keV for HXRBS and BATSE), where thermal emission can dominate in large flares, causing a flatter powerlaw slope ($\alpha_T \approx 1.4$). The average slope value is $\alpha_T = 2.08 \pm 0.10$.

7.3.2 Solar Flare Soft X-rays

Soft X-ray emission in solar flares mostly originates from free-free bremsstrahlung emission of heated flare plasma, which typically reaches temperatures of $T \approx 10$–35 MK. A pragmatic relationship between soft and hard X-ray emission of flare plasmas is characterized with the so-called *Neupert effect* (e.g., Dennis and Zarro 1993), which essentially states that the time profile of hard X-ray emission corresponds to the heating rate produced by nonthermal particles bombarding the chromosphere, while soft X-ray emission represents the chromospheric response of flare plasma heating. This model is called the *chromospheric evaporation* scenario. The thermal energy of the heated plasma and thus the time profile of emitted soft X-ray emission consequently approximately follows the time integral of the hard X-rays, until cooling by thermal conduction and radiative loss

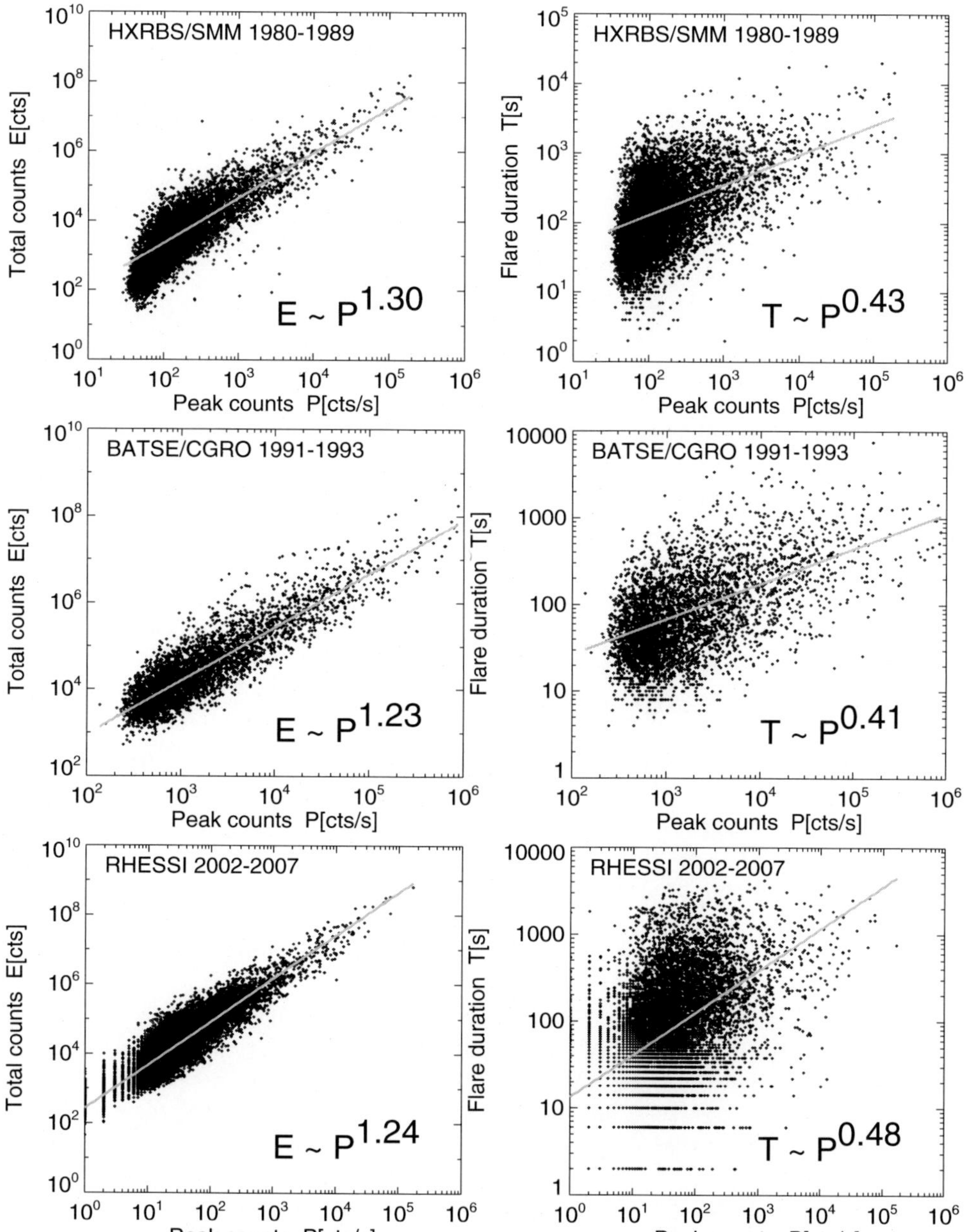

Fig. 7.10 Scatterplots between for the total counts $E(P)$ (left panels) or flare duration $T(P)$ (right panels) versus the peak count rate P for solar flares with HXRBS/SMM (1980–1989) (top), BATSE/CGRO (1991–1993) (middle), and RHESSI (2002–2007) (bottom). Linear regression fits are applied to all datapoints above a threshold of five times the minimum value in each parameter. All data are subject to a flux threshold P_{min}, which causes a truncation at $P \le P_{min}$, but does not affect linear regression fits of the form $y(x)$.

overcomes the heating rate in the late flare phase. From this scenario we expect the relationship,

$$F^{SXR}(t) \approx \int_0^t F^{HXR}(t')\, dt' \ . \tag{7.3.5}$$

If we define a SOC event by the energy release as observed in hard X-rays, characterized with a flare start time t_s, end time t_e, total duration $T = (t_e - t_s)$, peak energy flux $P^{HXR} = F(t = t_p)$, and total flux or fluence $E = \int_{t_s}^{t_e} F^{HXR}(t)\, dt$, then the peak time t_p^{HXR} of the hard X-rays corresponds to the inflection point with the steepest rise in the soft X-ray time profile, while the end time t_e^{HXR} corresponds to the peak time in soft X-rays. In order to obtain a consistent flare duration T in the two wavelengths, we have therefore to define,

$$T = (t_e^{HXR} - t_s^{HXR}) = (t_p^{SXR} - t_s^{HXR}) \ , \tag{7.3.6}$$

and to calculate the time derivative of the soft X-ray light curve $F^{SXR}(t)$ according to the Neupert effect (Eq. 7.3.5),

$$F^{proxi}(t) = \frac{dF^{SXR}(t)}{dt} \ , \tag{7.3.7}$$

to obtain a proxy $F^{proxi}(t)$ for the hard X-ray-like flare light curve where we can measure the peak energies P and total energies E. If we do not correct for this Neupert effect, we expect some significantly different frequency distributions and correlation parameters for flare event statistics in soft X-rays and hard X-rays.

First frequency distributions of flare peak fluxes in soft X-rays were reported from OSO-3 observations in the energy range of 7.7–12.5 keV (Hudson et al. 1969), where a cumulative distribution with a powerlaw tail with a slope of $\beta \approx 0.8$ was found which corresponds to a slope of $\alpha \approx \beta + 1 = 1.8$ for the differential frequency distribution. Further data in the 2–12 Å range (1–6 keV) with *Explorer 33* and *Explorer 35* satellites yielded solar flare statistics for ≈3,000 events during July 1966 and September 1968, from which powerlaw distributions of the peak flux ($\alpha_P = 1.75$) and the fluence ($\alpha_E = 1.44$) were reported (Drake 1971).

The Yohkoh mission (1991–2002) provided imaging observations of solar flares with the *Soft X-ray Telescope (SXT)* at temperatures of $T > 1.5$ MK (>0.13 keV). Shimizu (1995) analyzed small active region transient brightenings (small flares) during August 1992 and inferred from a sample of some 5,000 events in a single active region frequency distributions of soft X-ray peak fluxes with powerlaw slopes in the range of $\alpha_P = 1.64$–1.89, depending on the spatial area used in the sampling. The thermal energy of these events were estimated in the range of $E \approx 10^{27}$–10^{29}, and the powerlaw slope for energies was calculated to $\alpha_E \approx 1.5$–1.6 (Shimizu 1995). A similar study was performed by Shimojo and Shibata (1999), who analyzed 92 microflares during the lifetime of a single bright point (i.e., a miniature active region) and found a power law slope of $\alpha_P = 1.7 \pm 0.4$ for the soft X-ray peak flux.

The difference between frequency distributions sampled in hard X-rays and soft X-rays was modeled by Lee et al. (1995). For this purpose, flare statistics in hard X-rays (HXRBS, ISEE-3) and in soft X-rays (SMM/BCS, GOES) were reanalyzed (Fig. 7.11), but similar powerlaw slopes were found for the two wavelength ranges, which could only

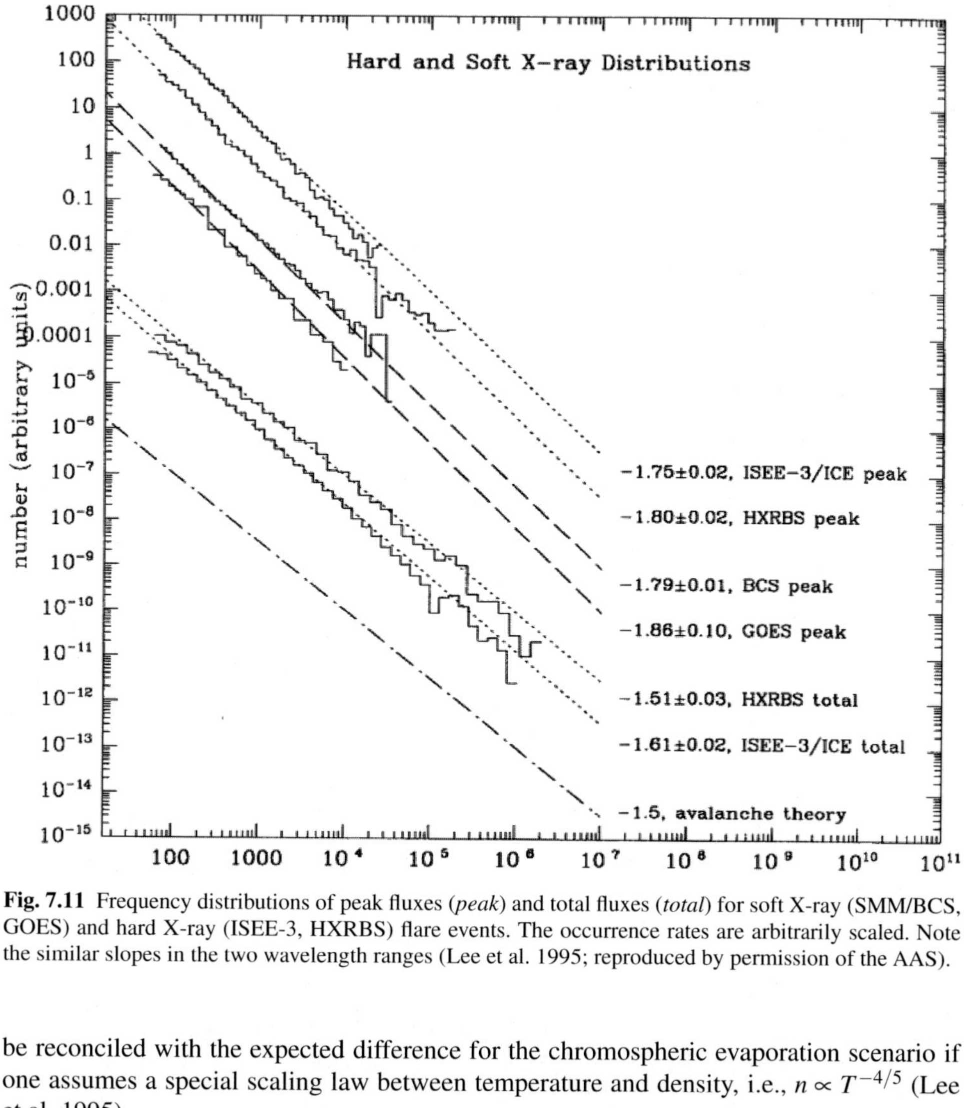

Fig. 7.11 Frequency distributions of peak fluxes (*peak*) and total fluxes (*total*) for soft X-ray (SMM/BCS, GOES) and hard X-ray (ISEE-3, HXRBS) flare events. The occurrence rates are arbitrarily scaled. Note the similar slopes in the two wavelength ranges (Lee et al. 1995; reproduced by permission of the AAS).

be reconciled with the expected difference for the chromospheric evaporation scenario if one assumes a special scaling law between temperature and density, i.e., $n \propto T^{-4/5}$ (Lee et al. 1995).

Using soft X-ray light curves from the *Geostationary Operational Environmental Satellites (GOES)*, which observe the Sun uninterrupted thanks to multiple spacecraft, complete flare statistics can be gathered. Feldman et al. (1997) sampled during 1993–1995 some 1,000 flare events in the 1–8 Å (0.08–0.67 keV) range and inferred a soft X-ray peak flux distribution with a powerlaw slope of $\alpha_P = 1.88 \pm 0.21$. A more comprehensive study of 50,000 soft X-ray flares observed with GOES during 1976–2000 was performed by Veronig et al. (2002a,b). The obtained frequency distributions exhibit significantly steeper slopes than previously found, i.e., $\alpha_P = 2.11 \pm 0.13$ for the peak flux, $\alpha_E = 2.03 \pm 0.09$ for the fluence, and $\alpha_T = 2.93 \pm 0.12$ for durations. This discrepancy with previous statistics (Table 7.3) most likely arises from two facts: (1) no pre-event background flux was sub-

Table 7.3 Frequency distributions measured from solar flares in soft X-rays. References: 1, Hudson et al. (1969); 2, Drake et al. (1971); 3, Shimizu (1995); 4, Lee et al. (1995); 5, Feldman et al. (1997); 6, Shimojo and Shibata (1999); 7, Veronig et al. (2002d); 8, Veronig et al. (2002a); 9, Yashiro et al. (2006).

Powerlaw slope of peak flux α_P	Powerlaw slope of total fluence α_E	Powerlaw slope of durations α_T	log range	Instrument	Reference
1.8			1	OSO-3	1
1.75	1.44		2	Explorer	2
1.64–1.89	1.5–1.6		2	Yohkoh	3
1.79			2	SMM/BCS	4
1.86			2	GOES	4
1.88 ± 0.21			3	GOES	5
1.7 ± 0.4			2	Yohkoh	6
1.98	1.88		3	GOES	7,8
$2.11 \pm 0.13^*$	$2.03 \pm 0.09^*$	$2.93 \pm 0.12^*$	3	GOES	8
$2.16 \pm 0.03^*$	$2.01 \pm 0.03^*$	$2.87 \pm 0.09^*$	3	GOES	9

* No background subtracted.

tracted, which substantially overestimates the peak flux and fluence of small events, and thus causes a steeper powerlaw slope, and (2) the Neupert effect could explain some discrepancy with respect to hard X-rays, but there seems to be a small difference (Fig. 7.11) according to the comparison of Lee et al. (1995). The Neupert effect has been investigated by correlating the soft X-ray peak flux with the hard X-ray fluence (Eq. 7.3.5) and a strong correlation was found, but it is not strictly proportional as predicted by the chromospheric evaporation model (Veronig et al. 2002c). Similar values ($\alpha_P = 2.16 \pm 0.03$, $\alpha_E = 2.01 \pm 0.03$, and $\alpha_T = 2.87 \pm 0.09$ were inferred by Yashiro et al. (2006), but these distributions suffer from the same lack of background subtraction as the study of Veronig et al. (2002a,b), which causes a bias in overestimating the flux of weak events and thus steepens the powerlaw slopes. Moreover, the study of Yashiro et al. (2006) demonstrated that a subset of flare events with simultaneous CME events exhibited flatter powerlaw distributions, which is to be expected for any subset that contains preferentially larger events.

In summary, for flare events observed in soft X-rays, we can group the results into two categories: (1) event statistics with pre-flare background subtraction, and (2) without pre-flare background subtraction. From the compilation shown in Table 7.3 it is clear that each group produces quite consistent results among themselves, but they differ significantly due to the well-understood bias caused by neglecting background subtraction, especially in GOES data, where every light curve contains also the total soft X-ray emission from all other active regions on the solar surface besides a particular flare event (e.g., Bornmann 1990). Thus, ignoring the second group (Veronig et al. 2002a,b; Yashiro et al. 2006) in Table 7.3, we obtain the following averages from the first group,

$$\begin{array}{ll} N(P_{SXR}) \propto P^{-\alpha_P} & \alpha_P = 1.79 \pm 0.06 \\ N(E_{SXR}) \propto E^{-\alpha_E} & \alpha_E = 1.50 \pm 0.05 \end{array} , \qquad (7.3.8)$$

which is not much different from the values obtained in hard X-rays (Eq. 7.3.3), i.e., $N(P_{HXR}) \propto P^{-1.75\pm0.05}$ and $N(E_{HXR}) \propto E^{-1.61\pm0.04}$. Although we expect some difference due to the Neupert effect (Eq. 7.3.5) in the relationship between soft and hard X-rays, the dissimilarity in the frequency distributions of SOC parameters is apparently not large, either because the time profiles are close to self-similar, or because the soft X-ray light curves are not purely thermal emission, but contain also significant nonthermal emission as observed at higher energies in hard X-rays. Insight into these problems could be obtained by comparing SOC statistics obtained from GOES light curves directly versus event detection from the time derivative of the GOES light curve (Eq. 7.3.7).

7.3.3 Solar Flare Extreme Ultraviolet Emission

The evolution of a solar flare in different wavelengths can be best understood by their temperature dependence. In a large flare, plasma becomes heated to $T \approx$ 20–35 MK, which produces bright emission in soft X-rays. Once the plasma cools down in the post-flare phase, soft X-ray emission fades and extreme ultraviolet (EUV) emission becomes brighter, which is produced by free-free and bound-bound emission at temperatures of $T \approx$ 1–2 MK. The systematic delay in the peak of the emission in different wavelengths can best be seen in multi-wavelength observations of a large flare, such as during the Bastille-Day (14 July 2000) flare shown in Fig. 7.12, where the timing of the peak emission in each wavelength is exactly ordered according to the temperature peak sensitivity of the different instruments. It peaks first in the *Yohkoh/HXT* 14–23 keV channel, which is sensitive to the highest temperature that occurred in the flare ($T \approx 35$ MK), then in the soft X-ray channels (GOES, Yohkoh/SXT), and finally in the EUV channels (TRACE 195, 171 Å), which are sensitive in the temperature range of $T \approx$ 1–2 MK. Therefore, SOC statistics of flare events can in principle be performed in all these wavelengths, but the frequency distributions of peak flux (P), fluence (E), and durations (T) are not necessarily identical, unless the time profiles in the different wavelengths are self-similar. The comparison of the light curves from 7 different wavelength ranges shown in Fig. 7.12 suggests that the durations become systematically longer in wavelengths corresponding to cooler temperatures, which implies a temperature-dependent scaling between peak flux and duration, $P \propto T^{\beta}$. Large flares with total energies of $E \approx 10^{27}$–10^{32} erg are visible in hard X-rays and soft X-rays (Fig. 1.14), but tiny flares with energies of $E \approx 10^{24}$–10^{26}, dubbed *nanoflares*, can only be detected in EUV, because they seem not to exceed temperatures of $T \lesssim 2$ MK, and thus lack soft or hard X-ray emission. Frequency distributions reported in solar EUV have mostly concentrated on these nanoflares, but EUV statistics on larger flares are strangely lacking completely.

A first systematic study on EUV nanoflares was carried out by Krucker and Benz (1998), using images from the *Extreme-ultraviolet Imaging Telescope (EIT)* onboard the *SOlar and Heliospheric Observatory (SOHO)*. The detection of events in the EUV images was performed with a code similar to the one described in Section 6.9, using the 171 and 195 Å filters ($T \approx$ 1.1–1.9 MK), and the energy was calculated based on a special physical model of the flare volume, assumed to be proportional to the area, i.e., $V \propto A$. This was the first study that reported significantly steeper powerlaw slopes ($\alpha_P \approx$ 2.3–2.6) than

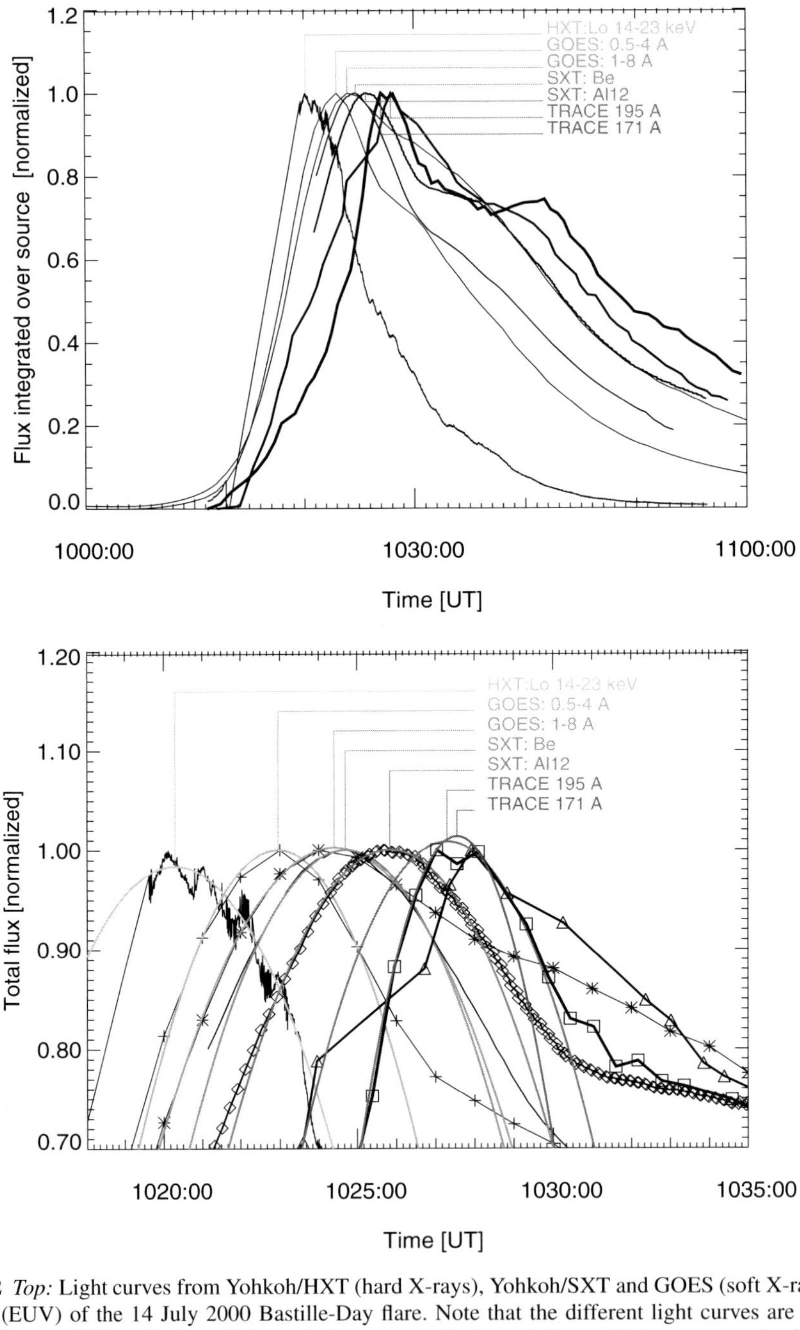

Fig. 7.12 *Top:* Light curves from Yohkoh/HXT (hard X-rays), Yohkoh/SXT and GOES (soft X-rays), and TRACE (EUV) of the 14 July 2000 Bastille-Day flare. Note that the different light curves are not self-similar. *Bottom:* Enlarged view of the emissions during their peak fluxes. Note a systematic delay that occurs in order of the decreasing temperature sensitivity of the instruments, due to the cooling of the flare plasma (Aschwanden and Alexander 2001).

Table 7.4 Frequency distributions measured in small-scale events in EUV, UV, and Hα. References: 1, Krucker and Benz (1998); 2, Aletti et al. (2000); 3, Parnell and Jupp (2000); 4, Aschwanden et al. (2000a,b); 5, Benz and Krucker (2002); 6, Aschwanden and Parnell (2002); 7, Georgoulis et al. (2002); 8, Greenhough et al. (2003); 9, McIntosh and Gurman (2005); 10, Nishizuka et al. (2009).

Powerlaw slope of peak flux α_P	Powerlaw slope of total fluence or energy α_E	Powerlaw slope of durations α_T	log range	Waveband λ (Å)	Reference
	$2.3-2.6$		1.3	171, 195	1
1.19 ± 1.13			2	195	2
	$2.0-2.6$		1.5	171, 195	3
$1.68-2.35$	1.79 ± 0.08		1.5	171, 195	4
	$2.31-2.59$		1.3	171, 195	5
	$2.04-2.52$		1.5	171, 195	5
1.71 ± 0.10	2.06 ± 0.10		2	171	6
1.75 ± 0.07	1.70 ± 0.17		2	195	6
1.52 ± 0.10	1.41 ± 0.09		1.5	AlMg	6
	1.54 ± 0.03		4	171+195+AlMg	6
2.12 ± 0.05			1	6563	7
$1.5-3.0$			1.5	1–500	8
		$1.4-2.0$	1	171,195,284	9
1.5		2.3	1.5	1550	10

previously reported in soft and hard X-rays. In Table 7.4 we list the powerlaw slopes α_E of energies (which are model-dependent), when the fluence was not reported. A similar study was done independently with another, but similar, event detection code, and powerlaw slopes in the range of $\alpha_P = 2.4$–2.6 were reported for the same volume model $V \propto A$, but a different range of $\alpha_P = 2.0$–2.1 for a modified volume model, i.e., $V \propto A^{3/2}$ (Parnell and Jupp 2000). The same data from these first two studies were reanalyzed with both volume models and powerlaw slopes of $\alpha_e = 2.52$–2.59 were inferred for the model $V \propto A$, and $\alpha_e = 2.04$–2.31 for the model $V \propto A^{3/2}$ (Benz and Krucker 2002), so we learned that the choice of the flare volume model changes the powerlaw slope of energies by about $\Delta\alpha \approx 0.4$.

A third study was conducted with the automated event detection code described in Section 6.9, which was designed to discriminate flare events (defined by impulsively heated and cooling loops) from non-flare events. For the frequency distributions of peak fluxes, a broken powerlaw was found at 171 Å with a slope varying from $\alpha_P = 1.68$ to $\alpha_P = 2.35$, but a single powerlaw at 195 Å with a slope of $\alpha_P = 1.85$ (Aschwanden et al. 2000a,b). Thermal flare energies were also determined using a cylindrical loop geometry for the flare volume, leading to a powerlaw slope of $\alpha_E = 1.79 \pm 0.08$ in the energy range of $E = 10^{24}$–$10^{26.5}$ erg (Fig. 1.14). This study also demonstrated that the selection of flare events can change the powerlaw slope by $\Delta\alpha \approx 0.3$. The next, more detailed, study was conducted using the combined EUV (TRACE) and soft X-ray data (Yohkoh), which allowed synthesize of a more complete temperature range, yielding more reliable total flare energies than previous studies in a single waveband (Aschwanden and Parnell 2002). The peak flux

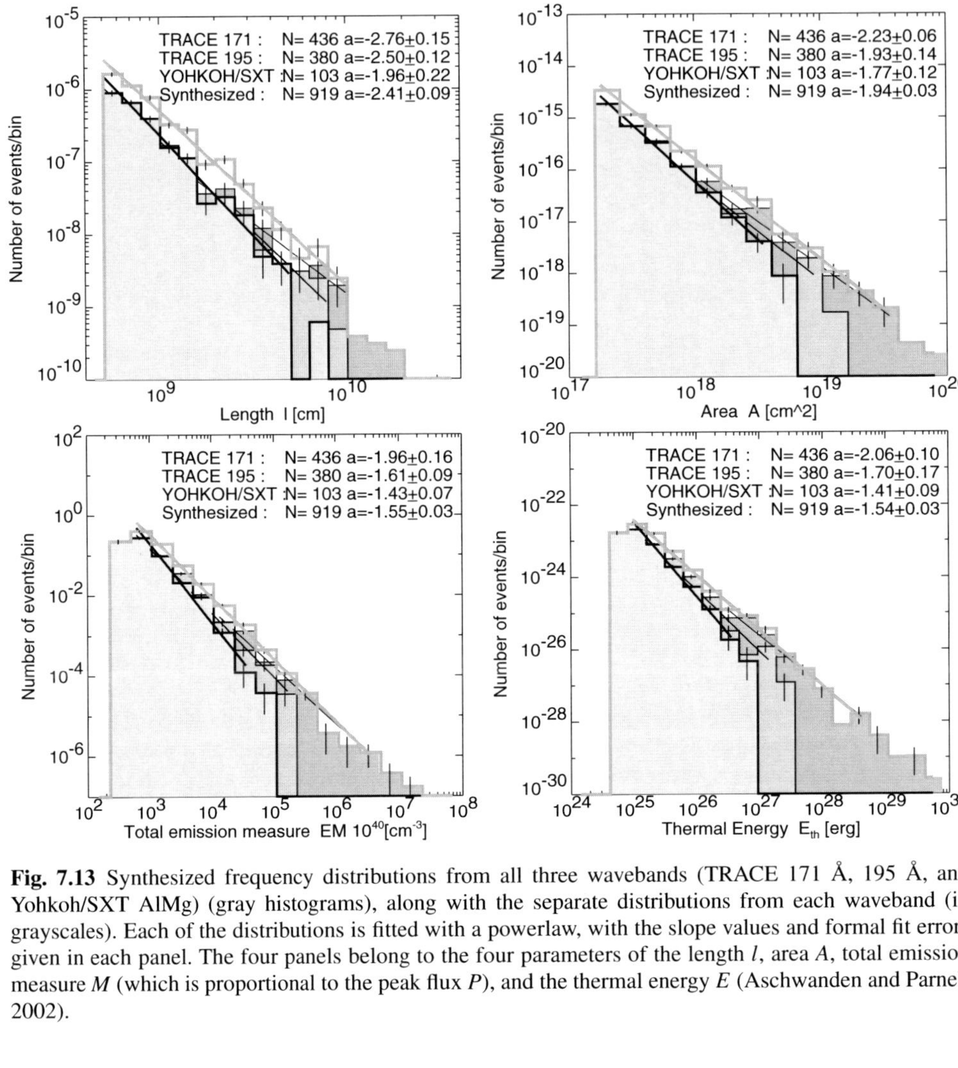

Fig. 7.13 Synthesized frequency distributions from all three wavebands (TRACE 171 Å, 195 Å, and Yohkoh/SXT AlMg) (gray histograms), along with the separate distributions from each waveband (in grayscales). Each of the distributions is fitted with a powerlaw, with the slope values and formal fit errors given in each panel. The four panels belong to the four parameters of the length l, area A, total emission measure M (which is proportional to the peak flux P), and the thermal energy E (Aschwanden and Parnell 2002).

distributions were found to have powerlaw slopes of $\alpha_P = 1.71 \pm 0.10$ (TRACE, 171 Å), $\alpha_P = 1.75 \pm 0.07$ (TRACE 195 Å), and $\alpha_P = 1.52 \pm 0.10$ (Yohkoh/SXT, AlMg, $T \gtrsim 2.4$ MK). Thermal flare energies were computed by taking the synthesized full temperature range as well as the fractal volume geometry (Chapter 8) of the flares into account, which yielded powerlaw slopes as steep as $\alpha_E = 2.06 \pm 0.10$ for the filter with the lowest temperature ($T \approx 1.0$ MK; TRACE 171 Å), or as low as $\alpha_E = 1.41 \pm 0.09$ for the filter with the highest temperature ($T \gtrsim 2.4$ MK; Yohkoh/SXT), while the synthesized distribution yields a powerlaw slope of $\alpha_E = 1.54 \pm 0.03$. Thus, this study demonstrated that there is also a temperature bias that steepens the powerlaw slope up to $\Delta\alpha \approx 0.9$, if the statistics is limited to a single narrowband filter of the lowest EUV temperature band. In Fig. 7.13 we show the frequency distributions of various parameters (length, area, total emission measure, and thermal energy) of the same event set measured in three different wavebands. Note the systematic flattening of the powerlaw slopes when including data with higher temperatures.

Further studies on frequency distributions were performed on EUV brightenings in the quiet Sun ($\alpha_P = 1.19 \pm 0.09$; Aletti et al. 2000), on full-disk EUV/XUV solar irradiance ($\alpha_P = 1.5$–3.0; Greenhough et al. 2003), or on EUV bright points over 9 years ($\alpha_T = 1.4$–2.0; McIntosh and Gurman 2005). Frequency distributions in flares were also evaluated for substructures that occur during a flare, such as UV brightenings of flare kernels observed in the C IV line ($\alpha_P = 1.5$, $\alpha_T = 2.3$; Nishizuka et al. 2009). Besides the EUV waveband, frequency distributions of small-scale variability events were also analyzed in the Hα line, which originates in the photosphere and chromosphere, such as in short-lived and small-scale events called *Ellerman bombs* ($\alpha_P = 2.12$; Georgoulis et al. 2002).

In summary we can say that the frequency distribution of nanoflares observed in EUV exhibit approximately the same powerlaw distributions of peak fluxes P and total energies E as observed in hard X-rays and soft X-rays, if the event definition is restricted to flare-like phenomena and if sufficiently broad temperature coverage is ensured to capture emission at the peak temperature of each event. However, several biases in the measurement of powerlaw slopes have been identified that are more severe in the cooler EUV waveband than in hotter soft X-ray wavebands, resulting from the event selection, narrowband temperature filters, and the geometric model of the flare volume, which enters the calculation of the thermal energy.

7.3.4 Solar Radio Emission

Solar radio bursts can be detected from ground-based instruments, and have thus been observed since their discovery by Hey and Southworth in 1942. Most solar radio bursts occur during solar flares, but they display a rich morphological variety that point to a number of different emission mechanisms, such as gyrosynchrotron emission of relativistic particles, electron beam-driven instabilities, loss-cone instabilities, or free-free (bremsstrahlung) emission. Radio emission at decimetric and microwave frequencies originate at the flare site and thus may show a detailed temporal co-evolution with the hard X-ray emission, while radio emission at metric and decametric wavelengths originate in the upper corona and heliosphere (Fig. 7.14), where they originate from local plasma instabilities or CME-driven shocks detached from the flare energy release process in the lower corona. This splits the statistics of solar radio bursts into two different realms, depending on the connectivity with the flare site, and consequently we expect possibly different frequency distributions for the two types. Solar radio bursts also span a large range of frequencies, from millimeter (≈ 300 GHz) to hectometer (≈ 3 MHz) wavelengths, and thus we might expect quite different frequency distributions depending on the wavelength or emission mechanism. A compilation of reported frequency distributions of solar radio bursts is given in Table 7.5.

The earliest frequency distributions of solar radio bursts were reported by Akabane (1956), who recorded solar radio bursts during 1951–1956 at 3 GHz and found powerlaw distributions with slopes of $\alpha_P \approx 1.8$. Further observations were reported in microwaves

Table 7.5 Frequency distributions measured from solar radio bursts, classified as type I storms, type III-like bursts, decimetric pulsation types (DCIM-P), decimetric millisecond spikes (DCIM-S), microwave bursts (MW), and microwave spikes (MW-S). References: 1, Akabane (1956); 2, Kundu (1965); 3, Kakinuma et al. (1969); 4, Fitzenreiter et al. (1976); 5, Aschwanden et al. (1995); 6, Aschwanden et al. (1998b); 7, Mercier and Trottet (1997); 8, Das et al. (1997); 9, Nita et al. (2002); 10, Ning et al. (2007).

Powerlaw slope of peak flux α_P	Powerlaw slope of total flux or total energy α_E	Powerlaw slope of durations α_T	log range	Waveband frequency f	Reference and type
1.8			2	3 GHz	1, MW
1.5			2	3, 10 GHz	2, MW
1.8			2	1, 2, 3.75, 9.4 GHz	3, MW
1.9–2.5			2	3.75, 9.4 GHz	3, MW
1.26–1.69			3	110 kHz–4.9 MHz	4, type III
1.28			2	100 MHz–3 GHz	5, type III
1.45 ± 0.31			3	100 MHz–3 GHz	6, type III
1.33 ± 0.11			3	100 MHz–3 GHz	6, DCIM-P
1.22–1.65			2.5	0.245–17 GHz	8, III, MW
1.71–1.91			4	0.100–2 GHz	9, III, MW
2.99 ± 0.63			3	100 MHz–3 GHz	6, DCIM-S
2.9–3.6			1.5	164, 237 MHz	7, type I
7.4 ± 0.4		5.4 ± 0.9	0.5	4.5–7.5 GHz	10, MW-S

(most likely to be produced by gyrosynchrotron emission) at 3 and 10 GHz during 1958–1959 with values of $\alpha_P = 1.5$ (Kundu 1965), at 1.2, 3.75, and 9.4 GHz during 1957–1962, with values of $\alpha = 1.8$ (Kakinuma et al. 1969), and at 3.75, and 9.4 GHz during 1957–1962, with values of $\alpha = 1.9$–2.5 (Kakinuma et al. 1969). Fitzenreiter et al. (1976) observed interplanetary type III bursts (produced by an electron beam instability) with the IMP-6 satellite during May–July 1971 at frequencies from 110 kHz to 4.9 MHz and found powerlaw distributions of their fluxes with slopes in the range of $\alpha_P = 1.26$–1.69. Interestingly, the value of the powerlaw slope systematically increases toward higher frequencies, which tells us something about the ratio of growth time τ_G to the saturation time t_S of the radio emission-producing instability, according to our model of exponentially growing instabilities, i.e. $\alpha_P = (1 + \tau_G/t_S)$ (Eq. 3.1.28). Statistics of flare-associated metric type III bursts yielded powerlaw slopes of $\alpha_P = 1.28$ (Aschwanden et al. 1995). Statistics on different types of flare-associated decimetric radio bursts included *decimetric type III types* (produced by electron beams) with $\alpha_P = 1.45 \pm 0.31$, *decimetric pulsation types* (produced by an oscillating instability) with $\alpha_P = 1.33 \pm 0.11$, and *decimetric millisecond spikes* (conceivably produced by an electron-cyclotron maser instability) with $\alpha_P = 2.99 \pm 0.63$ (Aschwanden et al. 1998b). There are always multiple radio bursts per flare, but when the distribution of peak fluxes or durations is investigated among the bursts occurring during a single flare, both powerlaw-like and exponential-like distributions are found. The relatively more restricted parameter space for a single flare could explain the exponential frequency distributions, which are not scale-free but define a dominant tem-

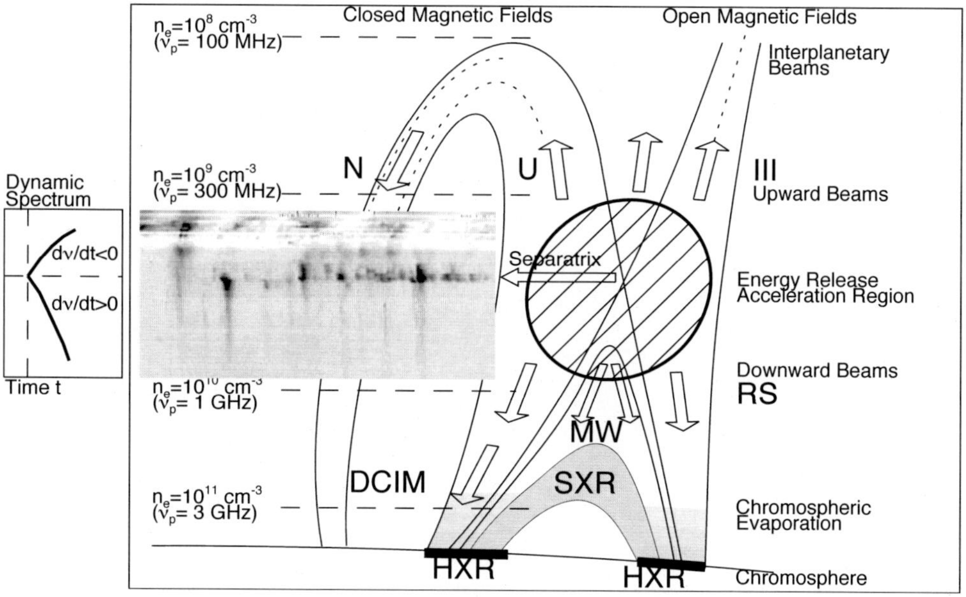

Fig. 7.14 Schematic overview of solar flare-related radio bursts: plasma emission excited by an electron beam instability produces radio bursts along open field lines escaping the acceleration region in upward direction (type III bursts), along upward escaping closed field lines (type U and N bursts), or in downward direction (reverse-slope drift [RS] bursts). Various decimetric (DCIM) radio bursts are produced by a losscone-type instability, sometimes with oscillatory patterns. Microwave emission (MW) produced by incoherent gyrosynchrotron emission mostly originates in flare loops where particles are injected from the acceleration region and subsequently become trapped (Aschwanden 2004).

poral or spatial scale. Mercier and Trottet (1997) sampled radio bursts from type I noise storms, which are produced above solar active regions without flares, probably associated with gentle continuous electron acceleration and found powerlaw distributions with slopes of $\alpha_P = 2.9$–3.6. Das et al. (1997) sampled radio bursts at frequencies from 245 MHz to 17 GHz and found some deviations from a strict powerlaw, which mostly affects the rollover at the low end of the distribution. A statistical analysis of decimetric millisecond spikes observed during single flares between 237 and 610 MHz exhibited both powerlaw-like and exponential-like flux distribution functions (Meszarosova et al. 1999, 2000). Isliker and Benz (2001) investigated how insufficient spatial and temporal resolution of these fast millisecond spikes affects the peak flux distribution function and found a tendency toward exponential behavior at large flux values. The most comprehensive statistics of solar radio bursts recorded over 40 years (1960–1999) compiled in NOAA catalogs was undertaken by Nita et al. (2002), finding powerlaw distributions with slopes in the range of $\alpha = 1.71$–1.91, using two different peak detection methods and sampling radio bursts in 8 frequency bands from 100 MHz to >2 GHz. Two examples of cumulative frequency distributions are shown in Fig. 7.15, measured at 2 GHz, which have a powerlaw slope (of the differential frequency distribution) of $\alpha = 1.82 \pm 0.01$ during the solar maximum and $\alpha = 1.81 \pm 0.02$ during the solar minimum, so there is no significant variation during the

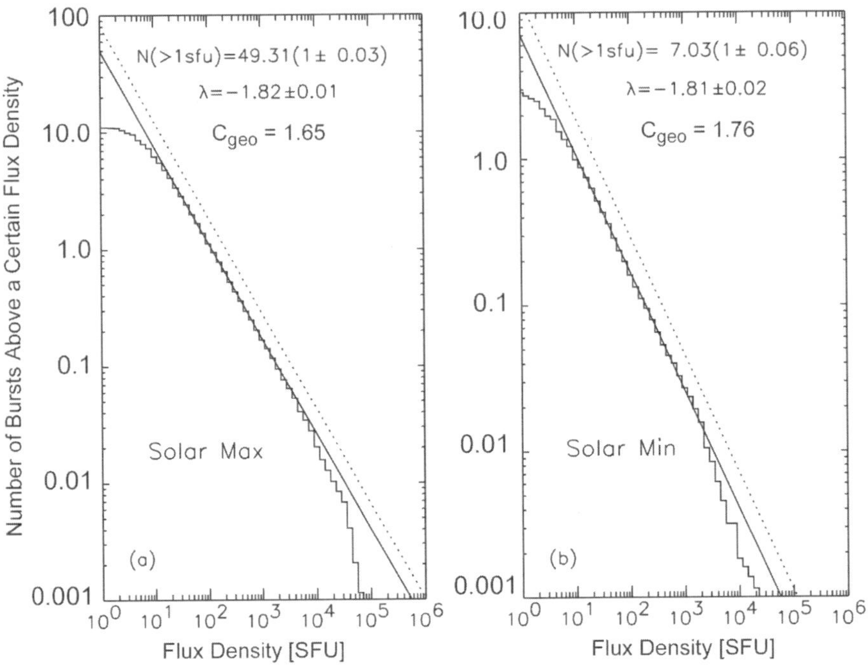

Fig. 7.15 Cumulative frequency distributions of radio bursts measured at 2 GHz during the solar maximum (left) and solar minimum (right) (Nita et al. 2002; reproduced by permission of the AAS).

solar cycle. Ning et al. (2007) reported very steep powerlaw slopes of $\alpha = 7.4 \pm 0.4$ for microwave bursts during a single event.

Inspecting Table 7.5 we see a clear pattern of two groups. The first group includes statistics of radio bursts occurring in many flare events, such as type III bursts and microwave bursts, which show similar powerlaw-like distributions as hard X-ray and soft X-ray flares, in the range of $\alpha_P \approx 1.3$–1.9. The second group includes statistics of radio burst fine structure during single flare events, such as decimetric millisecond spikes, type I sub-bursts, or microwave sub-bursts, which all exhibit very steep powerlaw distributions $\alpha \gtrsim 3$ or exponential distribution functions. The dissimilarity of statistical distributions of sub-bursts sampled during single flare events and the overall statistics sampled from many flares is clearly evident when they are juxtaposed in the same diagram (e..g, see examples in Aschwanden et al. 1998b). We conclude that physical parameters are more restricted during a single flare, and thus reveal a dominant temporal or spatial scale in the statistics of finestructure or sub-bursts, while a large statistical ensemble of many flares involves a much larger parameter range and produces the scale-free powerlaw distributions that are typical for flare statistics observed in other wavelength domains.

7.3.5 Solar Energetic Particle (SEP) Events

The highest particle energies detected in our heliosphere, mostly by in-situ detectors on spacecraft or by ground-based neutron monitors, can reach energies up to $\gtrsim 100$ MeV for electrons and $\gtrsim 1$ GeV for protons. While such high-energy particles were associated with cosmic rays earlier on, the current understanding is that they are accelerated either by shocks in coronal mass ejections, at typical distances of $R \approx 1$–5 solar radii, or in magnetic reconnection regions of solar flares in the lower corona. There is evidence for both scenarios, based on the timing inferred from the velocity dispersion of the detected particles: about half of the events have the time of their origin coincident with the flare peak times, while the other half originate with some significant delay, as expected for a CME-associated acceleration source. Nevertheless, whatever the origin of SEP events is, they represent very energetic phenomena and thus are expected to exhibit much flatter frequency distributions, like a subset of the largest and most energetic flare events.

An early frequency distribution of the intensity of 20–80 MeV protons (in units of protons cm^{-2} s^{-1} MeV^{-1}) was reported by Van Hollebeke et al. (1975), based on measurements of 185 SEP events during 1967–1972 with the *Interplanetary Monitoring Platform (IMP)* 4 and 5 spacecraft, who find a powerlaw distribution with a slope of $\alpha_P = 1.10 \pm 0.05$. Cliver et al. (1991) reported a powerlaw slope of $\alpha_P = 1.13 \pm 0.04$ for 24–43 MeV proton fluxes and $\alpha_P = 1.30 \pm 0.07$ for 3.6–18 MeV electron fluxes, based on 92 SEP events detected with the IMP-8 spacecraft during 1977-1983. Gabriel and Feynman (1996) collected data from the IMP 1, 2, 3, 5, 6, 7, 8, and the *Orbiting Geophysical Observatory (OGO)* 1 spacecraft observed during 1956–1990 and sampled frequency distributions of time-integrated particle fluxes (fluences), finding powerlaw slopes of $\alpha_E = 1.32 \pm 0.05$ for >10 MeV protons, $\alpha_E = 1.27 \pm 0.06$ for >30 MeV protons, and $\alpha_E = 1.32 \pm 0.07$ for >60 MeV protons, with little variation during the three solar cycles. A comprehensive compilation of the size distribution of >10 MeV solar proton events is provided by Miroshnichenko et al. (2001) for different datasets, based on the IMP spacecraft ($\alpha_P = 2.12 \pm 0.03$), the NOAA list ($\alpha_P = 1.47 \pm 0.06$), or SPE catalogues ($\alpha_P = 1.00$–1.43). The large range of powerlaw slopes in the latter dataset results from different threshold intensities, time ranges, or subsets with *sudden storm commencement (SSC)* associated events. The flattest slope $\alpha_P = 1.00 \pm 0.03$ was found for the lowest threshold, which corresponds to the most complete dataset, while the steepest powerlaw slope $\alpha_P = 2.12 \pm 0.03$ was found for the highest threshold and smallest data subset, and thus may be affected by the upper cutoff of the distribution. Looking at the temporal occurrence of SEP events, they are not well-correlated with the solar cycle, and thus unpredictable on time scales longer than the lifetime of an active region that has the necessary complex magnetic pattern (Xapsos et al. 2006; Hudson 2007).

In summary, the compilation in Table 7.6 shows that the frequency distributions of both the peak fluxes and fluences of SEP events are significantly flatter ($\alpha_P \approx \alpha_E \approx 1.1$–1.3) than for a comprehensive set of solar flares ($\alpha_P = 1.75 \pm 0.05$ and $\alpha_E = 1.61 \pm 0.04$). Since virtually all SEP events are also accompanied by a flare (unless the flare was occulted at the solar limb), the dataset of SEP events is essentially the most energetic subset of a flare distribution, regardless whether the acceleration of the high-energy particles occurred at

Table 7.6 Frequency distributions of solar energetic particle (SEP) events. References: 1, Van Hollebeke et al. (1975); 2, Belovsky and Ochelkov (1979); 3, Cliver et al. (1991); 4, Gabriel and Feynman (1996); 5, Smart and Shea (1997); 6, Mendoza et al. (1997); 7, Miroshnichenko et al. (2001); 8, Gerontidou et al. (2002).

Powerlaw slope of peak flux α_P	Powerlaw slope of total flux or total energy α_E	Powerlaw slope of durations α_T	log range	Energy range f	Reference and type
1.10 ± 0.05			3	20–80 MeV protons	1
1.40 ± 0.15				>10 MeV protons	2
1.13 ± 0.04			4	24–43 MeV protons	3
1.30 ± 0.07			4	3.6–18 MeV electrons	3
	1.32 ± 0.05		4	>10 MeV protons	4
	1.27 ± 0.06		4	>30 MeV protons	4
	1.32 ± 0.07		4	>60 MeV protons	4
1.47–2.42				>10 MeV protons	5
1.27–1.38				>10 MeV protons	6
1.00–2.12				>10 MeV protons	7
1.35			3.5	>10 MeV protons	8

the flare site in the lower corona or in associated CMEs further out in the heliosphere. The selection criterion for SEP events, e.g., the threshold for detecting > 10 MeV protons, includes all largest events of a peak flux distribution, but includes gradually less of the flare events with lower fluxes, which explains that their frequency distribution is flatter than for a complete set of flare events. Applying our exponential-growth model (Section 3.1), a powerlaw slope of $\alpha_P = (1+\tau_G/t_S) \approx 1.1$–$1.3$ corresponds to a mean ratio of $t_S/\tau_G = 3$–10 growth times, or mean amplification factors of $\exp(t_S/\tau_G) \approx 30$–$20{,}000$, which is hugely larger than the mean for average flares ($\exp(t_S/\tau_G) = \exp[1./(\alpha_E - 1)] \approx 5$. If we relate this mean amplification factor to the energy gain of the acceleration process, we expect that SEP events (with a powerlaw slope of $\alpha_E \approx 1.1$) produce a factor of $20{,}000/5 = 4{,}000$ higher energies, which explains the detection of > 10 MeV protons and >4 MeV electrons in SEP events.

7.4 Frequency Distributions in Astrophysics

While the Sun represents our local SOC laboratory that provides us abundant statistics and spatial information on each SOC event, observations of astrophysical sources offer only a few glimpses with sparse event statistics (due to the limited observing time allocation with expensive large telescopes) and no spatial information at all. This crucial limitation severely limits the characterization of frequency distributions, which requires ample statistics, but on the other hand, we can study exciting new phenomena that do not exist in our local solar system. The most-studied extra-solar SOC phenomena are stellar flares and accretion-disk or black-hole objects.

7.4.1 Stellar Flares

Here we review a few observations with published occurrence frequency distributions of stellar flares, mostly in UV wavelengths, which probe the transition regions at the base of stellar coronae. Robinson et al. (1999) observed the YZ Canis Minoris (YZ CMi) flare star (spectral type dM4.5e) on 1993 Nov 10 with the *High-Speed Photometer (HSP)* on the *Hubble Space Telescope (HST)* for 2.5 hrs and identified 54 flare events, finding a cumulative frequency distribution of the (time-integrated) flux with a slope of $\beta_E \approx 1.25 \pm 0.10$, approximately corresponding to a slope of $\alpha_E \approx 2.01 \pm 0.13$ for the differential frequency distribution, using Eq. (7.1.14) which includes the steepening effect of the cumulative frequency distribution near the upper cutoff, based on the logarithmic range $E_{max}/E_{min} \approx 10$ (see Fig. 3 in Robinson et al. 1999). The cumulative frequency distribution exhibits two bumps, so it is not well-characterized by a powerlaw function, which is often the case for small samples.

Audard et al. (2000) sample the flare activity of 12 (late-type) cool stars (spectral type F to M) from *Extreme Ultraviolet Explorer (EUVE)* Deep Survey observations. The cumulative frequency distributions of their total radiative energies E (which is assumed to be proportional to the total number of photon counts observed in the energy range of 0.01–10 keV) of these 12 stars is shown in Fig. 7.16. The cumulative frequency distributions are shown in form of rank-order plots, since there are only about 5–15 datapoints (flare events) measured for each star. Fitting a powerlaw distribution in the log-log plane they find the cumulative powerlaw indices β and estimate the powerlaw indices of the differential frequency distribution with the approximate relation $\alpha = \beta + 1$, listed in the first column of Table 7.7 (which corresponds to the value α^b in the fourth column of their Table 2). We estimate the corresponding powerlaw slopes α of the differential frequency distributions with the relationship $\alpha(\beta)$ given in Eq. (7.1.14), based on a powerlaw fit in the lower half of the cumulative distribution (marked with a thick line in Fig. 7.16) and the logarithmic range $\Delta E_{log} = \log(E_{max}/E_{min})$, with $q_2 = 10^{(-\Delta E_{\log}/2)}$, listed as values α_E^P in Table 7.7. We also fit the exact cumulative distribution function $N^{cum}(>x)$ as defined in Eq. (7.1.10), which includes the steepening at the upper end (marked with thin curves in Fig. 7.16), listed as values α_E^C in Table 7.7. We see that discrepancies between the three methods mostly arise where the logarithmic range of the sampled energies is small (listed in parenthesis in the column ΔE_{log} in Table 7.7), say $\lesssim 0.7$ decades (a factor of 5), excluding the rightmost energy bin containing the largest event. Powerlaw fits over such small ranges are not reliable because they could fit the gradual exponential-like cutoff without constraining the powerlaw part at lower energies. Thus, if we focus on the more reliable events with energy ranges of $\Delta E_{log} \geq 0.8$ (excluding the uppermost bin), we find 7 cases (out of the 12 stars analyzed by Audard et al. 2000) which have the following mean powerlaw slopes for each of the three methods: $\beta + 1 = 2.01 \pm 0.15$, $\alpha_E^P = 1.65 \pm 0.18$, and $\alpha_E^C = 1.75 \pm 0.26$. Thus, we find a significantly flatter slope of $\alpha_E \approx 1.7 \pm 0.2$ based on the two methods (α_E^P and α_E^C) that include the upper cutoff effect in the cumulative frequency distribution than the method $\alpha = \beta + 1$ that neglects this effect, used in Audard et al. (2000). A summary of various biases in the derivation of frequency distributions from stellar data is given in Güdel et al. (2003).

Table 7.7 Frequency distributions observed in stellar flares. The powerlaw slope α of the differential frequency distribution of energies is calculated with three methods: $\beta+1$ is from a powerlaw fit to the cumulative distribution (reported by authors); α_E^P is from a powerlaw fit to the lower half of the bins with correction Eq. (7.1.14), and α_E^C by fitting the cumulative distribution (Eq. 7.1.10). The logarithmic ranges $\Delta E_{log} = \log(E_{max}/E_{min})$ are given, where the numbers in parentheses give the range excluding the uppermost bin in the cumulative distribution that contains the largest flare. The values flagged with an asterisk were obtained from fitting photon arrival time distributions using Monte-Carlo simulations. References: 1, Robinson et al. (1999); 2, Audard et al. (2000); 3, Kashyap et al. (2002); 4, Güdel et al. (2003); 5, Arzner and Güdel (2004); 6, Arzner et al. (2007); 7, Stelzer et al. (2007).

Powerlaw slope of total flux $\beta+1$	α_E^P	α_E^C	Logarithmic range ΔE_{log}	Object	Instrument	Reference
2.25 ± 0.10	2.01		1.0	YZ Cmi	HSP/HST	1
1.89	2.40	2.15	1.6 (0.6)	HD 2726	EUVE	2
1.98	1.79	1.95	1.3 (0.8)	47 Cas	EUVE	2
2.27	1.71	1.93	1.0 (0.9)	EK Dra	EUVE	2
1.90	1.67	2.14	0.8 (0.4)	κ Cet 1994	EUVE	2
2.21	2.22	2.18	1.0 (0.6)	κ Cet 1995	EUVE	2
1.97	1.64	1.83	1.4 (1.1)	AB Dor	EUVE	2
2.50	2.00	2.68	0.5 (0.4)	ε Eri	EUVE	2
1.96	1.65	1.69	1.3 (1.1)	GJ 411	EUVE	2
1.85	1.72	1.81	1.5 (1.2)	AD Leo	EUVE	2
1.90	1.75	1.84	1.3 (1.0)	EV Lac	EUVE	2
1.91	2.20	1.79	1.6 (0.7)	CN Leo 1994	EUVE	2
2.14	1.26	1.19	1.0 (0.8)	CN Leo 1995	EUVE	2
$2.60 \pm 0.34^*$				FK Aqr	EUVE	3
$2.74 \pm 0.35^*$				V1054 Oph	EUVE	3
2.03–2.32*				AD Leo	EUVE	3
2.0–2.5*				AD Leo	EUVE	4
$2.3 \pm 0.1^*$				AD Leo	EUVE	5
1.9–2.5*				HD 31305	XMM	6
2.4 ± 0.5				TMC	XMM	7

Kashyap et al. (2002) analyzed also observations from the EUVI Deep Survey and inferred the frequency distribution $N(E) \propto E^{-\alpha}$ in an indirect way by Monte–Carlo simulations of photon arrival times, where the numerical model has three free parameters: α the powerlaw index of the energy distribution, r_F the average count rate due to flares, and r_C the average background count rate. So, the value α is found from the best fit of the modeled to the observed distribution of photon arrival times. The obtained values in the range of $\alpha \approx 2.2$–2.7 are significantly steeper than previously inferred values from similar stars. A similar value was obtained for AD Leo using the same method (Güdel et al. 2003; Arzner and Güdel 2004). Arzner et al. (2007) applied the same Monte-Carlo simulation technique to a sample of 22 stars observed with the *XMM-Newton* Extended Survey of the Taurus Molecular Cloud (XEST), but could constrain the powerlaw slope of the flare energy distribution with an acceptable fit ($\alpha_E = 2.0_{1.9}^{2.5}$) only for one case (HD 31305). It would be interesting to test the validity of this novel simulation method (that infers powerlaw slopes from fitting photon arrival time distributions) by comparing with powerlaw

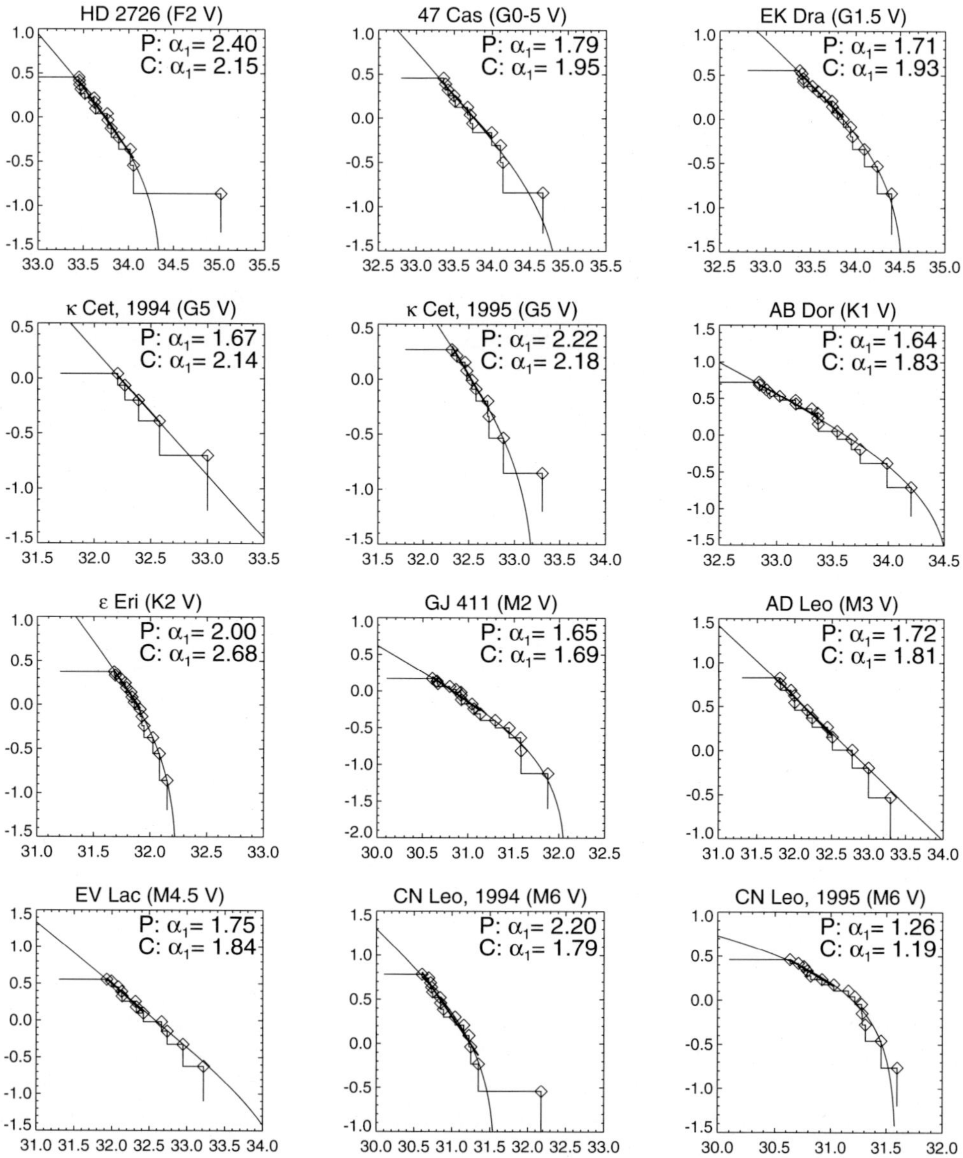

Fig. 7.16 Cumulative frequency distributions of flare energies (total counts) observed for 12 cool (type F to M) stars with EUVE (Audard et al. 2000). The flare events are marked with diamonds, fitted with a powerlaw fit in the lower half (P; thick line), and fitted with a cumulative frequency distribution (C; curved function).

slopes obtained from cumulative frequency distributions using the same data sets. Systematic biases of this method are not known yet, but it is conceivable that some assumptions (e.g., self-similar flare time profile) could influence the inferred powerlaw slopes (Arzner et al. 2007).

The inference of powerlaw slopes of frequency distributions from stellar flares cannot be obtained in the same way as for solar flares, where abundant statistics is available and a powerlaw slope can directly be fitted to the differential frequency distribution. Instead, the very small samples of flaring events per star require either the inversion of a rank-order plot (or cumulative frequency distribution) or a Monte Carlo simulation technique that fits a distribution of observed photon arrival times. Both methods have their own bias that need to be determined. For cumulative frequency distributions, the effect of the upper cutoff needs to be taken into account in small samples, which changes the powerlaw slope in the order of $\Delta\alpha \approx 0.3$. If this effect is taken into account, we infer values of $\alpha_E \approx 1.7 \pm 0.2$, which is similar to solar flares $\alpha_E \approx 1.61 \pm 0.04$ (Fig. 7.8), although the energies of the detected stellar flares are up to two orders of magnitude higher than the largest solar flares (Fig. 1.15).

7.4.2 Pulsar Glitches

Pulsars exhibit glitches in pulse amplitudes and frequency shifts that correspond to large positive spin-ups of the neutron star, probably caused by sporadic unpinning of vortices that transfer momentum to the crust. Conservation of the momentum produces then an increase of the angular rotation rate, like a twirling ice skater who draws the hands closer. The pulse height distribution of the Crab pulsar (NGC 0532 or PSR B0531+21) observed at 146 MHz was found to have a powerlaw slope of $\alpha_P \approx \beta + 1 = 3.5$ over a range of 2.25 to 300 times the average pulse size (Argyle and Gower 1972). Similar values were measured by Lundgren et al. (1995), with $\alpha_P \approx 3.06$–3.36 (Fig. 7.17). While the Crab pulsar is the youngest known pulsar (born in the year 1054), PSR B1937+21 is an older pulsar with a 20 times faster period (1.56 ms) than the Crab pulsar (33 ms). Cognard et al. (1996) measured a powerlaw distribution with a slope of $\alpha_P \approx \beta + 1 = 2.8 \pm 0.1$ from its occasional giant pulses. A theoretical model of the inertial momentum change in pulsar macro-glitches predicts a frequency distribution of $N(E) \propto E^{-1.14}$ (Morley and Garcia-Pelayo 1993), which is much flatter than previously observed. However, statistics on nine pulsars found powerlaw slopes in a range of $\alpha_E = -0.13, ..., 2.4$ (see Table 7.8) for the size distribution of pulse glitches (Melatos et al. 2008), but no correlation between the powerlaw slope and the spin-down age was found.

Interestingly, while most pulsars have a Gaussian or exponential pulse-amplitude distribution, only few pulsars, including the Crab pulsar, exhibit a powerlaw (Lundgren et al. 1995), which could be interpreted in terms of a SOC phenomenon (Young and Kenny 1996). Assuming a SOC model would also imply a powerlaw distribution for the pulse duration. Turbulence in neutral non-ionized fluids were considered as a possible mechanism that exhibit spatial and temporal scale invariance (Young and Kenny 1996). Alternative models in terms of modulational instabilities with stochastic growth and wave collapse were also proposed, which produce log-normal energy distributions with a steep power-

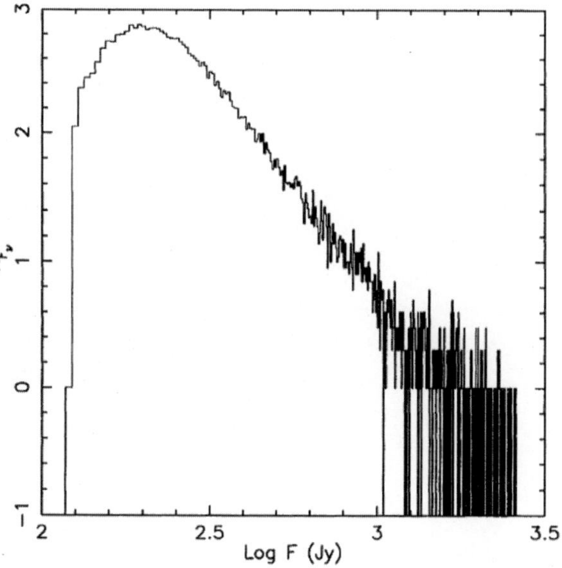

Fig. 7.17 Frequency distribution of giant-pulse flux densities measured from the Crab pulsar, observed during 15–27 May 1991 with the Green Bank 43-m telescope at 1,330, 800, and 812.5 MHz. The tail can be represented by a powerlaw distribution $N_F \propto F^{-\alpha}$ with $\alpha = 3.46 \pm 0.04$ for fluxes $F > 200$ Jy (Lundgren et al. 1995; reproduced by permission of the AAS).

Table 7.8 Frequency distributions observed from pulsar (giant-pulse) glitches (PSR), soft gamma-ray repeaters (SGR), black-hole object Cygnus X-1, and blazar GC 0109+224. Uncertainties in terms of one standard deviation are quoted in parentheses [...] for some cases. References: 1, Argyle and Gower (1972); 2, Lundgren et al. (1995); 3, Cognard et al. (1996); 4, Melatos et al. (2008); 5, Gogus et al. (1999); 6, Gogus et al. (2000); 7, Chang et al. (1996); 8, Mineshige and Negoro (1999); 9, Ciprini et al. (2003).

Powerlaw slope flux α_P	Powerlaw slope fluence α_E	Waveband	Object	Ref.
3.5		146 MHz	Crab pulsar	1
3.06-3.36		813–1330 MHz	Crab pulsar	2
2.8 ± 0.1		430 MHz	PSR B1937+21	3
2.4 [1.5,5.2]			PSR 0358+5413	4
1.2 [1.1,1.4]			PSR 0534+2200	4
0.42 [0.39,0.43]			PSR 0537-6910	4
1.8 [1.2,2.7]			PSR 0631+1036	4
−0.13 [−0.20,+0.18]			PSR 0835-4510	4
1.4 [1.2,+2.1]			PSR 1341-6220	4
1.1 [0.98,1.3]			PSR 1740-3015	4
0.57 [0.092,1.1]			PSR 1801-2304	4
0.36 [-0.30,1.0]			PSR 1825-0935	4
	1.66	>25 keV	SGR 1900+14	5
	1.43, 1.76, 1.67	20.8 keV	SGR 1806-20	6,7
7.1		1.2–58.4 keV	Cygnus X-1	8
1.55		optical	GC 0109+224	9

law tail $\alpha_E \approx 4$–7 at high energies, which could correspond to the "giant pulses" (Cairns 2004; Cairns et al. 2004).

Finally, a detailed cellular automaton SOC model was proposed for pulsar glitches, which could reproduce powerlaw slopes of $\alpha_E \approx 2.0$–4.3 for pulse sizes and $\alpha_T = 2.2$–5.5 for pulse durations (Warzawski and Melatos 2008). The underlying theoretical model of pulsar glitches is summarized by Warzawski and Melatos (2008) as follows: *The neutron superfluid in the stellar interior is threaded by many ($\approx 10^{16}$) vortices, approximately one per cent of which are pinned to the stellar crust at grain boundaries and/or nuclear lattice sites. As the pulsar crust spins down electromagnetically, a lag builds up between the velocity of the pinned vortex lines (corotating with the crust) and the superfluid. When the transverse Magnus force (directly proportional to the lag) surpasses a threshold value (equal to the strength of the pinning force), a catastrophic unpinning of vortices occurs, transferring angular momentum to the crust. In order for this mechanism to generate glitches on the scale observed, it requires up to 10^{12} vortices to unpin simultaneously, exhibiting a high level of collective, non-local behaviour.*

7.4.3 Soft Gamma-Ray Repeaters

Observations with the *Compton Gamma Ray Observatory (CGRO)* revealed a rare class of objects that show repetitive emission of low-energy gamma rays (>25 keV), termed *soft gamma-ray repeaters (SGR)*. In 1999, only four such SGR sources were known (three in our galaxy and one in the Magellanic Cloud), but at least three of them were identified to be associated with slowly rotating, extremely magnetized neutron stars, located in supernova remnants (Kouveliotou et al. 1998, 1999). They emit gamma-ray bursts with relatively soft spectra (like optically-thin bremsstrahlung at $k_B T \approx 20$–40 keV) and short duration of $\approx$0.1 s. Thompson and Duncan (1996) suggested that these gamma-ray bursts occur from neutron star crust fractures driven by the stress of an evolving, ultrastrong magnetic field ($B \gtrsim 10^{14}$ G).

Gogus et al. (1999) analyzed a database of 187 gamma-ray bursts (at energies of $\geq$25 keV) from SGR 1900+14 during the 1998–1999 active phase and found that the fluence or energy distribution of the bursts follows a powerlaw distribution over 4 orders of magnitude (Fig. 7.18, left). Also a correlation between the energy and duration was found, $E \propto T^{1.13}$, similar to solar flares (Eq. 7.3.2). Gogus et al. (2000) analyzed 290 events of SGR 1806-20 using data from the *Rossi X-Ray Timing Explorer (RXTE)*, 111 events detected with *CGRO/BATSE*, and 134 events detected with the *International Cometary Cometary Explorer (ICE)*, and found powerlaw slopes of $\alpha_E = 1.43$, 1.76, and 1.67 for the fluences, respectively (Fig. 7.18, right). The results were interpreted in support of the neutron star crustquake model of Thompson and Duncan (1996), in analogy to the SOC interpretation of earthquakes.

Soft gamma-ray repeaters with pulses originating from the same object are the exception rather than the rule, while most gamma-ray bursts detected with CGRO are non-repetitive, and thus come sporadically from different objects. Statistics of the temporal properties of those gamma-ray bursts has been gathered for several hundreds of events (e.g., Norris 1995; Norris et al. 1996; Quilligan et al. 2002), but it is not clear whether any

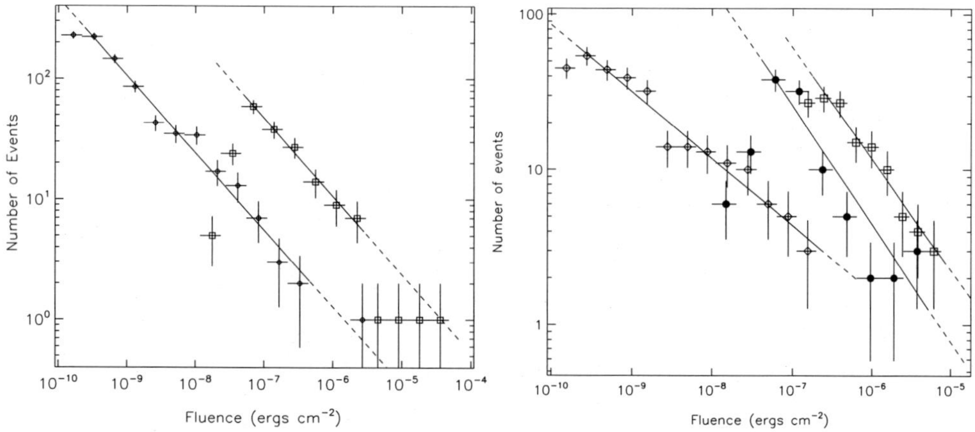

Fig. 7.18 Differential frequency distributions of the fluences of soft gamma-ray repeater sources: SGR 1900+14 (left), and SGR 1806-20 (right), observed with CGRO, RXTE, and ICE (Gogus et al. 1999, 2000; reproduced by permission of the AAS).

SOC characteristic could be retrieved from an observational sample that contains only one event per SOC system, where each SOC system has vastly different (unknown) distances to the observer.

7.4.4 Black Hole Objects

We discussed observations of black-hole candidates such as Cygnus X-1 in Section 1.9, numerical cellular automaton models of the surrounding accretion disks in Section 2.7.1, a shot-noise model of their power spectra in Section 4.8.4, and their waiting-time distributions in Section 5.7.2. In Fig. 7.19 (right) we show an observed occurrence frequency distribution of the peak intensity of the shots from a light curve of Cygnus X-1 (Negoro et al. 1995; Mineshige and Negoro 1999), along with a theoretical distribution (Fig. 7.19, left), simulated according to the cellular automaton model of Mineshige et al. (1994a,b). The observed peak-intensity distribution has a steep slope of approximately $\alpha_P \approx 7.1$. The cellular automaton model can accommodate a range of powerlaw slopes, depending on what fraction of mass m' (Eq. 2.7.4) is transferred by gradual diffusion in addition to the avalanche-like shots, e.g., simulations with $m' = m/100, m/10$, or $m/5$ produce powerlaw slopes of $\alpha_P \approx 5.6, 7.7$, and 11.5 (Mineshige and Negoro 1999; Takeuchi et al. 1995). Initial simulations without gradual diffusion produced energy distributions with powerlaws of $N(E) \propto E^{-2.8}$ and time scale distributions of $N(T) \propto T^{-1.4}$ (Mineshige et al. 1994b). Whatever the detailed scaling of the mass transfer in the accretion disk is, the fact of a powerlaw distribution of peak fluxes in the light curve is thought to support a SOC interpretation in terms of an avalanching system in a self-organized critical state.

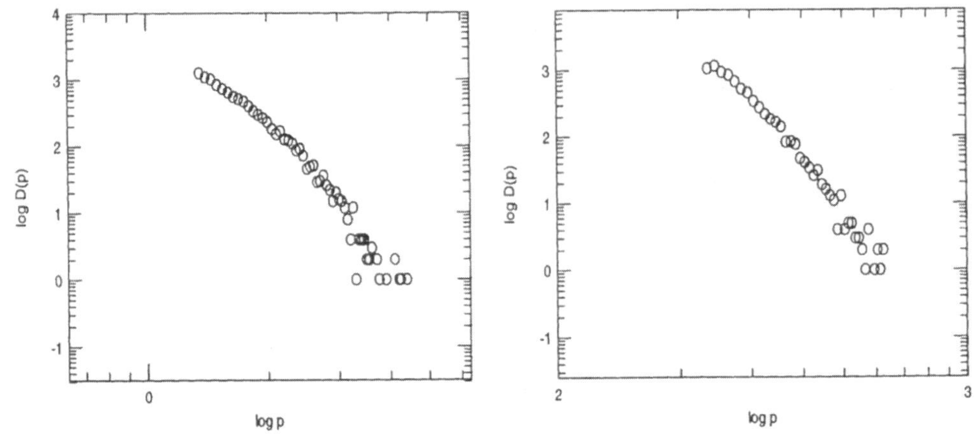

Fig. 7.19 *Left:* Numerically simulated frequency distribution of a cellular automaton model of mass avalanches in an accretion disk (Mineshige and Negoro 1999). *Right:* Observed frequency distribution of the peak intensities of pulses in the light curve of the black-hole object Cygnus X-1, exhibiting a powerlaw slope of $\alpha_P \approx 7.1$ (Negoro et al. 1995; Mineshige and Negoro 1999).

7.4.5 Blazars

Blazars (blazing quasi-stellar objects) are very compact quasars (quasi-stellar objects) associated with super-massive black holes in the center of active, giant elliptical galaxies. They represent a sub-group of active galactic nuclei (AGNs) which emit a relativistic jet in the direction of the Earth. Because of this particular geometry, where the jets are co-aligned with the line-of-sight to the observer, rapid variability and apparent super-luminous features are the paramount characteristics of these objects.

The optical variability of blazar GC 0109+224 was monitored from 1994 and the light curve was found to exhibit an intermediate behavior between flickering and shot noise, with a power spectrum of $P(\nu) \propto \nu^{-p}$ with $1.57 < p < 2.05$ (Ciprini et al. 2003). A combination of two modes between flickering (pink noise with $p > 0.8$) and pure shot noise (Brownian random walk or brown noise with $p \geq 2$) seems to be common in blazars (Hufnagel and Bregman 1992). Ciprini et al. (2003) constructed an occurrence frequency distribution of the peak fluxes of flare events and found a powerlaw distribution $N(P) \propto P^{-\alpha}$ with a slope of $\alpha \approx 1.55$, within a range of about one order of magnitude, and excluding the largest flares (Fig. 7.20). The powerlaw distribution of peak fluxes, along with the $1/f$ flicker noise spectrum of the light curve, was considered as an indication that blazars also represent a SOC phenomenon (Ciprini et al. 2003).

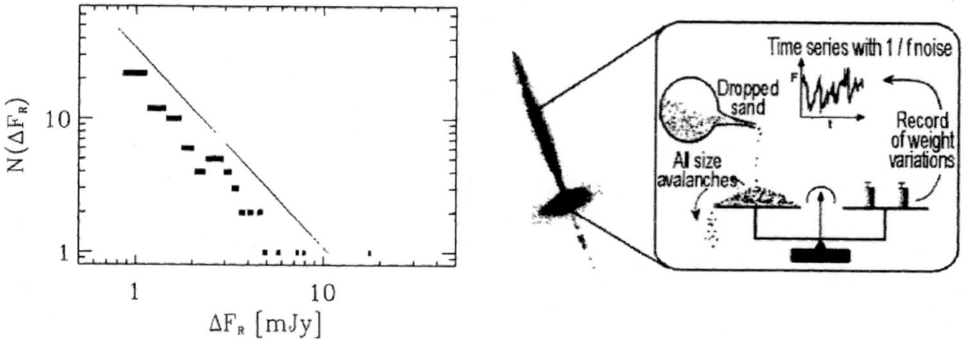

Fig. 7.20 *Left:* Frequency distribution of peak fluxes of flaring events in blazar GC 0109+224, including fluxes above of a 3σ-threshold, fitted with a powerlaw $N(P) \propto P^{-1.55}$. *Right:* A cartoon that illustrates the analogy of toppling avalanches in SOC sandpiles with jets emerging out of a blazar (Ciprini et al. 2003).

7.5 Summary

We have reviewed most of the occurrence frequency distributions observed in astrophysical event sets: magnetospheric substorms, solar and stellar flares, solar energetic particle events, solar radio bursts, pulsar glitches, soft gamma-ray repeaters, black hole objects, and blazars. Many frequency distributions of peak fluxes or fluences are found to be close to powerlaw distributions, with slopes varying in a considerable range of $\alpha \approx 1, ..., 10$, with a preference around $\alpha \approx 1.5$–2.0 for most phenomena. Statistics of the same phenomenon type exhibit their own characteristic value, such as (in increasing order): $\alpha_E \approx 1.1$–1.3 for magnetospheric substorm events, $\alpha_P \approx 1.1$–1.5 for solar energetic particle events (SEP), $\alpha_E \approx 1.4$–1.8 for soft gamma-ray repeaters, $\alpha_P \lesssim 1.5$ for blazars, $\alpha_P \approx 1.5$–1.8 for solar radio bursts, $\alpha_E \approx 1.6$–1.8 for solar (and probably stellar) flares, $\alpha_E \approx 3$ for pulsar glitches, or $\alpha_P \approx 7$ for black-hole objects. We identified a number of measurement biases that entered the published values, such as: (1) the "instrumental waveband bias" and "incomplete temperature coverage bias", which can lead to an overestimate of the powerlaw slope (e.g., for solar nanoflares detected with narrowband EUV filters); (2) the "big-event selection bias", which can lead to an underestimate of the powerlaw slope for event subsets that select larger events with a higher probability (e.g., SEP or CME events are not representative subsets of solar flares); or (3) the "upper-cutoff bias of cumulative frequency distributions", which leads to an overestimate of the powerlaw slope for small samples (e.g., stellar flares). All these biases in the measurement of powerlaw slopes and derivation of power indices in parameter correlations can be systematically studied with Monte-Carlo simulations (Section 7.1.4) and forward-fitted to the observed data. A self-consistent determination of the powerlaw slopes of peak fluxes (α_P), fluences (α_E), and durations (α_T) can quantify the correlations between the observables (P, E, T) that are most important for infering the scaling laws of underlying physical processes (see Chapter 9).

7.6 Problems

Problem 7.1: Simulate a distribution of random values that obey an exponential frequency distribution and verify that the histogrammed differential frequency distribution matches the analytical exponential function (follow Eqs. 7.1.28–7.1.30 and Fig. 7.3, left).

Problem 7.2: Simulate a distribution of random values that obey a powerlaw frequency distribution and verify that the histogrammed differential frequency distribution matches the analytical powerlaw function (follow Eqs. 7.1.31–7.1.33 and Fig. 7.3, right).

Problem 7.3: Use the numerically generated values of Problem 7.2 to construct the cumulative frequency distribution function and a rank-order plot (Fig. 7.2). What powerlaw slope do you infer from the cumulative frequency distribution or rank-order plot and how do you explain the difference to the original powerlaw slope of the numerically generated values?

Problem 7.4: Simulate the third case ($n = 10^4$ events) shown in Fig. 7.4 with different sets of random numbers and quantify the average accuracy or reproducibility or the powerlaw slopes and power indices of the parameter correlations.

Problem 7.5: What are necessary and sufficient conditions that the frequency distributions of solar flare energies observed in soft X-ray and hard X-ray wavelengths be identical?

Problem 7.6: Simulate a small sample of 15 random events that obey a powerlaw distribution function with a slope of $\alpha_E = 1.5$ to mimic a dataset of stellar flares (Fig. 7.16) and determine the powerlaw slope with three different methods: (1) with an overall powerlaw fit, (2) with a half powerlaw fit and the correction given in Eq. (7.1.14), and (3) with fitting the cumulative distribution function given in Eq. (7.1.10). How much different are the values determined with the three methods? Do you find a systematic bias when repeating the same experiment with different random number sets?

8. Fractal Geometry

Clouds are not spheres, mountains are not cones, coastlines are not circles, and bark is not smooth, nor lightenings travel in a straight line.

Benoit Mandelbrot

What really interests me is whether God had any choice in the creation of the world.

Albert Einstein

Fractals in nature originate from self-organized critical dynamical processes.

Per Bak and Kan Chen (1989)

Since Euclid (300 BC) we have been used to perceiving nature with the concept of a three-dimensional (3-D) geometry. We measure linear structures in one dimension, area-like structures in two dimensions, and volume-like structures in three dimensions. However, when we measure an object in terms of these three dimensions, we are aware that the geometric model describes a solid body, while a natural object may be inhomogeneous, porous, or even mostly empty, if we think on atomic scales. The counterpart to Euclidean geometry, the set theory with discrete elements, has been introduced by mathematicians like Georg Cantor, Karl Weierstrass, and Augustin-Louis Cauchy. The mathematical concept of discrete, irregular, inhomogeneous structures has then been discovered in the real world by Benoit Mandelbrot, who coined the definition of a *fractal dimension*, which represents a generalization (in terms of rational or irrational numbers) to the *Euclidean dimension* (which is restricted to integer values of 1, 2, 3, or n). A fractal dimension is a scale-free quantity that describes the fractional filling of a structure over some scale range, but usually does not extend to infinite microscopic or macroscopic scales. Popular examples are the coastline of Norway, ferns, trees, mountain landscapes, snowflakes, or clouds. The reason why we dedicate a chapter to fractal geometry here is, of course, because self-organized criticality also is governed by scale-free powerlaw distributions of observable parameters. Therefore, fractal geometry is nothing else than the spatial counterpart of self-organized criticality processes observed in the temporal and energy domain. In a paper entitled "The physics of fractals", Bak and Chen (1989) succinctly summarized *Fractals in nature originate from self-organized critical dynamical processes.*

General introductions to fractal geometry can be found in textbooks like *Fractals* (Mandelbrot 1977), *The Fractal Geometry of Nature* (Mandelbrot 1983), *The Beauty of Fractals* (Peitgen and Richter 1986), *Fractals Everywhere* (Barnsley 1988), *The Science of Fractal Images* (Peitgen and Saupe 1988), *Fractals, Chaos, Power Laws* (Schroeder 1991), *Critical Phenomena in Natural Sciences: Chaos, Fractals, Selforganization and Disorder* (Sornette 2004), *Discovery of Cosmic Fractals* (Baryshev and Teerikorpi 2002), or *Fractals and Chaos in Geology and Geophysics* (Turcotte 1997). Related articles can also be found in *The Physics of Fractals* (Bak and Chen 1989) and in popular articles like *The Language of Fractals* (Juergens et al. 1990) or *Chaos and Fractals in Human Physiology* (Goldberger et al. 1990). In the following we focus mostly on measurements of fractal dimensions or related spatial parameter distributions from astrophysical observations, which we relate to other SOC parameter distributions. Fractal structures were found in magnetospheric phenomena, solar flares, planetary systems, stardust, galactic structures, and cosmology.

8.1 1-D Fractals

In the next three sections we divide the discussion of fractal dimensions by their approximate spatial dimension, but this should not be taken too literally, because it is an intrinsic property of fractal structures that they deviate from a strict Euclidean dimension. One-dimensional structures are lines, segments of lines, contours, which can be straight, curved, intermittent, discrete, folded, intertwined, or deformed by any conceivable transform. If there is a repetitive pattern on different scales, such structures can be self-similar and fractal. Fractal structures are most naturally generated by a replication process that works in a self-similar way at different scales. For instance the growth of crystals occurs in subsequent layers that replicates the original molecular grid structure. Therefore, also the mathematical definition of fractal geometries usually makes use of a simple transformation rule that is repeated on successive size scales.

8.1.1 The Cantor Set and Koch Curve

A Cantor set, also called "Cantor dust", is a subdivision of a set into smaller pieces with a fixed fraction in each subsequent step. With progressive iterations, the number of elements increases to infinity, but their total length approaches zero. For instance, in the Cantor set shown in Fig. 8.1, a bar is subdivided into two bars by erasing the middle third, so the number of elements increases as $N = 2^i$ with every iteration i, while the length decreases as $\varepsilon = (1/3)^i$. In mathematical language, the set is uncountable but has a measure of zero. The Hausdorff dimension D of a one-dimensional fractal structure is defined as a powerlaw relation between the number N of elements and the length scale ε of an element,

$$N(\varepsilon) \propto \varepsilon^{-D} \qquad \text{for } \varepsilon \mapsto 0 \,, \tag{8.1.1}$$

Fig. 8.1 Construction of the "middle-third-erasing" Cantor set: The replication rule is to eliminate the middle third of every bar in subsequent subdivisions. The total length converges to zero, while the fractal dimension is $D = \log(2)/\log(3) = 0.630930....$

and quantifies how the number N of elements depends on the size scale ε. Thus, we can obtain the Hausdorff dimension D from $N = 2^i$, $\varepsilon = (1/3)^i = 3^{-i}$, and Eq. (8.1.1),

$$D = -\frac{\log N}{\log \varepsilon} = \frac{\log 2}{\log 3} \approx 0.630930.... , \qquad (8.1.2)$$

for this Cantor set. The fact that the fractal dimension D is smaller than the Euclidean value of $D = 1$ means that the 1-D structure has less than solid filling, and if the fractal process is continued to infinitely microscopic scales, it even approaches a total length of zero. Perhaps our whole universe has this fractal property if we probe matter down to microscopic scales, or even down to atomic and sub-atomic scales.

A classical example of a fractal one-dimensional structure is the Koch curve (Fig. 8.2). The initiator function is a straight line. A generator function is constructed by replacing the middle third by an equilateral triangle, so that the length of the fractal generator function is 4 units, while the size of the initiator function is 3 units. In subsequent iterations, each straight segment is replaced by another generator function (Fig. 8.2). The number of segments thus increases a factor of 4 with each iteration ($N = 4^i$), while the length of each segment becomes a factor 3 smaller each time ($\varepsilon = (1/3)^i = 3^{-i}$), which yields the Hausdorff dimension

$$D = -\frac{\log N}{\log \varepsilon} = \frac{\log 4}{\log 3} \approx 1.26186.... , \qquad (8.1.3)$$

Note that the fractal dimension is now larger than the Euclidean dimension $D = 1$ of a straight or smooth line, which indicates that the line is increasingly folded in a meandering pattern with smaller scales. Famous examples of this fractal structure is the coastline of Norway or Great Britain, which both became eroded by many fjords, valleys, rivers, streams, and creeks, so that the ragged coastal length increases the finer the spatial resolution of the topographical map is.

8.1.2 Irregularity of Time Series

Every type of one-dimensional data can be investigated in terms of fractal analysis. For instance, a string of binary data (e.g., 0 0 1 0 1 1 1 0 1 0 1 1 1 0 1 0 0 1 0 1 1 0) or a Morse code (e.g., ...—...–.-....–.—.-.–..) resemble pretty much the Cantor set shown in

Fig. 8.2 Construction of the "Koch curve": The generator function consists of three segments, with an equilateral triangle in the middle third, forming four straight segments of equal length. Six successive iterations are shown, where each straight segment is replaced by the fractal generator function.

Fig. 8.1, and thus a Hausdorff dimension (Eq. 8.1.1) can be determined, regardless whether the pattern is regular or irregular. However, a structure is only fractal, when the value of the Hausdorff dimension is found to be invariant at different scales, which means that the ratio of $\log(N)$ to $\log(\varepsilon)$ is constant, and thus implies a powerlaw behavior.

While the Cantor set (Fig. 8.1) represents a binary structure, 1-D data are generally multi-valued, such as a time series $f_i = f(t_i)$ with values in a range of $f_{min} \le f_i \le f_{max}$. A technique to measure the fractal dimension of a set of points $[t_i, f_i = f(t_i)]$ forming a graph or time profile of a function f has been developed by Higuchi (1988). The technique is normalized in such a way that a fractal dimension of $D = 1$ corresponds to a completely regular time profile (such as a constant or slowly-varying smooth time profile), but approaches the value of $D = 2$ for a completely irregular time series. Thus, the fractal dimension is a measure of the irregularity or complexity of a time profile. Let us consider a time series of values $f_i = f(i)$ as a function of the time step $i = 1, ..., N$,

$$f(1), f(2), f(3), ..., f(N) . \tag{8.1.4}$$

Then we generate subsets of time series with different time steps $k = 1, 2, 3, ...$, starting at all possible phases $m = 1, 2, 3, ..., k$,

$$f(m), f(m+k), f(m+2k), f(m+3k), ..., f(m+[(N-m)/k]k) . \tag{8.1.5}$$

For each time step k and phase m we can now define a length $L_m(k)$,

$$L_m(k) = \frac{1}{k}\left[\left(\sum_{i=1}^{|(N-m)/k|} |f(m+ik) - f(m+(i-1)k)|\right) \frac{N-1}{[(N-m)/k]k}\right] \tag{8.1.5}$$

Since we are interested in a time scale spectrum, but not in the phases m, we average the length $L_m(k)$ over all phases m and obtain a mean value $\langle L_m(k)\rangle$ for every time step k. If the length $\langle L_m(k)\rangle$ shows a powerlaw dependence on the time step k, the time series has a fractal dimension D,

$$\langle L_m(k)\rangle \propto k^{-D} . \tag{8.1.6}$$

Higuchi (1988) applied this algorithm to a time series of a fractional Brownian function, which has the property of self-similarity on all scales, and determined with this method the precise value of its fractal dimension $D = 2$.

8.1.3 Variability of Solar Radio Emission

The fractal analysis of Higuchi (1988) has been applied to time series of solar radio emission by Watari (1996a). The analyzed data consist of nine time series of daily solar radio fluxes at different frequencies from $\nu = 245$ MHz to $\nu = 15.4$ GHz, observed during the years 1976–1990, published in the *Solar-Geophysical Data* catalog, as well as the time series of the sunspot number. Thus the time resolution of the data is 1 day and the length is 15 years (i.e., 5,479 days or datapoints for each set). The time series are shown in Fig. 8.3, which all represent measures of the solar cycle variability observed at different wavelengths, different physical conditions, and different physical emission mechanisms. The measurement of the fractal dimension (Eq. 8.1.6) requires a time scale spectrum $\langle L_m(k)\rangle$, which are shown in Fig. (8.4), calculated in a range from $k = 1$ day to $k = 40$ days. Since a half solar rotation represents the longest possible time interval during which a solar radio source can be observed contiguously, the time series is expected to change its behavior at $k \lesssim 13$ days. The time scale spectra $\langle L_m(k)\rangle$ shown in Fig. 8.4 clearly show a powerlaw behavior at all frequencies in the range of $k \approx 1$–10 days, while a drop-off is visible in the range of $k \approx 10$–40 days, as expected from the solar rotation effect.

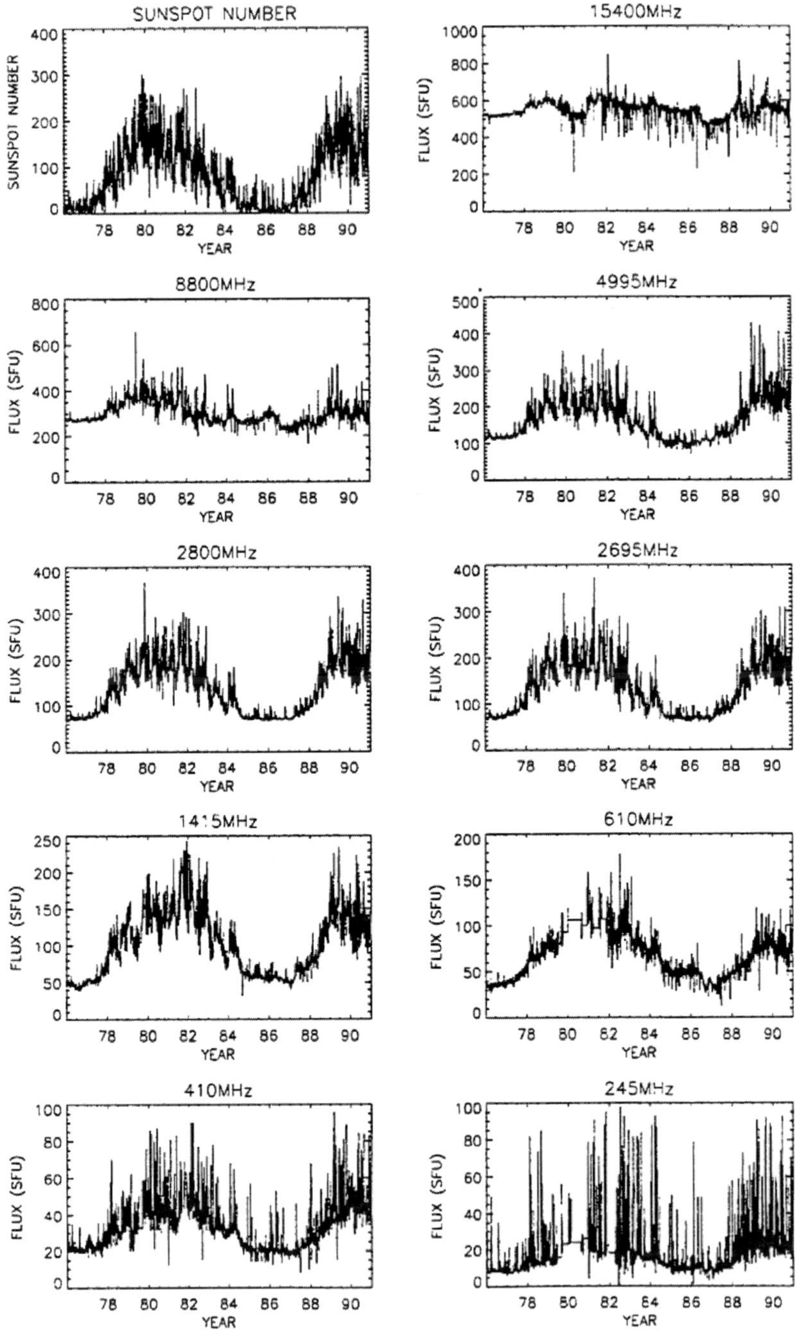

Fig. 8.3 Time profiles of the daily sunspot number (top left) and daily solar radio fluxes at frequencies of 245, 410, 610, 1,415, 2,695, 2,800, 4,995, 8,800, and 15,400 MHz (Watari 1996a).

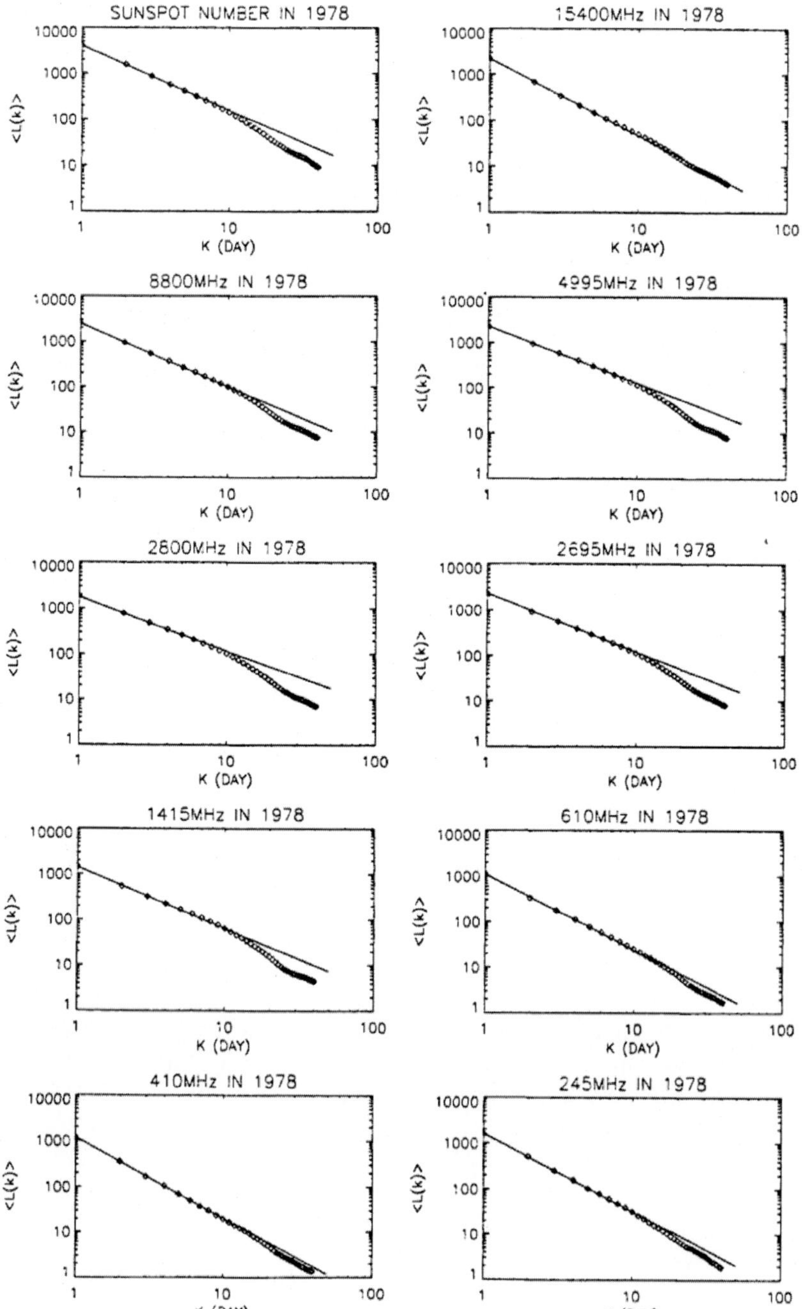

Fig. 8.4 Time scale spectrum $\langle L_m(k) \rangle$ of the 10 time series shown in Fig. 8.3 (Watari 1996a).

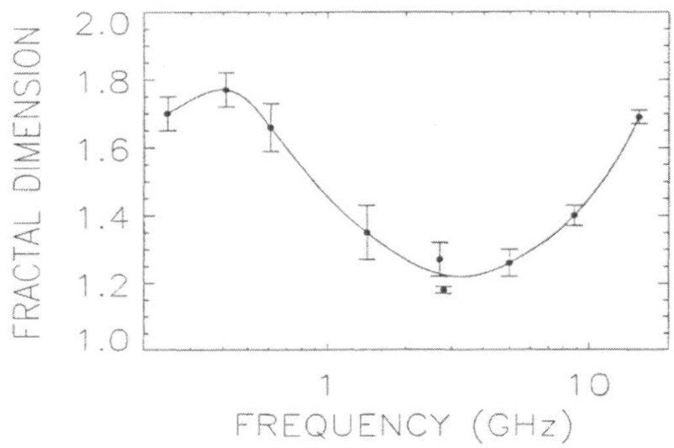

Fig. 8.5 Variation of the fractal dimension $D(\nu)$ as a function of the radio frequency ν from the year 1978, at the beginning of the solar cycle 21 (Watari 1996a).

An interesting result that came out of this study, besides the fractality of solar radio emission, is the dependence of the fractal dimension $D(\nu)$ on the radio frequency ν, which is shown in Fig. 8.5. There is a variation from a lowest fractal dimension of $D \approx 1.2$ at frequencies of $\nu \approx$ 2–5 GHz, to the highest fractal dimension with values of $D \approx 1.8$ at frequencies of 400 MHz, as well as near 15 GHz. This difference in the fractal dimension is likely to be a consequence of different radiation mechanisms. At decimetric frequencies ($\nu \approx 0.3 - 3.0$ GHz), solar radio emission is dominated by so-called *decimetric type III bursts*, which are caused by a beam-driven bump-in-tail instability producing plamsa emission. Such type III-like bursts occur very sporadically and irregularly due to the nonlinear nature of plasma instabilities, and thus can explain the high value of the fractal dimension measured in the $\nu \approx$ 0.3–1.0 GHz range. At higher frequencies, gyroresonance emission in strong magnetic fields, such as above sunspots, is the most dominant radio emission (e.g., Dulk 1985). Since the strong magnetic field above sunspots has a slowly-varying time evolution, this could explain the lower fractal dimension of $D \approx$ 1.2–1.3 at radio frequencies of $\nu \approx$ 1–5 GHz. This is also corroborated by the fact that Watari (1996a) found a similar low fractal dimension of $D \approx 1.2$ for the variability of the sunspot number. The third frequency domain at $\nu \gtrsim 10$ GHz, is too high to contain significant gyroresonance emission, and thus could be dominated by free-free bremsstrahlung from flare events, which occur very sporadically (see monthly averages in hard X-rays in Fig. 7.6), which could explain the upturn to a higher fractal dimension (Fig. 8.5) observed by Watari (1996a). In conclusion, the fractal dimension of the time series seems to provide a sensible diagnostic of physical emission mechanisms with different time variability characteristics.

8.2 2-D Fractals

By 2-D fractals we mean structures that can be measured from 2-D data, such as a flat or slightly curved image, as they are produced in abundance from CCD readouts of astronomical telescopes. If an image is strictly flat, any extracted structure can have a fractal dimension in the range of $D = 0,...,2$. Essentially, solid blobs appearing in an image have an Euclidean dimension of $D = 2$, curvilinear structures a dimension near $D = 1$, and dots a dimension near $D = 0$. A nice selection of fractal structures sorted by their dimension can be viewed on the wikipedia website (http://en.wikipedia.org/wiki/List_of_fractals_by_Hausdorff_dimension).

8.2.1 Hausdorff Dimension and Box-Counting Method

For Euclidean structures, the area A of a square is a quadratic function of the length scale or size L, i.e., $A = L^D$ with Euclidean dimension $D = 2$. If we cover the area A of linear size L with n squares, we have $n = L^D$ and can define an Euclidean dimension D by

$$D = \frac{\log n(L)}{\log L} , \tag{8.2.1}$$

which is also valid for other Euclidean dimensions $D = 1$ or $D = 3$. The same definition is extended to fractal structures, called the *Hausdorff dimension*, where D generally is a non-integer number,

$$D = \lim_{\varepsilon \mapsto 0} \frac{\log n(\varepsilon)}{\log(1/\varepsilon)} , \tag{8.2.2}$$

where $n(\varepsilon)$ is the number of self-similar structures of linear size $\varepsilon = 1/L$ that are needed to cover the whole structure. In Fig. 8.6 we show the iterative generation of the Sierpinski triangle, which is constructed by subdividing an equilateral triangle into four smaller triangles of half the size in each iteration step. Thus the fractal dimension of the Sierpinski

Fig. 8.6 Construction of the Sierpinski triangle in six iterative steps. Each triangle is subdivided into four triangles of half the size, with the middle one taken out. The Hausdorff dimension of the Sierpinski triangle is $D = \log(3)/\log(2) \approx 1.585$.

triangle can directly be calculated with Eq. (8.2.2),

$$D = \lim_{\varepsilon \mapsto 0} \frac{\log n(\varepsilon)}{\log(1/\varepsilon)} = \lim_{i \mapsto \infty} \frac{\log(3^i)}{\log(2^i)} = \frac{\log 3}{\log 2} \approx 1.58496... , \qquad (8.2.3)$$

The definition of the Hausdorff dimension (Eq. 8.2.2) leads directly to a practical measurement method. If we grid a 2-D image with a cartesian grid of size $L \times L$, where each macropixel has a size ε, the number of pixels $n(\varepsilon)$ that cover a fractal structure can be directly counted and set into relation with the linear extension of the structure $L = 1/\varepsilon$. If we define the number of pixels that cover a fractal structure as the *area* $A = n(\varepsilon)$, the fractal or Hausdorff dimension D can be obtained by

$$D = \frac{\log n(\varepsilon)}{\log(1/\varepsilon)} = \frac{\log A}{\log L} . \qquad (8.2.4)$$

Of course, a structure is only fractal when the same value D holds for a range of spatial resolutions $\varepsilon = 1/L$, so the box-counting has to be repeated for a range of spatial resolutions ε. For pixelized astronomical images with a size of $N_x \times N_x$, such as digital images from a CCD readout, it is often convenient to rebin the image by factors of 2^i, i.e., $\varepsilon = 1, 2, 4, 8, 16, ..., N_x$, which mimics the asymptotic limit $\varepsilon \mapsto 0$ in the definition of Eq. (8.2.3).

Sometimes, the fractal dimension D is also evaluated from the perimeter P or an area A, which is related as,

$$P \propto A^{D/2} . \qquad (8.2.5)$$

Note that the perimeter would scale as $P \propto A^{1/2}$ for linear features ($D = 1$), while it scales as $P \propto A$ for area-filling, meandering curves ($D = 2$).

The reader should be cautioned that the fractal dimension measured from a given observation depends very much on the definition of the measurement method. Thus, different computation methods may give differing values. The value of the Hausdorff dimension D (defined with Eq. 8.2.4) does not necessarily need to be identical with the fractal dimension D measured with the perimeter method (defined with Eq. 8.2.5), even when they are measured from an identical data set. Different methods used are specified in Table 8.1 (second column).

We show an example of the determination of the Hausdorff dimension for a solar EUV image recorded during the 1998 July 14 flare in Fig. 8.7 (called the Bastille-Day event because it occurred during the French national holiday). The image is rebinned into macropixels of size $\varepsilon = 1, 2, 4, 8, 16, 32, 64$ and the fractal dimension is determined by counting the macropixels with a brightness above some flux threshold, which yields the values of $D(\varepsilon = 1) = 1.607$, $D(\varepsilon = 2) = 1.563$, ..., $D(\varepsilon = 64) = 1.503$. The mean and standard deviation of the dimension determined with different macropixel sizes is $D = 1.55 \pm 0.03$, so it is approximately constant and thus the structure can be called fractal. A more accurate method would be to obtain D from the graph $\log(n)$ vs. $\log(1/\varepsilon)$ (Eq. 8.2.4).

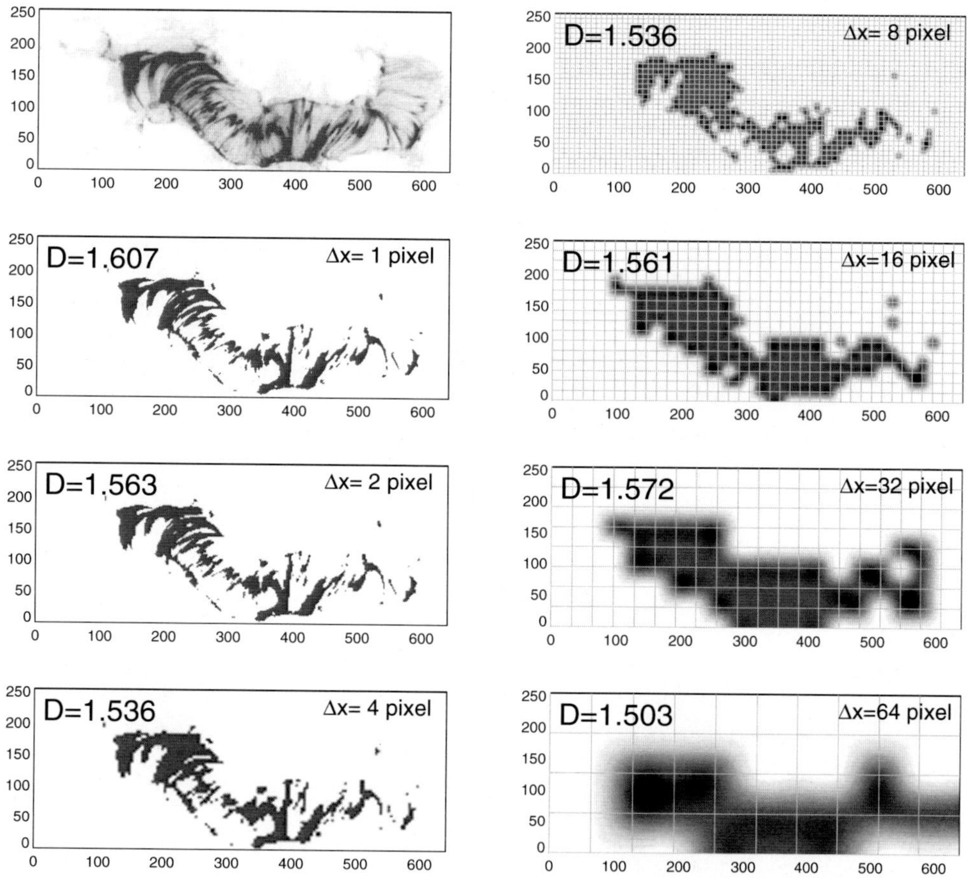

Fig. 8.7 Measurement of the fractal area of the Bastille-Day flare, observed by TRACE 171 Å on 2000-Jul-14, 10:59:32 UT. The Hausdorff dimension is evaluated with a box-counting algorithm for pixels above a threshold of 20% of the peak flux value, with a mean of $D_2 = 1.55 \pm 0.03$ for the 7 different spatial scales shown here. Note that the Hausdorff dimension is nearly invariant when rebinned with different scales (macropixel sizes of $\Delta x = \varepsilon$ = 1, 2, 4, 8, 16, 32, 64, indicated with a mesh grid). The original image with full resolution image ($N_x \times N_y = 640 \times 256$ pixels) is shown on a logarithmic greyscale in the top left frame, with a pixel size of $\Delta x = 0.5''$. The fractal dimension $D = \log(A)/\log(L)$ is simply evaluated from the number of rebinned macropixels $A(L)$ above the flux threshold and the rebinned image size $L = \sqrt{N_x \times N_y}/\Delta x$ (Aschwanden and Aschwanden 2008a).

8.2.2 Solar Photosphere and Chromosphere

The solar surface exhibits various features related to the magneto-convection (granulation, meso-granulation, super-granulation, network) or to areas of concentrated magnetic flux (sunspot umbrae, penumbrae, active regions, pores), which all have irregular geometries that have been subjected to fractal analysis (Table 8.1).

Table 8.1 Area fractal dimension D_2 of scaling between length scale L and fractal area $A(L) \propto L^{D_2}$ of various solar phenomena observed in white light, magnetograms, H-α, EUV, and soft X-rays. References: 1, Roudier and Muller (1987); 2, Hirzberger et al. (1997); 3, Bovelet and Wiehr (2001); 4, Paniveni et al. (2005); 5, Janssen et al. (2003); 6, Lawrence (1991); 7, Lawrence and Schrijver (1993); 8, Balke et al. (1993); 9, Meunier (1999); 10, Meunier (2004); 11, Lawrence et al. (1993); 12, Cadavid et al. (1994); 13, Lawrence et al. (1996); 14, McAteer et al. (2005); 15, Gallagher et al. (1998); 16, Georgoulis et al. (2002); 17, Aschwanden and Parnell (2002); 18, Aschwanden and Aschwanden (2008a,b).

Wavelengths regime and phenomenon (reference in superscript)	Method	Area fractal dimension D
Photosphere		
White-light of granules[1]	perimeter area	1.25, 2.15
White-light of granules[2]	perimeter area	1.3, 2.1
White-light of granular cells[2]	perimeter area	1.16
White-light of granules[3]	perimeter area	1.09
Magnetogram super-granulation[4]	perimeter area	1.25
Magnetograms of small scales[5]	perimeter area	1.41 ± 0.05
Magnetograms of active regions[6,7]	linear size area	1.56 ± 0.08
Magnetograms of plages[8]	linear size area	1.54 ± 0.05
Magnetograms of active regions[9]	linear size area	1.78–1.94
	perimeter area	1.48–1.68
Magnetograms of active regions[10]	perimeter area	
– Total		1.71–1.89
– Cycle minimum		1.09–1.53
– Cycle rise		1.64–1.97
– Cycle maximum		1.73–1.80
Magnetograms quiet Sun, active regions[11]	box-counting	multifractal
Magnetograms of active regions[12,13]	box-counting	multifractal
Magnetograms of active regions[14]	box-counting	1.25–1.45
Chromosphere		
EUV of quiet Sun network[15]	box-counting	1.30–1.70
H-α of Ellerman bombs[16]	box-counting	1.4
Corona, Flares		
EUV 171 Å of nanoflares[17]	box-counting	1.49 ± 0.06
EUV 195 Å of nanoflares[17]	box-counting	1.54 ± 0.05
Yohkoh SXT of nanoflares[17]	box-counting	1.65
EUV 171 Å of Bastille-Day flare[18]	box-counting	1.57–1.93

The solar granulation has a typical spatial scale of $L = 1{,}000$ km, or a perimeter of $P = \pi L \approx 3{,}000$ km. Roudier and Muller (1987) measured the areas A and perimeters P of 315 granules and found a powerlaw relation $P \propto A^{D/2}$ (Eq. 8.2.5), with $D = 1.25$ for small granules (with perimeters of $P \approx 500$–$4{,}500$ km) and $D = 2.15$ for large granules (with $P = 4{,}500$–$15{,}000$ km). The smaller granules were interpreted in terms of turbulent origin, because the predicted fractal dimension of an isobaric atmosphere with isotropic and homogeneous turbulence is $D = 4/3 \approx 1.33$ (Mandelbrot 1977). Similar values were found by Hirzberger et al. (1997). Bovelet and Wiehr (2001) tested different pattern recognition algorithms (Fourier-based recognition technique FBR and multiple-level tracking MLT) and found that the value of the fractal dimension strongly depends on the measurement method. The MLT method yielded a fractal dimension of $D \approx 1.1$, independent of the spa-

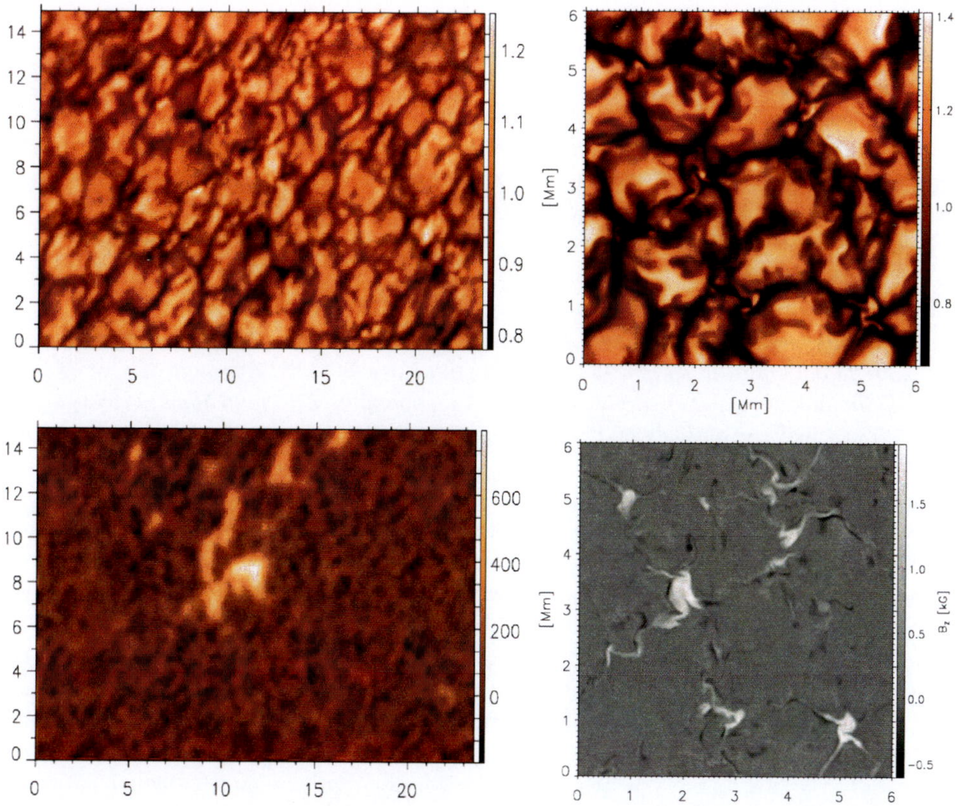

Fig. 8.8 *Left:* A speckle-reconstructed broadband image (top) and magnetogram (bottom) obtained with the Göttingen Fabry-Perot spectrometer at the Vacuum Tower Telescope on Tenerife. Tickmarks are given in arcseconds. *Right:* Snapshot of a numerical simulation of magneto-convection with the MURAM code, tuned to an average vertical field of 50 G. The upper panel shows the frequency-integrated intensity, while the lower panel shows the vertical magnetic field component B_z at a height with opacity $\tau_{5000} = 1$. The pixel size is 21 km, and the full image has a size of 6,000 km. Both the data and the numerical simulations were found to have a very similar fractal dimension of $D \approx 1.4$ (Janssen et al. 2003).

tial resolution, the heliocentric angle, and the definition in terms of temperature or velocity. Meunier (1999) evaluated the fractal dimension with the perimeter–area method and found $D = 1.48$ for supergranular structures to $D = 1.68$ for the largest structures, while the linear size-area method yielded $D = 1.78$ and $D = 1.94$, respectively. In addition, a solar cycle dependence was found by Meunier (2004), with the fractal dimension varying from $D = 1.09 \pm 0.11$ (minimum) to $D = 1.73 \pm 0.01$ for weak-field regions ($B_m < 900$ G), and $D = 1.53 \pm 0.06$ (minimum) to $D = 1.80 \pm 0.01$ for strong-field regions ($B_m > 900$ G), respectively. A fractal dimension of $D = 1.41 \pm 0.05$ was found by Janssen et al. (2003), but the value varies as a function of the center-to-limb angle and is different for a speckle-reconstructed image that eliminates seeing and noise. An example of data and numerical simulations with a time-dependent magneto-convection code is shown in Fig. 8.8, which both were found to have a very similar fractal dimension.

A completely different approach to measuring the fractal dimension D was pursued in terms of a 2-D diffusion process, finding a fractal diffusion with dimensions in the range of $D \approx 1.3$–1.8 (Lawrence 1991) or $D = 1.56 \pm 0.08$ (Lawrence and Schrijver 1993) by measuring the dependence of the mean square displacement of magnetic elements as a function of time. Similar results were found by Balke et al. (1993), The results exclude Euclidean 2-D diffusion but are consistent with percolation theory for diffusion of clusters at a density below the percolation threshold (Lawrence and Schrijver 1993; Balke et al. 1993).

Fractal dimensions were also evaluated with a box-counting method, finding a range of $D \approx 1.30$–1.70 for chromospheric network structures in a temperature range of $T = 10^{4.5} - 10^{6}$ K (Gallagher et al. 1998), a value of $D \approx 1.4$ for so-called *Ellerman bombs* (Georgoulis et al. 2002), which are short-lived brightenings seen in the wings of the Hα line from the low chromosphere, or a range of $D \approx 1.25$–1.45 from a large survey of 9,342 active region magnetograms (McAteer et al. 2005),

The physical understanding of solar (or stellar) granulation has been advanced by numerical convection models and N-body dynamic simulations, which predict the evolution of small-scale (granules) into large-scale features (meso or supergranulation), which is organized by surface flows which sweep up small-scale structures and form clusters of recurrent and stable granular features (Hathaway et al. 2000; Berrilli et al. 2005; Rieutord et al. 2008, 2010). The fractal structure of the solar granulation is obviously a self-organizing pattern that is created by a combination of subphotospheric magneto-convection and surface flows, which is a turbulence-type phenomenon, but is not in a critical state. The fractal structure of magnetic features, however, such as sunspots, active regions, magnetic pores), originate from magnetic flux emergence by buoyancy from the solar interior, which occur at independent places and times, and thus could possibly be attributed to a SOC system. The finding of a fractal dimension in magnetic features thus represents a necessary condition for scale-free (spatial) parameters that is typical for SOC, but not a sufficient condition. If the distributions of lifetimes, peak energies, and total energies of magnetic features also reveal powerlaw distributions, we can consider the driving system, i.e., the solar dynamo at the bottom of the tachocline (in a depth of $\approx$0.3 solar radii below the surface), to operate in a self-organized critical state. Instead of trickling sand grains on top of a SOC sandpile, the solar dynamo generates buoyant magnetic fluxtubes down in the tachocline, which cluster into small or large magnetic filament bundles when bubbling up to the solar surface in an avalanche-like fashion. The question is whether it is a SOC phenomenon or percolation. We will discuss percolation theory in Section 10.6.

8.2.3 Solar Flares

We have already extensively established that solar flares fulfill all criteria of a SOC system, regarding powerlaw distributions of total energies, peak energies, durations (Section 7.3), and waiting-time distributions in terms of a nonstationary Poisson process (Section 5.6).

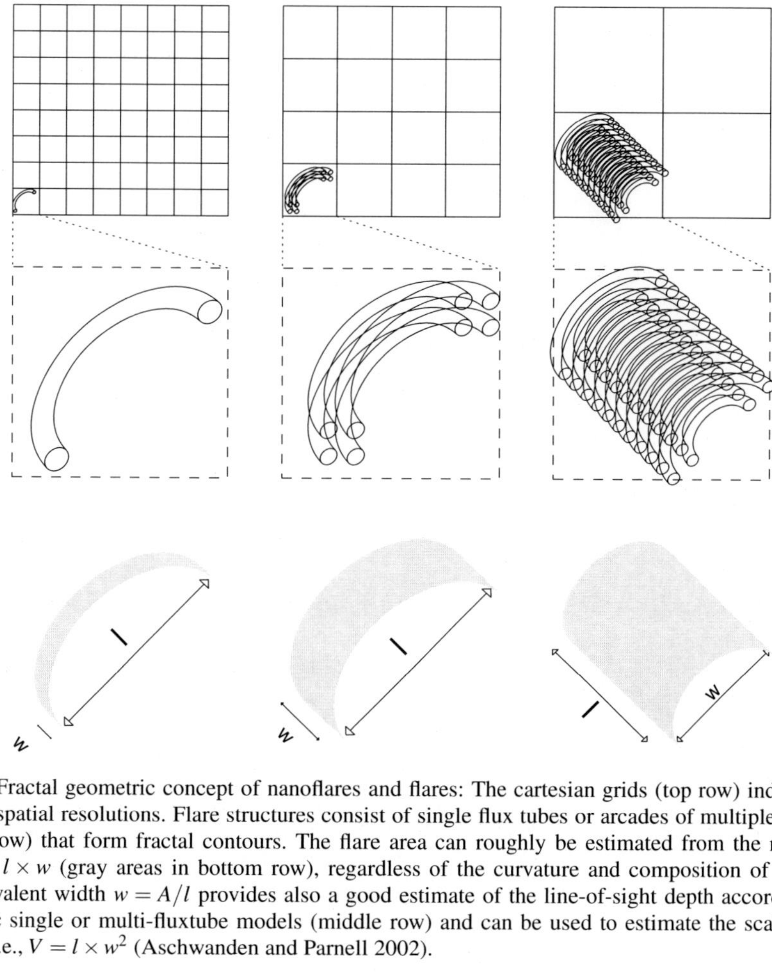

Fig. 8.9 Fractal geometric concept of nanoflares and flares: The cartesian grids (top row) indicate three different spatial resolutions. Flare structures consist of single flux tubes or arcades of multiple flux tubes (middle row) that form fractal contours. The flare area can roughly be estimated from the rectangular area $A = l \times w$ (gray areas in bottom row), regardless of the curvature and composition of the shape. The equivalent width $w = A/l$ provides also a good estimate of the line-of-sight depth according to the geometric single or multi-fluxtube models (middle row) and can be used to estimate the scaling of the volume, i.e., $V = l \times w^2$ (Aschwanden and Parnell 2002).

Consequently we expect also scale-free (powerlaw) distributions of spatial scales (lengths, areas, volumes) with fractal properties.

A fractal geometric concept of a solar flare is shown in Fig. 8.9, which consists of arcades of semi-circular flux tubes that generally are expected to have fractal (i.e., less than solid area-filling) contours above some flux level. For small flares (e.g., nanoflares observed in EUV), the fractal structure may not be resolved even in high-resolution ($\lesssim 1''$) images, but a crude characterization of their projected area A would at least show some asymmetry in their shape, which can be measured from the length l and width w of their elliptical shape (Fig. 8.9, bottom panels). Scaling the width w to the length l with a powerlaw index b, and characterizing the occurrence frequency distribution $N(l)$ of lengths with a powerlaw index a, we expect the following scaling relations and frequency distributions for fractal flare areas,

$$\begin{array}{ll} w(l) & \propto l^b \\ l(w) & \propto w^{1/b} \\ A(l) & \propto lw = l^{1+b} = l^D \\ N(l)\,dl & \propto l^{-a}\,dl \\ N(w)\,dw & \propto w^{-[1+(a-1)/b]}\,dw \\ N(A)\,dA & \propto A^{-(a+b)/(1+b)}\,dA \end{array} . \qquad (8.2.6)$$

The corresponding Hausdorff dimension $D = \log(A)/\log(l)$ is

$$D = (1+b) < 2 , \qquad (8.2.7)$$

Data analysis of $\approx$1000 nanoflares observed in EUV (TRACE and soft X-rays (Yohkoh) yielded values of $a = 2.5 \pm 0.2$ and $b = 0.5 \pm 0.2$, which corresponds to a Hausdorff dimension of $D = 1.5 \pm 0.2$ and an area distribution of $N(A) \propto A^{-2.0}$ (Aschwanden and Parnell 2002).

The geometric flare concept shown in Fig. 8.9 visualizes small flares that consist of only one single or a few loops (Fig. 8.9, left and middle), which is typical for EUV nanoflares, but also large flares, which consist of hundreds of loops, geometrically arranged in near-concentric arcades (Fig. 8.9, right). The fractal structure of such large flares has been investigated in detail for the Bastille-Day flare of 2000 July 14 (Aschwanden and Aschwanden 2008a). The story is not simple. Measuring the fractal dimension as a function of time, but normalizing it to the same flare area A_{max} defined around the peak time of the flare, the fractal area varies in the range of $A(t)/A_{max} = 0.08$–0.67, corresponding to a Hausdorff dimension of $D(t) = 1.57$–1.93. The time evolution is shown in Fig. 8.10, which exhibits some correlation of the fractal dimension with the EUV flux, which essentially tells us that more and more fractal structures (flare loops) brighten up before the flare peak. Typically, a flare starts when a first loop brightens up, which is a nearly linear feature and thus has a dimension of $D \gtrsim 1$, while more and more loops come into play as the flare progresses, until the flare area is almost solidly filled with $D \lesssim 2$. Moreover, the determination of the fractal dimension depends also on the flux threshold. Data as well as simulations show a variation of the fractal dimension of $D \approx 1.4$–1.9 depending on the chosen flux threshold, say in the range of $F_{th} = 10$–50 DN s^{-1} as shown in Fig. 8.11. Generally, the value of the fractal dimension drops with higher thresholds. In the same study, a total of 20 large (GOES X-class and M-class) flares were investigated from TRACE observations, which all have very complex fractal finestructure, as shown in Fig. 8.12, and the fractal dimensions cover a substantial range during the flare evolution. A summary of the fractal areas ranges $A(L)$ versus the length scale L is shown in Fig. 8.13, which has a mean fractal dimension of $D = 1.89 \pm 0.05$ during the flare peak, but covers a range of lower values of $D \gtrsim 1.0$–1.5 at the beginning of the flare. Thus, a complete SOC theory should also include the time evolution of the fractal geometry. Our simplest SOC work model (Section 3.1) quantifies a SOC avalanche in terms of an exponential growth phase and a linear decay phase, which implies a multiplicative pattern in energy release and spatial structures. In order to predict the temporal evolution of the 2-D fractal dimension, the 3-D evolution of spatial structures has to be mapped into a 2-D plane (see Section 8.3).

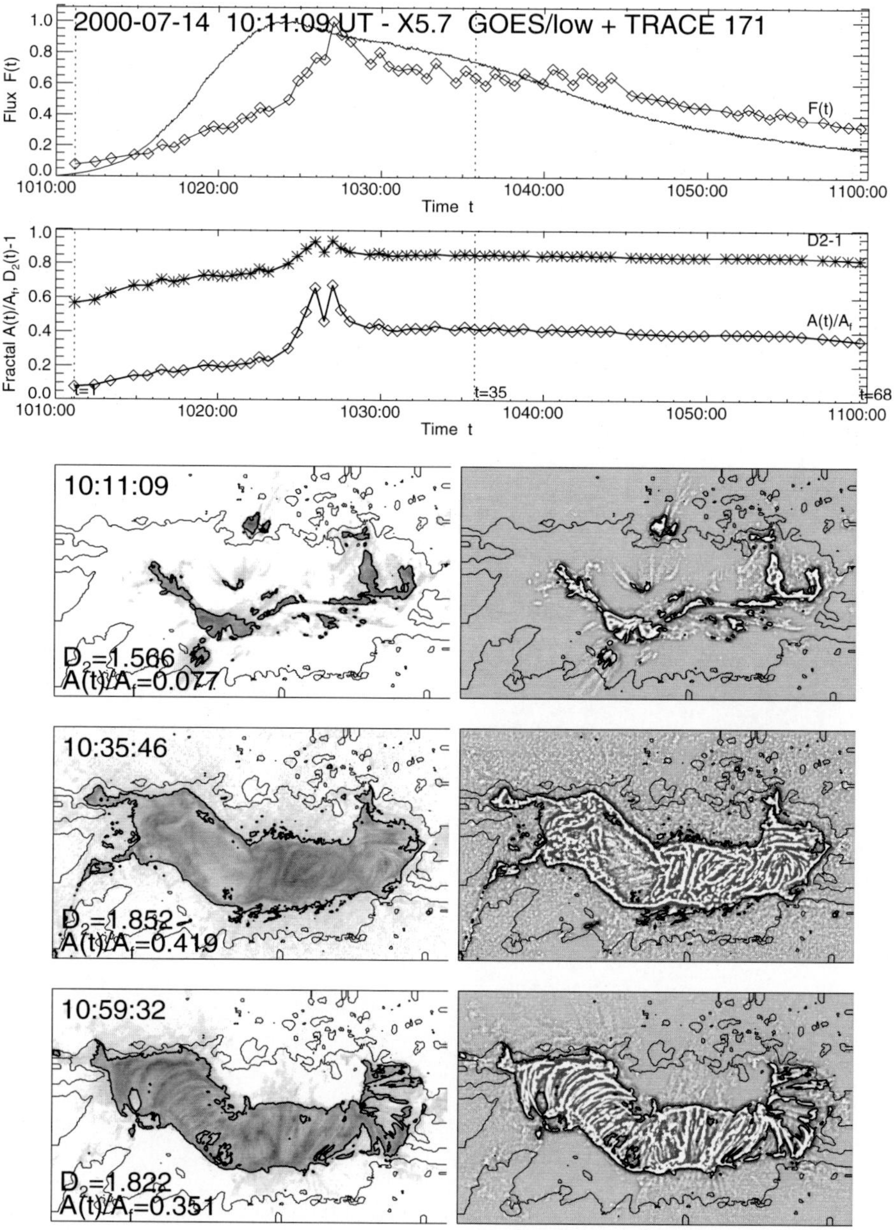

Fig. 8.10 The time evolution of the normalized EUV 171 Å flare flux $F(t)$ (diamonds) and soft X-ray flux from GOES (smooth curve) are shown for the Bastille-Day flare (top panel), along with the fractal area $A(t)/A_f$ and fractal dimension $D_2(t)$ (second panel). The TRACE 171 Å images at start (t_1), middle (t_{35}), and end time (t_{68}) are shown in the three lower panels on a logarithmic flux scale (three lower left panels) and high-pass filtered (three lower right panels). The instantaneous flare areas $A(t)$ are marked with thick black contours, while the time-integrated flare area A_f is marked with thin contours (Aschwanden and Aschwanden 2008a).

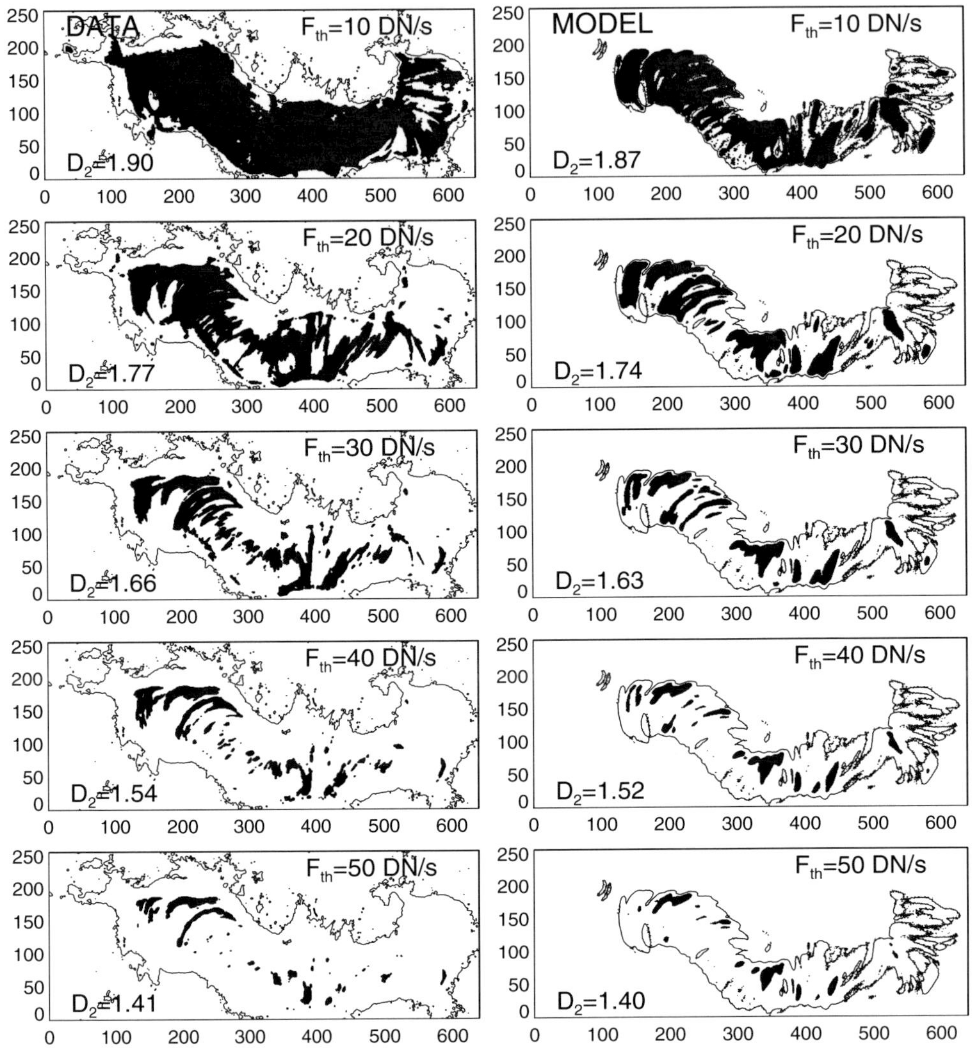

Fig. 8.11 The fractal area $A(F > F_{th})$ is shown for the same data image (left panels) and model image (right panels) as given in Fig. 8.10 for different flux thresholds $F_{th} = 10, 20, ..., 50$ DN s^{-1}. The flare area is contoured at a flux threshold of $F_{th} = 5$ DN s^{-1}. Note the similar dependence of the fractal dimension D_2 (indicated at bottom left corner of each panel) on the flux threshold for data and model (Aschwanden and Aschwanden 2008a).

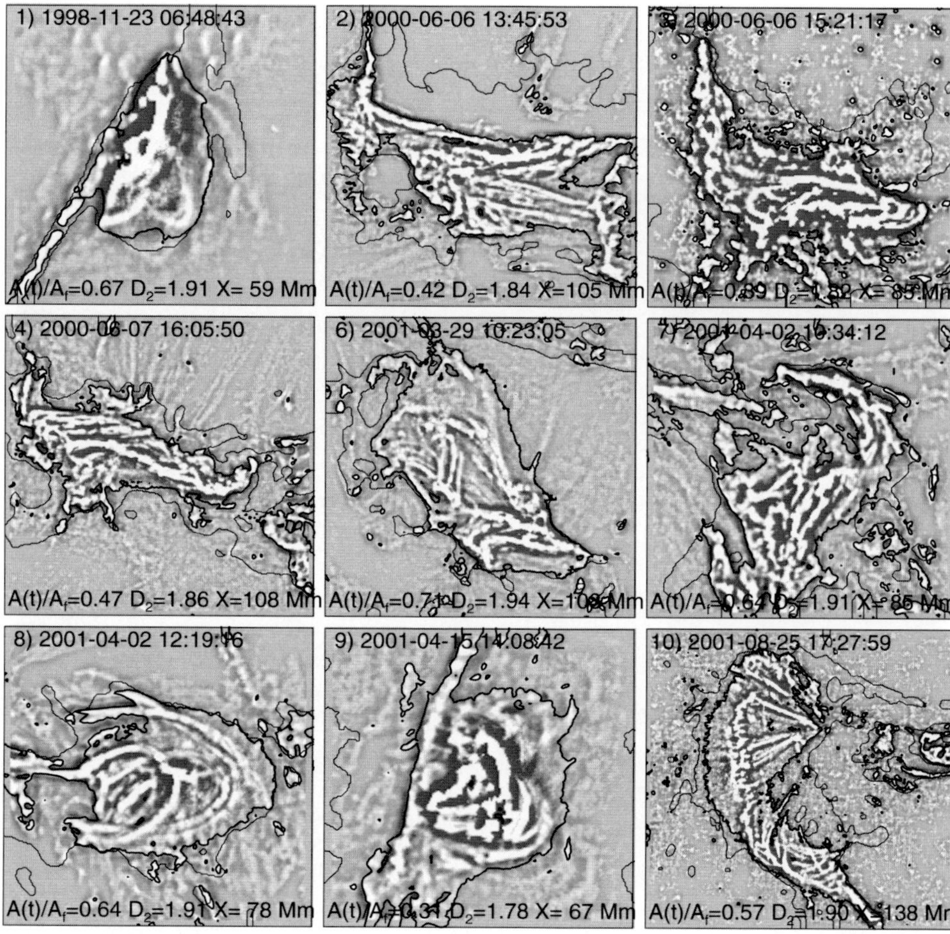

Fig. 8.12 Highpass-filtered images of nine X-class flares are shown, which enhance the fractal finestructure of flare loops (Aschwanden and Aschwanden 2008a).

8.3 3-D Fractals

The theoretical extension of 2-D to 3-D fractal dimension is straightforward. In the definition of the Hausdorff dimension we have to replace the area A by the volume V, and the number $n(\varepsilon)$ of elements that cover a fractal structure are 3-D voxels, rather than 2-D pixels,

$$D_V = \lim_{\varepsilon \mapsto 0} \frac{\log n(\varepsilon)}{\log(1/\varepsilon)} = \frac{\log V}{\log L} \ . \tag{8.3.1}$$

The practical measurement of a 3-D fractal dimension D_V, however, is not straightforward, but can be inferred with help of tomographic 2-D projections and computer simulations. Especially in astrophysical applications, only 2-D data are available in general, and thus

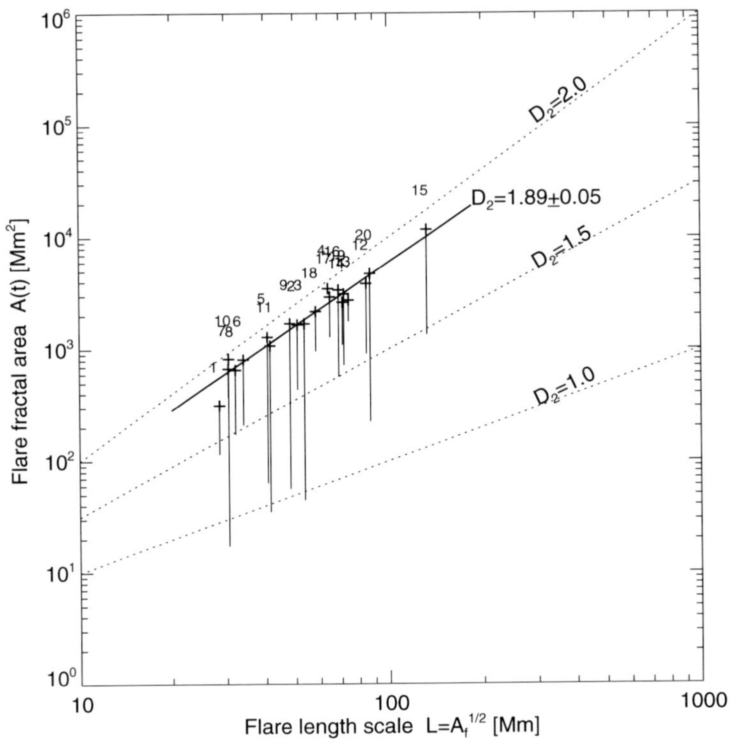

Fig. 8.13 Fractal areas $A(t)$ of flares as a function of the spatial length scale $L = \sqrt{A_f}$. For each flare there is an evolution of the fractal area $A(t)$ as a function of time (vertical range). The cross symbols mark the maximum of the fractal dimension reached during the entire flare duration. The fractal dimensions of $D_2 = 1.0, 1.5, 2.0$ are indicated with dotted lines, and the average maximum fractal dimension D_2 is indicated with a thick solid line, having a mean of $D_2 = 1.89 \pm 0.05$ (Aschwanden and Aschwanden 2008a).

the inference of a 3-D fractal dimension requires a spatial model, stereoscopic observations, or tomographic reconstructions. The scale invariance in terms of 3-D fractal geometry, however, has been probed from microscopic structures such as snow crystals (e.g., Westbrook et al. 2004), all the way to clustering of galaxies, cosmic voids, and dark matter (e.g., Gaite 2007).

8.3.1 Cellular Automaton Simulations

Cellular automaton simulations of SOC models have generally been performed in both 2-D and 3-D geometries (e.g., Bak et al. 1987, 1988; Lu and Hamilton 1991; Charbonneau et al. 2001). Special attention to the relationship between the 2-D and 3-D fractal dimension has been paid in the studies of Charbonneau et al. (2001), McIntosh and Charbonneau (2001), and McIntosh et al. (2002). An example of a 3-D avalanche with 2-D projections in a cellular automaton run is shown in Fig. 8.14. For their largest simulated datacubes

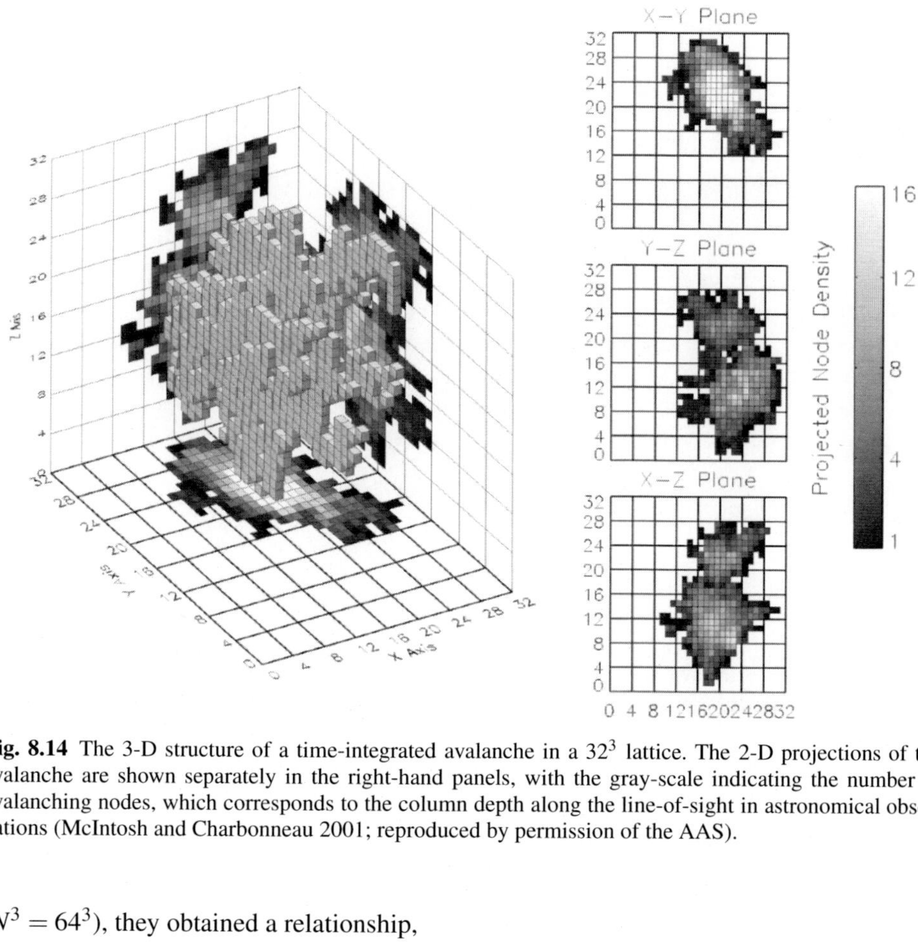

Fig. 8.14 The 3-D structure of a time-integrated avalanche in a 32^3 lattice. The 2-D projections of the avalanche are shown separately in the right-hand panels, with the gray-scale indicating the number of avalanching nodes, which corresponds to the column depth along the line-of-sight in astronomical observations (McIntosh and Charbonneau 2001; reproduced by permission of the AAS).

($N^3 = 64^3$), they obtained a relationship,

$$V(A) \propto A^{1.41 \pm 0.04} \,, \tag{8.3.2}$$

while the Euclidean scaling would be $V \propto A^{3/2}$, so the relationships are not identical for fractal and solid bodies.

There are different ways to define the linear size L of a fractal structure. One method is to define a *radius of gyration R*,

$$R^2 = \frac{1}{M} \sum_{i=1}^{M} |\mathbf{r}_i - \mathbf{R}_0|^2 \,, \tag{8.3.3}$$

where the sum runs over the M nodes that are part of the avalanche cluster, and $\mathbf{R}_0 = (1/M) \sum \mathbf{r}_i$ is the cluster's center of mass. Physically, R is the radius of the thin spherical shell (circular ring in 2-D) that has the same "mass" and moment of inertia as the original cluster (Stauffer and Aharony 1994; Charbonneau et al. 2001). Using this definition for the length scale ($L = R$), Charbonneau et al. (2001) find the following scaling for their simulation with the largest 3-D cube ($N^3 = 128^3$),

$$A(L) \propto L^{1.78 \pm 0.02} , \qquad (8.3.4)$$

while the Euclidean scaling would be $A \propto L^2$. Combining these two fractal scaling laws (Eqs. 8.3.2 and 8.3.4) we infer the relationship between the fractal volume V and the length scale L,

$$V(L) \propto L^{2.51 \pm 0.06} , \qquad (8.3.5)$$

which also differs from the Euclidean scaling $V \propto L^3$. Of course, these scaling laws apply to the particular setup of cellular automaton models we described in Chapter 2, but slightly different values are expected for different avalanche models or length scale definitions. We have also to be aware that these simulated fractal structures (as shown in Fig. 8.14) represent time-integrated structures, while the fractal dimensions of instantaneous snapshots are generally smaller.

8.3.2 Solar Flares

Since geometric 3-D models of solar flares are unavoidable in calculating electron densities and thermal energies from the observed volume-integrated emission measures in soft X-rays and EUV, which are necessary parameters to infer occurrence frequency distributions of flare energies for SOC models, the 3-D fractal dimension D_V is a fundamental parameter. Alternatively, one can specify a volume-filling factor q_V, which is the ratio of the fractal V to the Euclidean volume V_0,

$$q_V = \frac{V}{V_0} = \frac{L^{D_V}}{L^3} = L^{D_V - 3} , \qquad (8.3.6)$$

while the area-filling factor q_A can be defined analogously in terms of the area fractal dimension D_A,

$$q_A = \frac{A}{A_0} = \frac{L^{D_A}}{L^2} = L^{D_A - 2} . \qquad (8.3.7)$$

Based on the geometric concept of flares introduced in Fig. 8.9, we can construct a volumetric model in terms of an arcade that contains a variable number of concentric half loops that fill the half-cylindric volume to some extent, parameterized by the arcade length l_a, arcade width w_a, and average loop width w_{loop}. While the 3-D volume is invariant to rotation, the projected area will depend on the aspect angle, longitude, or center-to-limb distance, as shown in Fig. 8.15. If we allow for fractal filling with n_{loop} loop structures, which has the limit of $n_{loop}^{max} \approx l_a w_a / 2 w_{loop}^2$ for the arcade model shown in Fig. 8.15, one can derive the following geometric filling factors (Aschwanden and Aschwanden 2008b),

$$q_V = \frac{n_{loop}}{n_{loop}^{max}} = n_{loop} \left(\frac{2 w_{loop}^2}{l_a w_a} \right) . \qquad (8.3.8)$$

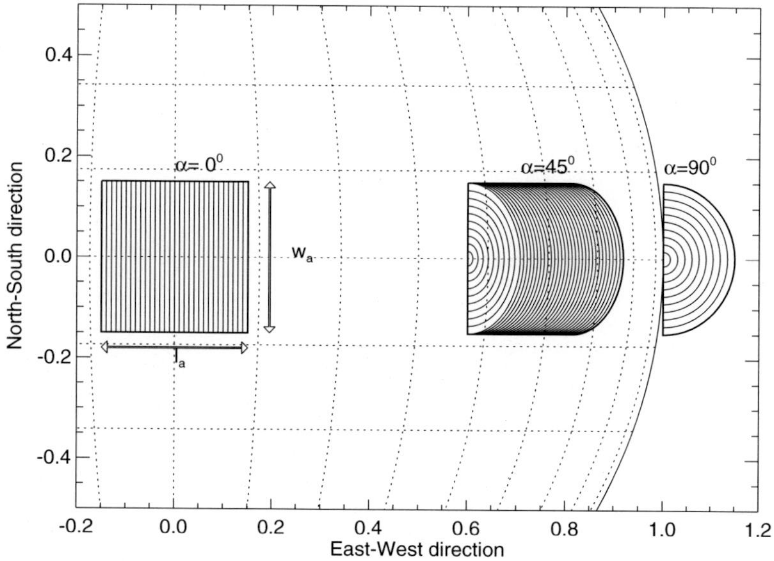

Fig. 8.15 The geometry of a semi-cylindrical flare arcade model is shown, quantified by the arcade length l_a in the east–west direction, the arcade width w_a in the north–south direction, and the line-of-sight angle α to the solar vertical (or relative longitude difference to solar disk center). The three cases correspond to $\alpha = 0°, 45°, 90°$ with an aspect ratio of $w_a/l_a = 1$. The total (Euclidean) flare area is outlined in thick linestyle, while the loop quantization is indicated with thin lines (Aschwanden and Aschwanden 2008b).

$$q_A = \frac{A(n_{loop})}{A_0} = \left[1 - \exp\left(-n_{loop}\frac{A_1}{A_0}\right)\right] , \qquad (8.3.9)$$

where A_1 is the Euclidean area that depends on the aspect angle α,

$$A_1(\alpha) \approx w_{loop}\frac{w_a}{2}\left[1 + \left(\frac{\pi}{2} - 1\right)\sin^{3/2}(\alpha)\right] . \qquad (8.3.10)$$

Using the definitions of Eqs. (8.3.6) and (8.3.7), the area and volume fractal dimensions can then be calculated from the area- and volume-filling factors,

$$D_V = 3 + \frac{\ln q_V}{\ln L} . \qquad (8.3.11a)$$

$$D_A = 2 + \frac{\ln q_A}{\ln L} . \qquad (8.3.11b)$$

where the length scale L can be defined from the Euclidean volume V_0,

$$V_0 = \frac{\pi}{2}\left(\frac{w_A}{2}\right)^2 l_a , \qquad (8.3.12)$$

$$L = V_0^{1/3} . \qquad (8.3.13)$$

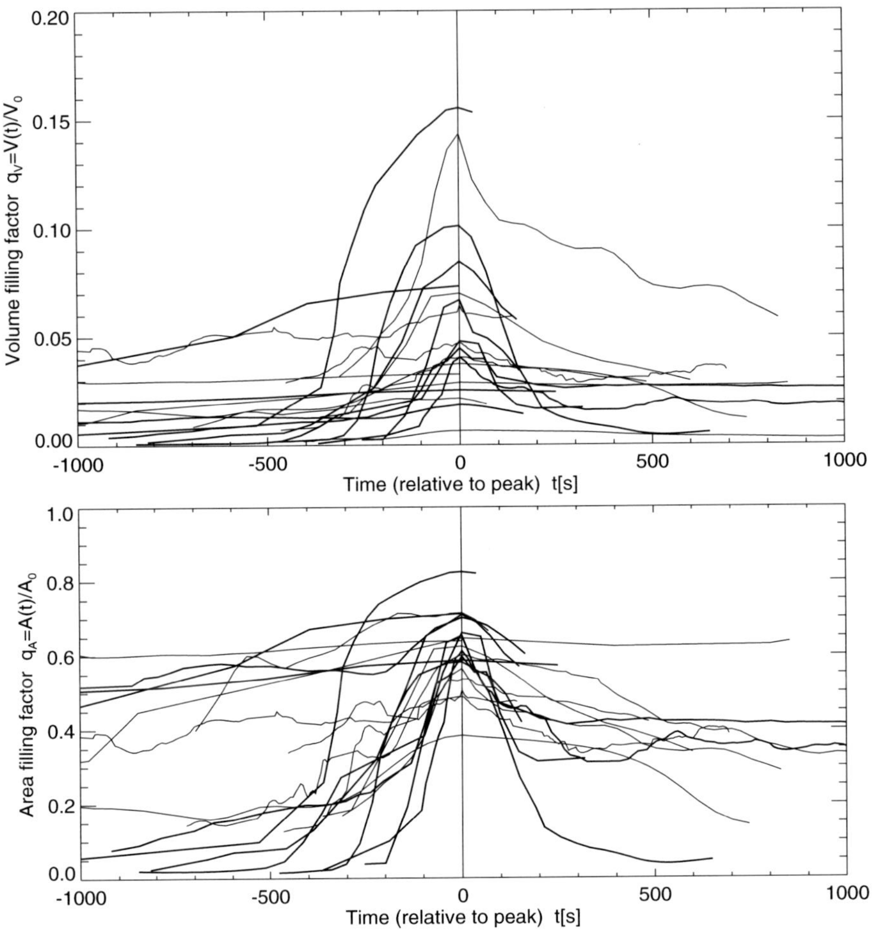

Fig. 8.16 The area-filling factor $q_A(t)$ (bottom panel) and the inferred volume-filling factor $q_V(t)$ (top panel) are shown for 20 flares, as a function of the time relative to the peak in the maximum fractal area. Flares which have an increase of more than 0.5 in the fractal area during the rise time are outlined with thick linestyle. Note that maximum area-filling factors do not exceed 0.8, while maximum volume-filling factors do not exceed 0.15 (Aschwanden and Aschwanden 2008b).

We have already shown how the observed area fractal dimension varies as a function of time during a flare (Fig. 8.10), and consequently also the area- and volume-filling factors do. In Fig. 8.16 the results of the time evolution of flare-filling factors q_A and q_V are shown for 20 large flares, varying typically in the range of $q_V \approx 0.001$–0.03 at flare start, $q_V \approx 0.03$–0.08 at flare peak, and $q_V \approx 0.01$–0.06 at flare end. These filling factors are very important, because they constrain the true mean electron density n_e. If an average electron density $\langle n_e \rangle = \sqrt{EM/V_0}$ is estimated for a unity filling factor (solid filling of the flare volume), the correct mean electron density in the fractal flare volume scales as.

$$n_e = \sqrt{EM/V} = \langle n_e \rangle \sqrt{V_0/V} = \langle n_e \rangle \sqrt{1/q_V} \ . \tag{8.3.14}$$

This correction is important in deriving correct thermal energies of flares,

$$E_{th} = \int 3n_e(T) k_B T V(T) \, dT \approx 3 n_e k_B T_e V = \frac{3 k_B EM T_e}{n_e} \ , \tag{8.3.15}$$

where EM is the total emission measure, T_e the electron temperature, and n_e the electron density at the peak time of the flare.

8.4 Multifractal Analysis

The geometric concepts we described so far are all *monofractal*, which contain self-similar and scale-invariant structures that can be characterized by a single fractal dimension, such as the Hausdorff dimension D. However, there is no structure in the universe that exhibits the same fractal dimension at all scales from the microscopic to the macroscopic limit. Geometric structures are generated by different physical processes that operate within a preferred scale range each, and thus the resulting structures have a different degree of inhomogeneity or fractality at different scales. The concept of *multifractals* attempts to characterize the degree of geometric complexity with multiple scaling exponents or fractal dimensions, which in the continuum limit results into a spectrum of fractal dimensions. While the fractal dimension D is defined by $n(\varepsilon) \approx \varepsilon^{-D}$ for monofractals in the framework of the box-counting method (Eq. 8.2.2), there is a spectrum $f(\alpha)$ of exponents for multifractals, also called *singularity spectrum*,

$$n(\varepsilon) \propto \varepsilon^{-f(\alpha)} \ , \tag{8.4.1}$$

where α is the relative strength or significance. Examples of the singularity spectrum $f(\alpha)$ are shown in Fig. 8.17 for a monofractal (Sierpinski carpet with Hausdorff dimension $D = log(8)/log(3) \approx 1.89279$), for a theoretical multifractal image (Cadavid et al. 1994), and for observational data from solar magnetogram data (Hewett et al. 2008; Conlon et al. 2008). The latter example shows a typical singularity spectrum, which has a peak of $f(\alpha)_{max}$ and a minimum of $f(\alpha)_{min}$, which is also characterized by the terms *contribution diversity* $C_{div} = \alpha_{max} - \alpha_{min}$ and *dimensional diversity* $D_{div} = f(\alpha)_{max} - f(\alpha)_{min}$, both being measures of the geometric complexity and richness of a fractal structure. Related measures of complexity are also multiscaling of Kadanoff and Lipshitz–Hölder exponents (e.g., see Georgoulis et al. (1995) and references therein).

The magnetic field seen at the solar surface reveals a richness of morphological structures that are termed sunspots, plages, network, intranetwork, magnetic knots and pores, etc. A quiet-Sun photospheric magnetogram was first analyzed in terms of multifractal analysis by Lawrence et al. (1993), who modeled the singularity spectrum with a Gaussian random process. More detailed modeling was done by Cadavid et al. (1994) by adding Gaussian white noise to theoretical self-similar and multifractal structures, finding that the degree of multifractality is enhanced for more intermittent distributions and strong correlations between cells. The influence of finite spatial resolution on the determination of multi-

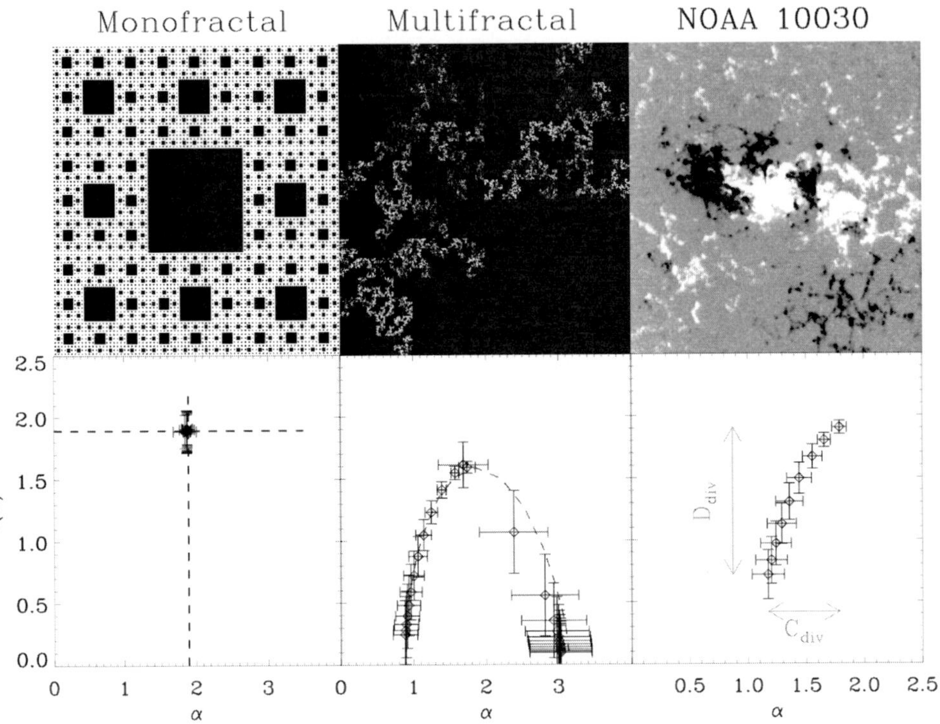

Fig. 8.17 A monofractal image of the Sierpinski carpet (left), a theoretical multifractal image (middle), and an observed multifractal solar magnetogram of active region NOAA 10030 (right), along with the singularity spectra $f(\alpha)$ (bottom panels) determined for these structures (Conlon et al. 2008).

fractal scaling was investigated by Lawrence et al. (1996) who found that the box-counting method is unreliable if it does not fill the embedding Euclidean dimension ($D < 2$). The multifractal singularity spectrum was also applied in time sequences of photospheric magnetograms to study the evolution of active regions (Conlon et al. 2008), see a snapshot in Fig. 8.17. It was found that active regions that evolved into large-scale coherent structures show a decrease of dimensional diversity D_{div}, and a relationship was found between the flaring rate in an active region and the multifractal properties (Conlon et al. 2008). The multifractal complexity was also found to vary as a function of the solar cycle, or between the northern and southern hemisphere (Sen 2007).

Multifractal analysis appears to be a sensitive tool for characterizing complexity and changes in complexity of spatial morphological structures, either as a function of space, or as a function of time, similar to the Bayesian statistics of nonstationary Poisson processes used in the time domain (Section 5.2).

Another multi-scale method that is related to multifractal analysis is the *structure function*, which has been developed to describe the statistical behavior of fully developed turbulence (Kolmogorov 1941). Structure functions express the degree of correlation at different length scales, equivalent to the correlation function of velocity fluctuations as a function of the spatial distance, which has a similar scaling behavior as the singularity spectrum of a multifractal structure. The scaling behavior of structure functions has been studied in magnetograms of the solar photosphere (Abramenko et al. 2002, 2003; Abramenko 2005), revealing significant changes of the structure function before and during solar flares, similar to the evolutionary changes of the multifractal singularity spectrum measured for active regions (Conlon et al. 2008).

8.5 Spatial Power Spectrum Analysis

A more traditional multi-scale method is the *spatial 2-D Fourier power spectrum* of a 2-D spatial image, which quantifies the correlated intensity as a function of spatial scales. If we take an image with a size of $N \times N$ pixels and denote the image coordinates with the indices (n,m), the intensity of a particular pixel is $I_{n,m}$. 2-D power spectra I_k (with complex Fourier coefficients) can then be calculated (e.g., Gomez et al. 1993a),

$$I_k = \sum_{n=1}^{N} \sum_{m=1}^{N} I_{n,m} \exp\left[\frac{2\pi i}{N}(nn' + mm')\right] , \tag{8.5.1}$$

where the Fourier component or wave vector $(k = 2\pi/\lambda)$ in the (k_x, k_y) plane is,

$$k = \frac{2\pi}{N\Delta x}(n',m') , \quad n',m' = 0,1,...,(N-1) , \tag{8.5.2}$$

and Δx is the linear pixel size. The 2-D power spectrum $P(k_x,k_y)$ is then defined as

$$P(k_x,k_y) = \left(\frac{\Delta x}{2\pi}\right)^2 |I_k|^2 . \tag{8.5.3}$$

An example of such a 2-D Fourier power spectrum of a solar image recorded in soft X-ray wavelengths is shown in Fig. 8.18. The presence of a broad-band spectrum (in contrast to a δ-function peak for non-fractal large-scale spatial structures) indicates spatial structures over a large scale range, down to the image resolution Δx. From the 2-D power spectra, 1-D omnidirectional power spectra can be computed, which average the spectra in all radial directions. Such omnidirectional Fourier spectra have been found to scale with $P(k) \propto k^{-3}$ for some solar active regions (Martens and Gomez 1992; Gomez et al. 1993a), which was explained in terms of a turbulent Kolmogorov spectrum $P(k) \propto k^{-5/3}$, combined with the spectral modifications resulting from the velocity distribution of photospheric granulation and the emission mechanism observed in soft X-rays (Gomez et al. 1993b).

Power spectra analysis in other regions of the solar surface and in other wavelengths were performed in a number of studies. The power spectra are very wavelength-dependent. In the quiet Sun, power spectra of $P(k) \propto k^{-2.7}$ were measured in soft X-rays (Benz et al. 1997), and $P(k) \propto k^{-2.5}$ in EUV Fe XII (Berghmans et al. 1998). Power spectra derived from photospheric magnetograms, after correction for the seeing (modulation transfer function), yielded $P(k) \propto k^{-1}$ for the photospheric network, $P(k) \propto k^{-3.5}$ for the non-

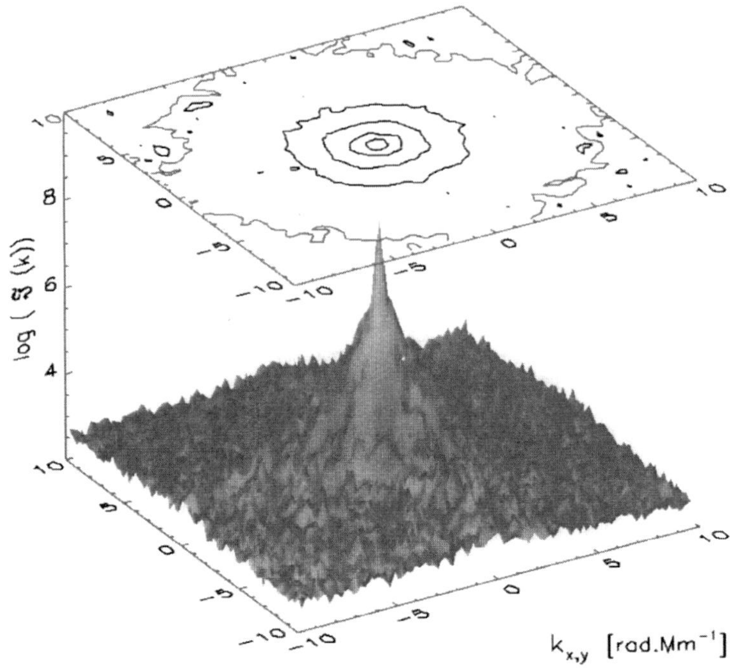

Fig. 8.18 2-D Fourier power spectrum of a soft X-ray image of a solar active region recorded with the *Normal Incidence X-ray Telescope (NIXT)* telescope during a rocket flight in 1989, with a spatial resolution of 0.75″. The 2-D power spectrum is nearly isotropic (Gomez et al. 1993a; reproduced by permission of the AAS).

network (Lee et al. 1997), $P(k) \propto k^{-1.7}$ for active regions, and $P(k) \propto k^{-1.3}$ for quiet-Sun regions (Abramenko et al. 2001). Observations in extreme ultraviolet, which probe the lower corona rather than the photosphere, yield power spectra of $P(k) \propto k^{-2.0}$ for bright points, $P(k) \propto k^{-2.1}$ for loops, $P(k) \propto k^{-1.9}$ for the background corona, $P(k) \propto k^{-1.6}$ for dark lanes (network), while power spectra in the transition region (in the He II line) yield values of $P(k) \propto k^{-1.5}$ for the same structures (Berghmans et al. 1998). Similar power spectra were measured for full-Sun images in EUV, $P(k) \propto k^{-1.57}$ in the S VI (933 Å) and $P(k) \propto k^{-1.74}$ in the S VI (944 Å) (Buchlin et al. 2006). Power spectra measured with a highest resolution of 0.1″ (70 km on the solar surface) in the G-band were found to be as steep as $P(k) \propto k^{-4.0}$ in sunspot penumbrae, and $P(k) \propto k^{-3.6}$ in active granulation (Rouppe Van der Voort et al. 2004).

The variety of power spectra measured in solar data reflects a number of effects that affect the precise value of the slope: (1) the physical mechanism (e.g., MHD turbulence in subphotospheric granulation cells), (2) the wavelength of the observer (optical, EUV, soft X-rays), which mostly indicates different altitude levels (photosphere, chromosphere, transition region, corona), and (3) instrumental effects (seeing and spatial resolution). The powerlaw index p of a spatial power spectrum $P(k) \propto k^{-p}$ can be transformed into a distribution of spatial length scales $N(L)$ with $k = 2\pi/L$, with a similar formalism as we derived in Section 4.8.4 in the time domain. However, there are also other methods to measure the distribution $N(L)$ of length scales directly, as we describe in the next Section 8.6.

8.6 Statistics of Spatial Scales

The common denominator of fractal structures with SOC theory is the property of scale-free parameter ranges, which can be described with powerlaw relations between various geometric parameters (e.g., length, area, volume). The fractal property implies two important consequences for SOC theory: (1) it describes the geometry of the instantaneous internal microstructure of a SOC event, but also (2) describes the relationships of geometric size parameters between different SOC events. The two relationships may even influence each other during a SOC event. For instance, a landscape has a fractal structure as a static property, but leaves an imprint of its static fractality also on dynamic events, such as landslides, flooding, or show avalanches, which follow the channeling and ducting of the fractal terrain. Fractal landscapes (valleys, craters) may even be the witnesses of SOC-like processes (erosion, mountain slides, volcanic eruptions). Solar flares are magnetic reconnection events that occur in the environment of a fractal magnetic field, and thus the resulting energy of a magnetic instability released in a flare, which heats up chromospheric plasma and redistributes it throughout coronal fluxtubes, reflects a similar fractality as the previous static magnetic field. So, we can interpret the fractal geometry of static structures as ducts or remnants of dynamical SOC events, which exhibit the multiplicative imprints of exponentially-growing catastrophes. It is like the domino effect, where an avalanche chain reaction takes place in a pre-arranged fractal geometry.

In the following we focus on the second consequence of fractal geometries, namely the statistics of spatial size scales between different events. If the main SOC parameters we used so far (i.e., the peak energy P, the total (time-integrated) energy E, and the duration T) possess powerlaw frequency distributions, and if there is a simple powerlaw scaling law of the SOC parameters (P,E,T) with length scales L, we expect also occurrence frequency distributions $N(L)$ of length scales to exhibit a powerlaw-like functional form (Section 7.1.6). Hence, we study the length scale frequency distributions $N(L)$ for different astrophysical SOC processes in the following.

8.6.1 Solar Photosphere and Chromosphere

As we alluded to in previous sections on the (multi)fractal structure of magnetic structures seen on the solar surface and in the solar atmosphere, there is the notion that the internal solar dynamo is the driver and generator of magnetic features (sunspots, active regions, filaments, flares, coronal mass ejections), and thus could represent a dissipative nonlinear system in the state of self-organized criticality (in contrast to turbulence or percolation theories). The static magnetic features seen on the solar surface represent then the remnants of buoyant magnetic fluxtubes generated by the SOC state of the tachocline on one hand, while dynamic magnetic reconnection processes in the solar corona represent a secondary SOC process generated by the SOC state of the solar atmosphere on the other hand.

Early statistics (before SOC) on the distribution of the most prominent magnetic features on the solar atmosphere, namely active regions, was obtained by measuring the areas of over 1,000 active regions during 1967–1981 in daily magnetograms from the Mount Wilson Observatory, detected above a threshold of 10 G. The resulting area size distribution with sizes of $A = 3$–1,350 square degrees (1 square deg = 48.5×10^{-6} of the solar hemisphere), were found to fit an exponential distribution of $N(A) \propto \exp(-A/175)$ (Tang et al. 1984). A more extended study was conducted by Harvey and Zwaan (1993), which differed from the study of Tang et al. (1984) in several ways: (1) only bipolar regions that reach their peak size on the visible hemisphere were included, (2) regions are included only once, (3) each region was measured during the peak, and (4) corrections for visibility and data gaps were made. The resulting size distribution was characterized by a sum of exponential and logarithmic terms and was much different from the one obtained by Tang et al. (1984), but agreed in the invariance of the functional shape during the solar cycle. The size distribution of sunspots sampled over more than 100 years (from Greenwich Observatory 1874–1976) was found to follow a log-normal distribution (Baumann and Solanki 2005), which is powerlaw-like at the upper end, but exhibits a gradual flattening towards smaller sizes. Statistics on areas of magnetic features depend very much on the selection (active regions, sunspots, emerging bipoles), the time evolution (growth, peak, or decay phase), and the counting method (daily records, multiple countings per solar rotation). Moreover, since the total available magnetic energy per feature depends on both the area and the field strength, statistics on areas alone may not be most useful.

An area-related quantity is the magnetic flux $\Phi = \int B\, dA \approx BA$, which includes the magnetic field strength B and is largely independent on the instrumental resolution, because it represents a spatial integral. However, the relationship between the magnetic flux $\Phi \approx BA$ and the area A is not simple and seems to vary in an active region on time scales of days (Chumak and Zhang 2003). Statistics on the distribution of magnetic field strengths in the range of $B = 0$–1,800 G has been quantified in Dominquez Cerdena et al. (2006), which depends very much on the instrumental resolution and whether the Zeeman signal tends to cancel opposite polarization. A powerlaw-like distribution of magnetic fluxes was found for intranetwork (with a slope of $\alpha \approx 1.68$) and network flux (with a slope of $\alpha \approx 1.27$) in the a range of $\Phi \approx 10^{16}$–10^{18} Mx (Wang et al. 1995; Meunier 2003). A series of studies was conducted (Hagenaar et al. 1997, 2003; Hagenaar 2001; Hagenaar and Shine 2005) on the statistical distribution of cell sizes in the chromospheric network, ephemeral magnetic regions, and moving magnetic features around sunspots, and synthesized the different statistics into a single composite powerlaw-like distribution function that contains the magnetic fluxes of emerging bipoles at the lower end and entire active regions at the upper end, spanning a range of about four orders of magnitude ($\Phi \approx 5 \times 10^{18} - 5 \times 10^{22}$ Mx), shown in Fig. 8.19 (left). Parnell et al. (2009) used a "clumping algorithm" and extended this way the range of magnetic fluxes over about seven decades and found that the synthesized distribution of all magnetic features in the range of $\Phi = 10^{16} - 10^{23}$ Mx fit a powerlaw distribution with a slope of $\alpha \approx 1.85 \pm 0.14$ (Fig. 8.19, right). The statistical distributions of magnetic fluxes in active regions has been modeled in terms of percolation models (Wentzel and Seiden 1992; Seiden and Wentzel 1996; Fragos et al. 2004). We will discuss physical SOC models that involve the observed size distributions of magnetic areas and magnetic fluxes in Chapter 9.

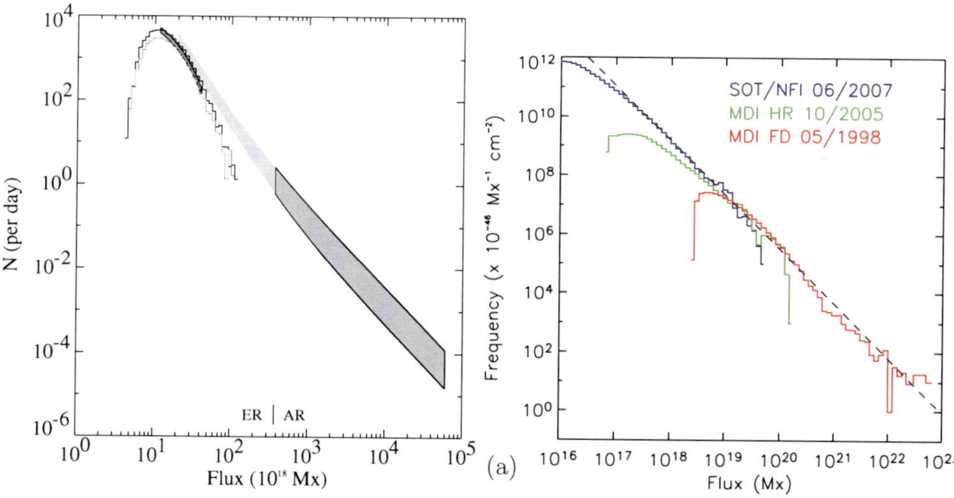

Fig. 8.19 *Left:* Composite distribution function of magnetic bipoles emerging on the Sun per day, per flux interval of 10^{18} Mx, and active regions (Hagenaar et al. 2003). *Right:* Synthesized histograms of magnetic features observed with SOT/Hinode and MDI/SOHO, identified with an automated clumping-feature algorithm. The combined powerlaw slope is $\alpha = 1.85 \pm 0.14$ (Parnell et al. 2009), (reproduced by permission of the AAS).

8.6.2 Solar Flares

There are only few studies that offer statistics on spatial scales of solar flares (Table 8.2). Area statistics of very small solar flares in the energy range of $E \approx 10^{24}$–3×10^{25} erg, called *EUV transient brightenings*, was sampled by Berghmans et al. (1998), finding an approximate powerlaw distribution with a slope of $\alpha_A = 2.7$ at a transition region wavelength (304 Å) and $\alpha_A = 2.0$ in a coronal wavelength (195 Å), measured with SOHO/EIT at spatial scales of $L \approx 3$–20 Mm. A similar SOHO/EIT study was conducted by Aletti et al. (2000), who measured the size of an EUV brightening from the number of pixels that have an intensity above a threshold of 2σ or 3σ, and obtained a (fractal area) size

Table 8.2 Frequency distributions of area sizes observed in solar flares. References: 1, Berghmans et al. (1998); 2, Aletti et al. (2000); 3, Aschwanden and Parnell (2002).

Events type	Wavelength of lengths	Range of areas L	Powerlaw slope α_A $N(A) \propto A^{-\alpha_A}$
EUV brightenings[1]	304 Å (He II)	3–20 Mm	2.7
EUV brightenings[1]	195 Å (Fe XII)	3–20 Mm	2.0
EUV brightenings (2σ)[2]	195 Å (Fe XII)	2–60 Mm	1.26 ± 0.04
EUV brightenings (3σ)[2]	195 Å (Fe XII)	2–20 Mm	1.36 ± 0.05
EUV nanoflares[3]	171, 195 Å	2–20 Mm	2.56 ± 0.23

distribution with powerlaw slopes of $\alpha_A = 1.26$–1.36 in the range of $A = 1$–$1{,}000$ pixels (corresponding to a length scale of $L \approx 2$–60 Mm). In a systematic study of EUV nanoflares detected with TRACE 171 and 195 Å, the (fractal) flare areas were measured with an elliptical area with length l and width w, yielding an area of $A = lw$ (Eq. 8.2.6). From a set of 281 automatically detected nanoflare events, size distributions of $N(l) \propto l^{-\alpha_l}$ with $\alpha_l = 2.10 \pm 0.11$ and $N(w) \propto w^{-\alpha_w}$ with $\alpha_w = 4.43 \pm 0.22$ and $N(A) \propto A^{-\alpha_A}$ with $\alpha_A = 2.56 \pm 0.23$ were found (Aschwanden and Parnell 2002), for a size range of $L \approx 2$–20 Mm.

Thus, there is very scarce statistics on distribution of spatial scales. A full-scale SOC model should also include geometric scaling laws, but little effort has been put into this aspect. What are our theoretical expectations for a geometric SOC model? One potential model is *Euclidean fragmentation*, which we envision simply by breaking down a solid structure into smaller space fragments. For instance, if we break a square-like chocolate into 16 equal pieces, each little square has a quarter length of the original size, so we have $N(L=1) = 1$ and $N(L=1/4) = 16$, and thus $N(L) \propto L^{-2}$. For solid structures, the expected scaling would then just be the reciprocal relationship of Euclidean scaling, e.g., $N(L) \propto L^{-1}$ for breaking a linear structure into smaller pieces, $N(L) \propto L^{-2} = A^{-1}$ for subdividing an area-like 2-D structure, and $N(V) \propto L^{-3} = V^{-1}$ for fragmenting a volume structure. For fractal geometries, we might expect a reciprocal scaling of the fractal dimension, but the definition of a length scale for fractal structures is more tricky. A comparison with the distributions measured in Table 8.2 shows at least some values are close to the expected scaling of $N(A) \propto A^{-1}$, but clearly more statistics is needed to narrow down more reliable values of the powerlaw slope based on a wider range of spatial scales.

There are some other area-related flare studies. The study of Sammis et al. (2000) investigated the flare peak fluxes as a function of the area of active regions and a trend was found that large active regions produce larger flares, but this general trend was found to be less significant than the dependence on the magnetic classification ($\alpha, \beta, \gamma, \delta$ classes of magnetic complexity of sunspots). In addition some studies explored whether the frequency distribution of peak fluxes in flares depends on the sizes of active regions and some systematic differences were found (e.g., Kucera et al. 1997; Sammis 1999) as expected for biased subsets, while the scale invariance was corroborated when compared among different active regions (Wheatland 2000c), which is expected for SOC models.

8.6.3 Lunar Craters

Craters can generally be produced either by volcanic eruptions or by meteoroid impacts, both representing violent catastrophic events that may exhibit SOC behavior. Many craters seen on the Moon or Earth appear to be the result of meteoroid impacts. Both the Moon and the Earth were subjected to intense bombardment between 4.6 and 4.0 billion years ago, which was the final stage of the sweep-up of debris left over from the formation of the solar system. The impact rate during that time was a thousand times higher than today's rate. Lunar craters, therefore, represent remnants or witnesses of catastrophic events that left a measurable imprint from which we can measure the size and perhaps even calculate the energy.

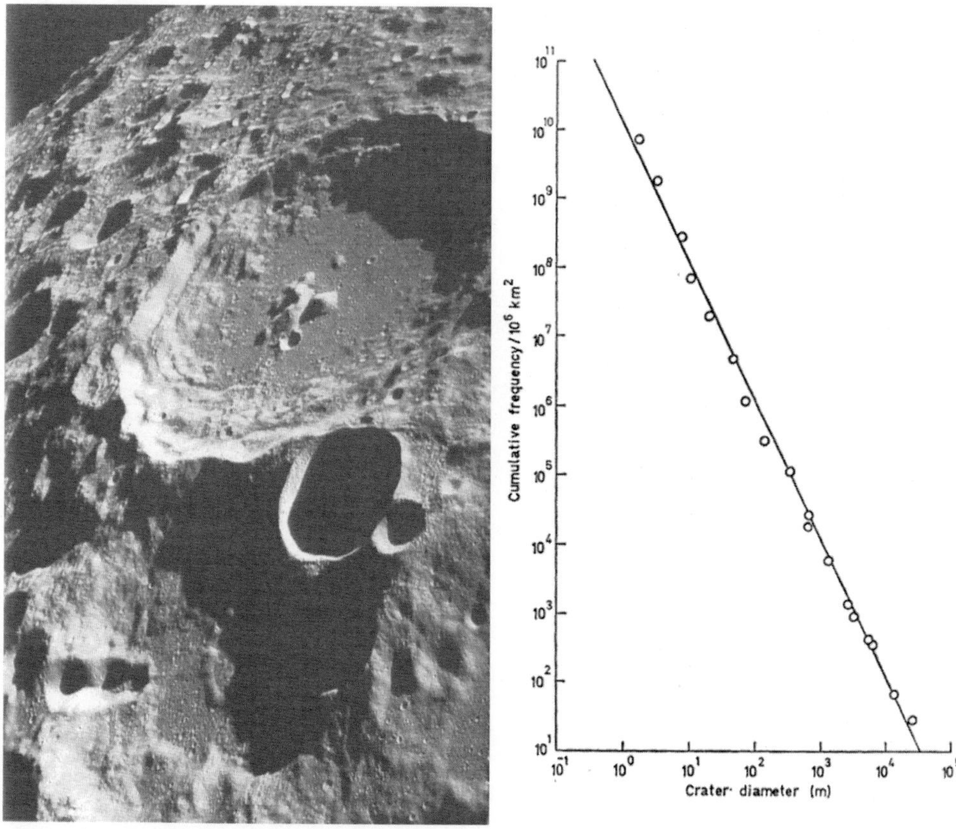

Fig. 8.20 *Left:* The lunar crater Daedalus, about 93 km in diameter, was photographed by the crew of Apollo 11 as they orbited the Moon in 1969 (NASA photo AS11-44-6611). *Right* Cumulative frequency distribution of crater diameters measured from *Ranger 8* in the lunar *Mare Tranquillitatis* (Cross 1966).

The size distribution of lunar craters was measured from pictures of the lunar orbiters *Ranger 7, 8, 9* by Cross (1966), who measured the diameters L from a total of 1,600 craters, ranging from 0.65 to 69,000 m, and found an approximate powerlaw function for the cumulative frequency distribution,

$$N^{cum}(>L) \propto L^{-2} \tag{8.6.1}$$

which corresponds to a differential frequency distribution of $N(L) \propto L^{-3}$ according to (Eq. 7.1.8). Cross (1966) conducted statistics of lunar craters for each Mare separately. One example of a cumulative frequency distribution of craters from the *Mare Tranquillitatis* using *Ranger 8* measurements is shown in Fig. 8.20 (right). A similar powerlaw index of 2.75 was also found for the size distribution of meteorites and space debris from man-made rockets and satellites (Fig. 3.11 in Sornette 2004).

Interpreting this result, we may think again of the concept of *Euclidean fragmentation*, for which we expect $N(L) \propto L^{-3}$. The more or less solid mass that was forming the solar system probably has been fragmented by collisions and tidal forces into smaller pieces. Conservation of mass and volume yields then the scaling law, $N(L) \propto L^{-3}$, since

$$V = N(L_0)L_0^3 = N(L)L^3 = \text{const} , \tag{8.6.2}$$

where L_0 is the original average size and L is the smaller average size of fragments at a later time of the fragmentation process. The fragmentation process leads naturally to a self-similar fractal geometry, since fragments from frequent collisions tend to grind spherical objects, and combined with the spherically propagating shock waves during an impact event, leads also to a self-similar distribution of circular craters. Do impact craters qualify for a SOC system? Both the Euclidean fragmentation process (driven by two-body collisions) as well as the impact of a fragment on the lunar surface are both highly nonlinear dissipation processes, occur with a random waiting-time distribution, and exhibit scale-free powerlaw distributions of energies and sizes, and thus possess all typical characteristics of a SOC process. However, we cannot measure the time history of the event to obtain the peak energy, total energy, and duration, but are left with the imprints of the spatial sizes only.

8.6.4 Asteroid Belt

The asteroid belt between the planets Mars and Jupiter contains a large number of irregular bodies or minor planets with sizes from about 1,000 km (Ceres 1,020 km; Pallas 538 km; Vesta 549 km; Juno 248 km) down to the size of dust particles. While most planetesimals from the primordial solar nebula formed bigger planets under the influence of self-gravitation, the gravitational perturbations from the giant planets Jupiter and Saturn prevented a stable conglomeration of planetesimals in the zone between Mars and Jupiter, and thus we still live with a fragmented soup of primordial planetesimals, called the asteroid belt (Fig. 8.21, left). The asteroid belt has evolved into the present configuration by dynamical depletion due to the gravitational disturbance from the giant planets (which pull planetesimals into highly eccentric orbits) and collisions (which fragment the planetesimals further).

The asteroid size distribution has been studied in the *Palomar Leiden Survey* (Van Houten et al. 1970) and *Spacewatch Surveys* (Jedicke and Metcalfe 1998), where a power law of $N^{cum}(>L) \propto L^{-1.8}$ was found for the cumulative size distribution of larger asteroids ($L > 5$ km), which corresponds to a differential powerlaw slope of $\alpha_L \approx 2.8$. In a *Sloan Digital Sky Survey* collaboration (Fig. 8.21, right), a broken powerlaw was found with $N(L) \propto L^{-2.3}$ for large asteroids (5–50 km) and $N(L) \propto L^{-4}$ for smaller asteroids (0.5–5 km) (Ivezic et al. 2001). In the *Subaru Main-Belt Asteroid Survey*, a cumulative size distribution $N^{cum}(>L) \propto L^{-1.29\pm0.02}$ was found for small asteroids with $L \approx 0.6$–1.0 km (Yoshida et al. 2003; Yoshida and Nakamura 2007), which corresponds to a differential powerlaw slope of $\alpha_L \approx 2.3$.

Interpreting these results, which specify powerlaw slopes of the differential size distribution in the range of $\alpha_L \approx 2.3$–4.0, the average is close to the value $\alpha_L \approx 3$ expected

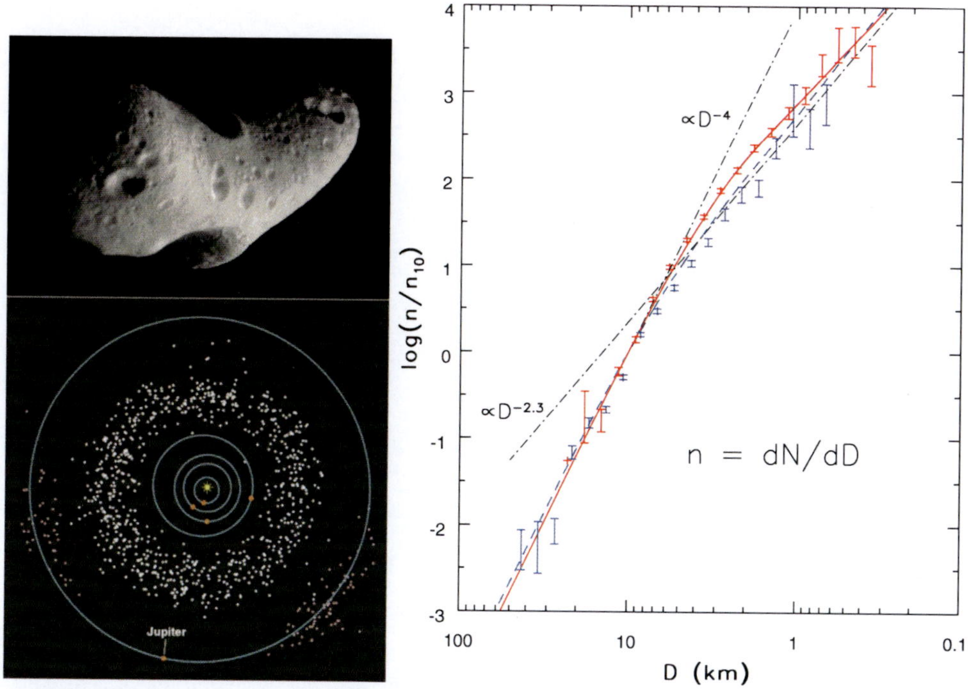

Fig. 8.21 *Left top:* A picture of the near-Earth asteroid *Eros* with a size of 30 km, pictured by a space probe. *Left bottom:* The main asteroid belt located between the Jupiter and Mars orbit. The subgroup of *Trojan asteroids* are leading and trailing along the Jupiter orbit. (Courtesy of NASA/Johns Hopkins University Applied Physics Laboratory). *Right:* Differential size distribution of asteroids observed in the *Sloan Digital Sky Survey* collaboration (Ivezic et al. 2001). (Reprinted with permission of Elsevier)

for Euclidean fragmentation, which predicts $N(L) \approx L^{-3}$, similar to the statistics of lunar craters (Section 8.6.3). However, the observational manifestation is quite different for these two phenomena, one observed before impact and the other after impact on a specific target. The fact that the size distribution exhibits a broken powerlaw could indicate that two different physical processes dominate in the two regimes, for instance dominant collisions with less gravitational orbit perturbation for the large asteroids ($L \gtrsim 5$ km), but stronger orbit perturbation and pre-dominant dynamic depletion for smaller asteroids. Nevertheless, a similar argument for asteroid formation as a SOC process can be made as for the creation of lunar craters.

8.6.5 Saturn Ring

Jupiter and Saturn are the two largest planets in our solar system, and thus it is no surprise that they also have numerous moons, rings, and ringlets thanks to their strong gravitational field. While the rings are located close to the planet (7,000 km to 80,000 km above Saturn's equator), the orbits of the moons are outside the rings. Mechanical resonances (i.e., in

Fig. 8.22 Photo of Saturn's B-ring with Huygens gap, Cassini division, and spoke structures, recorded by the *Cassini* spacecraft (credit: NASA, JPL, Space Science Institute).

orbits that have a period with a harmonic ratio to the outer moons' periods) destabilize inner rings, leading to gaps (e.g., Encke gap, Cassini division), or stabilize the zones in between (Figs. 1.11 and 8.22). The Saturn ring consists of particles ranging from 1 cm to 10 m, with a total mass of 3×10^{19} kg, just about a little less than the moon Mimas. The origin of the ring was hypothesized to come either from nebular material left over from the formation of Saturn itself or from the tidal disruption of a former moon.

The distribution of particle sizes in Saturn's ring was determined with radio occultation observations using data from the *Voyager 1* spacecraft and a scattering model, which exhibited a powerlaw distribution of $N(r) \propto r^{-3}$ in the range of 1 mm $< r <$ 20 m (Zebker et al. 1985; French and Nicholson 2000). This result, again, is consistent with Euclidean fragmentation, similar to the distribution of sizes of asteroids (Section 8.6.4) and lunar craters (Section 8.6.3). Can we consider the evolution of the Saturn ring as a SOC process? Events are caused by collisional encounters, which probably occur at random time intervals (though very rare on human time scales) and the energy release during a collisional impact is likely to be a nonlinear dissipative (fragmentation) process, leading to powerlaw distributions of energies with some scaling to the powerlaw size distribution of the projectile and target. The critical threshold is some minimal velocity difference Δv (between the projectile and target body) for inelastic impacts with subsequent fracturing, while small Δv merely cause elastic reflections without catastrophic disintegration. Hence the same argument for a SOC process can be made as for asteroids and lunar craters.

8.7 Summary

SOC processes produce scale-free powerlaw-like size distributions of their dynamical parameters (peak energy, total energy, duration), which also extends to their geometric parameters (length, area, volume). The powerlaw-like size distributions of geometric parameters then consequently imply also powerlaw-like scaling laws between geometric parameters, such as $A(L) \propto L^{D_A}$ and $V(L) \propto L^{D_V}$. These geometric scaling laws can be either Euclidean ($D_A = 2$ and $D_V = 3$) or fractal ($D_A < 2$ and $D_V < 3$). Cellular automaton models of SOC processes can reproduce fractal geometries in the spatial propagation of avalanches, and thus fractal scaling laws are expected for most SOC processes. We discussed 1-D fractals (Section 8.1), which can be applied to 1-D time series of astrophysical observations (e.g., variability of solar radio emission). Measuring 2-D fractals (Section 8.2) can be done most conveniently in astrophysical images inside our solar system (e.g., magnetospheric substorms, solar photosphere, or solar flares). The derivation of 3-D fractal dimensions (Section 8.3) is more tricky, because it requires either geometric models or lattice-based computer simulations. The measurement of fractal characteristics can be done either by box-counting algorithms (Section 8.2), multifractal analysis (Section 8.4), spatial power spectrum analysis (Section 8.5), or by statistics of spatial scales (Section 8.5). With the latter method we explored magnetic structures in the solar photosphere as well as during flares and found them all to be fractal. In contrast, the size distribution of lunar craters, asteroids, and Saturn ring particles all exhibit a Euclidean scaling law of $N(L) \propto L^{-3}$, as expected for a fragmentation process. In summary, fractal or Euclidean scaling laws of geometric parameters and their powerlaw-like size distributions are necessary conditions for SOC processes, but not sufficient to prove a SOC process, because non-SOC processes (such as intermittent turbulence) can also produce powerlaw-like distributions of spatial scales.

8.8 Problems

Problem 8.1: What is the 1-D ("Sierpinski dust") and 3-D analog ("Sierpinski tetahedron) of the 2-D Sierpinski triangle shown in Fig. 8.6. Calculate their fractal dimensions.

Problem 8.2: Construct a time series (say $N = 10{,}000$ points) with a random generator and measure its fractal dimension with the method of Higuchi (Section 8.1.2). Smooth the time series with a box-car of $n_{sm} = 10$ and 100 and determine its fractal dimension. How much smoothing is needed to obtain a near-Euclidean dimension of $D \approx 1$ within 1%?

Problem 8.3: Download digital astronomical images of a spiral galaxy, a globular cluster, and a star field. Measure their fractal dimensions for various thresholds (say 10%, 20% and 50% of the maximum intensity). In which cases do you obtain a near-Euclidean dimension of $D = 1$ as expected for dot-like stars. Which case shows the highest fractal dimension and what spatial structure is it associated with?

Problem 8.4: Verify the analytical expressions of the area- q_A and volume-filling factors q_V given Eqs. 8.3.8–8.3.10 for an aspect angle of $\alpha = 0°$ by means of a Monte-Carlo simulation for $n_{loop} = 10, 100$, and 1,000 loop elements in a half-cylinder configuration as shown in Fig. 8.15.

Problem 8.5: Design a simple Monte-Carlo simulation for the fragmentation of planetesimals, assuming that collisions occur in random time intervals and between random fragments, where each collision splits a planetesimal into two half volumes. Sample the size distribution $N(L)$ after 10^3, 10^4 and 10^5 events and fit a powerlaw distribution $N(L) \propto L^{-\alpha_L}$. Do you find a Euclidean dimension of $\alpha_L = 3$? Think of improvements that would make the model more realistic.

9. Physical SOC Models

In physics, you don't have to go around making trouble for yourself, – nature does it for you.

Frank Wilczek

All science is either physics or stamp collecting.

Ernest Rutherford

Why is the concept of self-organized criticality (SOC), such an interdisciplinary subject, being applied in geophysics, astrophysics, or financial physics with equal fervor? On the most general level, the common denominator of all SOC processes in different science disciplines is the statistics of nonlinear processes, which exhibit omnipresent powerlaw distributions. Nonlinear processes are characterized by a nonlinear growth phase, during which *coherent growth* is enabled, which has multiplicative characteristics, in contrast to linear processes with *incoherent* and additive characteristics. Thus, incoherent random processes exhibit binomial, Gaussian, Poissonian, or exponential distribution functions, while coherent processes exhibit powerlaw-like distributions. This is the fundamental trait that earthquakes, solar flares, or stock market crashes have in common, although the underlying physics could not be more different. Therefore, it is important to understand that the powerlaw feature does not require any particular physical model: it can all be explained by mathematical theory in terms of statistical probabilities, as we discussed in Chapters 3 and 4. SOC behavior can thus also be simulated by mathematical rules, as we illustrated in terms of cellular automaton models in Chapter 2. Consequently, our treatment of SOC systems has been entirely *physics-free* so far, not requiring any particular physical model to understand the observed statistical distributions and correlations. However, there are some free parameters we used in our analytical SOC models (Chapter 3) that can only be explained in terms of a physical model for a particular phenomenon, such as the value α_i of the powerlaw slope for each parameter distribution i, or the powerlaw indices β_{ij} between various correlated parameters i and j. At this point, SOC models become specific because the physics of solar flares is different from the physics of tectonic plates. In the following we will focus on specific physical models of astrophysical SOC processes.

9.1 A General (Physics-Free) Definition of SOC

Before we indulge in the manifold physical models of SOC phenomena, let us first summarize a physics-free definition of SOC phenomena, based on the general treatment we elaborated in the previous Chapters 1–8, in particular the exponential-growth model described in Section 3.1. This definition should enable us to identify SOC systems from observations and to discriminate SOC processes from other *non-SOC* processes, of which we will give a relevant selection in Chapter 10.

On the most general level, there are three necessary and (perhaps) sufficient criteria that define a SOC system, which we postulate here as a preliminary working definition:

1. **Statistical Independence:** The events that occur in a SOC system are statistically independent and not causally connected in space or time. The statistical independence can be verified from the waiting-time distribution in the time domain, and by spatial localization in the space domain (if imaging or in-situ observations are available). Waiting-time distributions should be consistent with a stationary or nonstationary Poisson process, in order to guarantee statistical independence by means of probabilities.
2. **Nonlinear Coherent Growth:** The time evolution of a SOC event has an initial nonlinear growth phase after exceeding a critical threshold. The nonlinear growth of dissipated energy, or an observed signal that is approximately proportional to the energy dissipation rate, exhibits an exponential-like or multiplicative time profile for coherent processes. (Incoherent random processes, in contrast, show a linear evolution and have additive characteristics.)
3. **Random Duration of Rise Times:** If a system is in a state of self-organized criticality, the rise time or duration of the coherent growth phase of a SOC event (avalanche) is unpredictable and thus exhibits a random time scale. The randomness of rise times can be verified from their statistical distributions being consistent with binomial, Poissonian, or exponential functions.

This is mostly a mathematical definition of a SOC system. The prototype of a SOC process is the BTW sandpile, and we can qualitatively verify that sandpile avalanches fulfill these three criteria: (1) subsequent sand avalanches occur at random, occasionally triggered by an infalling sand grain; (2) sand avalanches grow in a multiplicative manner once they get rolling; and (3) the growth phase (or rise time) of an avalanche lasts a random time interval, depending on the random path along which the avalanche propagates and encounters locations with slopes that are slightly steeper than the overall average critical value of a sandpile in SOC state. The numerical prototype of a SOC process is the cellular automaton (Section 2.1.3), which can easily be tested to see whether the numerically generated distributions of waiting times, growth rates, and rise times fulfill the three criteria of our mathematical SOC definition.

Verifying a SOC system with our three mathematical or physics-free criteria can most directly be accomplished by testing the following three relationships (for the simplest case of a stationary Poisson process):

$$\begin{array}{ll} N(\Delta t) \propto \exp(-\Delta t) & \text{random waiting times} \\ \log(P) \propto \tau_{rise} & \text{exponential growth} \\ N(\tau_{rise}) \propto \exp(-\tau_{rise}) & \text{random rise times} \end{array} \quad . \qquad (9.1.1)$$

With the analytical derivation of our standard model we have demonstrated that the criteria (2) and (3) lead to a powerlaw distribution $N(P) \propto P^{-\alpha_P}$ of peak energies, as well as to approximate powerlaw distributions of total energies and durations, $N(E)$ and $N(T)$, which is generally used as a test of SOC systems. Criterion (1) on the waiting-time distributions was often used to verify or disprove a SOC system, but it is not a sufficient condition to evaluate SOC, since waiting-time distributions can exhibit exponential (for stationary Poisson processes) or powerlaw-like distribution functions (for nonstationary Poisson processes). In the following review of physical SOC models we will discuss their compliance with our mathematical definition of SOC processes, and we will discuss also whether the criteria are sufficient to exclude non-SOC processes (in Chapter 10).

9.2 Astrophysics

The identification of physical mechanisms in nonlinear dissipative systems that exhibit SOC behavior is quite a new field that leaves a lot of room for new ideas and modeling in terms of existing theories. In fact, the literature on physical models of astrophysical SOC processes is very sparse, except for some applications in solar and magnetospheric physics. In Table 9.1 we give a tentative list of possible physical interpretations of SOC phenomena, which should be taken with a grain of salt and as a possible starting point for future modeling, rather than as a list of established results. We will briefly discuss the examples given in Table 9.1 in the following.

9.2.1 Galaxy Formation

Galaxies are observed at all sizes and there is a hierarchy of structures from dwarf galaxies (e.g., the Magellanic Cloud), single galaxies, groups, clusters, and superclusters of galaxies (Fig. 9.1). The standard bigbang model together with the inflationary model describes the cosmological evolution of the universe over the last 13.75 ± 0.17 billion years. The formation of galaxies has been modeled in terms of two opposite scenarios, i.e., the *top-down* scenario that starts with a monolithic collapse of a large cloud (Eggen, Lynden-Bell, and Sandage 1962), versus the *bottom-up* scenario where smaller objects merge and form larger structures that ultimately turn into galaxies (Searle and Zinn 1978), which is more widely accepted now. In most models on galaxy formation, thin, rotating galactic disks result as a consequence of clustering of dark matter halos, gravitational forces, and conservation of angular momentum. The fractal-like patterns of the universe from galactic down to solar system scales is thought to be a consequence of the gravitational self-organization of matter (Da Rocha and Nottale 2003). Whether the whole universe is in a state of self-organized criticality has not been clearly addressed in literature (see Section 1.10 and the textbook by Baryshev and Teerikorpi 2002), but it is conceivable that the driving forces of

Table 9.1 Examples of physical processes with SOC behavior.

SOC phenomenon	Source of free energy or physical mechanism	Instability or trigger of SOC event
Galaxy formation	gravity, rotation	density fluctuations
Star formation	gravity, rotation	gravitational collapse
Blazars	gravity, magnetic field	relativistic jets
Soft gamma ray repeaters	magnetic field	star crust fractures
Pulsar glitches	rotation	Magnus force
Blackhole objects	gravity, rotation	accretion disk instability
Cosmic rays	magnetic field, shocks	particle acceleration
Solar/stellar dynamo	magnetofriction in tachocline	magnetic buoyancy
Solar/stellar flares	magnetic stressing	magnetic reconnection
Nuclear burning	atomic energy	chain reaction
Saturn rings	kinetic energy	collisions
Asteroid belt	kinetic energy	collisions
Lunar craters	lunar gravity	meteoroid impact
Magnetospheric substorms	electric currents, solar wind	magnetic reconnection
Earthquakes	continental drift	tectonic slipping
Snow avalanches	gravity	temperature increase
Sandpile avalanches	gravity	super-critical slope
Forest fire	heat capacity of wood	lightning, campfire
Lightning	electrostatic potential	discharge
Traffic collisions	kinetic energy of cars	driver distraction, ice
Stockmarket crash	economic capital, profit	political event, speculation
Lottery win	optimistic buyers	random drawing system

gravitation in an expanding universe lead to sporadic density fluctuations that initiate a locally nonlinear growth phase of self-gravitating matter like an avalanche in a sandpile SOC model. The spatial and temporal independence of SOC events throughout the universe is somewhat guaranteed by the cosmological flatness and horizon problem.

9.2.2 Star Formation

Star formation is initiated by the local collapse of a molecular cloud under self-gravity. In the *triggered star formation* scenario, a gravitational collapse of a molecular cloud is initiated by a collision between two clouds, by a nearby supernova explosion that ejects shocked matter, or even by galactic collisions that cause compression and tidal forces. If there is sufficient mass available (the *Jeans mass criterion*), which depends on the initial size of the unstable galactic fragment, the collapsing cloud will build up a dense core that forms into a star with nuclear burning, otherwise it ends up as a brown dwarf. Considering star formation as a SOC process, it is conceivable that it fulfills the three SOC criteria of (1) statistical independence (if there are many independent sites of star-forming molecular clouds throughout the galaxies), (2) nonlinear coherent growth (gravitational collapse), and (3) randomness of formation time (if there is a large variation of accretion rates). However, some molecular clouds may be triggered externally by shock waves from nearby supernovae, which would correspond to "sympathetic flaring" and would violate

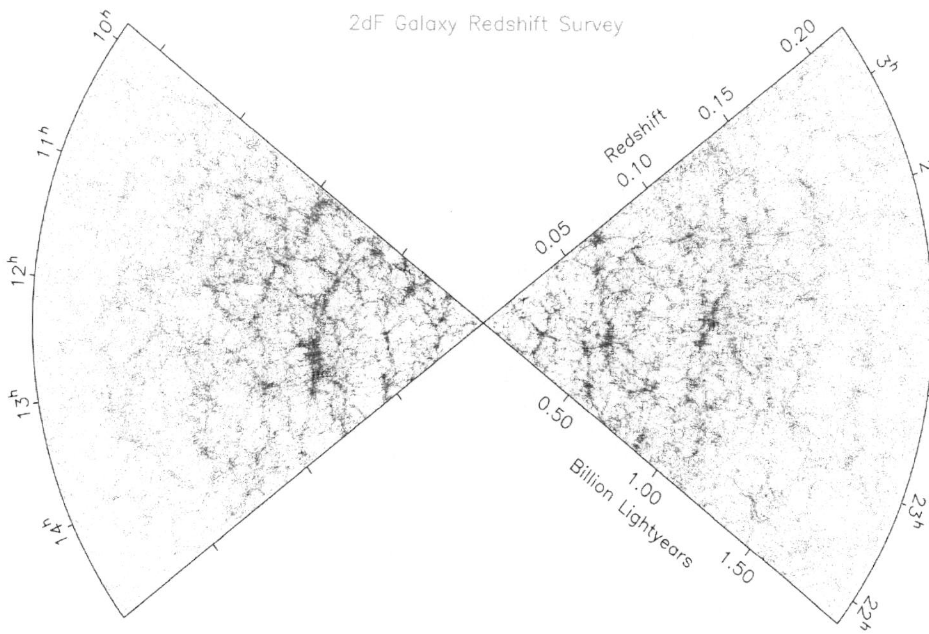

Fig. 9.1 The 2dF galaxy redshift survey (2dFGRS), conducted at the Anglo-Australian Observatory, shows a map of the galaxy distribution out to redshifts of $z = 0.23$ or approximately 2 billion lightyears, which includes approximately 250,000 galaxies. Note the fractal large-scale structure of the universe that makes up the galaxy density (Colless et al. 2001).

the first SOC criterium. Observational tests of the SOC criteria are obviously required. One pioneering study explored the scaling relations of molecular clouds and their fractal structure with observations of the starburst cluster in *30 Doradus* (Fig. 9.2) under the aspect of self-organized criticality (Melnick and Selman 2000).

9.2.3 Blazars

We discussed blazars (blazing quasi-stellar objects) briefly in Section 7.4.5, since pulses from such an object (blazar GC 0109+224) exhibit a powerlaw distribution $N(P) \propto P^{-1.55}$ in the intensity of the optical pulses (Ciprini et al. 2003). Blazars are a group of *active galactic nuclei (AGNs)* that have the special geometry of their relativistic jet pointing towards the observer on Earth. These relativistic jets are thought to be produced by matter that spirals toward the central black hole of the host galaxy, where the accumulated matter forms a hot accretion disk with a relatively compact size of $\approx 10^{-3}$ parsecs (Fig. 9.3). In the center of the torus-like accretion disk, strong magnetic fields are believed to produce axial relativistic jets that eject plasma away from the AGNs over distances of $\approx 10^4$–10^5 parsecs. The relativistic jet produces synchrotron radiation in radio and X-rays, as well as inverse Compton emission in X-rays and gamma rays, while the thermal emission produces also ultraviolet and strong optical emission lines.

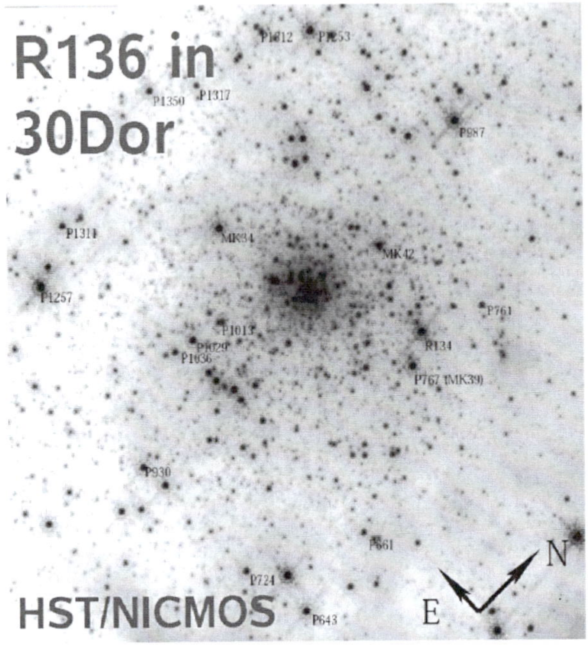

Fig. 9.2 The central star cluster R136 (30 Doradus) in the extragalactic giant HII region in the Large Magellanic Cloud (LMC), photographed with the Hubble Space Telescope (HST) NICMOS camera. This starburst cluster was analyzed in terms of SOC statistics by Melnick and Selman (2000).

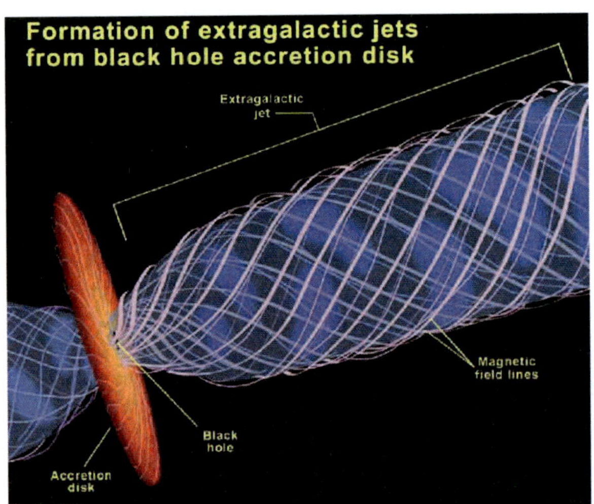

Fig. 9.3 Schematic diagram of a blazar, containing an active supermassive black hole in its core, surrounded by an accretion disk that accumulates infalling matter. The magnetic field wraps around the rotation axis and forms a relativistic extragalactic jet along the rotation axis. (Credit: NASA).

Can blazars be considered to be SOC systems? The observed high variability of pulses from blazars is mostly explained by the relativistic beaming, which has a strong dependence on the small angle between the forward direction of the relativistic jet and the line-of-sight to the Earth's observer. Synchrotron emission has a strong angular dependence of its emissivity. Thus, if a relativistic electron beam, which has maximum emissivity in forward direction, spirals around the rotation axis of the accretion disk, the small directional changes of the steep spiral cause large fluctuations in synchrotron emissivity for an observer at a fixed angle. If the mass infall into an AGN occurs stochastically, the first SOC criterion is fulfilled. One particular mass inflow could produce multiple pulses, depending on the complexity of its spiral-like trajectory, which may cause quasi-periodic brightness fluctuations (at each helical turn) or more randomized fluctuations if the trajectory is more complex than a symmetric spiral. The second SOC criterion of nonlinear growth could be attributed to the nonlinear change of emissivity as a function of the angular change during the spiraling orbit. The third SOC criterion of randomness of pulse rise times can easily be satisfied by the irregularity of the relativistic jets that are caused by the random mass and angular distributions of infalling blobs. Hence, pulses from blazars can fulfill all three SOC criteria. The analogy to a sandpile SOC model is even more striking when we think of the randomized input of dropped sand that causes sand avalanches of all sizes, similar to the infalling matter in a blazar (Fig. 7.20). Gravity provides the free energy in both systems, and the gravitational acceleration has a multiplicative effect on the growth of avalanches when they propagate and accumulate (or accrete) more ambient mass. Ciprini et al. (2003) also analyzed the variability of the blazar light curve by calculating the structure function for unevenly sampled data (i.e., the squared flux differences) and found an approximate $1/f$ flicker noise spectrum, which essentially corroborates the third SOC criterion of random pulse rise times.

9.2.4 Neutron Star Physics

The physics of neutron star crusts involves nuclear physics, condensed matter physics, superfluid hydrodynamics, and general relativity, which is reviewed, e.g., in Chamel and Haensel (2008). There are accreting neutron stars in low-mass binaries (Fig. 1.16), where a binary star is sufficiently tight for the companion to fill its Roche lobe, and mass is transferred through the inner Langrangian point via an accretion disk towards the neutron star surface. Accretion onto a neutron star releases $\approx$200 MeV per accreted nucleon, and thus energy is radiated in X-rays. The accreted material is hydrogen-rich, which fuels hydrogen burning into helium in the outer envelope of the neutron star. The helium burning is unstable for some range of accretion rates, which can ignite triggers of thermonuclear flashes, producing X-ray bursts with energies of $\approx 10^{39}$–10^{40} erg, which represent one class of so-called *soft X-ray transients*. Other soft X-ray transients are produced by unstable accretion rates in accretion disks. Some soft X-ray bursts are quasiperiodic with typical recurrence times of hours to days. These soft X-ray transients are a possible SOC phenomenon, because they supposedly fulfill the three SOC criteria of (1) statistical independence (of their recurrence), (2) nonlinear coherent growth (of thermonuclear flashes), and (3) random rise times (of unstable accretion rate fluctuations).

Pulsars are fast-rotating neutron stars (e.g., Crab or Vela pulsar) that emit extremely periodic signals like a clock, but occasional sporadic glitches revealed irregularities in their rotational frequencies. There are two types of irregularities: (1) timing noise that might result from irregular transfers of angular momentum between the neutron star crust and the liquid (superfluid) interior of the neutron star, and (2) sudden glitches of the rotational frequency (with typical amplitudes of $\Delta\Omega/\Omega \approx 10^{-9}$–$10^{-6}$), which is now mostly interpreted in terms of neutron starquakes. The starquake model assumes that neutron stars are not perfectly spherical, but slightly deformed because of centrifugal forces. Because the neutron star crust is solid rather than fluid, the star stays oblate and cannot adjust to a more spherical shape, which builds up stresses in the crust while the star spins down. When the stress reaches a critical threshold, the neutron star crust cracks and the neutron star adjusts its shape to reduce its deformation (Fig. 9.4). Thus, pulsar glitches are very likely a SOC system, as indicated by the powerlaw distribution of their giant-pulse fluxes (Fig. 7.17), as described in Section 7.4.2. Pulsar glitches most likely fulfill our three SOC criteria of (1) statistical independence (of thresholded stress releases in the neutron star crust), (2) the nonlinear coherent growth (during the rise time of giant pulses), and (3) randomness of rise times (of the giant-pulse time profiles).

Soft Gamma Repeaters are believed to be strongly magnetized neutron stars (also called magnetars) possessing the strongest magnetic fields ($B \approx 10^{14}$–10^{15} G) known in the universe. Similar to the interpretation of pulsar glitches, soft gamma repeaters are believed to be produced by crust quakes induced by magnetic stresses in the central neutron star

Fig. 9.4 The neutron starquake model involves a spinning neutron star with the strongest known magnetic fields in the universe (magnetars), which occasionally release energy by catastrophic unpinning of vortices, manifested in pulsar glitches and soft gamma-ray repeaters. The artists rendering depicts the neutron star SGR J1550-5418, which has a rotation period of 2.07 s and holds the record for the fastest-spinning magnetar (Credit: NASA, GSFC, Swift, Fermi).

(Fig. 9.4). Neutron star quakes are giant catastrophic events like earthquakes and are likely to be accompanied by global seismic vibrations or oscillations. One likely oscillation mode is the torsional shear mode, which could be responsible for the detected oscillations in the frequency range of $\nu \approx 10$–1000 Hz. An alternative model for supergiant flares of soft gamma-ray repeaters is energy release during a starquake of a solid quark star, which can free up energies up to 10^{48} erg (Xu et al. 2006). As argued above, the starquakes responsible for the observed giant pulses of soft gamma repeaters, magnetars, and pulsars are all suitable candidates for nonlinear dissipative systems in a SOC state.

9.2.5 Blackhole Objects and Accretion Disks

Accretion disks represent circumstellar mass accumulations in the shape of disks or halos that form in a natural way as a consequence of the rotation-induced angular momentum and the attractive gravitational force of massive objects, such as a young star or protostar in a molecular cloud, a white dwarf, a neutron star, or a black hole. Accretion may start initially at large radii, while the gravitational force causes the loose material to spiral inward, and conservation of the angular momentum will increase the rotation speed the closer the mass approaches the central object. The gravitational force compresses also the material and causes electromagnetic emission in the infrared for accretion disks of young stars and protostars. For more massive accretion disks around neutron stars and black holes, charged particles produce free-free bremsstrahlung in X-rays, as well as gyrosynchrotron emission in radio and X-rays, if there exists a sufficiently strong magnetic field.

The details of accretion disk physics are complicated. To first order, conservation of angular momentum in a gravitational field is expected to lead to elliptical orbits (Keplerian disk), and thus mass infalling towards the center of an accretion disk requires loss of angular momentum or momentum transport outwards. In addition, a hydrodynamic solution with laminar flows is not possible due to the Rayleigh–Taylor instability, which causes an interchange instability at the interface between two fluid layers of different densities. Consequently, turbulence-enhanced viscosity was invoked to explain the angular-momentum transport (Shakura and Sunyaev 1973). Balbus and Hawley (1991) established that a weakly magnetized accretion disk around a compact central object would be highly unstable and provided this way a mechanism for angular momentum transport.

A number of studies modeled accretion disks in terms of a SOC cellular automaton model (e.g., Mineshige et al. 1994a,b; Takeuchi et al. 1995; Takeuchi and Mineshige 1996; Xiong et al. 2000; Pavlidou et al. 2001), as we described in Section 2.7. In the original model of Mineshige et al. (1994a), a SOC avalanche is simply thought to occur as a multiplicative chain reaction of adjacent cells with mass concentrations that start to "coagulate" (like the formation of blood clots) as a consequence of some unknown instability (which could be the Balbus–Hawley instability according to our current thinking). Such a mechanism can easily fulfill our three SOC criteria of (1) statistical independence (for spontaneous occurrence of the instability), (2) nonlinear coherent growth (to next neighbor cells of mass concentrations), and (3) random durations of rise times (since the accretion disk is highly inhomogeneous). The earlier (Mineshige et al. 1994a) and later model (Mineshige et al. 1994b) differ in the assumption of gradual diffusion on top of the avalanching "mass clumping", which yields different powerlaw distributions of time scales that satisfy the third SOC criterion.

While the original accretion disk models of Mineshige et al. (1994a,b) are purely mechanical, a more recent cellular automaton model includes the magnetic field in the accretion disk (Pavlidou et al. 2001). Magnetic loop structures (similar to the solar corona) are thought to exist in accretion disks, which are subject to magnetic reconnection forced by magnetic stressing, and this way can lead to avalanching mass infall (an analog of solar coronal mass ejections, though in the opposite direction). The process of magnetic reconnection is further enhanced by the Balbus–Hawley instability and magnetic buoyancy of magnetic fields inside accretion disks. It has been suggested that a sufficiently radially extended distribution of magnetic loops in accretion disks could provide the anomalous viscosity needed to enable the outward transport of angular momentum for mass infall to the central object (Kuijpers 1995). A specific cellular automaton model with this physical scenario was constructed in Pavlidou et al. (2001), formulated in terms of three free parameters (probabilities of spontaneous, stimulated generation, and diffusive disappearence of magnetic flux) to infer the probabilistic power spectra of energy release times, which seems to fulfill our three requirements of a SOC system. Numerical simulations of this cellular automaton process were also performed (Fig. 2.24), which corroborate the SOC model further.

9.2.6 Cosmic Rays

Cosmic rays are high-energy particles (protons, helium nuclei, or electrons) that originate from within as well as from outside of our galaxy, usually detected when they hit the Earth's atmosphere and produce a shower of particles. The energy spectrum shown in Fig. 9.5 covers an amazing large energy range of $E \approx 10^9$–10^{21} eV. In comparison, the highest energy particles accelerated in our solar system, during solar flares and coronal mass ejections, called *solar energetic particle events (SEP)*, reach maximum energies of $\approx$1 GeV, which is at the low end of the cosmic-ray spectrum. Interestingly, the cosmic ray energy spectrum can almost perfectly be fit by a powerlaw with a slope of $\alpha \approx 2.7$, although a more detailed examination reveals a double powerlaw with a "knee" at $E \approx 10^{16}$ eV. The interpretation is that those particles with smaller energies originate from various sources within our galaxy, from supernova remnants, pulsars, pulsar-wind nebulae, and gamma-ray burst sources. The particles with higher energies have a uniform distribution over the sky and are speculated to come from outside of our galaxy, possibly from active galactic nuclei (AGN) jets, but an accurate localization is elusive.

For the acceleration of cosmic rays, diffusive (Fermi) shock acceleration, collisionless shock acceleration in relativistic perpendicular shocks (also called "shock surfing acceleration"), and stochastic cyclotron-resonance acceleration mechanisms are considered. Whatever the detailed acceleration mechanism is, the powerlaw spectrum of energies could be interpreted as a manifestation of a SOC system. If the time evolution of the energy gained during the acceleration process has a nonlinear growth profile (our second SOC criterion) and the acceleration time lasts for a random time interval (our third SOC criterion), the resulting energy spectrum will be a powerlaw. Different cosmic ray particles are likely to be accelerated independently (our first SOC criterion). Using this scenario, the system that is responsible for the acceleration of cosmic rays, e.g., supernova shocks, magnetic fields in pulsars or active galactic nuclei, are in a state of self-organized criticality in the sense that particles get randomly accelerated and leave the system with unpredictable energies, regardless what their initial condition (i.e., the thermal distribution) was. Of course, the total energy spectrum of all observed cosmic ray events as shown in Fig. 9.5 is a convolution of accelerated spectra from different locations and the superposition from many cosmic sources, multiple shock crossings, and thus may not be representative of the energy spectrum from a single accelerator. It is like adding up many SOC systems with different scales and maximum energy cutoffs. The "knee" in the spectrum clearly indicates different maximum energies obtained within and outside of our galaxy. The powerlaw slope of $\alpha \approx 1.7$ for energies below the "knee" can be explained by a scaling law between the volume and

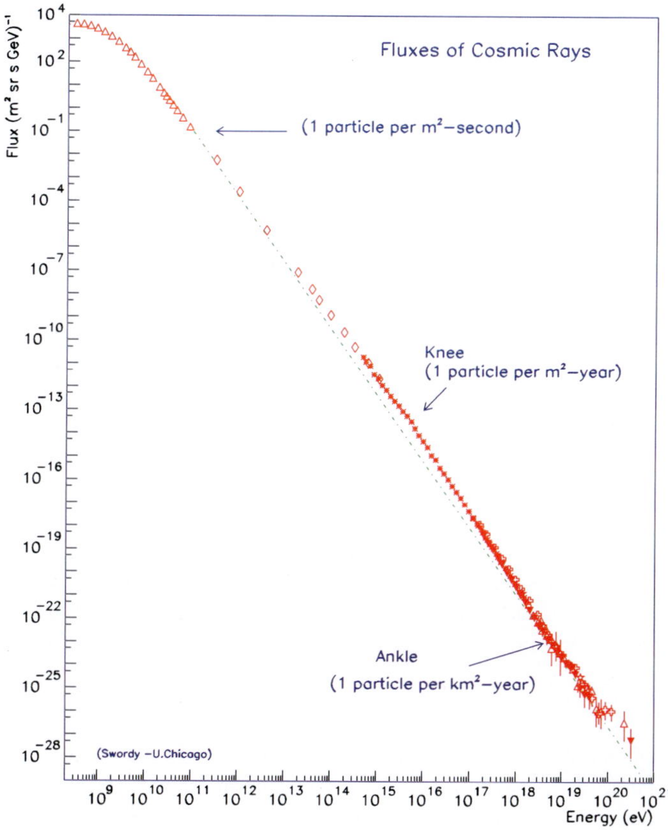

Fig. 9.5 Cosmic ray spectrum in the energy range of $E = 10^9$–10^{21} eV, covering over 12 orders of magnitude. There is a "knee" in the spectrum around $E \approx 10^{16}$ eV, which separates cosmic rays originating within our galaxy (at lower energies) and those from outside the galaxy (at higher energies) (Credit: Simon Swordy, University of Chicago).

magnetic field energy density (e.g., Golitsyn 1997). So, there are considerable degrees of freedom to model the universal cosmic ray spectrum in terms of multi-SOC systems.

9.3 Solar and Stellar Physics

Physical models for solar flares have often also been applied to stellar flares, which have all spatial information concealed by distance, but reveal similar statistical distributions of temporal parameters during flare events and have similar physical conditions in their (solar and stellar) coronae.

9.3.1 Maxwell's Electrodynamics

Per Bak's paradigm of a SOC model, the famous BTW sandpile model, is a purely mechanical model that can in principle be modeled in terms of gravitational and kinematic forces. For solar or stellar flares, in contrast, we have overwhelming evidence that electromagnetic forces are in play, and thus physical modeling of these astrophysical SOC phenomena can only be accomplished in terms of Maxwell's electrodynamic equations, which more generally, turn into the framework of magnetohydrodynamic (MHD) equations in the case of highly ionized plasmas. This approach of electrodynamic modeling for SOC phenomena was first postulated in Lu (1995a) and is reviewed in Charbonneau et al. (2001).

The starting point of SOC modeling in terms of MHD was initiated with cellular automaton models, where a *discretization of the MHD equations* was attempted (Section 2.6.3). Each cell (i,j) or node in a 2-D or 3-D lattice grid (i,j,k) was characterized by the quantity of a magnetic field strength B_{ij} (or B_{ijk}), rather than by a mechanical mass element m_{ijk} in the generic BTW sandpile model. Some models assigned the perpendicular magnetic field component B_k to each cell (Vassiliadis et al. 1998; Isliker et al. 1998a; Takalo et al. 1999a), which generally does not fulfill Maxwell's divergence-free condition,

$$\nabla \cdot \mathbf{B} = 0 , \tag{9.3.1}$$

when the standard cellular automaton redistribution rule (Eq. 2.6.1) is applied, while others assigned the vector potential quantity $\mathbf{A}$ (Isliker et al. 2000; 2001), which defines the magnetic field $\mathbf{B}$ by

$$\mathbf{B} = \nabla \times \mathbf{A} , \tag{9.3.2}$$

which trivially fulfills Maxwell's equation, i.e., $\nabla \cdot \mathbf{B} = \nabla \cdot (\nabla \times \mathbf{A}) = 0$.

In order to calculate an energy for a SOC event, the magnetic energy density integrated over the volume (i.e., the number of unstable cells in a discretized grid) was generally used,

$$E_B = \int_V \frac{B^2}{8\pi}\, dV , \tag{9.3.3}$$

which can also be computed in terms of the vector potential $\mathbf{A}$ from each cell (Galsgaard 1996). However, since the magnetic configuration in a potential field is stable, has no currents, and does not produce instabilities resulting into flares or other SOC events, it is more meaningful to calculate the difference of the nonpotential magnetic energy E_{NP} (e.g., calculated by a linear or nonlinear force-free field extrapolation) and the potential magnetic energy E_P,

$$\Delta E_B = \int_V \frac{B_{NP}^2}{8\pi}\, dV - \int_V \frac{E_P^2}{8\pi}\, dV \ , \tag{9.3.4}$$

as it was applied to study the evolution of a solar active region (e.g., Vlahos and Georgoulis 2004), or individual coronal loops (Morales and Charbonneau 2008a). The threshold for an instability or SOC event can then be formulated in terms of a minimum current **j** according to Ampère's law,

$$\mathbf{j} = \frac{c}{4\pi}(\nabla \times \mathbf{B}) \ , \tag{9.3.5}$$

which is a physically meaningful threshold for many plasma instabilities and magnetic reconnection processes.

9.3.2 The Solar Dynamo

The discretization of Maxwell's equation is a first step towards a physical SOC model that involves magnetic instabilities and electric currents, but there are many conceivable physical scenarios. A cellular automaton model captures only the most essential elements of the evolution of a nonlinear dissipative system without solving the exact solutions of the underlying MHD equations. The essential elements of a nonlinear dissipative system in a SOC state are: (1) a driver or source of free energy, (2) a critical threshold for an instability, (3) the nonlinear growth phase, and (4) saturation of the instability after a random time interval. Thus we can build a variety of SOC models for almost every kind of free energy reservoir and possible instabilities.

The source of free energy in magnetically-driven convective stars is the internal magnetic dynamo, which generates magnetic structures probably at the bottom of the convection zone (in the so-called tachocline), which then rise due to their magnetic buoyancy to the solar (or stellar) surface, appearing as sunspots or starspots. The magneto-convection below the surface as well as the differential rotation constantly deform the topology of magnetic features, which leads to twisting, stressing, and braiding of magnetic field lines in solar (and stellar) coronae, ultimately leading to magnetic instabilities that relax and resolve by the process of magnetic reconnection events, which are thought to be a paradigm of SOC events. We visualize this generic physical scenario in Fig. 9.6, which can be broken down into two SOC processes and one non-SOC process: (1) generation of magnetic flux tubes by the solar dynamo in the solar interior and subsequent emergence to the solar surface, possibly being a SOC process, (2) magneto-convection below the solar surface that produces self-organizing fractal structures (granulation) but is driven by turbulence (which is a non-SOC process), and (3) magnetic reconnection events manifested as flares and CMEs, which is a widely-accepted SOC process.

The solar dynamo is the ultimate source of free energy and driver of most observable phenomena in the magnetized atmosphere. So, it is worthwhile to consider whether the solar dynamo itself is a SOC system. An analogy would be the hot interior of our planet

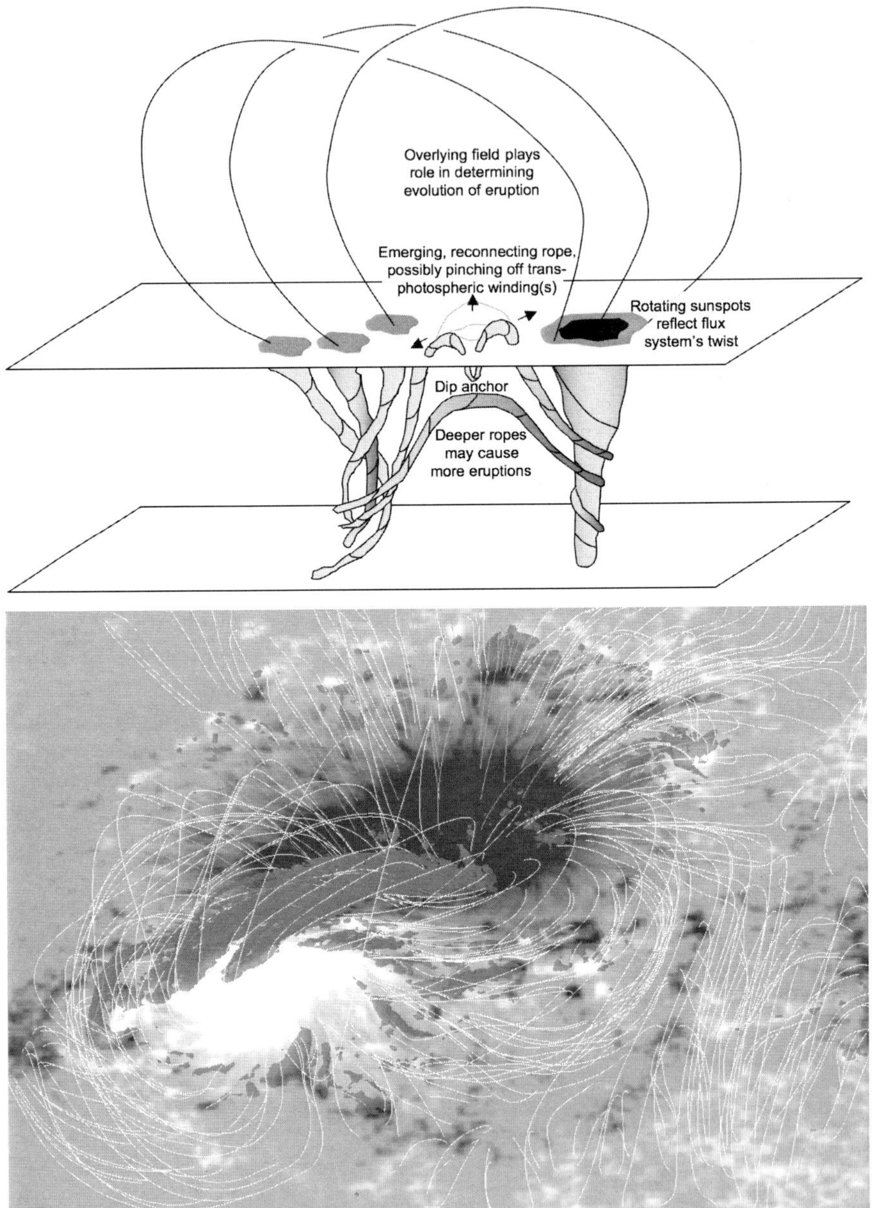

Fig. 9.6 *Top:* Schematic representation of an emerging field configuration generated by the solar dynamo at the bottom of the tachocline, with subsequent emergence at the solar surface due to magnetic buoyancy, creating a twisted coronal magnetic field. *Bottom:* A nonlinear force-free field (NLFFF) calculation of an active region prior to an X3.4 (GOES-class) flare. The two magnetic polarities (black and white) are connected by a twisted flux rope with strong electrical currents (gray). The vector magnetograph data (gray scale) were observed with Hinode (Schrijver 2009). (Reprinted with permission of Elsevier)

Earth, which occasionally produces volcanic eruptions at the Earth's surface, which are believed to be SOC events (Table 1.4). Although there are different physical scenarios of the solar dynamo, ranging from shallow sub-surface turbulent convection down to magnetic instabilities in the tachocline, they all would produce buoyant magnetic structures that emerge at the solar surface as bipoles, sunspots, and active region complexes, which manifest fractal geometries and powerlaw-like size and magnetic flux distributions (Fig. 8.19). The emergence of magnetic dipoles in bright points as well as the formation of active regions is distributed all over the solar surface (though concentrated in low latitudes), and thus seem to fulfill our first SOC criterion of statistical independence. Individual magnetic structures within a single active region, however, are temporally and spatially connected and should not be treated as independent SOC events. The second SOC criterion of nonlinear coherent growth could easily be tested by plotting the size or magnetic flux of many emerging active regions as a function of time, but we are not aware of a large statistical study focused on this SOC aspect. Also the third criterion of random duration of the growth phase can be tested straightforwardly. If the solar dynamo represents a SOC system, powerlaw-like distributions of the peak energy, total energy, and lifetime of active regions, bright points, transient ephemeral regions, or emerging bipoles are predicted, as well as random distributions of waiting times between subsequent phenomena, although modulated as a nonstationary Poisson process with a quasi-periodic solar cycle period of $\approx$11 years.

9.3.3 Magnetic Field Braiding

A key mechanism of magnetic instabilities that trigger magnetic reconnection events in the solar (or stellar) corona is the braiding of magnetic field lines by subphotospheric magneto-convection, which may lead to coronal heating and flare events. Photospheric granular and supergranular flows advect the footpoints of coronal magnetic field lines towards the network, which can be considered as a flow field with a random walk characteristic (Fig. 9.7). This process twists coronal field lines by random angles, which can be modeled by helical twisting of cylindrical fluxtubes. The rate of build-up of nonpotential energy (dW/dt) integrated over the volume $V = \pi r^2 l$ of a cylindrical fluxtube is

$$\int \frac{dW}{dt}\, dV = \frac{\Phi B_0 \langle \mathrm{v}^2 \rangle \tau_c}{4\pi l}\ , \tag{9.3.6}$$

where $\Phi = \pi r^2 B_0$ is the magnetic flux, B_0 is the photospheric magnetic field strength, l the length of the fluxtube, r its radius, $\langle \mathrm{v} \rangle$ the mean photospheric random velocity, and τ_c the correlation time scale of random motion. Sturrock and Uchida (1981) estimate that a correlation time of $\tau_c \approx$ 10–80 min is needed, whose lower limit is comparable with the lifetimes of granules, to obtain a coronal heating rate of $dW/dt \approx 10^5$ (erg cm^{-2} s^{-1}), assuming small knots of unresolved photospheric fields with $B_{ph} \approx 1200$ G.

The idea of topological dissipation between twisted magnetic field lines that become wrapped around each other (Fig. 9.7) has already been considered by Parker (1972). Similarly to Sturrock & Uchida (1981), Parker (1983) estimated the build-up of the magnetic stress energy $B_0 B_t / 4\pi$ of a field line with longitudinal field B_0 and transverse component

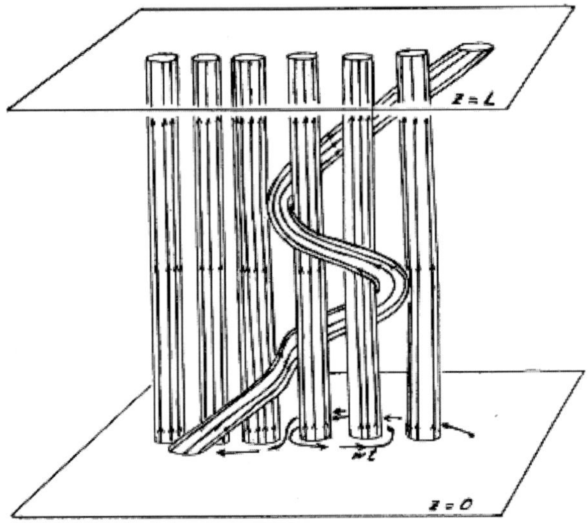

Fig. 9.7 Topology of magnetic fluxtubes that are twisted by random walk footpoint motion, leading to a state where fluxtubes are wound among their neighbors (Parker 1983; reproduced by permission of the AAS).

$B_t = B_0 \mathrm{v} t/l$,

$$\frac{dW}{dt} = \frac{B_0 B_t}{4\pi} \mathrm{v} = \frac{B_0^2 \mathrm{v}^2 t}{4\pi l} , \tag{9.3.7}$$

and estimated an energy build-up rate of $dW/dt = 10^7$ (erg cm^{-2} s^{-1}), based on $B_0 =$ 100 G, v $= 0.4$ km s^{-1}, $l = 10^{10}$ cm, and assuming that dissipation is sufficiently slow that magnetic reconnection does not begin to destroy B_t until it has accumulated random motion stress for 1 day. The manifestation of such sporadic dissipation events of tangential discontinuities in the coronal magnetic field in the form of tiny magnetic reconnection events is then thought to be detectable as nanoflares in the soft X-ray corona, whenever the twist angle

$$\tan\theta(t) \approx \frac{\mathrm{v}t}{l} \tag{9.3.8}$$

exceeds some critical angle. Parker (1988) estimates, for a critical angle given by a moderate twist of $B_t = B_z/4$, corresponding to $\theta = 14°$, for fluxtubes with length L that are braiding within a characteristic horizontal scale of ΔL, that the typical energy of such a nanoflare would be

$$W = \frac{l^2 \Delta L\, B_t^2}{8\pi} \approx 6 \times 10^{24} \quad (\mathrm{erg}) , \tag{9.3.9}$$

based on $l = \mathrm{v}\tau = 250$ km, v $= 0.5$ (km s^{-1}), $\tau = 500$ s, $\Delta L = 1{,}000$ km, and $B_t = 25$ G. Thus, the amount of released energy per dissipation event is about nine orders of magnitude smaller than in the largest flares, which defines the term nanoflare.

There are several variants of random stressing models. A spatial random walk of footpoints produces random twisting of individual fluxtubes and leads to a stochastic build-up of nonpotential energy that grows linearly with time, with episodic random dissipation events (Sturrock & Uchida 1981; Berger 1991). The random walk step size is short compared with the correlation length of the flow pattern in this scenario, so that field lines do not wrap around each other. The resulting frequency distribution of processes with linear energy build-up and random energy releases is an exponential function, which is not consistent with the observed powerlaw distributions of nanoflares. On the other hand, when the random walk step size is large compared with the correlation length, the field lines become braided and the energy builds up quadratically with time, yielding a frequency distribution that is close to a powerlaw. In this scenario, energy release does not occur randomly, but is triggered by a critical threshold value (e.g., by a critical twist angle; Parker 1988; Berger 1993), or by a critical number of (end-to-end) twists before a kink instability sets in (Galsgaard and Nordlund 1997).

The first analytical SOC avalanche model of twisted and braided magnetic field lines, thought to mimic solar nanoflares and coronal heating of active regions, was conceived by Zirker and Cleveland (1993a), following the generic magnetic field braiding model of Parker (1988), which could reproduce the observed powerlaw distribution of flare energies over some energy range. However, the results depend on details such as whether the nonpotential magnetic energy is calculated from (rotationally) twisted or braided (random-walk) structures (Zirker and Cleveland 1993a,b), or whether the threshold criterion for a SOC avalanche is defined in terms of a critical angle, a critical (nop-potential) energy, or a critical current (Podladchikova et al. 1999; Krasnoselskikh et al. 2002). Also the conservation of relative helicity plays a role, which appears to be a necessary condition in some SOC models to produce powerlaw distributions of event sizes (Chou 1999, 2001). Numerical SOC simulations based on the magnetic braiding model aimed to explain the quiet Sun coronal heating (Podladchikova et al. 1999; Krasnoselskikh et al. 2002), energy releases in emerging and evolving active regions (Vlahos 2002; Vlahos et al. 2002; Vlahos and Georgoulis 2004), temperature fluctuations of coronal loops caused by unresolved random heating events (Walsh et al. 1997), or the coherence length of braided coronal loops (Berger and Asgari-Targhi 2009).

An important detail in coronal heating models is the spatial distribution of heating events. The original nanoflare scenario of Parker (1988) assumes a homogeneous plasma along braided or twisted fluxtubes (Fig. 9.7), which predicts a uniform distribution of nanoflares along the loops, which is a tacit assumption in most numerical nanoflare models. Observational data (e.g., Aschwanden et al. 2007) and numerical MHD simulations (e.g., Gudiksen and Nordlund 2005a,b), however, yield strong evidence for a higher nanoflaring rate in the non-force-free and more tangled transition region than in the force-free upper corona. The classical Parker (1988) model thus should be modified to implement the higher degree of magnetic field misalignments in the transition region, in order to provide a realistic framework for numerical SOC models and simulations. Realistic assumptions of the spatial distribution of SOC events affect the resulting occurrence frequency distributions of length scales and volumes, and thus also the volume-dependent flare energies.

9.3.4 Magnetic Reconnection in Solar/Stellar Flares

There is overwhelming evidence that solar (and by inference stellar) flares are triggered by a magnetic reconnection process, during which magnetic energy is released and subsequently heats up the flare plasma and accelerates particles to relativistic (nonthermal) energies. Thus the energy of a flare, which we consider as a SOC event, can be estimated from the magnetic, thermal, or nonthermal energy. Lu et al. (1993) specified a generic physical model of an elementary magnetic reconnection process that quantifies the three observables (E, P, T) of a SOC event in physical quantities.

The total magnetic energy E_B released during an elementary reconnection process in an elementary volume L^3 with an average magnetic energy density $B^2/8\pi$ is

$$E_B = L^3 \left(\frac{B^2}{8\pi} \right) . \quad (9.3.10)$$

The magnetic reconnection process starts when a stressed field becomes unstable and ends after it relaxes into a lower energy with a new stable magnetic configuration. Relaxation happens at the Alfvén speed $v_A = B/(4\pi\rho)^{1/2}$, where ρ is the mass density. Thus, the time scale T of a reconnection process is given by,

$$T = \frac{L}{v_A} \xi , \quad (9.3.11)$$

where ξ is a constant factor that depends on the geometry of the current sheet in the reconnection region, estimated to be of order $\xi \approx 10^1$–10^2 for solar flare conditions (Parker 1979). Combining Eqs. (9.3.10) and (9.3.11), we thus have for the peak energy release rate P,

$$P = \frac{E_B}{T} = L^2 \frac{B^2}{8\pi} \frac{v_A}{\xi} . \quad (9.3.12)$$

The average magnetic field strength B is likely to decrease with a slightly negative power with size L, because larger flares extend to higher altitudes and the coronal field strength falls off with height. If we assume an (empirical) scaling of,

$$B(L) \propto L^{-1/4} , \quad (9.3.13)$$

neglecting other dependencies (i.e., square root of mass density $\rho^{1/2}$ and geometry factor ξ constant), we find (using Eqs. 9.3.10–9.3.13) the following scaling laws as a function of size L,

$$\begin{aligned} E(L) &\propto L^3 B(L)^2 \propto L^{2.5} \\ T(L) &\propto L^1 B(L)^{-1} \propto L^{1.25} \\ P(L) &\propto L^2 B(L)^3 \propto L^{1.25} \end{aligned} \quad (9.3.14)$$

which exactly reproduces the correlations predicted by our simple analytical exponential-growth model (Eq. 3.1.27) described in Section 3.1,

$$\begin{aligned} E &\propto P^2 \\ E &\propto T^2 \\ T &\propto P \end{aligned} \tag{9.3.15}$$

and is close to the values of the correlations found by Lu et al. (1993) from cellular automaton simulations ($E \propto P^{1.82}$, $E \propto T^{1.77}$, and $P \propto T^{0.90}$). Thus, this simple magnetic reconnection scenario can approximately reproduce the observed parameter correlations between the observables (E, P, T), and provides us in addition a scaling law of the average magnetic field strength with the size of the system, i.e., $B \propto L^{-1/4}$.

A slightly different approach was pursued by involving separators in the reconnection geometry (Wheatland 2002; Craig and Wheatland 2002; Wheatland and Craig 2003). The scaling of the flare energy is assumed to scale with the area of a current sheet, $E \propto L^2$ (rather than with the volume L^3 in Eq. (9.3.10)), and a duration $T \propto L$ corresponding to the Alfvénic transit time, which also yields the same correlations as derived in Eq. (9.3.15). Moreover, a probability distribution $N(L) \propto L^{-2}$ of separator lengths (or probability $N(L) \propto L^{-1}$ in one dimension) was assumed, which corresponds to solid (Euclidean) filling, and yields a frequency distribution of flare energies,

$$N(E)\, dE = N[L(E)] \left| \frac{dL}{dE} \right| dE = L^{-2}(E) E^{-1/2}\, dE = E^{-3/2}\, dE \tag{9.3.16}$$

that is consistent with observations (Table 7.2). Implicitly, this model assumes no dependence of the average magnetic field on the flare energy. If we include the empirical scaling given in Eq. (9.1.13), $B \propto L^{-1/4}$, the predicted flare frequency distribution would be (using Eq. 9.3.14),

$$N(E)\, dE = N[L(E)] \left| \frac{dL}{dE} \right| dE = E^{-4/5} E^{-3/5}\, dE = E^{-8/5}\, dE = E^{-1.6}\, dE \ , \tag{9.3.17}$$

which is also consistent with observations, e.g., $N(E) \propto E^{-1.61 \pm 0.04}$ for total counts (fluences) in hard X-rays (Fig. 7.8). Similar combinations of possible scaling laws are discussed in Litvinenko (1998b).

There exists a number of more sophisticated magnetic reconnection models that quantify the frequency distribution of flare energies. Litvinenko (1996) uses a time-dependent continuity equation that takes the dynamical evolution and mutual interaction of multiple reconnecting current sheets by coalescence into account and derives a frequency distribution $N(E) \propto E^{-\alpha}$ of flare energies E with a powerlaw slope in the range of $3/2 < \alpha < 7/4$. Longcope and Noonan (2000) use a scenario of a coronal magnetic field that is stressed by photospheric shear, where currents flow along the photospheric network and magnetic separators. Continuous driving triggers occasional ("stick–slip") reconnection along separators and avalanche-like releases of magnetic energy, producing similar powerlaw distributions as observed.

9.3.5 Thermal Energy of Flare Plasma

The peak energy release rate P or total energy E are key parameters in the evaluation of SOC systems. For solar and stellar flares, photon count rates of fluxes are often used as a proxy for the energy release rate P, and total (time-integrated) counts or fluences for the total energy E. The observables (flux, fluence) are generally approximately proportional to the physical quantities (peak energy release rate and total energy), but the exact relationship requires physical models and is wavelength-dependent. There are essentially three different energy quantities that are modeled in the context of solar (and stellar) flares: (1) the magnetic energy E_B, (2) the thermal energy E_T, and (3) the nonthermal energy N_{NT}. We dealt with the magnetic energy in the last two sections (9.3.3 and 9.3.4), and consider now the thermal energy in the following, defining the relationships between the observables (peak count rate) and physical parameters (emission measure, density, and temperature) in particular (following Aschwanden et al. 2008c).

A solar or stellar coronal flare is usually detected from light curves in extreme ultraviolet (EUV) or soft X-ray wavelengths, from which a (background-subtracted) peak count rate c_p [cts s^{-1}] at the flare peak time $t = t_p$ can be measured. The count rate $c(t)$ for optically-thin emission (as it is the case in EUV and soft X-rays) is generally defined by the temperature integral of the total (volume-integrated) differential emission measure distribution $dEM(T)/dT$ [cm^{-3}] and the instrumental response function $R(T)$ (in units of [cts s^{-1} cm^3]),

$$4\pi d^2\, c(t) = \int \frac{dEM(T)}{dT}\, R(T)\, dT\ , \tag{9.3.18}$$

where the factor $(4\pi d^2)$ comes from the total emission over the full celestial sphere at a stellar distance d (in parsecs). The differential emission measure distribution (DEM) of flares shows usually a single peak at the flare peak temperature T_p, so that the emission measure peak at the flare peak time, $EM_p = dEM(t = t_p, T)/dT \approx dEM(t = t_p, T = T_p)$, can be approximated with a single temperature (which corresponds to an emission measure-weighted average value),

$$4\pi d^2\, c_p = 4\pi d^2\, c(t = t_p) \approx EM_p\, R(T_p)\ . \tag{9.3.19}$$

The total (volume-integrated) emission measure EM_p at the flare peak is defined as the squared electron density n integrated over the source volume V,

$$EM_p = \int n^2\, dV \approx n_p^2 V\ , \tag{9.3.20}$$

where the right-hand approximation implies that $n_p^2 = n^2(t = t_p, T = T_p)$ is the squared electron density at the flare peak time averaged over the volume V of the flare plasma, assuming a unity filling factor. Integrating the count rate $c(t)$ over the flare duration τ_f yields the total counts C, which in the case of a single-peaked DEM can also be approximated (with Eq. 9.3.19) as

$$4\pi d^2\, C = 4\pi d^2 \int c(t)\, dt \approx 4\pi d^2\, c_p\, \tau_f = EM_p\, R(T_p)\, \tau_f\ . \tag{9.3.21}$$

The radiative loss rate for optically thin plasmas is a function of the squared density and the radiative loss function $\Lambda(T)$,

$$\frac{dE_R}{dV\ dt} = n_e n_i \Lambda(T) \approx n_e^2 \Lambda(T) \ , \tag{9.3.22}$$

(in the coronal approximation of fully ionized plasma, i.e., $n_e \approx n_i$) where the radiative loss function has a typically value of $\Lambda(T) \approx 10^{-23...-22}$ [erg cm^3 s^{-1}] in the temperature range of $T \approx 10^{6...8}$ K. From this we can define a peak luminosity L_X in soft X-rays by integrating over the volume and temperature range,

$$L_X = V \int n^2(t = t_p, T)\Lambda(T)\ dT \approx EM_p\ \Lambda(T_p) \ . \tag{9.3.23}$$

The total radiated energy E_X integrated over the flare duration is then

$$E_X = \int\int\int n^2(t,T)\Lambda(T)\ dV\ dT\ dt \approx EM_p\ \Lambda(T_p)\ \tau_f \ . \tag{9.3.24}$$

This yields a convenient conversion from observed total counts $4\pi d^2\ C$ (Eq. 9.3.21) into total radiated energy E_X (Eq. 9.3.24),

$$E_X = \frac{\Lambda(T_p)}{R(T_p)} 4\pi d^2\ C = f(T_p)\ 4\pi d^2\ C \ , \tag{9.3.25}$$

which involves a temperature-dependent conversion factor $f(T_p) = \Lambda(T_p)/R(T_p)$.

For comparison we calculate also the total thermal energy E_T of the flare volume at the flare peak time $t = t_p$,

$$E_T = \int 3n(t = t_p, T) k_B T(t = t_p) V(t = t_p)\ dT \approx 3 n_p k_B T_p V = \frac{3 k_B EM_p T_p}{n_p} \tag{9.3.26}$$

where $n_p = n(t = t_p, T = T_p)$ represents the electron density at the flare peak time $t = t_p$ and DEM peak temperature $T = T_p$. The relation between the total thermal energy E_T and the total radiated energy E_X is then

$$E_T \approx E_X \frac{3 k_B T_p}{n_p(T_p)\ \Lambda(T_p)\ \tau_f(T_p)} \ , \tag{9.3.27}$$

where the peak electron density $n_p(T_p)$ and the flare duration $\tau_f(T_p)$ may have a statistical dependence on the flare peak temperature T_p, and this way define the temperature dependence in the correlation between the thermal energy E_T and the total radiated energy E_X.

The occurrence frequency distributions of solar (or stellar) flares can be carried out simply with observables (peak counts c_p, total counts C, and durations τ_f), or with physically derived quantities (peak luminosity L_X, total radiated energy E_X, and duration $\tau_f[T_p]$), using the relations Eqs. (9.3.18–27). To obtain distributions of thermal energies, the flare volume V has to be estimated, which can be fractal with a filling factor (Section 8.3.2) and can only be measured for solar flares, while stellar flares remain unresolved point sources for current instruments. With the observed peak counts c_p, the emission measure EM_p can be determined (Eq. 9.3.19) and the peak density $n_p \approx \sqrt{(EM_p/V)}$ (Eq. 9.3.20). In addition, a peak temperature measurement T_p is needed in order to obtain the thermal energy $E_T \approx 3n_p k_B T_p V$ (Eq. 9.3.26). Some care needs to be exercised in evaluating the peak temperature T_p from narrowband filters, in order to avoid instrumental biases of the temperature coverage. The two main critical issues of temperature bias and fractal volumes in the evaluation of flare energies are discussed in Aschwanden and Parnell (2002). Flare detection in EUV generally underestimates the flare temperature ($T_{EUV} < T_p$), which leads to steeper powerlaw slopes in the occurrence frequency distribution of flare energies (e.g., Parnell and Jupp 2000; Benz and Krucker 2002; see also Section 7.3.3).

As we have seen in the derivation of relationships between observed fluxes and thermal energies, physical variables such as electron densities and temperatures are involved, which require a physical model. While we included only the process of radiative loss in the derivation above (Eq. 9.3.22), the processes of heating and conductive losses may also be included in hydrodynamic models. The assumption of energy balance in 1-D hydrodynamic coronal loops or flare loops leads to scaling laws between the physical parameters of the electron temperature T_p, the electron density n_p, and the length scale L. Solar and stellar flares can be modeled in terms of a superposition of multiple 1-D hydrodynamic loops (e.g., Aschwanden et al. 2008c). Additional inclusion of the magnetic field yields "universal scaling laws" for solar and stellar flares (Shibata and Yokoyama 1999, 2002; Cassak et al. 2008). Such scaling laws provide the physical foundation for observed correlations between SOC parameters.

9.3.6 Nonthermal Energy of Flares

A generic energy spectrum of a large flare is shown in Fig. 9.8, which exhibits dominantly thermal emission in soft X-rays ($\approx$1–10 keV), nonthermal bremsstrahlung emission in hard X-rays ($\approx$10 keV–1 MeV), nuclear de-excitation lines in gamma rays ($\approx$1–10 MeV), relativistic electron bremsstrahlung at $\approx$10–100 MeV, and pion radiation at $\gtrsim$100 MeV. Theoretically we would expect that the total thermal energy is approximately equal to the nonthermal energy in hard X-ray producing electrons, because the thermal flare plasma is heated in the chromosphere by the precipitating nonthermal electrons and ions, according to the chromospheric evaporation scenario (also called "thick-target model"). It is therefore customary to estimate the nonthermal flare energy E_{NT} from hard X-ray observations, as alternative to the thermal flare energy E_T obtained from soft X-ray and EUV observations. A comparison of frequency distributions of nonthermal flare energies and active region sizes shows also a good correspondence (Wheatland and Sturrock 1996). In the following we outline the relationship between hard X-ray counts C and nonthermal flare energy E, which can be used to derive occurrence frequency distributions $N(E)$ of flare energies from the observed distributions $N(C)$ of hard X-ray counts.

The standard derivation of the *thick-target model* (e.g., see Chapter 13 in Aschwanden (2004)) approximates the observed hard X-ray spectrum $I(\varepsilon_x)$ as a powerlaw function of the photon energy ε_x) (Brown 1971),

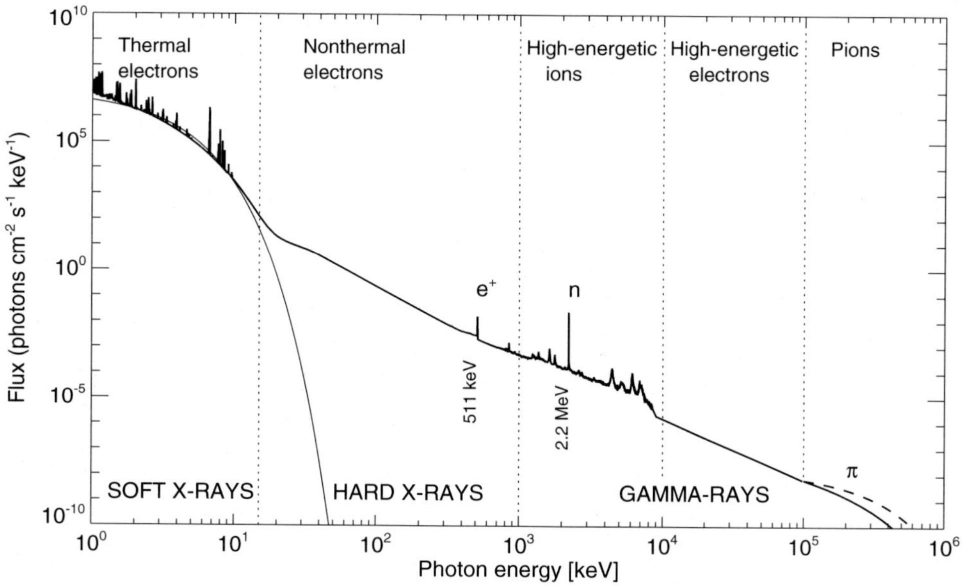

Fig. 9.8 Composite photon spectrum of a large flare, extending from soft X-rays (1 – 10 keV), hard X-rays (10 keV – 1 MeV), to gamma rays (1 MeV – 100 GeV). The energy spectrum is dominated by different processes: by thermal electrons (in soft X-rays), bremsstrahlung from nonthermal electrons (in hard X-rays), nuclear de-excitation lines (in $\approx$ 0.5–8 MeV gamma rays), by bremsstrahlung from high-energetic electrons (in $\approx$10–100 MeV gamma rays), and by pion decay (in $\gtrsim$100 MeV gamma-rays). Note also the prominent electron-positron annihilation line (at 511 keV) and the neutron capture line (at 2.2 MeV).

$$I(\varepsilon_x) = I_1 \frac{(\gamma - 1)}{\varepsilon_1} \left(\frac{\varepsilon_x}{\varepsilon_1} \right)^{-\gamma} \qquad (\text{photons cm}^{-2}\ \text{s}^{-1}\ \text{keV}^{-1}) \,, \tag{9.3.28}$$

where ε_1 is a reference energy, above which the integrated photon flux is I_1 (photons cm^{-2} s^{-1} keV^{-1}), and γ is the powerlaw slope. The parameters ε_1 and γ of the hard X-ray spectrum are time-dependent. The total number of photons above a lower cutoff energy ε_1 is the integral of Eq. (9.3.28),

$$I(\varepsilon_x \geq \varepsilon_1) = \int_{\varepsilon_1}^{\infty} I(\varepsilon_x)\, d\varepsilon_x = I_1 \qquad (\text{photons cm}^{-2}\ \text{s}^{-1}) \,. \tag{9.3.29}$$

Brown (1971) solved the inversion of the photon spectrum for the Bethe–Heitler bremsstrahlung cross-section and found the following *instantaneous nonthermal electron spectrum* $n_e(\varepsilon)$ present in the X-ray-emitting region,

$$n_e(\varepsilon) = 3.61 \times 10^{41} \gamma (\gamma - 1)^3\, B\left(\gamma - \frac{1}{2}, \frac{3}{2} \right) \frac{I_1 \sqrt{\varepsilon}}{n_0 \varepsilon_1} \left(\frac{\varepsilon}{\varepsilon_1} \right)^{-\gamma}$$

$$(\text{electrons keV}^{-1}) \,, \tag{9.3.30}$$

with the associated *electron injection spectrum* $f_e(\varepsilon)$,

$$f_e(\varepsilon) = 2.68 \times 10^{33} \gamma^2 (\gamma - 1)^3 \, B\left(\gamma - \frac{1}{2}, \frac{3}{2}\right) \frac{I_1}{\varepsilon_1^2} \left(\frac{\varepsilon}{\varepsilon_1}\right)^{-(\gamma+1)}$$

$$(\text{electrons keV}^{-1} \text{ s}^{-1}) , \tag{9.3.31}$$

with n_0 (cm^{-3}) the mean electron or proton density in the emitting volume, ε_1 [keV] the lower cutoff energy in the spectrum, I_1 (photons cm^{-2} s^{-1} keV^{-1}) the total X-ray photon flux at energies $\varepsilon \gtrsim \varepsilon_1$, and $B(p,q)$ is the *Beta function*,

$$B(p,q) = \int_0^1 u^{p-1} (1-u)^{q-1} \, du , \tag{9.3.32}$$

which is calculated in Hudson et al. (1978) for a relevant range of spectral slopes γ and is combined in the auxiliary function $b(\gamma)$,

$$b(\gamma) = \gamma^2 (\gamma - 1)^2 \, B\left(\gamma - \frac{1}{2}, \frac{3}{2}\right) \approx 0.27 \, \gamma^3 . \tag{9.3.33}$$

So the powerlaw slope of the electron injection spectrum ($\delta = \gamma + 1$) is steeper than that (γ) of the photon spectrum in the thick-target model. With this notation we can write the electron injection spectrum as

$$f_e(\varepsilon) = 2.68 \times 10^{33} \, (\gamma - 1) b(\gamma) \, \frac{I_1}{\varepsilon_1^2} \left(\frac{\varepsilon}{\varepsilon_1}\right)^{-(\gamma+1)} \qquad (\text{electrons keV}^{-1} \text{ s}^{-1}) . \tag{9.3.34}$$

The total number of electrons above a cutoff energy ε_c is then

$$F(\varepsilon \geq \varepsilon_c) = \int_{\varepsilon_c}^{\infty} f_e(\varepsilon) \, d\varepsilon = 2.68 \times 10^{33} \, b(\gamma) \frac{(\gamma - 1)}{\gamma} \frac{I_1}{\varepsilon_1} \left(\frac{\varepsilon_c}{\varepsilon_1}\right)^{-\gamma} \qquad (\text{electrons s}^{-1}) . \tag{9.3.35}$$

The power in nonthermal electrons above some cutoff energy ε_c is

$$P(\varepsilon \geq \varepsilon_c) = \int_{\varepsilon_c}^{\infty} f_e(\varepsilon) \, \varepsilon \, d\varepsilon = 2.68 \times 10^{33} \, b(\gamma) I_1 \left(\frac{\varepsilon_c}{\varepsilon_1}\right)^{-(\gamma-1)} \qquad (\text{keV s}^{-1}) . \tag{9.3.36}$$

or a factor of (keV/erg) $= 1.6 \times 10^{-9}$, smaller in cgs units,

$$P(\varepsilon \geq \varepsilon_c) = \int_{\varepsilon_c}^{\infty} f_e(\varepsilon) \, \varepsilon \, d\varepsilon = 4.3 \times 10^{24} \, b(\gamma) I_1 \left(\frac{\varepsilon_c}{\varepsilon_1}\right)^{-(\gamma-1)} \qquad (\text{erg s}^{-1}) . \tag{9.3.37}$$

Solar flares have typical photon count rates in the range of $I_1 = 10^1$–10^5 (photons s^{-1} cm^{-2}) at energies of $\varepsilon \geq 20$ keV and slopes of $\gamma \approx 3$. Thus, for $\varepsilon_c = \varepsilon_1 = 20$ keV, and using $b(\gamma) \approx 0.27\gamma^3 \approx 7$ (Eq. 9.3.33), we estimate using Eq. (9.3.37) a nonthermal power of $P(\varepsilon \geq 20 \text{ keV}) \approx 3 \times 10^{25} - 3 \times 10^{30}$ erg s^{-1}. Integrating this power over typical flare durations of $\tau_{flare} \approx 10^2$ s yields a range of $W = P(\varepsilon \geq 20 \text{ keV}) \times \tau_{flare} \approx 3 \times 10^{27} - 3 \times 10^{32}$ [erg] for flare energies. A frequency distribution of total nonthermal flare energies in electrons (>25 keV) which covers this range has been determined in Crosby et al. (1993), see Fig. 1.14.

Applying this thick-target model to hard X-ray data observed with *HXRBS/SMM*, using a lower energy cutoff of 25 keV, the following correlations were found between the observed peak count rate P, the peak hard X-ray flux I at 25 keV, the spectrally-integrated hard X-ray flux I above 25 keV, the peak energy flux F in electrons, and the total energy E in electrons,

$$\begin{aligned} I(25 \text{ keV}) &\approx P^{1.01} \\ I(>25 \text{ keV}) &\approx P^{1.07} \\ F(>25 \text{ keV}) &\approx P^{0.94} \\ E(25 \text{ keV}) &\approx P^{1.25} \\ F \times D(>25 \text{ keV}) &\approx E^{1.18} \end{aligned} \tag{9.3.38}$$

Thus, the simply observed peak count rate P is a good proxy for the nonthermal flare energy E or the time-integrated total flare energy $F \times D$. The powerlaw slope of the occurrence frequency distribution for any of these parameters can then easily be calculated from the slope α_P of the count rate distribution $N(P) \propto P^{-\alpha_P}$ and the correlation coefficients β given in Eq. (9.3.38) using the relation Eq. (7.1.42). For instance, the average powerlaw slope of peak counts in hard X-rays is $\alpha_P = 1.75$ (Fig. 7.7). Using the correlation $E(>25 \text{ keV}) \propto P^{\beta}$ with $\beta = 1.25$, we estimate $\alpha_E = 1 + (\alpha_P - 1)/\beta = 1.60$ for the powerlaw slope of the energy distribution, which indeed agrees well with $\alpha_E = 1.61$ of the frequency distribution of total counts (Fig. 7.5).

9.3.7 Particle Acceleration

A typical energy spectrum of a solar flare (Fig. 9.8) can be characterized by an exponential-like thermal spectrum at low energies ($E \lesssim 10$ keV) and by a powerlaw-like nonthermal spectrum at high energies ($E \gtrsim 10$ keV), sometimes extending up to $\lesssim 100$ MeV in large flares. These two spectral components strikingly display the dual nature of incoherent and coherent random processes. The thermal spectrum is produced by collisional interactions, which operate as an incoherent random process that has additive characteristic and an exponential-like random distribution. The nonthermal spectrum, in contrast, is produced by nonthermal particles that were accelerated coherently in an essentially collisionless plasma, either by electric fields, stochastic wave–particle interactions, or shocks. The coherent energy gain has a nonlinear dependence as a function of time. If (1) individual charged particles are accelerated independently, (2) the nonlinear energy gain is close to an exponential function, and (3) the acceleration time is a random time interval, all three criteria of a SOC process (Section 9.1) are fulfilled, the resulting energy spectrum is consequently a powerlaw, and we can consider the particle acceleration region of a flare as an individual SOC system. The independence of individual acceleration trajectories is certainly fulfilled for stochastic wave–particle interactions, diffusive (second-order) Fermi, or diffusive shock acceleration processes. Note that flare statistics from the whole Sun, where the entire solar corona is considered to be a SOC system, is then a "SOC system of

SOC systems". In other words, if the solar corona is the analog of a sandpile and flares are individual sand avalanches, we can also consider every sand avalanche as a SOC system itself, where each sand grain gains different amounts of energy according to a powerlaw distribution. If we proceed in the hierachy of SOC systems further, from stars to galaxies and the entire universe, we end up at the cosmic-ray spectrum shown in Fig. 9.5, which still has a powerlaw-like functional shape.

Let us consider the physical basis of how a particle acceleration region in a coronal plasma can fulfill the assumed SOC characteristics. A simple model is electric DC-field acceleration, where a particle (say an electron with electric charge e) gains energy proportional to the electric field strength $\mathscr{E}_\parallel$ (in the mildly relativistic regime),

$$m_e \frac{dv_\parallel}{dt} = e\,\mathscr{E}_\parallel\ , \tag{9.3.39}$$

which leads to a quadratic dependence of energy gain as a function of time,

$$E(t) = \frac{1}{2} m_e v^2(t) = \frac{e^2\,\mathscr{E}_\parallel^2}{2m_e} t^2\ . \tag{9.3.40}$$

The quadratic dependence is not exactly exponential (only to the second order), but sufficiently close for a small number of growth times. If the particles are accelerated for a random time interval,

$$N(t)dt \propto \exp(-t/t_A)\ dt \tag{9.3.41}$$

with t_A the e-folding value of random acceleration times, the resulting energy spectrum is

$$N(E)dE = N(t[E]) \left| \frac{dt}{dE} \right| dE = N_0 \exp\left(-\sqrt{\frac{E}{E_0}}\right) E^{-1/2}\ dE\ , \tag{9.3.42}$$

with the reference energy E_0,

$$E_0 = \frac{e^2\,\mathscr{E}_\parallel^2}{2m_e} t_A^2\ . \tag{9.3.43}$$

This energy spectrum (Eq. 9.3.42) has a powerlaw-like function in the low energy part and falls off exponentially at higher energies, which could be consistent with some observations, but it would not explain flares with a powerlaw spectrum over a large energy range, as shown in Fig. 9.8. Of course, the electron injection spectrum has also to be convolved with a (e.g., Bethe–Heitler) bremsstrahlung cross-section (see Section 9.3.6), in order to predict the observed hard X-ray spectrum shown in Fig. 9.8. However, we used two essential assumptions that can be modeled in different ways. First, we assumed a uniform constant DC electric field, which may not exist in coronal conditions, while dynamical and spatially inhomogeneous fields are more likely and would produce a different acceleration time profile than the quadratic one assumed in Eq. (9.3.40). Secondly, we assumed that particles are accelerated during a random time interval. This could be the case in a thin current sheet, where particles are randomly scattered out of the acceleration region due to their chaotic orbits and different initial pitch angles. In general, the detailed distribution of acceleration times depends on the initial pitch angle distribution as well as on the particular 3-D geometry of the current sheet. Similar modeling could be discussed for alternative acceleration processes, such as stochastic gyroresonant wave-particle interactions or diffusive shock acceleration. However, whatever the details of the physical models are, every coherent acceleration mechanism that leads to a systematic energy gain of a particle above the thermal energy level, will produce a powerlaw-like energy spectrum if the acceleration times are random. Thus, many particle acceleration mechanisms in astrophysical plasmas can be considered as a SOC process and be modeled as such.

Let us mention a few relevant studies on particle acceleration in solar flare conditions. Electric DC field acceleration of field-aligned currents in (time-varying) shear flow (vortices) in flare loops leading to a powerlaw-like energy spectrum of the form of Eq. (9.3.42) has been considered by Tsuneta (1995). A number of leading particle acceleration models have been reviewed in Miller et al. (1997), including sub-Dreicer and super-Dreicer electric DC field acceleration, stochastic MHD turbulence, and shock acceleration. These models all produce a powerlaw-like energy spectrum at mildly relativistic energies, but exhibit an exponential-like fall-off at highly relativistic energies. A series of particle acceleration simulations have been conducted in the spirit of the SOC concept: with random shocks (Anastasiadis and Vlahos 1991, 1993, 1994), with random DC electric fields (Anastasiadis et al. 1997), with random magnetic fields (Dauphin 2007), 3-D MHD turbulence (Dmitruk et al. 2003), or in terms of a cellular automaton model (Anastasiadis et al. 2004), which all contain the elements of independent acceleration time histories for each particle, random acceleration times, and produce powerlaw-like energy spectra, thus essentially fulfilling the basic requirements for a SOC system. The ratio of the acceleration time to the e-folding growth time (of energy gain) predicts the flattest powerlaw slope for subsets with high energy gain (Eq. 3.1.10), which has been applied to the threshold effect of proton acceleration in solar flares (Miroshnichenko 1995). A SOC state of first-order Fermi acceleration was also considered in astrophysical shocks (Malkov et al 2000).

9.3.8 Coherent Radio Emission

Solar radio emission can be subdivided into the two categories of incoherent emission (e.g., free-free bremsstrahlung, gyroresonance, or gyrosynchrotron) and coherent emission (e.g., plasma emission or electron-cyclotron maser emission). For an overview see, e.g., Benz 1993, or chapter 15 in Aschwanden 2004). The category of coherent emission has exactly the exponential-growth characteristics we expect for SOC events. Some plasma instabilities, such as the bump-in-tail instability of electron beams, or a loss-cone instability driven by an anisotropic particle distribution, exhibit an exponential growth of electrostatic or electromagnetic waves, which saturate at some point once the unstable particle distribution flattens out to a stable plateau. The beam-driven instability produces plasma emission at the density-dependent plasma frequency of $\nu_{pe} = 9{,}000\sqrt{n_e}$. Plasma emission is a multi-stage process, which includes, e.g., (1) formation of an (unstable) particle beam distribution by velocity dispersion, (2) generation of Langmuir turbulence, and (3) its nonlinear evolution and conversion into escaping (electromagnetic) radiation (plasma emission). This basic process is responsible for a variety of solar radio burst types, which have a different morphology in an observed dynamic spectrum depending on the magnetic configuration and local density structure in which they are generated. Solar radio bursts with plasma emission include (Fig. 7.14) type I storms (Langmuir turbulence), type II bursts (beams from shocks), type III bursts (upward propagating beams), reverse-slope bursts (downward propagating beams), type J and U bursts (beams along closed loops), type IV continuum (trapped electrons), and type V bursts (slow electron beams). Another category of coherent radio emission is produced by loss-cone particle distributions, which have an enhancement of particles at large pitch angles and thus provide free energy for gyroresonant waves by quasi-linear diffusion to a lower energy state at lower pitch angles. A prominent representative of the latter category is the electron-cyclotron maser emission, which is believed to operate in solar flare loops, auroral kilometric radiation (AKR), Jupiter's decametric emission, and in stellar flares (e.g., Dulk 1985).

In essence, we can consider every plasma environment as a SOC system, if it is capable of producing coherent radio emission. Since most of the plasma instabilities occur very fast (on sub-second time scales) and since most astrophysical plasmas are quite extended, individual radio bursts are most likely to be generated independently in the time and space domain, and thus fulfill our first SOC criterion of statistical independence (Section 9.1). The second SOC criterion of an exponential-like growth phase is fairly characteristic for coherent wave-particle interactions. The third criterion of a random rise time is also easily to satisfy, because the criticality is often given by a gradient in the particle distribution ($\partial f/\partial v_{\parallel}$ for beam instabilities, or $\partial f/\partial v_{\perp}$ for loss-cone instabilities), which are subject to large fluctuations in various temporal and spatial domains. Thus, coherent radio bursts are likely to originate in a SOC system and thus are expected to exhibit powerlaw-like frequency distributions of their peak fluxes or fluences, as it was indeed found for numerous datasets (Table 7.5 and Section 7.3.4).

While statistics of solar radio bursts gathered from many flare events dominantly exhibit powerlaw distributions of their peak fluxes, this is not necessarily the case for statistics of radio bursts during a single flare episode (e.g., Aschwanden et al. 1998b; Isliker et al. 1998b). An interpretation of solar radio bursts in terms of SOC models is also discussed in Vilmer and Trottet (1997) and Bastian and Vlahos (1997). Some detailed theoretical models of type III bursts involve the stochastic growth evolution of Langmuir waves (e.g., Robinson 1993), rather than the exponential growth evolution predicted by quasi-linear diffusion theory, which leads to "clumpy" Langmuir emission (e.g., Cairns and Robinson 1999) and might introduce some modification of the powerlaw-like frequency distributions of peak fluxes.

9.3.9 Master Equation

Our basic analytical model of a SOC system is the exponential-growth model (Section 3.1), which is characterized by a growth time τ_G of an instability, an e-folding saturation time τ, and a linear decay time t_D. Assuming a random distribution of saturation times, $N(t_S) \propto \exp(-\tau/t_S)$, this model predicts a powerlaw distribution function $N(E) \propto E^{-\alpha}$ for the released energy E of SOC events. An alternative approach to derive the occurrence frequency distribution of energies is a balance equation between the energy build up rate dE/dt and the energy release rate $E/\Delta t$, which occurs in time intervals we called *waiting times* Δt. Such a *steady-state transport equation* was proposed by Litvinenko (1994),

$$\frac{d}{dE}\left(\frac{dE}{dt}N(E)\right)+\frac{N(E)}{\Delta t}=0\,, \qquad (9.3.44)$$

where dE/dt is the mean rate of energy increase available for a flare, $N(E)$ is the flare probability function, and Δt is the mean waiting time between flares. Inserting some scaling laws that apply to reconnecting current sheets, Litvinenko (1994) derived a frequency distribution of $N(E) \propto E^{-7/4}$, which is close to the observed ones in solar flares.

Wheatland and Glukhov (1998) expanded this steady-state transport equation (Eq. 9.3.44) into a more general probability equation that is also called *master equation*,

$$\frac{d}{dE}\left(\frac{dE}{dt}N(E)\right)+N(E)\int_0^E \alpha(E,E')\,dE' - \int_E^\infty N(E')\alpha(E,E')\,dE' = 0 \qquad (9.3.45)$$

which describes the rate of change in the probability distribution $N(E)$ with three terms, including the energy build-up (first term), the number of active regions that fall out of the energy interval $(E, E+\Delta E)$ due to flaring (second term), and those active regions that fall from a higher energy state into the interval $(E, E+\Delta E)$ due to flaring (third term). The transition rate from energy state E to E' is denoted by the coefficient $\alpha(E,E')$. In a steady state situation, the sum of the three terms should balance out to zero. The master equation (Eq. 9.3.45) cannot easily be solved to obtain a general solution, but for some special assumptions Wheatland and Glukhov (1998) could arrive at a powerlaw shape for the energy distribution $N(E)$. With the master equation approach, a differential equation could also be derived that describes how the free energy in the corona changes as a function of the driving and flaring rate (Litvinenko and Wheatland 2001; Wheatland and Litvinenko 2001). Monte-Carlo simulations of this model demonstrated that the behavior of waiting-time distributions can significantly deviate from simple Poisson statistics (Wheatland 2009).

9.4 Magnetospheric Physics

9.4.1 Coronal Mass Ejections and Magnetospheric Storms

Major disturbances in the Earth's magnetosphere are caused by space weather events triggered by geoeffective solar flares and coronal mass ejections (CMEs), which are called *magnetic storms*. A CME produces a shock wave in the heliospheric solar wind that can strike the Earth's magnetosphere typically 1–1.5 days later.

The phenomenon of a CME occurs with a frequency of few events per day, carrying a mass in the range of $m_{CME} \approx 10^{14}$–10^{16} g, which corresponds to an average mass loss rate of $m_{CME}/(\Delta t \cdot 4\pi R_\odot^2) \approx 2\times10^{-14}$–$2\times10^{-12}$ (g cm^{-2} s^{-1}), which is $\lesssim 1\%$ of the solar wind mass loss in coronal holes, or $\lesssim 10\%$ of the solar wind mass in active regions. The transverse size of CMEs can cover from a fraction up to more than a solar radius, and the ejection speed is in the range of $v_{CME} \approx 10^2$–2×10^3 (km s^{-1}). Ambiguities from line-of-sight projection effects make it difficult to infer the geometric shape of CMEs. Possible interpretations include fluxropes, semi-shells, or bubbles. There is a general consensus that a CME is associated with a release of magnetic energy in the solar corona, but its relation to the flare phenomenon is controversial. Even big flares (at least GOES M-class) have no associated CMEs in 40% of the cases. A long-standing debate focused on the question of whether a CME is a by-product of the flare process or vice versa. This question has been settled in the view that both CMEs and flares are quite distinctly different plasma processes, but related to each other by a common magnetic instability that is controlled on a larger global scale. A CME is a dynamically evolving plasma structure, propagating outward from the Sun into interplanetary space, carrying a frozen-in magnetic flux and expanding in size. If a CME structure travels from a sub-solar point radially towards the Earth, it is called a *halo-CME*, an *Earth-directed*, or *geo-effective* event. CME-accelerated energetic particles reach the Earth most likely when a CME is launched in the western solar hemisphere, since they propagate along the curved *Parker spiral* interplanetary magnetic field. The solar wind pressure varies according to solar activity and the occurrence of CME shock waves, which induce currents in the ionosphere, cause *geomagnetic storms* in the Earth's magnetosphere, and can disrupt global communication and navigation networks, or can cause failures of satellites and commercial power systems.

We already discussed some SOC aspects of CMEs in earlier sections, i.e., the waiting-time distribution of CMEs in Section 5.6.3, and the frequency occurrence distributions of CME-associated *solar energetic particle (SEP)* events in Section 7.3.5. Since CMEs and solar flares are intimately connected, both CMEs and flares are primary SOC phenomena with similar frequency distributions and waiting-time distributions, while geomagnetic storms are secondary SOC phenomena, representing a subset of the large solar flare and CME events that are geo-effective. This subset essentially includes events that originate on the western side of the solar disk and are magnetically connected with the Earth. However, the SOC statistics of geo-effective CMEs is expected to be similar to that of all CMEs. In analogy, earthquake statistics on a particular continent are expected to be similar to the global statistics from the entire planet.

9.4.2 Heliospheric Field and Magnetospheric Substorms

The *heliospheric 3-D magnetic field* is defined by the flow of the solar wind. The field in the regions between the planets near the ecliptic plane is more specifically called the *interplanetary magnetic field*. The basic geometry of the interplanetary magnetic field has the form of an Archimedean spiral, as inferred by Parker (1963) from the four assumptions: (1) the solar wind moves radially away from the Sun at a constant speed; (2) the Sun rotates with a constant period (i.e., with a synodic period of 27.27 days at the prime meridian defined by Carrington); (3) the solar wind is azimuthally symmetric with respect to the solar rotation axis; and (4) the interplanetary magnetic field is frozen-in the solar wind and anchored at the Sun. The solar wind stretches the global, otherwise radial field into spiral field lines with an azimuthal field component. The resulting Archimedean spirals leave the Sun near-vertically to the surface and cross the Earth orbit at an angle of $\approx 45°$. Measurements of the magnetic field direction at Earth orbit reveal a *two-sector pattern* during the period of declining solar activity and a *four-sector pattern* during the solar minimum,

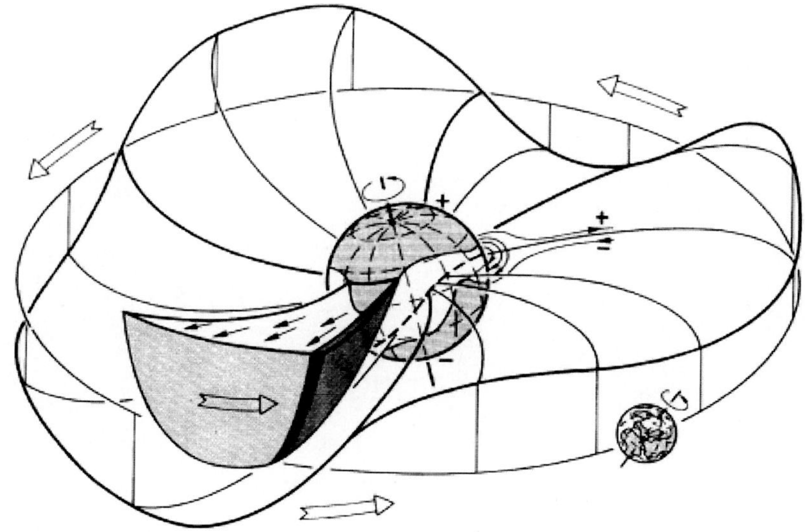

Fig. 9.9 The interplanetary magnetic field has a spiral-like radial field and the boundary layer between the two opposite magnetic polarities in the northern and southern hemispheres is warped like a *"ballerina skirt"*. This concept was originally suggested by Hannes Alfvén in 1977.

with oppositely directed magnetic field vectors in each sector. From this ecliptic cut, a warped heliospheric current sheet can be inferred that has the shape of a *"ballerina skirt"* (Fig. 9.9). The solar axis is tilted by 7.5° to the ecliptic plane, and the principal dipole magnetic moment of the global field can be tilted by as much as $\approx 20°–25°$ at activity minimum, and thus the warped sector zone extends by at least the same angle in northerly and southerly direction of the ecliptic plane. The strength of the interplanetary magnetic field, of course, depends on the solar cycle, varying between $B \approx 6$ nT and 9 nT ($\approx 10^{-5}$ G) at a distance of 1 AU. The interplanetary magnetic field can be heavily disturbed by CME-related shocks and propagating CMEs. The magnetic field is near-radial near the Sun and falls off with $B(R) \approx B_r(R) \propto R^{-2}$ there, while it becomes more azimuthal at a few AU and falls off with $B(R) \approx B_\varphi(R) \propto R^{-1}$ at larger heliocentric distances according to the model of Parker.

Besides the major disturbances caused by CMEs, the Earth's magnetosphere is also affected by smaller disturbances of the solar wind and the interplanetary magnetic field, that wrap around the Earth's magnetopause. As can be easily imagined from the magnetic configuration shown in Fig. 9.9, the occurrence of a new active region on the Sun, which is the "footprint" of the interplanetary magnetic field on the solar surface, can easily flip the warped heliospheric current sheet (ballerina skirt) at the location of the Earth. Brief magnetospheric disturbances occur when the interplanetary magnetic field (IMF) flips southward, which triggers magnetic reconnection at the dayside magnetopause and transfers energy from the solar wind to the magnetosphere. Part of the transferred energy is stored in the magnetotail, where also magnetic reconnection and field relaxation events can occur, which are termed *magnetospheric substorms*. A magnetospheric substorm has

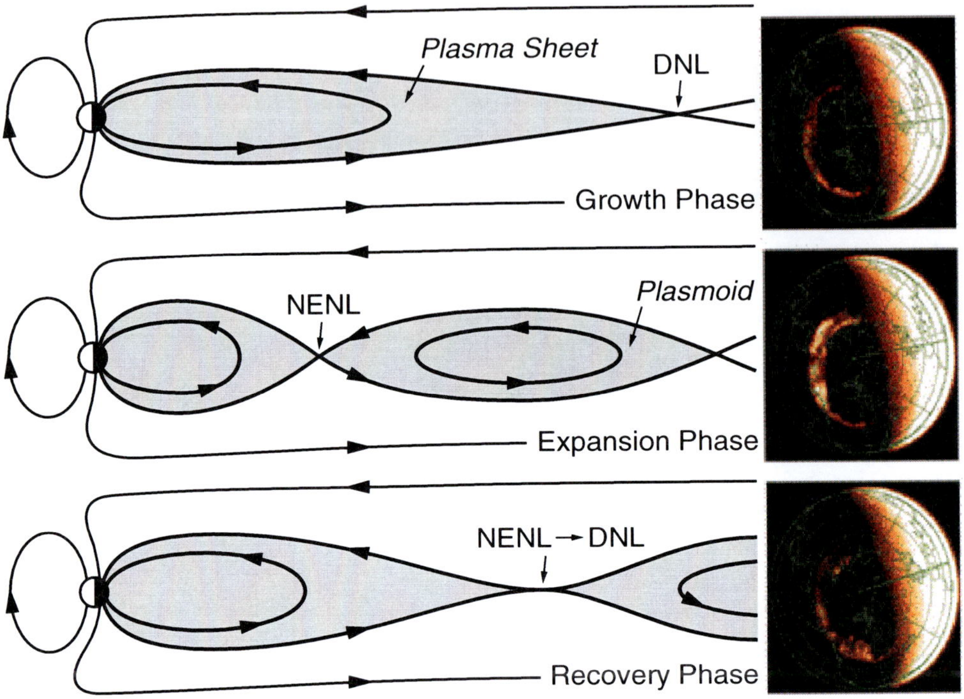

Fig. 9.10 The three phases of a geomagnetic substorm are shown: the growth phase (top), the expansion phase (middle), and the recovery phase (bottom) (Baumjohann and Treuman 1996). The accompanying three auroral images were obtained with the IMAGE WIC instrument (credit: NASA).

three phases (Fig. 9.10): (1) the growth phase (when energy from the solar wind is transferred to the dayside magnetosphere), (2) the substorm expansion phase (when the energy stored in the magnetotail is released, the inner magnetosphere relaxes from the stretched tail, and the tail snaps into a more dipolar configuration and energizes particles in the plasma sheet), and (3) the recovery phase (during which the magnetosphere returns to its quiet state). The whole process causes changes in the auroral morphology (Fig. 1.9) and induces currents in the polar ionosphere. Part of the transferred energy is dissipated by particle precipitation into the ionosphere, which produces auroral displays and magnetic disturbances. The frequency of substorms is about 6 per day on average, but larger during geomagnetic storms.

One indicator of geomagnetic activity is the so-called *Auroral Electrojet Index (AE)*, which provides a global, quantitative measure of the enhanced ionospheric currents that flow below and within the auroral oval. Ideally, the AE index measures deviations from quiet day values of the horizontal magnetic field around the auroral oval. The AE index was found to be correlated with substorm morphologies, the behavior of communication satellites, radio propagation, radio scintillation, and the coupling between the interplanetary magnetic field (IMF) and the Earth's magnetosphere. Low-frequency stochastic fluctuations of the geomagnetic AE-index with a $1/f$ spectrum have been interpreted in terms of a SOC system (Consolini 1997; Chapman et al. 1998; Uritsky and Pudovkin 1998), as well as the lifetime distributions of magnetospheric disturbances as measured from the AE index (Takalo 1993; Takalo et al. 1999a). The SOC events are believed to be triggered by sudden changes of the energy input, such as the southward turning and subsequent northward turning of the IMF, or pressure pulses from the solar wind (Takalo et al. 1999a). It is suggested that spatially localized current instabilities, current disruptions by kinetic instabilities (Lui 1996), or the merging of coherent structures around Alfvénic resonances (Chang 1999a,b) lead to the initiation of magnetospheric substorms.

Numerical simulations with 2-D resistive MHD models that involve anomalous resistivity of a current-driven kinetic instability have been performed by Klimas et al. (2004), which revealed some novel results which we quote here: *Under steady loading of plasma containing a reversed magnetic field topology, an irregular loading-unloading cycle is established in which unloading is due primarily to annihilation at the field reversal. Following a loading interval during which the current-sheet supporting the field reversal thins and intensifies, an unloading event originates at a localized reconnection site that then becomes the source of waves of unstable current sheets. These current sheets propagate away from the reconnection site, each leaving a trail of anomalous resistivity behind. An expanding cascade of field line merging results. Some statistical properties of this cascade are examined. It is shown that the diffusive contribution to the Poynting flux in these cascades occurs in bursts, whose duration, integrated size, and total energy content exhibit scale-free power law probability distributions over large ranges of scales. Although not conclusive, these distributions do provide strong evidence that the model has evolved into SOC* (Klimas et al. 2004).

There are also simple analytical models for magnetospheric substorms. We described one minimal substorm model in Section 5.5, which could explain the waiting-time distributions expected for a SOC system (Freeman and Morley 2004). However, there are also alternative interpretations to the SOC model, which we discuss in Chapter 10.

9.5 Summary

On the most general level, what SOC phenomena have in common are the powerlaw-like occurrence frequency distributions that express the scale-free nature of dissipative nonlinear processes without preferred temporal or spatial scales. However, the powerlaw behavior is a mathematical or numerical property only, and thus can be described in terms of entirely physics-free statistics. At the beginning of this chapter we established a physics-free definition of SOC phenomena, which is aimed to provide necessary and (perhaps) sufficient criteria to define and identify a SOC system. The three SOC criteria include: (1) statistical independence of SOC events (in the temporal and spatial domain), (2) a nonlinear coherent growth phase (above some threshold level), and (3) the randomness of rise times of the nonlinear growth phase. The latter criterion implicitly is a consequence of the criticality of the system. In the remainder of this Chapter we discussed the physics of SOC processes in astrophysical, solar/stellar, and magnetospheric applications. The physical mechanisms, although all involving a nonlinear instability (Table 9.1), are completely different for each SOC phenomenon, involving mechanical, electromagnetic, or other instabilities. While we discussed only the basic physical nature of individual SOC processes, such as galaxy formation, star formation, neutron star physics, accretion disk physics, cosmic ray physics, solar and stellar dynamos, magnetic reconnection, particle acceleration, solar radio emission mechanisms, kinetic and current instabilities, evidence for SOC phenomena requires quantitative modeling of these physical mechanisms that ultimately leads to detailed predictions of statistical parameter correlations and the analytical form of the resulting parameter distributions. This more advanced step in our understanding of SOC phenomena has not yet been reached for most astrophysical applications, except for modeling of magnetospheric and solar data to some extent, featuring SOC manifestations from our closest astrophysical neighbors.

9.6 Problems

Problem 9.1: Find examples of incoherent and coherent physical processes. Identify the linear and nonlinear nature of the physical parameters involved in these processes. How is the additive and multiplicative nature of these processes manifested?

Problem 9.2: Find SOC phenomena described in this book where all three SOC criteria given in Section 9.1 can be verified. Which parameters need to be measured for a full verification?

Problem 9.3: Is the powerlaw shape of occurrence frequency distributions a sufficient criterion to identify SOC processes? How can a powerlaw distribution function be modeled with incoherent (non-SOC) processes.

Problem 9.4: Identify which of the three basic parameter distributions (peak energy, total energy, duration) has been measured for each SOC phenomenon listed in Table 9.1, using the information from Chapter 7. For which phenomena can the three SOC criteria given in Section 9.1 be verified?

Problem 9.5: Predict the correlations between the parameters of (1) peak counts C and peak energy P, (2) peak energy P and total energy E, and (3) total energy E and total duration T for a solar/stellar flare with a temperature of $T \approx 10$ MK, based on the model described in Section 9.3.5.

Problem 9.6: Derive a frequency distribution of flare energies E based on dimensional arguments, using scaling relations between length scales L, mass M, and time scales M. Hint: Derive first the relationship for energies, $E \propto ML^2T^{-2}$. Compare your result with the distribution $N(E) \propto E^{-3/2}$ obtained in Litvinenko and Wheatland (2001).

10. SOC-Like Models

A new scientific truth does not triumph by convincing its opponents and making them see the light, but rather because its opponents eventually die, and a new generation grows up that is familiar with it.

Max Planck, (*Scientific Autobiography*, 1949).

It is a capital mistake to theorize before one has data.

Sir Arthur Conan Doyle (1859–1930), *Sherlock Holmes*.

Is it SOC or not? asked Hendrik Jeldtoft Jensen in the final chapter of his book *Self-Organized Criticality: Emergent Complex Behavior in Physical and Biological Systems.* Let us quote the beginning of that chapter that is still unanswered: *Ever since the term "self-organized criticality" was introduced by Bak, Tang, and Wiesenfeld (BTW) in their 1987 paper for Physical Review Letters, the concept has been surrounded by a hectic air of controversy. There are a number of reasons for this. One reason is the bold and optimistic claims that were made. The attitude was that here finally is a line of thinking that will allow us to bring the statistical physics of Boltzmann and Gibbs in touch with the exciting real world of nonequilibrium physics, and that SOC is powerful enough to explain everything from mountain formation to stock-market variation. Super-general theories always meet a certain amount of skepticism from expert scientists working in the specific fields. It is difficult to draw a precise line between the general and the specific. It might not appear likely to the geologist that the many specific details or earthquakes can be understood in terms of a simple numerical cellular automaton. The biologist working on the immensely complicated interconnected web of evolving species might not find it anything but a bad joke to represent evolution in terms of a string of random numbers with nearest neighbor interaction only. So what then is SOC good for? Let us consider some important questions. (1) Can we identify SOC as well-defined distinct phenomenon different from any other category or behavior? (2) Can we identify a certain construction that can be called a theory of self-organized critical systems? (3) Has SOC taught us anything about the world that we did not know prior to BTW's seminal 1987 paper? (4) Is there any predictive power in SOC – that is, can we state the necessary and sufficient conditions a system must fulfill in order to exhibit SOC? And, if we are able to establish that a system belongs to the category of SOC systems, does then actually help us to understand the behavior of the system?* Jensen answered these questions, 12 years ago, with a cautious affirmation. In the meantime, we have collected a vast amount of more data, especially from astrophysical observations, which can be interpreted in terms of SOC systems, but at the same time, critical papers have been published that challenged the SOC interpretation for a number of phenomena with powerlaw-like frequency distributions and proposed alternative interpretations, such as intermittent turbulence or forced self-organized criticality. In this final book chapter we deal with SOC-like processes that exhibit SOC-consistent properties (such as powerlaw distributions), but violate other necessary SOC criteria, as we defined in Section 9.1. This way we hope to establish a clear dividing line between SOC and non-SOC processes, at least within the framework of our SOC definition derived here.

10.1 Hierarchical SOC Systems

In our universe, all astrophysical objects are coupled somehow in a hierarchical order. The big bang is supposedly responsible for the common origin of the universe, in which galactic structures formed (Fig. 9.1), galactic arms nurture molecular clouds and trigger star formation (Fig. 9.2), a solar or stellar system forms planets and small bodies, and so on. On an astronomical scale it is not a priori clear how to place a BTW sandpile model into the whole system. We identified a number of likely SOC processes, but some of them are clearly coupled. We might therefore generalize the simple BTW sandpile paradigm into a hierarchy of SOC systems, as visualized in Fig. 10.1.

Let us place some SOC systems, whose physics we discussed in Chapter 9, into this generalized picture of hierarchical SOC systems. Cosmic rays (Section 9.2.6), for instance, seem to be accelerated outside of the galaxy, as well as inside of our galaxy. Thus, some cosmic rays are initially accelerated in extragalactic systems and enter our own galaxy already with a powerlaw spectrum. When they enter our galaxy they might be further accelerated during reflections at magnetic clouds in our galaxy by the first-order Fermi acceleration process, which is a secondary sandpile process and could explain the broken (double) powerlaw spectrum (Fig. 9.5). In Section 9.3.2 we discussed the solar dynamo, which produces buoyant magnetic flux tubes that rise from the tachocline and emerge at the solar surface, which is a primary SOC process. The solar surface acts as a filter and randomizes the emergent magnetic flux tubes by magneto-convection. As a secondary process, the photospheric magneto-convection stresses and twists the coronal magnetic field, which eventually leads to magnetic reconnection and produces flares and coronal mass ejections, which represent secondary SOC processes. A subset of the launched coronal mass ejections are geo-effective and hit the Earth, where they induce ring currents in the ionosphere, which causes particle precipitation in auroral zones and display auroral ovals, causes magnetospheric substorms, which are possibly SOC processes of a third stage. Some events have a direct coupling, such as solar energetic particles (SEP), which can hit the Earth and trigger *ground-level enhancements (GLEs)* and neutron showers, possibly another SOC process.

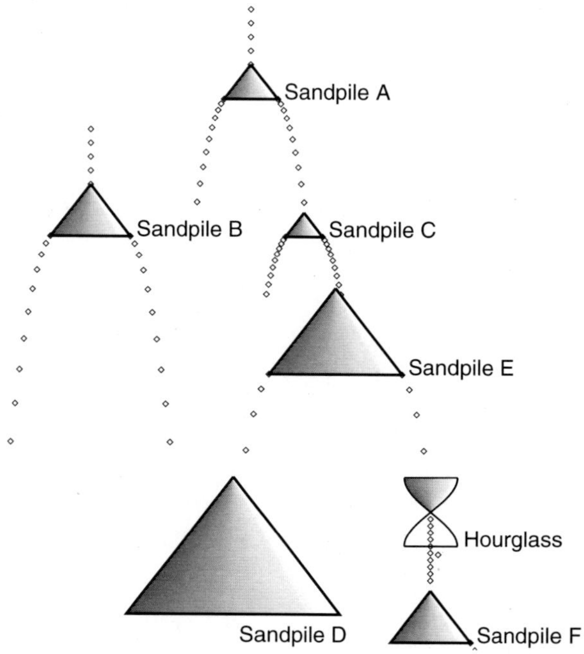

Fig. 10.1 Schematic concept of hierarchical SOC systems, illustrated with the analog of multiple sandpiles that feed each other from top to bottom. The top sandpiles A and B have a constant input rate of dripping sand grains, while the secondary sandpiles C, D, and E at the bottom of the system receive avalanche-like inputs. The avalanche-like output of sandpile E drips into an hourglass, so that the input rate is constant for sandpile F.

The concept of a hierarchical sandpile model is a novel aspect that has not been discussed in literature yet, but there is mounting pressure from astrophysical observations to conceive such a model. A first question is whether a hierarchical sandpile model is consistent with the standard SOC scenario. The classical BTW sandpile model assumes a constant time-averaged input of dropping sand grains, while the output of sporadic avalanches display a highly nonlinear response. If we couple multiple sandpiles, all secondary sandpiles experience a bursty input from the avalanche output of the primary sandpiles, which seems to violate the assumption of a constant time-averaged input rate. In order to avoid this inconsistency, some filters are required that smooth out the bursty output from primary sandpiles before they feed secondary sandpiles, otherwise the input and output of secondary sandpiles could be highly correlated and contradict the nonlinearity of the standard BTW sandpile model. In Fig. 10.1 we visualize such a randomization filter with the analog of an hourglass, which releases a steady stream of sand grains regardless how intermittent the input is. In some sense, the hourglass is the opposite to a sandpile, because it converts a highly nonlinear input into a constant output, while sandpiles in a critical state convert a linear input into a highly nonlinear output. Nature indeed seems to provide such filters that randomize bursty and avalanche-like outputs of nonlinear dissipative systems, such as the solar or stellar convection zone that decouples emergent buoyant flux tubes from flare loops, or the Earth's bowshock that smoothes out solar wind disturbances before they produce magnetospheric substorms. So, is a hierarchical SOC system consistent with a standard (single) SOC system? The answer is that it is only consistent with our definition of a SOC system (Section 9.1) if there is a randomization filter (i.e., an hourglass in the analogy of sandpiles) in between subsequent SOC systems, otherwise it would produce "sympathetic flaring" and violate the first SOC criterion of statistical independence. It is beyond the scope of this book to develop a complete generalization of hierarchical SOC models, but this concept may help to model the obvious coupling of some multiple SOC systems.

10.2 Self-Organization without Criticality

There may be some confusion between the terms *self-organization (SO)* and *self-organized criticality (SOC)*. Let us make a simple analogy. If the wind blows over a sand desert, it might create a wavy pattern of ripples on sand dunes that have a highly regular geometry (Fig. 10.2, top left), which is an example of a self-organization process. However, this geometric pattern is quasi-static and has no criticality, like a sand beach (Fig. 1.2, left), in contrast to a sandpile that has been raised up to a critical slope, so that sporadic avalanches occur along the critical slope, which is a dynamic process.

There are many self-organization processes in physics, such as structural first-order phase transitions and spontaneous symmetry braking, second-order phase transitions at which the system exhibits scale-invariant structures (critical opalescence of fluids at a critical point), structure formation in thermodynamic systems away from equilibrium (Bénard cells), or self-organization of electromagnetic waves or solitons into vortices in a magnetized electron–positron plasma (e.g., Kaladze and Shukla 1987). Self-organization in chemistry includes molecular self-assembly, reaction–diffusion systems (Belousov–Zhabotinsky reaction with spiral-like patterns) and oscillating chemical reactions, autocatalytic networks, or liquid crystals. Self-organization in biology include spontaneous folding of proteins and biomacromolecules, geometric patterns on the skin of animals (zebra, giraffe, tiger, various tropical fishes), formation flight of birds, etc. For an extensive description see the textbook *Self-Organization in Biological Systems* by Scott Camazine (Camazine et al. 2001).

In astrophysics, what comes to mind is the granulation pattern on the solar (and supposedly stellar) surfaces, which is created by magneto-convection, similar to the hexagonal cell pattern of Bénard cells that occur in the boiling water of a frying pan. Also some geometric patterns in galaxy formation (Kalapotharakos et al. 2004), star formation, and planetary systems might be a result of self-organization (for instance the spiral structure of our galaxy: Fig. 10.2, bottom left). Other self-organizing structures result when the governing process is self-similar expansion, diffusion-limited aggregation, or percolation. Saturn's rings, the asteroid belt, or lunar craters, in contrast, can be considered to be SOC processes, because the result is a statistical distribution of rare catastrophic SOC events (via collisional next-neighbor interactions). Planetary orbits follow Bode's law, which is probably also the end-product of a self-organization process, as well as the oligarchic growth of protoplanets. Self-organization occurs in systems with fully developed turbulence, such

Fig. 10.2 Examples of self-organizing patterns: Top left: Sand dunes in Namibia (credit: Jef Maion); *Top right:* Jupiter atmosphere with white bands or ammonia ice clouds (credit: Jeff Root); *Bottom left:* Top-view of our Milky way galaxy according to an artist's rendering (Credit: NASA); *Bottom right:* The cracked ice plains of Jupiter's moon Europa (Credit: Galileo Project, JPL, NASA).

as in Jupiter's atmosphere (Fig. 10.2, top right), giving rise to cyclone-like whirls and eddies in equatorial bands, featuring the so-called *Jupiter red spot* as the most prominent landmark. This red spot has been modeled in terms of Rossby autosolitons, which is an undamped solitary vortex that is self-organized in axisymmetric zonal counterflows in a rotating parabolic "shallow-water" layer and is sustained by flows (Antipov et al. 1985). Even the evolution of our planet Earth went through stages of great environmental disequilibrium forming self-organized structures from bubbles at the sea–air interface to tectonic plates, a process that apparently was necessary for conditions of life, and did not happen on Mars, for instance (Chang, 1988). An extensive discussion of self-organized patterns in astrophysics can be found in the textbook *The Discovery of Cosmic Fractals* by Baryshev and Teerikorpi (2002).

How can we distinguish SO from SOC processes? Most SO patterns are quasi-stationary, while SOC always involves dynamic (catastrophic) processes. If dynamic processes are involved in the formation of SO patterns, they usually involve system-wide processes, such as diffusion, turbulence, convection, magneto-convection, which essentially operate with long-range interactions (via pressure, streams, flows), while SOC processes occur spontaneously with an explosive evolution and multiplicative growth via next-neighbor interactions. The restriction to next-neighbor interactions in SOC processes essentially guarantees the statistical independence of individual events, while SO patterns exhibit a close coupling over a large range. For instance, a Bénard cell convection pattern can be quasi-stationary, but if one cell disappears, the surrounding pressure will cause an adjustment of all adjacent cells and over an arbitrary distance, until pressure is balanced again. In other words, one sandpile avalanche does not know what another avalanche does, but convection cells adjust to each other until a new equilibrium is reached. So, since SO patterns can exhibit scale-free powerlaw distributions of spatial scales, such as the Kolmogorov spectrum in turbulent MHD cascades, we do not use the powerlaw feature as a distinction criterion between SO and SOC processes, but rather the three SOC criteria defined in Section 9.1, which should disqualify most SO processes to be considered as SOC systems.

10.3 Brownian Motion and Diffusion

There are some common properties between SOC avalanches and Brownian motion: both have power spectra $P(\nu) \propto \nu^{-p}$ with a powerlaw function, have similar fractal dimensions, and involve nearest-neighbor interactions. So, let us test whether Brownian motion could be a SOC process.

Brownian motion describes the random motion of atoms or molecules in a gas, which is governed by frequent collisions with nearest neighbors, termed after the Scottish botanist Robert Brown. Although we cannot resolve atomic scales, the Brownian motion can vividly be demonstrated by watching the vibrations of a dust particle suspended in a fluid under the microscope. This already tells us something about the scale-free properties and self-similarity of Brownian motion on different spatial scales. The physical model of Brownian motion includes some simplifications of the real process and is defined by the following three assumptions: (1) additional degrees of freedom like rotation are neglected; (2) the time span between subsequent collisions is constant; and (3) the velocities before and after a collision are assumed to be statistically independent. The latter assumption implies a Gaussian distribution of velocities, because the kinetic energies of the particles follow a Boltzmann distribution in thermodynamic equilibrium, which is equivalent to a Gaussian distribution of velocities. The motion of a Brownian particle is thus described by,

$$x(t_n) = \sum_{i=1}^{n} v_i \, \delta t \ , \tag{10.3.1}$$

where the time $t_n = n\delta t$ is discretized with a constant time step δt and v_i is drawn from a Gaussian random distribution (Eq. 4.2.1). An example of such a 1-D random walk is shown in Fig. 10.3 (left), as well as a 2-D diffusion process in Fig. 10.3 (right) by simulating a similar (independent) random motion $y(t_n)$ in a second space coordinate, for $n = 10{,}000$ time steps with $\delta t = 1$. Thus, $x(t)$ is a linear combination of independent, Gaussian-distributed random values, which implies statistically also a Gaussian distribution for the superposition of Gaussians, with a mean and standard deviation of,

$$\langle x(t)\rangle = \sum_i^n \langle v_i\rangle \delta t = 0 , \tag{10.3.2}$$

$$\sigma(t)^2 = \left\langle [x(t) - \langle x(t)\rangle]^2 \right\rangle = \sum_{i=1}^n \sum_{j=1}^n \langle v_i v_j\rangle \delta t^2 = n\sigma_n^2 \delta t^2 = \sigma_v^2 \delta t^2\, t . \tag{10.3.3}$$

Thus, the standard deviation increases with the square root of time,

$$\sigma(t) \propto \sqrt{t} \tag{10.3.4}$$

In Fig. 10.3 we marked this expected time dependence of the standard deviation (Fig. 10.3, left).

This model of the Brownian motion is also applied to a number of diffusive transport phenomena, such a heat transport, diffusion of gases, meridional flows of magnetic features on solar and stellar surfaces, etc. If we define a general 1-D diffusion equation,

$$\frac{\partial f(x,t)}{\partial t} = \kappa \frac{\partial^2 f(x,t)}{\partial x^2} , \tag{10.3.5}$$

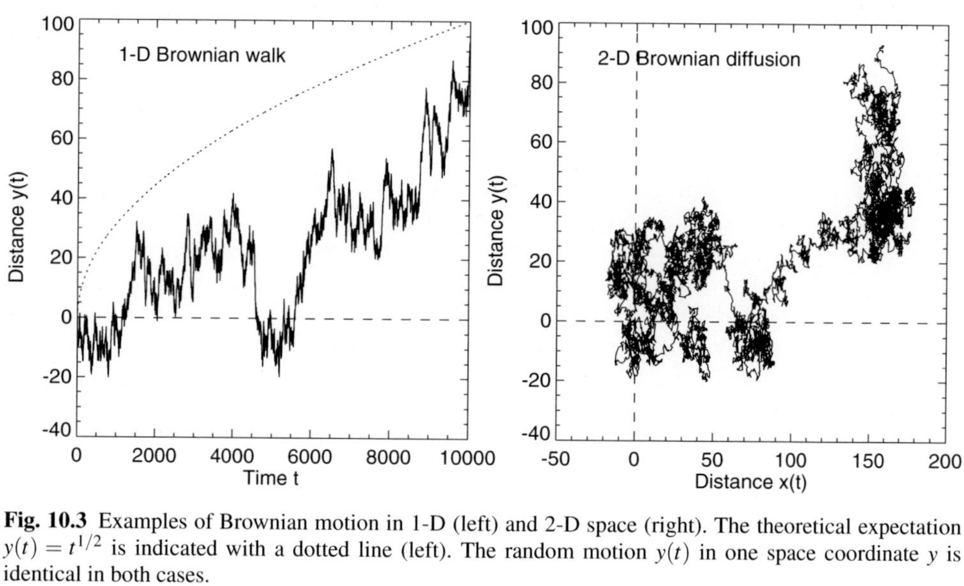

Fig. 10.3 Examples of Brownian motion in 1-D (left) and 2-D space (right). The theoretical expectation $y(t) = t^{1/2}$ is indicated with a dotted line (left). The random motion $y(t)$ in one space coordinate y is identical in both cases.

for a density function $f(x,t)$ with a diffusivity coefficient κ, and insert an initially Gaussian random function $f(x,t)$, with the time dependence of a Brownian random walk, $\sigma(t)^2 = ct$ (Eq. 10.3.4),

$$f(x,t) = \frac{1}{\sqrt{2\pi}\sigma(t)} \exp\left(-\frac{x^2}{2\sigma(t)^2}\right) = \frac{1}{\sqrt{4\pi ct}} \exp\left(-\frac{x^2}{4ct}\right) , \qquad (10.3.6)$$

into the diffusion equation (10.3.5), we see immediately that the function satisfies the diffusion equation and yields the solution $\kappa = c/2$. Thus the diffusion equation is equivalent to the Brownian motion.

The self-similarity of Brownian motion can be defined by the fractal dimension. If we use the definition of the Hausdorff dimension D in 2-D space (Section 8.2.1), with $n(L)$ the number of area elements that are covered in a square area with length L, we have for Brownian random walk a dimension of,

$$D = \frac{\log n(L)}{\log L} = \frac{\log L^2}{\log L} = 2 . \qquad (10.3.7)$$

This result appears to be strange on the first glance, because it is the same fractal dimension as Euclidean filling, but when we look at the area covered by a random walk (as shown in Fig. 10.3, right), the covered area seems to be very fractal, and thus we would estimate a smaller dimension of $D < 2$. However, the truth is that many locations are revisited many times during a random walk, so that the number of encountered locations is as much as the solid Euclidean area. In other words, if we think of the infamous drunkard who searches for his penny on the ground with a random walk, it would take him about the same time to search systematically square by square, so both search strategies are statistically equivalent.

Another important property of Brownian random walk is the power spectrum. It can be shown with the Fourier transform (e.g., Hergarten 2002) that the resulting power spectrum has a slope of 2,

$$P(\nu) \propto \nu^{-2} , \qquad (10.3.8)$$

which is also called *Brownian noise* (Section 4.7).

So, since Brownian motion is fractal, has a $1/f$-type power spectrum (in the sense of the generalized nomenclature described in Section 4.7), and propagates via next-neighbor (collisional) interactions, does it qualify to be a SOC process? Let us imagine sand avalanches where each sand grain follows a Brownian random walk and test our three SOC criteria given in Section 9.1. The first SOC criterion of temporal and spatial independence between different avalanches could be fulfilled in a scenario where the avalanches are triggered independently. Also multiple diffusion processes can be started independently, such as acid rain drops that fall on a water surface. The second SOC criterion of nonlinear coherent growth is clearly violated, since a nonlinear evolution of the energy dissipation rate, $P(t) \propto t^p$, requires a positive power index larger than unity ($p > 1$), while the average diffusion velocity actually decreases according to $\langle v_{diff} \rangle = \langle x^2 \rangle^{1/2}/t \propto t^{1/2}/t \propto t^{-1/2}$ for Brownian motion, which also implies a decreasing kinetic energy with time, $E_{kin} \propto \langle v_{diff} \rangle^2 \propto t^{-1/4}$. Also the third SOC criterion of a random rise time cannot be fulfilled, especially since diffusion processes go on forever, though with asymptotically slower speed. Thus, Brownian motion clearly does not qualify as a SOC process, despite the scale-free similarities.

While we discussed the basic classical diffusion here, more complex concepts have been considered (Section 3.6), such as a coupled equation system including a 1-D diffusion process that can mimic a cellular automaton simulation (Eqs. 3.6.7; Lu 1995c), a diffusion equation that can mimic a 3-D cellular automaton redistribution rule (Isliker et al. 1998a), or (anomalous) hyper-diffusion equations that can mimic the continuum limit of cellular automaton simulations (Liu et al. 2002; Charbonneau et al. 2001). In principle, it is conceivable that the combination of a nonlinear-growth process with a diffusion process can produce SOC-like avalanches, but diffusion alone seems not to qualify as a SOC process. Other anomalous diffusion processes include *quasi-linear diffusion* in tokamak plasma confinement (Dendy and Helander 1997), or an anomalous diffusion process called *Lévy flight*, that has been applied to earthquake statistics (Sotolongo-Costa et al. 2000), or the *Hurst effect*, which was applied to the fluctuations of the AE-index of magnetospheric substorms (Uritsky and Pudovkin 1998), to the fluctuations of solar activity (Lepreti et al. 2000), or to the bursty time profile of solar flare hard X-rays (McAteer et al. 2007). Hurst exponents above the value of $H = 1/2$ expected for Brownian random motion, reveal some hidden memory or persistence that controls the observed fluctuations, leading also to steeper power spectra (such as black noise, with $P(\nu) \propto \nu^{-3}$, see discussion in Section 7.4).

10.4 MHD Turbulence

10.4.1 Solar Corona

Convection and turbulence are important in fluids with high Reynolds numbers. Since the *magnetic Reynolds number* $R_m = l_0 v_0 / \eta_m$ is high in the coronal plasma ($R_m \approx 10^8$–10^{12}), turbulence may also develop in coronal loops (although there is the question whether turbulence could be suppressed in coronal loops due to the photospheric line-tying). Theoretical models and numerical simulations that study MHD turbulence include the *kinematic viscosity* or *shear viscosity* ν_{visc} in the MHD momentum equation, and the *magnetic diffusivity* $\eta_m = c^2/4\pi\sigma$ in the MHD induction equation,

$$\rho \frac{D\mathbf{v}}{Dt} = -\nabla p - \rho \mathbf{g} + (\mathbf{j} \times \mathbf{B}) + \nu_{visc}\rho \left[\nabla^2 \mathbf{v} + \frac{1}{3}\nabla(\nabla \cdot \mathbf{v}) \right] , \qquad (10.4.1)$$

$$\frac{\partial \mathbf{B}}{\partial t} = \nabla \times (\mathbf{v} \times \mathbf{B}) + \eta_m \nabla^2 \mathbf{B} . \qquad (10.4.2)$$

Similar to the models of stress-induced current cascades, random footpoint motion is assumed to pump energy into a system at large scales (into eddies the size of a granulation cell, $\approx$1000 km), which cascade due to turbulent motion into smaller and smaller scales, where the energy can be more efficiently dissipated by friction, which is quantified by the kinematic or shear viscosity coefficient ν_{visc}. Friction and shear are dynamical effects resulting from the nonlinear terms $(v_{1,i}v_{1,j})$, $(v_{1,i}B_{1,j})$, $(B_{1,i}v_{1,j})$, and $(B_{1,i}B_{1,j})$ in Eqs. (10.4.1–10.4.2) and are only weakly sensitive to the detailed dynamics of the boundary conditions. Analytical (3-D) models of MHD turbulence have been developed by Heyvaerts and Priest (1992), Inverarity et al. (1995), Inverarity and Priest (1995), and Milano et al. (1997), where the nonlinear viscosity terms are specified as diffusion coefficients. These turbulent diffusion coefficients are free parameters, which are constrained self-consistently by (1) assuming that the random footpoint motion has a turbulence power spectrum (e.g., a Kolmogorov spectrum $P(k) \propto k^{5/3}$); and (2) by matching the observed macroscopic parameters (i.e., velocity of footpoint motion, density, and magnetic field). Heyvaerts and Priest (1992) predict turbulent velocities of $v_{turb} \approx 20$–30 km s^{-1}, which are consistent with the excess broadening of lines observed with SOHO/SUMER, which shows a peak of $\xi = 30$ km s^{-1} at a transition region temperature of $T \approx 3 \times 10^5$ K (e.g., Chae et al. 1998).

Analytical models of turbulent heating are applied to sheared arcades (Inverarity et al. 1995) and twisted fluxtubes (Inverarity & Priest 1995). Turbulent heating has been numerically simulated in a number of studies, which exhibit a high degree of spatial and temporal intermittency (Einaudi et al. 1996a,b; Galsgaard and Nordlund 1996; Dmitruk and Gomez 1997). An example of such a simulation is shown in Fig. 10.4, where it can be seen how larger eddies fragment into smaller ones, forming current sheets and triggering magnetic reconnection during this process. Heating occurs by Ohmic dissipation in the thinnest current sheets. Milano et al. (1999) emphasize that the locations of heating events coincide with quasi-separatrix layers. The formation of such current sheets has also been analytically studied in the context of turbulent heating by Aly and Amari (1997). Numerical simulations reveal intermittent heating events with energies of $E_H = 5 \times 10^{24}$ to 10^{26} erg and a frequency distribution with a powerlaw slope of $\alpha \approx 1.5$, similar to observed nanoflare distributions in EUV (Dmitruk and Gomez 1997, 1999; Dmitruk et al. 1998). Continuous slow fluctuating footpoint driving leads to a steady state with a random superposition of current sheets (Longcope and Sudan 1994).

On the observational side, the distribution of spatial structures, Fourier spectra, and structure functions of solar UV emission was associated with intermittent MHD turbulence (Buchlin et al. 2006). The time variability of solar magnetic activity has been associated with a turbulent cascading process based on its multi-fractal properties (Lawrence et al. 1995), or based on its power spectra, structure functions, or correlation lengths (e.g., Abramenko et al. 2001, 2002, 2003). Radio emission in the form of decimetric narrowband spikes was also interpreted as a manifestation of coronal MHD turbulence because of its power spectrum $P(\nu) \approx \nu^{-1.6}$ being close to a Kolmogorov-type spectrum (Karlicky et al. 1996).

Does MHD turbulence in the solar corona qualify as a SOC process? Among the common properties are, based on numerical MHD simulations: the frequency distributions of dissipated energies with a powerlaw slope of $\alpha_E \approx 1.5$ (Dmitruk and Gomez 1997; Dmitruk et al. 1998), powerlaw distributions for total dissipated energy, peak energy, and duration of events (Galtier and Pouquet 1998; Galtier 1999, 2001; Georgoulis et al. 1995; Einaudi and Velli 1999; Buchlin et al. 2005), or powerlaw distributions for waiting times (Galtier 2001). Despite of all these similarities, not all of our three SOC criteria (Section 9.1) are fulfilled: (1) MHD turbulence cascading exhibits a correlation length between larger and smaller spatial scales, and thus violates our first SOC criterion of statistical

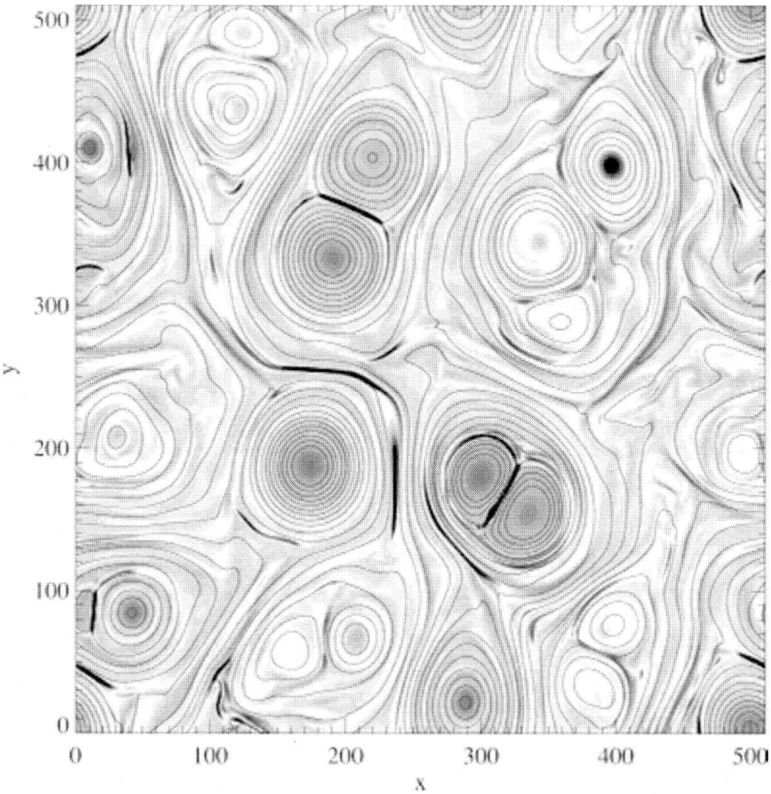

Fig. 10.4 Example of a numeric 2-D simulation of MHD turbulence with a Reynolds number of $R_m = 2000$, showing the magnetic field lines (contours) and electric current density (greyscale). Reconnection zones have formed between a number of adjacent islands that are coalescing, triggering localized nonsteady reconnection events throughout the simulation box (Matthaeus 2000). Reprinted from Zhou et al. (2004) with permission; Copyright by American Physical Society.

independence between individual energy dissipation events; (2) MHD turbulence can produce intermittent heating events that have a nonlinear growth characteristics, and thus may fulfill our second SOC criterion, which explains that both processes produce powerlaw distributions of dissipated energies; (3) the rise time of intermittent heating or energy dissipation events could possibly fulfill the randomness of our third SOC criterion. So, it is mostly the property of spatial and temporal correlation lengths that distinguishes energy dissipation events driven by MHD turbulence in the solar corona from SOC processes, such as energy dissipation events produced by stressing and braiding of magnetic fields leading to sporadic magnetic reconnection events. A consequence of the spatial and temporal correlation between subsequent energy dissipation events is manifested in the distribution of waiting times, which is expected to be a random distribution for a SOC process (i.e., exponential-like for a stationary Poisson process), while it is a powerlaw distribution for MHD turbulence. Therefore, this distinctive statistical feature was used to associate solar flares with the process of MHD turbulence rather than with a SOC process (e.g., Boffetta et al. 1999). However, this argument is only valid under the assumption of a stationary Poisson process, while in reality it is clear that solar flares are governed by a nonstationary Poisson process, and thus naturally produce a powerlaw distribution with slopes of $p \approx 2$–3 (e.g., Wheatland 2000a; Aschwanden and McTiernan 2010). Since the distinction between SOC processes and intermittent turbulence (IT) processes is somewhat uncertain and data are consistent with both, even a coexistence of the two processes was proposed, in the sense that both SOC and IT may be manifestations of a single complex dynamical process entangling avalanches of magnetic energy dissipation with turbulent particle flows (Uritsky et al. 2007; Watkins et al. 2009).

10.4.2 Solar Wind

MHD turbulence is thought to be ubiquitous in the Sun and heliosphere, and so in the solar wind, which is considered to be a turbulent magnetofluid (e.g., see review by Petrosyan et al. (2010)). Evidence for turbulence in the solar wind comes from in-situ measurements (from the *Mariner, Pioneer, Helios, ISEE-3, IMP, Voyager, ACE*, and *Ulysses* missions) of fluctuations in plasma velocity, magnetic field, and plasma density, which ripple the average large-scale solar wind that falls off relatively smoothly with radial distance from the Sun. Spatial scales of solar wind turbulence can be measured from an astronomical unit (1 AU) down to the thermal proton gyroradius (about 50 km), and recently with *CLUSTER* even down to the electron radius (Fig. 10.5). Measurements near Earth orbit (1 AU) seems to indicate that the solar wind exhibits fully developed MHD turbulence, a Kolmogorov power spectrum, and the presence of MHD waves. However, the input spectrum from the solar corona seems not to be of the Kolmogorov type with the fluctuating power proportional to $f^{-5/3}$ as usually found in interplanetary space, but rather of the f^{-1} type power spectrum (Matthaeus and Goldstein 1986), which can be modeled in terms of a steady-state of cascading Alfvénic waves (Vainio et al. 2003) and simulated with 3-D hydrodynamic and MHD (Dmitruk and Matthaeus 2007). The scenario of the MHD turbulent cascade is thought to begin with an MHD energy reservoir at the largest scales with a spectrum of f^{-1} fed by the lower solar corona, while turbulent interactions produce a cascade of energy through vortices and eddies to progressively smaller sizes (with spectrum $f^{-5/3}$), while energy dissipation becomes only significant at sufficiently small scales, where heating sets in. Alfvénic waves that escape the Sun have a dissipation length of a few solar radii, and the temperature of the solar wind indeed increases over the first few solar radii, especially for protons. Wave–particle interactions in the turbulent solar wind include also resonant absorption and cyclotron resonance, which leads to anisotropic particle distributions and temperature anisotropies.

Self-organization in the solar wind is thought to produce clustering of low-frequency waves in the solar wind, leading to a Hausdorff fractal dimension of $D \approx 4/3$ and a power spectrum of $f^{-5/3}$ (Milovanov and Zelenyi 1999). Self-organization and intermittency of magnetic turbulence in the solar wind is manifested in the high degree of correlation between magnetic field and velocity field fluctuations, especially at times of very low levels of fluctuations in mass density and magnetic field intensity (Veltri et al. 1999; Carbone

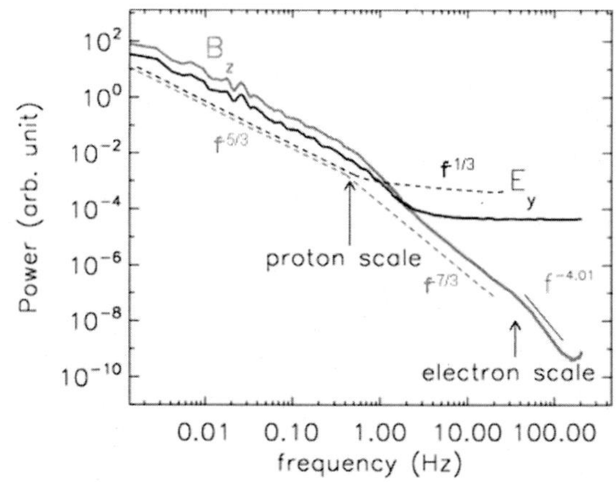

Fig. 10.5 A spectrum of the solar wind is shown, based on *CLUSTER* observations from large scales ($\approx$105 km) down to small scales ($\approx$3 km) is shown, with the proton and electron gyroradius scale indicated. The solar wind spectrum is interpreted in terms of a turbulent MHD cascade, with the theoretically predicted slopes of $f^{-5/3}$ and $f^{-7/3}$ from gyro-kinetic theory. The plot proves that the energy continues cascading below the proton scale down to the electron scale, where it is converted to heat (via electron Landau damping resonance) causing the steepening of the B_z spectrum to f^{-4} (Howes et al. 2008; Sahraoui et al. 2009; credit: ESA, CLUSTER).

et al. 2004). Powerlaw distributions have been found for the probability density function of burst energies and durations, as well as for the inter-burst waiting times, which could all be consistent with a SOC process governed by a nonstationary Poisson process (Freeman et al. 2000a). The distribution of waiting times could also be consistent with the Lévy flight (anomalous diffusion) model (Hnat et al. 2007). Numerical simulations with *reduced MHD* codes indeed produce similar powerlaw-like distributions of burst sizes (in mean-square current density), as well as powerlaw-like distributions of burst durations and waiting times, which is consistent with SOC models and nonstationary Poisson processes (Watkins et al. 2001a). However, the distributions of the kinetic energy density in the inertial range of solar wind turbulence was found to be self-similar only approximately (Podesta et al. 2006b, 2007; Podesta 2007). In addition, the presence of a background magnetic field not only introduces a symmetry breaking in interplanetary space but also organizes fluctuations in their large scale orientation (Bruno et al. 2007). The intermittent turbulence of the solar wind was also analyzed with a multi-fractal Cantor set, which provided evidence for a two-scale cascading process (Macek and Szczepaniak 2008; Macek and Wawrzaszek 2009). On the other side, a single generalized scaling function was found to characterize turbulent fluctuations consistently, independent of plasma conditions, even at very low levels of solar activity (Chapman and Nicol 2009). Recent analysis of the solar wind has shown a quasi-universal spectrum following the Kolmogorov law $\propto k^{-5/3}$ at MHD scales, a $\propto k^{-2.8}$ powerlaw at ion scales (Fig. 10.5), and an exponential law at the electron gyroradius, which for the first time demonstrates that the electron Larmor radius plays the role of a dissipation scale in space plasma turbulence (Alexandrova et al. 2009), although there is still some controversy about the dissipation scale in space plasma turbulence (see comment paper by Matthaeus et al. (2008)). Statistical analysis of waiting times of solar wind discontinuities and modeling with Hall-MHD simulations revealed a clusterization of discontinuities that is not Poissonian randomly distributed, giving further evidence for intermittent, anisotropic, and fully developed MHD turbulence (Greco et al. 2009a,b).

The bottom line of all these studies is that fluctuations in the solar wind share many common properties with (nonstationary) SOC processes as well as with intermittent MHD turbulence models, so that it is rather difficult to discriminate between these two processes. The only option that can be eliminated is a stationary SOC process, because waiting times of solar wind fluctuations are powerlaw-like rather than exponential. Consulting our basic three SOC criteria (Section 9.1), only the first SOC criterion of temporal and spatial independence is violated to some degree, because solar wind fluctuations reveal some correlation length and clusterization effects, which seems to be the only discrimination criterion from an ideal SOC process.

10.4.3 Magnetospheric Substorms

We discussed physical scenarios of magnetospheric substorms (Section 9.4.2) and described the statistical distribution of auroral blobs and AE bursts (Section 1.6), substorm durations (Section 7.2), waiting-time distributions (Section 5.5), and cellular automaton models (Section 2.5), that all support the interpretation in terms of a SOC process. However, an alternative model is *intermittent turbulence (IT)* (e.g., Angelopoulos et al. 1999). The most compelling evidence for turbulence processes operating in the plasma sheet in the geotail is the observation that the flow variability amplitude is many times larger than the average, and that the flow directions are at random, which challenges the laminar flow hypothesis (Angelopoulos et al. 1999; Kovacs et al. 2001). A schematic of the bursty flows in the geotail is shown in Fig. 10.6. Angelopoulos et al. (1999) conclude: *According to observations, the magnetotail is in a bi-modal state: nearly stagnant, except when driven turbulent by transport-efficient fast flows. The distributions of flows are in agreement with sporadic (intermittent) variability in the magnetotail. The variability may resemble hydrodynamic turbulence around a jet. The presence of turbulence alters the conductivity and the mass/momentum diffusion properties across the plasma sheet and may permit cross-scale coupling of localized jets into a global perturbation. Bursty-flow-driven turbulence is a physical process that may have an important role to play in the establishment of a state of self-organized criticality.*

It appears that it is currently not possible to discriminate between the proposed models for magnetospheric substorms, such as SOC versus IT, and that new tests are required (e.g., Watkins et al. 2001b; Watkins 2002; Chapman and Watkins 2001; Antonova 2004). The power spectrum of AE bursts is in the range of f^{-2} to $f^{-1.8}$ (Consolini et al. 1996) could potentially be consistent with a modified Kolmogorov spectrum and thus support a turbulent origin. The observed frequency distribution of substorm durations has a powerlaw shape, which could be consistent with both, a SOC model with nonstationary driver (Uritsky et al. 2007; Watkins et al. 2009), or with intermittent turbulence (Boffetta et al. 1999; Lepreti et al. 2004), but more realistic turbulence models are needed to enable

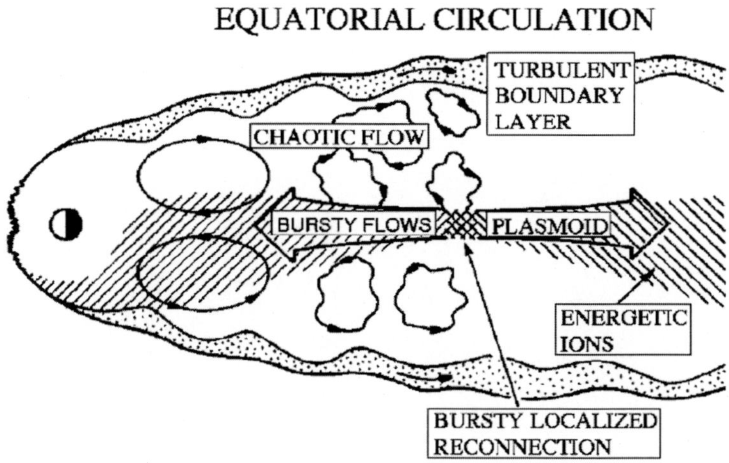

Fig. 10.6 Schematic representation of magnetospheric circulation at the equatorial plane, showing localized bursty flows that drive vortical (turbulent) flows. Magnetospheric substorms originate in the bursty reconnection region in the geotail (Angelopoulos et al. 1999; Kennel 1995).

more discriminative tests. Waiting times between substorms were found not to be Poissonian distributed, as expected in SOC models: calm periods prior to recurring high-speed-driven storms were found to be substantially greater than random occurrence probability (Borovsky and Steinberg 2006). New observations, such as the discovery of a distinct class of short-scale *drift-kinetic Alfvén (DKA)* vortex motions in the Earth's magnetospheric cusp region made by *CLUSTER* my shed new light on turbulent plasma and energy transport through the magnetospheric boundary layers (Sundkvist et al. 2005).

10.4.4 Interstellar Medium

The *interstellar medium (ISM)* is believed to be turbulent, from scales ranging from from 1 AU to kiloparsecs, with an embedded galactic magnetic field that controls star formation and acceleration and propagation of cosmic rays. Observational measurements of the turbulent state comes mostly from radio-interferometry, which quantifies scattering of radio waves (of point-like sources, such as quasars) and scintillations (in the time domain). A compilation of such radio measurements, combined with the latest measurements of Doppler velocity fluctuations using the *Wisconsin Hα Mapper (WHAM)* telescope is shown in Fig. 10.7, which displays a power spectrum over a grand scale of 12 orders of magnitude, following a powerlaw slope of approximately $P(L) \propto L^{-5/3}$, as expected for a Kolmogorov turbulence spectrum (Chepurnov and Lazarian 2010). This is probably the largest spatial range (from 10^6 to 10^{17} m) over which the spectrum of a Kolmogorov cascade was ever measured. It represents evidence that energy in injected at the largest scales of $\approx$100 pc in the galaxy and cascades all the way down to the size of solar systems. The leading ideas about the energy injection at these large scales are via supernova explosions and via the magneto-rotational instability.

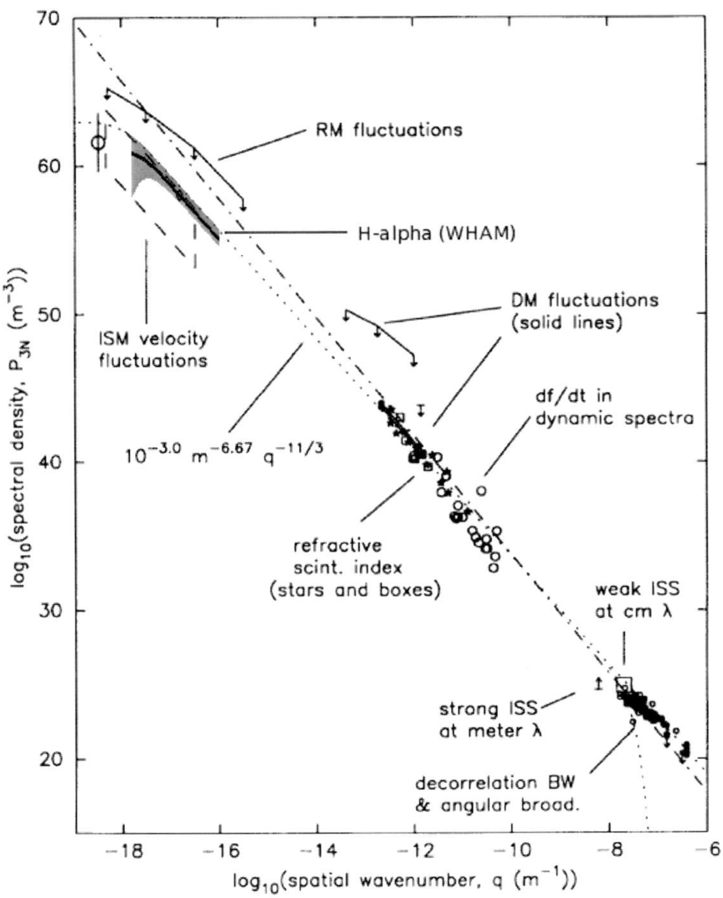

Fig. 10.7 A composite turbulence power spectrum of the electron density in the interstellar medium including the most recent data from the *Wisconsin Hα Mapper (WHAM)* (Chepurnov and Lazarian 2010; extending the previous figure of Armstrong et al. (1995), (reproduced by permission of the AAS).

In the context of our focus on SOC systems, density fluctuations in the interstellar medium reveal a perfect powerlaw distribution of their power spectrum, but obviously are not produced by a SOC process. Thus, the powerlaw shape of a power spectrum is clearly not a sufficient criterion to establish the presence of a SOC process, it can equally be created by turbulence. A turbulence cascade, however, violates our first SOC criterion of spatial and temporal independence (Section 9.1), because spatial structures in a turbulent cascade have some correlation length.

10.5 Forced Criticality Models

In the original BTW model, the critical state arises from self-organization within the system itself, for instance by the critical slope of the sandpile. When a little disturbance occurs (e.g., in form of a dripping sand grain), an avalanche is formed along those paths that have a slightly steeper slope than the average critical slope. Once the avalanche proceeds, it leaves a coarse surface behind that has the same average critical value, but with slight local deviations that will play a role for the next avalanche. So, the slope of the sandpile is self-organizing in the sense that it always stays near the critical value. Even big avalanches will neither make the slope sub-critically flat, nor leave a super-critical steep slope behind, which is the essence of self-organized criticality. In the terminology of nonlinear dynamics, a system that returns to the same state is also said to have an "attractor".

An alternative concept that shares all the avalanche phenomenology (of powerlaw distributions), but is not necessarily self-organized, was introduced by Chang (1992, 1998a,b, 1999a,b), called *forced criticality (FC)* or *FSOC* in combination with self-organized criticality. The key aspect of this FSOC model is that some external dynamics exerts forces on a system to produce powerlaw-like distributions of avalanches without internal self-organization.

10.5.1 Magnetospheric Physics

The FSOC concept was mostly motivated by the physics of magnetospheric substorms, which seems to require a continuous loading process in order to drive it into a critical or near-critical state (Horton and Doxas 1996). Chang et al. (2003) envision the magnetosphere to be filled with coherent magnetic structures that approach each other, merge, or scatter. If they merge, current sheets are formed and instabilities and turbulence initiate magnetic reconnection. The merging produces larger and larger structures, until a distribution of various sizes is generated, with a powerlaw probability distribution of the scale sizes of fluctuations, as well as powerlaw frequency (ω) and mode number (k) spectra of the correlations of the associated fluctuations. The merging of structures introduces correlations over a large spatial and temporal range, in contrast to the generation of independent structures in a SOC system. Ultimately, these multi-scale coherent magnetic structures may be responsible for current instabilities that lead to the onset of magnetospheric substorms. Data analysis of observations in the intermittent turbulence region of the magnetotail and from the auroral electron jet index (AE) seem to support the predictions of the FSOC concept (Consolini 1997; Lui 1998; Angelopoulos et al. 1999; Lui et al. 2000; Uritsky et al. 2002; Consolini and De Michelis 2001, 2002).

The concept of forced criticality represents an intermediate concept between a classical SOC model (which has a steady input), a hierarchical SOC system (which has partially intermittent bursty input), and intermittent turbulence (which has a correlation length of spatial scales). In the FSOC model, structures are formed by merging, the opposite to the fragmentation in turbulent cascade models. Both merging and fragmenting have naturally some correlation length, which is opposite to the statistical independence of SOC models. So, we applied both SOC and FSOC models to magnetospheric substorms. SOC models with finite size effects (Section 2.5.1) can explain the bimodal frequency distribution of energies (Chapman et al. 1998). Cellular automaton models with discretized MHD (Section 2.5.2) can relate the random electric current avalanches to the powerlaw distributions of magnetic energies in substorms (Takalo et al. 1999a). The powerlaw-like waiting-time distributions of magnetospheric substorms could be explained with a SOC model that is driven by solar wind fluctuations (Freeman and Morley 2004; Sections 5.5 and 9.2.4). The powerlaw distributions of substorm durations could also be explained by SOC models (Section 7.2). And finally we discussed a scenario of magnetospheric substorms based on intermittent turbulence (Section 10.4.3), which is a non-SOC model. In summary, a number of physical models have been developed to describe various aspects of the complexity of magnetospheric phenomena, which introduce some ambiguity in the interpretation. Some models could possibly be combined in the framework of a hierarchical SOC model system that includes all coupled processes between the solar wind, the magnetosphere, the ionosphere, and the geotail.

10.6 Percolation Models

Percolation processes describe the movement and filtering of fluids through porous materials, which applies in physics, chemistry, material sciences, and geography. Mathematical theories have been developed to model the percolation phenomenon, based on combinatorial and statistical concepts of connectedness that exhibit universality in form of powerlaw distributions, similar to branching theory (Section 2.6.5). A prominent example is *coffee percolation*, which contains a solvent (water), a permeable substance (coffee grounds), and soluble constituents (aromatic chemicals). Percolation theory has also been applied to galactic spiral structures produced by randomly-propagating star formation (Seiden and Gerola 1982; Schulman and Seiden 1986), current systems in the magnetotail (Milovanov et al. 2001), the spread of medical diseases, the undergound seeping of rainfall, or the spread of forest fires. Percolation theory can be summarized in the following question: if a liquid is poured on top of some porous material, will it be able to make its way from hole to hole and reach the bottom? Statistically, the answer depends on the connectedness of next neighbors in a 3-D lattice grid. The connections between two next neighbors can be open and let the liquid pass with probability p, or they can be closed and the probability to pass is $(1-p)$. Completion of a pass from the top to the bottom can then be expressed as statistical probability of all combinations along each possible path. There is a critical value p_c for the probability that decides between the two outcomes. For systems with subcritical values $p < p_c$, percolation will die out exponentially. For two dimensions, the critical value is $p_c = 1/2$. From these basic properties we see that percolation is very much dependent on the initial conditions (i.e., whether the percolation probability is below or above the critical value of the system), which is fundamentally different from SOC processes, where the occurrence of avalanches is absolutely insensitive to the initial conditions.

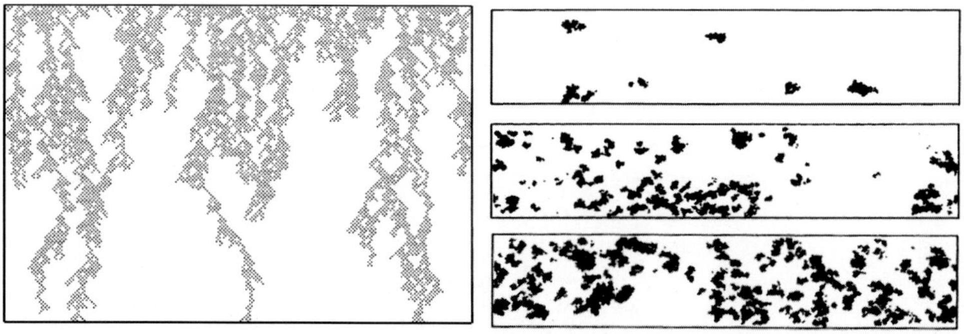

Fig. 10.8 *Left:* A 2-D simulation of a percolation process (credit CCL Northwestern University). *Right:* Three simulations of active region distributions on a full-Sun synoptic map with a 30° latitude strip, varying the hopping probability from $P_{st} = 0.1810$ (top), $P_{st} = 0.1818$ (middle), to $P_{st} = 0.1822$ (bottom), mimicking changes from the minimum to the maximum of the solar cycle (Wentzel and Seiden 1992; reproduced by permission of the AAS).

10.6.1 Solar Active Regions

We have already discussed the solar dynamo, which is a fundamentally important process for all solar activity phenomena, possibly being a candidate for a SOC process (Section 9.3.2). An interesting alternative model for the generation of active regions in terms of percolation theory was proposed by Wentzel and Seiden (1992), and the interpretation of the solar dynamo as a percolation concept is also discussed in Schatten (2007). The birth of active regions is thought to start deep in the solar interior, in the thin tachocline layer that separates the outer convection zone at $\approx 0.7\ R_\odot$ from the radiative zone, where magnetic fields of order $B \approx 10^5$ G are generated which produce magnetic flux tubes that emerge through the convection zone by buoyancy. The statistical distribution of emerging fluxtubes has been modeled as a percolation phenomenon (Wentzel and Seiden 1992; Seiden and Wentzel 1996). The percolation process can be parameterized with two free parameters: (1) the probability P_{st} that the release and rise of one flux tube stimulates the subsequent release and rise of a neighboring flux tube, and (2) the lifetime of magnetic flux once it has arrived at the solar surface. This model can explain (1) the longevity of active regions on the solar surface and the persistence of magnetic flux emergence at the same location, (2) the occurrence of persistent empty regions like coronal holes, (3) the substantial solar cycle variation as a phase-transition process near the critical threshold value, and (4) the observed exponential size distribution of active regions. However, recent studies have also shown powerlaw distributions of area-related parameters, such as for magnetic fluxes that represent the product of the area and magnetic field strength (Fig. 8.19; Parnell et al. 2009), or of active region energies (Wheatland and Sturrock 1996; Vlahos et al. 2002a,b; Vlahos and Georgoulis 2004). In a study of one particular active region it was found that the fractal dimension was close to the prediction $D = 1.56$ from percolation theory for clusters of tracers placed randomly on a lattice with a tracer density below a critical threshold (Balke et al. 1993).

Is the emergence of active regions a SOC or a percolation phenomenon? The answer depends very much on the physical model. Wentzel and Seiden (1992) assumed a simple two-parameter percolation model as stated above, while Vlahos et al. (2002a,b) add a new element by keeping track of the energy release through flux cancellation (reconnection) if flux tubes of opposite polarities collide. The magnetic polarity is certainly a basic physical property of emerging fluxtubes that determines the bipolar nature of active regions, which is not part of standard percolation theories with neutral fluids. Another open question is whether there is really a phase-transition process between the solar minimum and maximum, or is it just a gradual nonstationary Poisson process, as envisioned for solar SOC processes. It would be interesting to test whether percolation models fulfill our three basic SOC criteria (Section 9.1). While different active regions might meet the statistical independence criterion, subsequent emerging dipoles within an active region may not be created independently. Given the inaccessibility of the solar interior, physical models are required to decide whether the solar dynamo can be modeled as a SOC, percolation, or branching process.

10.7 Nonlinear Chaotic Systems

Self-organized criticality is a special state of nonlinear dissipative systems. Furthermore, SOC avalanches occur randomly and unpredictably, and thus share some common properties with chaotic systems, which seem to be governed by disorder and irregularity. However, nonlinear chaotic systems in modern usage are described by coupled differential equations that have a *deterministic*, and hence a (theoretically) predictable time evolution. Chaos theory is typically concerned with nonlinear equation systems of a few degrees of freedom, while SOC systems contain an infinite number of degrees of freedom (or metastable states). Some nonlinear systems which display deterministic chaos are (e.g., Schuster 1988): a forced pendulum, fluids near the onset of turbulence, lasers, nonlinear optical devices, Josephson junctions, chemical reactions, classical many-body systems (e.g., asteroid orbits), particle accelerators (e.g., in the plasma sheet boundary layer in the geotail), plasmas with wave–particle interactions, atomic spin flips (Ising models), harmonic oscillator in a fluid (Langevin equations), biological models of population dynamics (e.g., the Lotka–Volterra equation), medical heart pacemakers, or the meteorological butterfly effect (described by the Lorenz equation system; Lorenz 1963). Although the time evolution prescribed by a nonlinear differential equation system for a given initial condition is in principle deterministic and predictable, it is practically impossible to predict the long-time behavior of such a system because no explicit analytical solution exists and computer solutions are subject to numerical accuracy that diverge exponentially with time. Chaotic systems arise even without noise (such as the Lorenz equations), but the influence of external noise acting on chaotic systems, as it relevant for astrophysical data, smooths its fractal properties (e.g., Geisel 1985). The complexity of a nonlinear chaotic system is often characterized with the *dimension of strange attractors* (e.g., Grassberger 1985; Guckenheimer 1985), which approximately corresponds to the number of differential equations that are needed to describe the system dynamics. A low-dimensional attractor can be described with two equations (e.g., the Lotka–Volterra equation), or with three equations (e.g., the Lorenz equations). In the other extreme, Brownian motion could in principle be described by some 10^{26} coupled differential equations for every atom, but such high-dimensional systems are more practically characterized by concepts of random noise. Among the mechanical cellular automatons we described the model of coupled pendulums (Section 2.1.1) or the slider-block spring model (Section 2.4.1), which both can be described by n coupled differential equations and thus represent nonlinear chaotic systems. Computer simulations with $n \approx 10^2$–10^3 elements indeed revealed chaotic behavior (e.g., Turcotte 1999; Huang and Turcotte 1990), and at the same time displayed avalanche behavior typical for cellular automata. So, it is also said that SOC is a property of classical dynamical systems which have a critical point as an attractor. Their macroscopic dynamics exhibits spatial and temporal scale-invariance (powerlaw distributions) of the critical point of a phase transition, but are insensitive to the initial conditions. Despite of these similarities of nonlinear chaotic systems with SOC systems, a fundamental difference is the statistical independence of individual SOC events in the temporal and spatial domain (first SOC criterion, Section 9.1), while subsequent fluctuations in a nonlinear chaotic system are subject to a deterministic time evolution that is prescribed by a system of coupled differential equations, and thus are not statistically independent.

10.7.1 Astrophysics

Let us mention a few astrophysical observations that have been interpreted in terms of nonlinear chaotic systems, while some of them have also been modeled with SOC systems. The irregular X-ray variability of the neutron star Her X-1 has been subjected to a time series analysis with the method of Procaccia (1985) and a low-dimensional attractor ($D \approx 2.3$), while some higher-dimensional chaos was found for the accretion disk (Voges et al. 1987). However, this result was disputed by Norris and Matilsky (1989), who concluded that the insufficient signal-to-noise ratio does not allow us to distinguish it from an ordinary attractor contaminated with noise. The light curves of three long-period cataclysmic variable stars have been analyzed with the technique of Grassberger and Procaccia (1983a,b) in the search for an attractor dimension, but the light curves could be modeled with a periodic and a superimposed random component (Cannizzo et al. 1990). A low-dimensional attractor with a dimension of ≈ 1.5 was found in the Vela pulsar with a correlation sum technique (Harding et al. 1990). The *quasi-periodic oscillations (QPO)* of the *low-mass X-ray binary star (LMXB)* Scorpius X-1 was quantified with power spectrum analysis and with a wavelet technique, which were found to be consistent with a *dripping handrail* accretion model, a simple dynamical system that exhibits *transient chaos* (Scargle et al. 1993; Young and Scargle 1996). A time series from the R Scuti star, a RV Tau type star, was found to exhibit deterministic chaos (with an embedding dimension of 4), because it was not multi-periodic and could not be generated by a linear stochastic process (Buchler et al. 1996). The quasi-periodic light curve is shown in Fig. 9.10 (top), along with a synthetic light curve generated with a corresponding low-dimensional attractor (Fig. 9.10, middle and bottom).

The orbits of moons that are part of 3-body or N-body problems can show significant deviations from strict periods due to the aperiodic disturbances of other moons and planets.

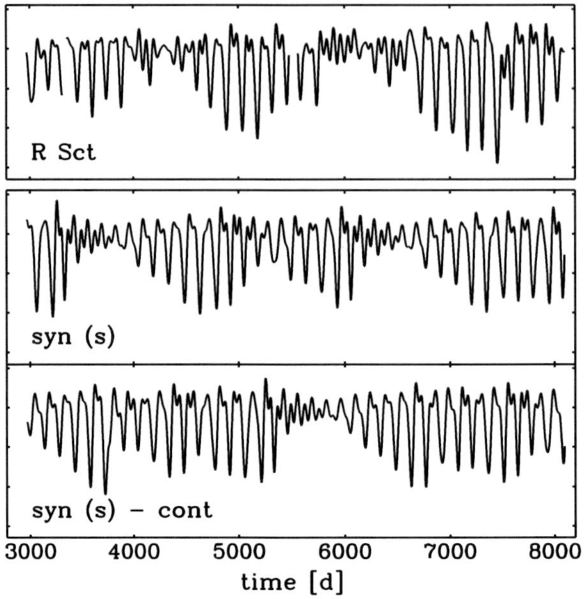

Fig. 10.9 The smoothed light curve observed from the RV Tau-type star R Scuti (top) and synthetically generated light curves with a model of a low-dimensional strange attractor (middle and bottom) (Buchler et al. 1996; reproduced by permission of the AAS).

The Saturnian satellite Hyperion, for instance, is believed to exhibit chaotic behavior, and thus has an unstable orbit (Boyd et al. 1994).

10.7.2 Solar Physics

In solar physics, time series analysis in terms of nonlinear chaotic systems and searches for strange attractors started at about the same time as in astrophysics, when chaos theory became popular after the publication of suitable nonlinear time series analysis methods (e.g., Grassberger and Procaccia 1983a). The first applications were carried out in time series of radio bursts, which were recorded with high time resolution and at times exhibited quasi-periodic signals. Low-dimensional attractors were identified in radio burst time series based on embedding dimensions, maximum Lyapunov exponents, and Kolmogorov entropy (Kurths and Herzel 1986, 1987; Kurths 1987; Kurths et al. 1990, 1991), a transition from regular to chaotic structures as part ot the Ruelle–Takens–Newhouse route to chaos was identified (Kurths and Karlicky 1989), novel nonlinear methods such as symbolic dynamics, correlation dimensions, and wavelet analysis were explored (Schwarz et al. 1993; Kurths and Schwarz 1994, 1995; Kurths et al. 1995), dimensional analysis was conducted with tests of stationarity and discrimination between deterministic and stochastic time signals (Isliker 1992a,b; Isliker and Kurths 1993; Isliker and Benz 1994a,b; Isliker 1994, 1996; Yurovsky and Magun 1996, 1998; Yurovsky 1997; Ryabov et al. 1997), and fractal dimensions, recurrence plots, and frequency distributions of solar radio bursts were derived (Higuchi 1988; Watari 1996a; Meszarosova et al. 1999, 2000; Veronig et al. 2000; Karlicky et al. 2000). However, all these radio data were obtained with spectrometers without any imaging information, and thus no spatial identification of the driving nonlinear system with chaotic behavior could be furnished.

Further studies on nonlinear systems with chaotic behavior in solar data were concerned with the chaotic dynamics in the solar wind (Polygiannakis and Moussas 1994), with a low-dimensional chaotic attractor in short-term solar ultraviolet time series (Chatterjee 1999), with modeling of the onset of Alfvén turbulence and transition to chaos in the solar corona (Chian et al. 2002), and with the Lyapunov exponents in hydrodynamic convection of the solar dynamo (Kurths and Brandenburg 1991).

Of course, the most obvious nonlinear dissipative system with chaotic behavior is the solar 11-year cycle (or magnetic 22-year Hale cycle), which shows quasi-periodic time behavior near a *limit cycle* (Fig. 10.10). A few relevant studies are concerned with the low-dimensional chaos of the solar cycle (Kremliovsky 1994), the chaotic behavior of the north–south asymmetry of sunspots (Watari 1996b), the nonlinear analysis of solar cycles (Serre and Nesme-Ribes 2000), the prediction of solar cycle 23 using nonlinear methods (Verdes et al. 2000), the multiperiodicity, chaos, and intermittency of the solar cycle (Charbonneau 2001; Spiegel 2009), the evidence for low-dimensional chaos in sunspot cycles (Letellier et al. 2006), or the interpretation in terms of high-dimensional convective turbulence and intermittency fluctuations (Mininni et al. 2002, Charbonneau et al. 2004).

What is the relationship of nonlinear chaotic systems to systems with self-organized criticality? The solar dynamo seems to be a quasi-periodic system with chaotic behavior that modulates the solar activity, which includes emergence of buoyant magnetic fluxtubes,

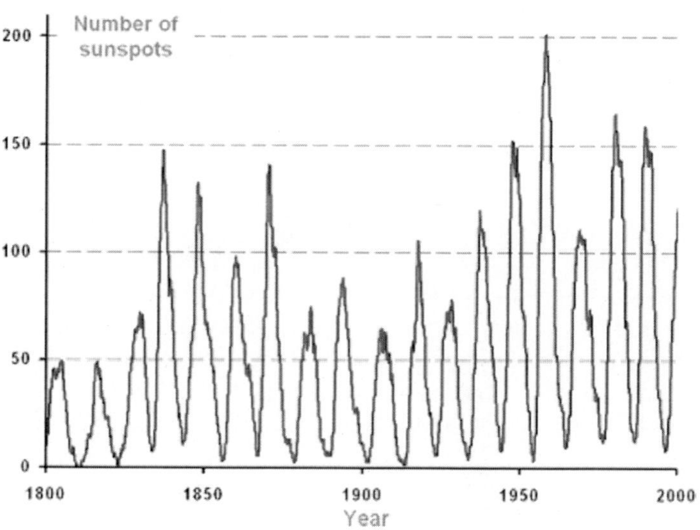

Fig. 10.10 The variation of the sunspot number from 1800 to 2000, showing the 11-year solar cycle of sunspot maxima. The solar cycle is believed to be a limit cycle of a nonlinear system with chaotic behavior (credit: NGDC).

flaring, and coronal mass ejections, which all are SOC phenomena. Thus, we can have a nonlinear chaotic system as a driver of SOC systems, which governs the nonstationary input rate on time scales that are generally much longer than the time scale of nonlinear SOC avalanches. Thus the time-dependent fluctuations of chaotic drivers do not determine the time scale distributions of secondary SOC processes, but they modulate the slowly-varying or intermittent energy input rate only. Therefore, the temporal behavior of the chaotic driver is decoupled from the temporal behavior (e.g., waiting-time distributions, occurrence frequency distributions of durations) of the SOC system. Low-dimensional attractors were identified with long limit cycles (i.e., 11 years for the solar dynamo), as well as with very fast limit cycles (on time scales of seconds during flaring radio emission). Can radio bursts driven by such fast chaotic cycles be consistent with SOC events? The answer is no, because radio bursts that are triggered by such a fast chaotic attractor would exhibit the same spatial and temporal correlation that is inherent to the low-dimensional attractor, and thus would violate the statistical independence of individual SOC avalanches we postulated for SOC systems (Section 9.1). In summary, nonlinear chaotic systems are not SOC systems, but can play a role as drivers of the energy input rate of SOC systems.

10.8 Summary

There exist many kinds of nonlinear dissipative systems that do not necessarily exhibit self-organized criticality. In this chapter we discussed a number of nonlinear systems that own SOC-like features, such as powerlaws for the distribution of some parameters, but do not qualify as SOC systems. The results of our discussion are summarized in Table 10.1, which includes, besides stationary, nonstationary, and hierarchical SOC processes, alternative nonlinear systems such as self-organization (without criticality), Brownian motion or diffusion, MHD turbulence, (externally) forced criticality, percolation, or chaotic systems. For the identification of a SOC system, powerlaw distributions in the size (or energy) of events are obviously not a sufficient criterion, since other non-SOC processes produce also powerlaws (e.g., the fractal geometry of patterns that result from self-organization, turbulence, or externally forced criticality). Also the waiting-time distribution is not a sufficient discriminator, because SOC processes can have exponential waiting-time distributions (for stationary Poisson processes) as well as powerlaw distributions (for nonstationary Poisson processes). Therefore we derived a definition of a SOC system based on three (necessary and sufficient) criteria (Section 9.1), which includes (1) the statistical independence of events (spatially and temporally), (2) the nonlinear (exponential-like) growth phase during the rise time of an event, and (3) the randomness of rise times, which implies a critical state of a system above some threshold level. Using these three criteria that mostly characterize the dynamic aspects of SOC events, we can discriminate self-organization, Brownian motion, MHD turbulence, forced criticality, percolation, or chaotic systems from SOC systems, as indicated in Table 10.1. Nonlinear systems wth chaotic behavior can often be described in terms of a coupled nonlinear equation system with a strange attractor and a quasi-periodic limit cycle. The nonlinear pulses of such chaotic systems do not fulfill the independence criterion of SOC events, but a chaotic system can modulate the energy input of a SOC system. In conclusion, characterization of nonlinear processes given in

Table 10.1 reflects only the typical behavior within the scope of astrophysical applications we discussed in this text here, and thus is rather tentative, while further corroboration with detailed parametric tests of observational data, numerical simulations, and theoretical modeling is envisioned in future endeavors.

Table 10.1 Discrimination criteria for SOC and non-SOC systems.

Process	Criterion 1: Statistical independence of events	Criterion 2: Nonlinear growth phase	Criterion 3: Random rise times	Occurrence frequency distribution of energy	Waiting time distribution
Stationary SOC	Yes	Yes	Yes	Powerlaw	Exponential
Nonstationary SOC	Yes	Yes	Yes	Powerlaw	Powerlaw
Hierarchical SOC					
−Coupled	No	Yes	Yes	Powerlaw	Powerlaw
−Filtered	Yes	Yes	Yes	Powerlaw	Powerlaw
Self-Organization	No	No	No	Powerlaw	...
Brownian Motion	No	No	No	...	...
MHD Turbulence	No	No	No	Powerlaw	Powerlaw
Forced Criticality	No	Yes	Yes	Powerlaw	...
Percolation	No	No	No	Exponential	...
Chaotic Systems	No	Yes	No	Exponential	Quasi-periodic

10.9 Problems

Problem 11.1: Discuss the physical origin of randomization filters in the input process of hierarchical SOC systems (i.e., the hourglass effect) for all SOC phenomena listed in Table 9.1.

Problem 11.2: Find more examples of self-organizing patterns (Section 10.2) in astrophysics, plasma physics, chemistry, and biophysics using a search engine on a web-browser.

Problem 11.3: Simulate Brownian motion with a random generator (as shown in Fig. 10.3 in 1-D or 2-D space) and sample a size distribution of avalanches that consist of a random number of particles undergoing Brownian diffusion. Do you obtain a powerlaw distribution?

Problem 11.4: Turbulence exhibits a powerlaw spectrum with a powerlaw function of $P(k) \propto k^{-5/3}$, with k being the wavenumber. How does this scaling law translate into a power spectrum $P(\nu)$ of a time series and into a distribution of time scales $N(T)$ (Hint: Use the relationship between a shot noise spectrum and the distribution of pulse durations described in Section 4.8.4.)

Problem 11.5: Use the Lotka–Volterra equation to describe a nonlinear system with quasi-periodic behavior. Can you simulate with this equation system similar time profiles as observed for the R Scuti star (Fig. 10.9) or the solar cycle (Fig. 10.10). Give a physical interpretation of the variables in the Lotka–Volterra equation system that would explain the (quasi-periodic) time scale of the limit cycle?

Appendices

Appendix A: Physical Constants

Physical quantity	Symbol	Value	cgs units
Speed of light in vacuum	c	$= 2.9979 \times 10^{10}$	cm s^{-1}
Elementary charge	e	$= 4.8023 \times 10^{-10}$	statcoulomb
Electron mass	m_e	$= 9.1094 \times 10^{-28}$	g
Proton mass	m_p	$= 1.6726 \times 10^{-24}$	g
Proton/electron mass ratio	m_p/m_e	$= 1.8361 \times 10^{3}$	
Gravitational constant	G	$= 6.6720 \times 10^{-8}$	dyne cm^2 g^{-2}
Boltzmann constant	k_B	$= 1.3807 \times 10^{-16}$	erg K^{-1}
Planck constant	h	$= 6.6261 \times 10^{-27}$	erg s
Rydberg constant	$R_H = me^4/4\pi\hbar^3 c$	$= 1.0974 \times 10^{5}$	cm^{-1}
Bohr radius	$a_0 = \hbar^2/m_e e^2$	$= 5.2918 \times 10^{-9}$	cm
Electron radius	$r_e = e^2/m_e c^2$	$= 2.8179 \times 10^{-13}$	cm
Stefan–Boltzmann constant	$\sigma = 2\pi^5 k_B^4/(15c^2h^3)$	$= 5.6774 \times 10^{-5}$	erg cm^{-2} s^{-1} K^{-4}
1 electronvolt	ε_{eV}	$= 1.6022 \times 10^{-12}$	erg
	T_{eV}	$= 1.1604 \times 10^{4}$	K
	λ_{eV}	$= 1.2398 \times 10^{-4}$	cm
	ν_{eV}	$= 2.4180 \times 10^{14}$	Hz
1 ångstrøm	(Å)	$= 10^{-8}$	cm
1 jansky	(Jy)	$= 10^{-23}$	erg s^{-1} cm^{-2} Hz^{-1}
1 solar flux unit	(SFU)	$= 10^{-19}$	erg s^{-1} cm^{-2} Hz^{-1}
1 astronomical unit	(AU)	$= 1.50 \times 10^{13}$	cm
Solar radius	$R_\odot$	$= 6.96 \times 10^{10}$	cm
Solar mass	$M_\odot$	$= 1.99 \times 10^{33}$	g
Solar gravitation	$g_\odot = GM_\odot/R_\odot^2$	$= 2.74 \times 10^{4}$	cm s^{-2}
Solar escape speed	v_∞	$= 6.18 \times 10^{7}$	cm s^{-1}
Solar age	$t_\odot$	$= 4.60 \times 10^{9}$	years
Solar radiant power	$L_\odot$	$= 3.90 \times 10^{33}$	erg s^{-1}
Solar radiant flux density	$F_\odot$	$= 6.41 \times 10^{10}$	erg cm^{-2} s^{-1}
Solar constant (flux at 1 AU)	$f_\odot$	$= 1.39 \times 10^{6}$	erg cm^{-2}
Solar solid angle (at 1 AU)	$\Omega_\odot = \pi R_\odot^2/AU^2$	$= 6.76 \times 10^{-5}$	ster
Photospheric temperature	T_{phot}	$= 5{,}762$	K

Appendix B: Plasma Parameters

Physical quantity	Definition	Numerical formula (cgs units)
Thermal pressure	$p_{th} = 2n_e k_B T_e$	$= 2.76 \times 10^{-16}\, n_e\, T$ (dyne cm^{-2})
Magnetic pressure	$p_m = B^2/(8\pi)$	$= 3.98 \times 10^{-2}\, B^2$ (dyne cm^{-2})
Plasma-β parameter	$\beta = (p_{th}/p_m)$	$= 6.94 \times 10^{-15}\, n_e T_e B^{-2}$
Thermal scale height	$\lambda_T = 2k_B T_e/(\mu_C m_p g_\odot)$	$= 4.73 \times 10^3\, T_e$ (cm)
Electron thermal velocity	$v_{Te} = (k_B T_e/m_e)^{1/2}$	$= 3.89 \times 10^5\, T_e^{1/2}$ (cm s^{-1})
Ion thermal velocity	$v_{Ti} = (k_B T_i/\mu m_p)^{1/2}$	$= 9.09 \times 10^3\, (T_i/\mu)^{1/2}$ (cm s^{-1})
Ion mass density	$\rho = n_i m_i = n_i \mu m_p$	$= 1.67 \times 10^{-24}\, \mu\, n_i$ (g cm^{-3})
Sound speed	$c_S = (\gamma p_{th}/\rho)^{1/2}$	$= 1.66 \times 10^4\, (T/\mu)^{1/2}$ (cm s^{-1})
Alfvén speed	$v_A = B/(4\pi \mu m_p n_i)^{1/2}$	$= 2.18 \times 10^{11}\, B\, (\mu n_i)^{-1/2}$ (cm s^{-1})
Electron plasma frequency	$f_{pe} = (n_e e^2/\pi m_e)^{1/2}$	$= 8.98 \times 10^3\, n_e^{1/2}$ (Hz)
Ion plasma frequency	$f_{pi} = (n_i Z^2 e^2/\pi \mu m_p)^{1/2}$	$= 2.09 \times 10^2\, Z(n_i/\mu)^{1/2}$ (Hz)
Electron gyrofrequency	$f_{ge} = eB/(2\pi m_e c)$	$= 2.80 \times 10^6\, B$ (Hz)
Ion gyrofrequency	$f_{gi} = ZeB/(2\pi \mu m_p c)$	$= 1.52 \times 10^3\, ZB/\mu$ (Hz)
Electron collision frequency	f_{ce}	$= 3.64 \times 10^0 n_e \ln\Lambda T_e^{-3/2}$ (Hz)
Ion collision frequency	f_{ci}	$= 5.98 \times 10^{-2}\, n_i \ln\Lambda Z^2 T_i^{-3/2}$ (Hz)
Electron collision time	$\tau_{ce} = 1/f_{ce}$	$= 2.75 \times 10^{-1}\, T_e^{3/2}/(n_e \ln\Lambda)$ (s)
Ion collision time	$\tau_{ci} = 1/f_{ci}$	$= 1.67 \times 10^1\, T_i^{3/2}/(n_i \ln\Lambda Z^2)$ (s)
Electron gyroradius	$R_e = v_{Te}/(2\pi f_{ge})$	$= 2.21 \times 10^{-2}\, T_e^{1/2} B^{-1}$ (cm)
Ion gyroradius	$R_i = v_{Ti}/(2\pi f_{gi})$	$= 9.49 \times 10^{-1}\, T_i^{1/2} \mu^{1/2} Z^{-1} B^{-1}$ (cm)
Debye length	$\lambda_D = (k_B T_e/4\pi n_e e^2)^{1/2}$	$= 6.90 \times 10^0\, T^{1/2} n_e^{-1/2}$ (cm)
Dreicer field	$E_D = Ze \ln\Lambda/\lambda_D^2$	$= 1.01 \times 10^{-11} Z \ln\Lambda n_e T_e^{-1}$ (statvolt cm^{-1})
Electrical conductivity	$\sigma = n_e e^2 \tau_{ce}/m_e$	$= 6.96 \times 10^7 \ln(\Lambda)^{-1} Z^{-1} T_e^{3/2}$ (Hz)
Magnetic diffusivity	$\eta = c^2/(4\pi\sigma)$	$= 1.03 \times 10^{12}\, \ln(\Lambda) Z\, T_e^{-3/2}$ (cm^2 s^{-1})
Magnetic Reynolds number	$R_m = lv/\eta$	$= 9.73 \times 10^{-13}\, lv\, T^{3/2}\, \ln\Lambda^{-1}$
Thermal Spitzer conductivity coeff.	κ	$= 9.2 \times 10^{-7}$ (erg s^{-1} cm^{-1} K$^{-7/2}$)
Thermal conductivity	$\kappa_\parallel = \kappa T^{5/2}$	$= 9.2 \times 10^{-7}\, T^{5/2}$ (erg s^{-1} cm^{-1} K^{-1})
Radiative loss rate	$\Lambda_0(T \approx 1$ MK$)$	$= 1.2 \times 10^{-22}$ (erg s^{-1} cm^3)
Coronal viscosity	ν_{visc}	$= 4.0 \times 10^{13}$ (cm^2 s^{-1})

– cgs units: length l (cm), mass m (g), time t (s), Temperature T (K), magnetic field B (G), densities n_i, n_e (cm^{-3}).
– Adiabatic index: $\gamma = c_p/c_v = (N+2)/N = 5/3 = 1.67$.
– Ion/proton mass ratio $\mu = m_i/m_p$: $\mu(H) = 1$, $\mu(He) = 4$, $\mu(Fe) = 56$.
– Mean molecular weight in corona (H:He = 10 : 1): $\mu_C = (10*1+1*4)/11 = 1.27$
– Coronal approximation (full ionization): $n_i = n_e$.
– Coulomb logarithm: $\ln\Lambda = 23 - \ln(n_e^{1/2} T_e^{-3/2}) \approx 20$ for $T_e \lesssim 10$ eV.
– Charge state: proton $\mapsto Z = 1$, Fe IX $\mapsto Z = 8$.

Notation

Physical Units Symbols

A	ampère, unit for electric current (SI)
Å	ångstrøm = 10^{-8} cm
AU	astronomical unit
C	coulomb, unit for electric charge (SI)
cm	centimeter, unit for length (cgs)
dyne	unit for force (cgs)
erg	unit for energy (cgs)
eV	electronvolt; keV, MeV, GeV
g	gram, unit for mass (cgs); kg (SI)
G	gauss, unit for magnetic field (cgs); kG
J	joule, unit for energy (SI)
Hz	hertz = s^{-1}, unit for frequency (SI); kHz, MHz, GHz
K	kelvin, unit for temperature (cgs, SI); MK
m	meter, unit for length (SI); μm, mm, cm, dm, km, Mm
N	newton, unit for force (SI)
rad	radian, unit angle π
s	second, unit for time (cgs, SI)
ster	steradian, unit for solid angle (ster=rad^2)
T	tesla, unit for magnetic field (SI)
V	volt, unit for electric potential (SI)
W	watt, unit for power (SI); kW, MW

Latin Symbols

$\mathbf{A}$	magnetic vector potential function
A	area (cm^2)
a	amplitude (cm)
$\mathbf{B}$	magnetic field vector, magnetic induction
B	magnetic field strength (G)
$B(p,q)$	beta function
C	count rate (s^{-1})
C	contour curve of surface integral
D	fractal dimension
D	decay time (s)
$D(x,t)$	diffusion constant (cm^2 s^{-1})
D	total derivative $\partial/\partial t + v\,\partial/\partial x$
d	distance (cm)
E	total energy (erg)
E_{kin}	kinetic energy (nonrelativistic $E_{kin} = \frac{1}{2}mv^2$)
E_m	magnetic energy ($E_m = B^2/8\pi$)
E_{th}	thermal energy ($E_{th} = k_B T_e$)
E_X	total radiated energy in X-rays (erg)
EM	emission measure $EM = n^2 z$ (cm^{-5})
$\mathscr{E}$	electric field strength (statvolt cm^{-1})
e	elementary electric charge

e	energy (erg)
F	photon flux (erg s^{-1} cm^{-2} keV^{-1})
F	force (dyne)
F_d	dynamic friction force (dyne)
F_s	static friction force (dyne)
f	frequency (Hz)
$f(x)$	function
G	gravitational constant
$\mathbf{g}$	gravitational acceleration (cm s^{-2})
h	height (cm)
h	Planck constant
I	current (statampere)
I	intensity of radiation (erg s^{-1} cm^{-2} Hz^{-1} ster^{-1})
$\mathbf{j}$	current density vector
k_B	Boltzmann constant
L, l	length (cm)
L_X	luminosity in X-rays
L	Laplacian
M, m	mass (g)
m	magnitude
m_e	electron mass
$N(x)$	differential frequency distribution of parameter x
$N^{cum}(>x)$	cumulative frequency distribution of parameter x
n	number
n_e	electron number density (cm^{-3})
P	peak energy dissipation rate (erg s^{-1})
P	time period (s)
P	perimeter (cm)
$P(x)$	probability distribution function of parameter x
$P(\nu)$	power spectrum versus frequency ν
p	powerlaw index of power spectrum $P(\nu) \propto \nu^{-p}$
p	powerlaw index of waiting time distribution $N(p) \propto (\Delta t)^{-p}$
p	pressure (dyne cm^{-2})
q	ratio
q	electric charge
R, r	radius or range (cm)
R_m	magnetic Reynolds number
$R_\odot$	solar radius
$R(T)$	instrumental temperature response function
r	rate (s^{-1})
S	surface (specifying a surface integral)
S	source function
S	size
s	path distance along curve (cm)
T	time duration (s)
T	temperature (K)
T_e	electron temperature (K)
t	time (s)
t_s	saturation time (s)
V	volume (cm^3)
$\mathbf{v}$, v	velocity (cm s^{-1})
v_A	Alfvén speed
W	energy release rate (erg s^{-1})
W_S	saturation energy rate (erg s^{-1})

w	width (cm)
x	spatial coordinate or position
y	spatial coordinate or position
z	spatial coordinate (along line-of-sight)
z	height difference

Greek Symbols

α	powerlaw index of differential frequency distribution
α_A	powerlaw index of area A
α_P	powerlaw index of peak energy rate P
α_E	powerlaw index of total energy E
α_S	powerlaw index of size S
α_T	powerlaw index of time duration T
α	angle (deg)
α	correlation coefficient
β	powerlaw index of cumulative frequency distribution
β	correlation coefficient
Γ	growth rate $(1/\tau_G)$
γ	powerlaw index of power spectrum
γ	powerlaw index of photon spectrum
γ	correlation coefficient
γ	damping constant
∇	nabla operator
Δ	Laplace operator
Δ	difference
Δt	waiting time between events ($\Delta t = t_{i+1} - t_i$)
δ	powerlaw index of electron spectrum
ε	infinitesimal length scale
ε	photon energy $\varepsilon = h\nu$ (keV)
ε_x	hard X-ray photon energy $\varepsilon = h\nu_x$
η	magnetic diffusivity
η	energy decay rate (erg s^{-1})
$\Theta(x)$	Heavyside step function
θ, ϑ	angle
κ	diffusion constant
$\Lambda(T)$	radiative loss function
λ	wavelength (cm)
λ	event occurrence rate $(1/\Delta t)$
μ	mean (of Gaussian distribution)
ν	frequency (s^{-1} = Hz)
ν_{visc}	coronal viscosity
ρ	mass density, $\rho = n\, m$
ρ	random number
σ	standard deviation (of Gaussian distribution)
σ	electrical conductivity
τ	time scale (s)
τ_G	growth time
τ_d	decay time
τ_{rise}	rise time
Φ	magnetic flux (Mx = G cm^2)
φ	azimuthal angle

Acronyms

1-D, 2-D, 3-D	one-, two-, three-dimensional
ACE	Advanced Composition Explorer
AE	Auroral Electron jet index
AGN	Active Galactic Nuclei
BATSE	Burst And Transient Source Experiment (on CGRO)
BCS	Bragg Crystal Spectrometer (on Yohkoh)
BCSW	Bak–Chen–Scheinkman-Woodford (1993) model
BTW	Bak–Tang–Wiesenfeld (1987) model
Cassini	Cassini orbiter, part of the Cassini–Huygens space probe
CCD	Charge Coupled Device (camera)
CME	Coronal Mass Ejection
DC	Direct Current
CCC	Cross-Correlation Coefficient
CEOF	Complex Empirical Orthogonal Function analysis (method)
CGRO	Compton Gamma Ray Observatory (spacecraft)
Cluster	Cluster (ESA space mission)
CV	Cataclysmic Variable stars (Canes Venatici type stars)
DCIM	DeCIMetric bursts
DEM	Differential Emission Measure (distribution)
DKA	Drift-Kinetic Alfvén vortex motions
DNA	DeoxyriboNuclei Acid
EIT	Extreme-ultraviolet Imaging Telescope (on SoHO)
ETH	Eidgenössische Technische Hochschule (Zurich, Switzerland)
EUV	Extreme UltraViolet
EUVE	Extreme UltraViolet Explorer (spacecraft)
EUVI	Extreme-UltraViolet Imager (on SECCHI/STEREO)
FBR	Fourier-Based Recognition (method)
Fermi	Fermi Gamma-ray Space Telescope (spacecraft)
FFT	Fast Fourier Transform
FWHM	Full Width Half Maximum
FSOC	Forced and/or Self-Organized Criticality model
FUV	Far UltraViolet imager (on IMAGE spacecraft)
GEOTAIL	magnetospheric satellite
GOES	Geostationary Orbiting Earth Satellite (spacecraft)
GRANAT	International Astrophysical Observatory (Russian spacecraft)
GRB	Gamma-Ray Burst spectrometer (on ULYSSES spacecraft)
GSFC	Goddard Space Flight Center (NASA)
Hα	hydrogen line (6562.8 Å)
HSP	High-Speed Photometer (on HST spacecraft)
HST	Hubble Space Telescope (spacecraft)
HXRBS	Hard X-Ray Burst Spectrometer (on SMM)
HXR	Hard X-Rays
HXT	Hard X-ray Telescope (on Yohkoh)
IBM	International Business Machines Corporation
ICA	Independent Component Analysis (method)
ICE	International Cometary Explorer (ISEE-3 spacecraft)
IMAGE	Imager for Magnetopause-to-Aurora Global Exploration (spacecraft)

IMF	Interplanetary Magnetic Field
IMP	Interplanetary Monitoring Platform (spacecraft)
ISEE-3	International Sun/Earth Explorer 3 (ICE spacecraft)
IT	Intermittent Turbulence
JPL	Jet Propulsion Laboratory (Pasadena, USA)
KLT	Karhunen–Loéve Transform (method)
LASCO	Large Angle Spectrometric COronagraph (on SOHO)
LMXB	Low-Mass X-ray Binary star
LMC	Large Magellanic Cloud (a galaxy)
MDI	Michelson Doppler Imager (on SoHO)
MHD	Magneto-HydroDynamics
MLT	Multiple Level Tracking (method)
MW	MicroWaves
MW-S	MicroWave Spike bursts
NASA	National Aeronautics and Space Administration
NGC	New General Catalogue (of nebulae and star clusters)
NGDC	National Geophysical Data Center (USA)
NICMOS	Near Infrared Camera and Multi-Object Spectrometer (on HST)
NIXT	Normal Incidence X-Ray Telescope (rocket instrument)
NOAA	National Oceanic and Atmospheric Administration (USA)
OFC	Olami–Feder–Christensen (1992) model
OSO-7	Orbiting Solar Observatory 7 (satellite)
PCA	Principal Component Analysis (method)
PHEBUS	Payload for High Energy BUrst Spectroscopy (on GRANAT)
POD	Proper Orthogonal Decomposition (method)
POLAR	Polar satellite
PSR	PulSaR
QPO	Quasi-Periodic Oscillations (in stellar data)
Ranger-8	lunar spacecraft
RCL	Resistor (R), Capacitor (C), inductor (L) circuit
RHESSI	Reuven Ramaty High Energy Solar Spectroscopic Imager (spacecraft)
RXTE	Rossi X-Ray Timing Explorer (spacecraft)
SDSS	Sloan Digital Sky Survey (ground-based telescope)
SECCHI	Sun Earth Connection Coronal and Heliospheric Investigation (on STEREO)
SEP	Solar Energetic Particle events
SGR	Soft Gamma Repeaters
SMM	Solar Maximum Mission (spacecraft)
SO	Self-Organization
SOC	Self-Organized Criticality
SOHO	SOlar and Heliospheric Observatory (spacecraft)
SSC	Sudden Storm Commencement (magnetospheric events)
SSW	Solar SoftWare (software package in IDL)
STEREO	Solar TErrestrial RElations Observatory (spacecraft)
SuperDARN	Super Dual Auroral Radar Network
SWAVES	STEREO/WAVES instrument (on STEREO spacecraft)
Swift	spacecraft to observe gamma-ray bursts (NASA)
SXR	Soft X-Rays
SXT	Soft X-ray Telescope (on Yohkoh)
TRACE	Transition Region And Coronal Explorer (spacecraft)
UCB	University of California, Berkeley
Ulysses	interplanetary spacecraft
UV	ultraviolet
UVI	UltraViolet Imager (onboard POLAR spacecraft)
Voyager	Voyager 1 and 2 (interplanetary spacecraft)

WATCH	Wide Angle Telescope for Cosmic Hard X-Rays (on GRANAT)
WHAM	Wisconsin Hα Mapper (ground-based telescope)
WIC	Wideband Imaging Camera (a FUV instrument on IMAGE)
WIND	interplanetary spacecraft
WTD	Waiting Time Distribution
XEST	XMM Extended Survey of the Taurus Molecular Cloud
XMM	X-ray Multi-Mirror Misson (spacecraft), also called Newton
XUV	eXtreme UltraViolet

Image Credit: Public Websites

Fig. 1.1a: http://members.virtualtourist.com/m/5f4b4/, Water-storage dam at Yaotsu, Gifu, Japan.

Fig. 1.1b: http://www.wholey.net/HST/HST.html, A large wet snow avalanche at Deadman Canyon (Jim and Louise Wholey).

Fig. 1.2a: http://www.vacation-rental-house.com/house-to-rent.html, Sand beach in northern California.

Fig. 1.2b: http://indianriverenterprises.com/6.html, Sandpile produced by conveyer belt of the Indian River Enterprises.

Fig. 1.4a: http://www.sliceofscifi.com/wp-content/uploads/2008/03/usa_night_ satellite.jpg, North America at night, satellite picture taken by International Space Station (ISS, NASA).

Fig. 1.8: http://books.google.com/, from book by Zehnder, J.N. 1974, Der Goldauer Bergsturz: Goldau, 207p. Fig.4 on p.19 of book shows a sketch of Rossberg after Arth-Goldau rockslide, drawn by Fritz Morach.

Fig. 1.9: http://sd-www.jhuapl.edu/Aurora/UVI_on_Earth.html, Global image of the auroral oval, observed with the *Ultraviolet Imager (UVI)* onboard the NASA satellite "Polar" (George Parks, University of California, Berkeley, and Ching Meng, JHU/APL).

Fig. 1.11a: http://www.lightandmatter.com/html_books/1np/ch10/figs/saturn.jpg, Saturn rings observed by the NASA spacecraft *"Voyager 2"* (JPL).

Fig. 1.11b: http://www.outer-space-art-gallery.com/images/bergeronsaturn.jpg, space art rendering of Saturn rings, posted at website of Cosmic Cafe and Outer Space Art Gallery.

Fig. 1.12a: http://trace.lmsal.com/POD/looposcillations/paperI/images/T171_010415_221512.gif, solar postflare arcade observed by the NASA spacecraft TRACE (LMSAL).

Fig. 1.12b: http://trace.lmsal.com/POD/images/arcade_9_nov_2000.gif, solar postflare arcade observed by the NASA spacecraft TRACE (LMSAL).

Fig. 1.16: http://antwrp.gsfc.nasa.gov/apod/image/0607/rsoph_pparc_big.jpg, Artistic rendering of white dwarf star RS Oph, Astronomy picture of the day 2006 July 26 (David A. Hardy, PPARC, and GSFC/NASA).

Fig. 2.6: http://www.ics.uci.edu/~ eppstein/ca/, John Conway's "Game of Life" cellular automatons (D. Eppstein).

Fig. 2.9: http://en.wikipedia.org/wiki/Bak–Sneppen_model, Bak-Sneppen model, Wikipedia, figure created by Claudio Rocchini.

Fig. 8.20a: http://meteorites.wustl.edu/lunar/crater_l.jpg, Lunar craters photographed by NASA's *Apollo 11* mission, posted by Randy L. Korotev at website of Dept. of Earth and Planetary Sciences, Washington University St. Louis.

Fig. 8.21a: http://www.astrobio.net/albums/xsolar/aci.sized.jpg, Asteroid Eros and asteroid belt, posted by Astrobiology Magazine, American Scientist Magazine, NASA and Johns Hopkins University Applied Physics Laboratory (JHU/APL).

Fig. 8.22: http://en.wikipedia.org/wiki/Rings_of-Saturn, http://photojournal.jpl .nasa.gov/catalog/PIA08955, Saturn B ring, photographed by NASA's spacecraft *Cassini*, NASA, JPL, Space Science Institute.

Fig. 9.1: http://www.mso.anu.edu.au/2dFGRS/Public/Pics/2dFzcone_big.gif, 2dF galaxy redshift survey, posted at website of The Australian National University (ANU), figure created by Matthew Colless.

Fig. 9.2: http://www.aip.de/image_archive/images/mandersen.jpg, R136 in 30 Doradus, photographed by the *NICMOS* instrument on NASA's *Hubble Space Telescope (HST)*, figure credit Morten Andersen, posted at website of Astrophysical Institute Potsdam.

Fig. 9.3: http://i.space.com/images/h_jet_schematic_02.jpg, http://www.tutorgig .com/ed/Black_hole, Artistic rendering of blazar, NASA website.

Fig. 9.4: http://www.nasa.gov/images/content/311187main_fermiswift_magnetar2 _HI.jpg, Artistic rendering of magnetar, NASA/GSFC website of SWIFT mission.

Fig. 9.5: http://astroparticle.uchicago.edu/archives.htm, University of Chicago, credit Simon Swordy.

Fig. 9.10: http://pluto.space.swri.edu/image/glossary/substorm.jpg, Substorm cartoon from W. Baumjohann and R.A. Treuman, Basic Space Plasma Physics, 1996. The three auroral images were obtained with the *WIC* instrument onboard NASA's *IMAGE* mission, posted at webpage of Southwest Research Institute.

Fig. 10.2a: www.maion.com/photography/_photos/nami1673.jpg, Sand dunes in Namibia, website of Jef Maion.

Fig. 10.2b: http://www.freemars.org/jeff/planets/Jupiter.jpg, Jupiter photographed by NASA's Voyager 2 mission, posted by Jeff Root at website of the Minnesota Space Frontier Society.

Fig. 10.2c: http://cdn.physorg.com/newman/gfx/news/hires/milkywaygala.jpg, Artist's rendering of Milky way galaxy, posted on website on Science News, Technology, Physics, Nanotechnology, Space Science, Earth Science, Medicine of Physorg.com.

Fig. 10.2d: http://www.sciencedaily.com/images/2007/02/070210172729.jpg, Jupiter's moon Europa, photographed by NASA and JPL *Galileo* mission, posted at website Science Daily: News and Articles in Science, Health, Environment and Technology.

Fig. 10.5: http://sci.esa.int/science-e-media/img/c1/solarwind-spectrum410.gif, Solar wind spectra of turbulence cascade, measured with ESA's *Cluster* mission, modeled with gyro-kinetic theory (Howes et al. 2008), posted at ESA/CLUSTER website.

Fig. 10.8a: http://ccl.northwestern.edu/netlogo/models/Percolation, Percolation code from NetLogo Models Library: Sample Models/Earth Science, posted at website of *Center for Connected Learning (CCL)* and Computer-Based Modeling at Northwestern University.

Fig 10.10: http://astronomy.swin.edu.au/cosmos/S/Sunspot+Cycle, Solar cycle sunspot number, credit National Geophysical Data Center (NGDC), posted at website of Swinburne University of Technology, Australia.

References

Abramenko, V.I., Yurchyshyn, V., Wang, H., and Goode, P.R. 2001, *Magnetic power spectra in the solar photosphere derived from ground and space based observations*, Solar Phys. **201**, 225-240.

Abramenko, V.I., Yurchyshyn, V.B., Wang, H., Spirock, T.J., and Goode, P.R. 2002, *Scaling Behavior of Structure Functions of the Longitudinal Magnetic Field in Active Regions on the Sun*, Astrophys. J **577**, 487-495.

Abramenko, V.I., Yurchyshyn, V.B., Wang, H., Spirock, T.J., and Goode, P.R. 2003, *Signature of avalanche in solar flares as measured by photospheric magnetic fields*, Astrophys. J. **597**, 1135-1144.

Abramenko, V.I. 2005, *Multifractal Analysis of Solar Magnetograms*, Solar Phys. **228**, 29-42.

Akabane, K. 1956, *Some features of solar radio bursts at around 3000 Mc/s*, Publ. Astron. Soc. Japan **8**, 173-181.

Aki, K. 1981, *Earthquake prediction*, (ed. D. Simpson and P. Richards), American Geophysical Union: Washington, DC, p. 566-574.

Aletti, V., Velli, M., Bocchialini, K., Einaudi, G., Georgoulis, M., and Vial, J.C. 2000, *Microscale structures on the quiet sun and coronal heating*, Astrophys. J. **544**, 550-557.

Alexandrova, O., Saur, J., Lacombe, C., Mangeney, A., Mitchell, J., Schwartz, S. J., and Robert, P. 2009, *Universality of Solar-Wind Turbulent Spectrum from MHD to Electron Scales*, Phys. Rev. Lett. **103**, CiteID 165003.

Alstrom, P. 1988, *Mean-field exponents for self-organized critical phenomena*, Phys. Rev. Lett. A **38/9**, 4905-4906.

Alstrom, P. and Leao, J. 1994, *Self-organized criticality in the Game of Life*, Phys. Rev. E **49**, R2507.

Aly, J.J. and Amari, T. 1997, *Current sheets in two-dimensional potential magnetic fields. III. Formation in complex topology configurations and application to coronal heating*, Astron. Astrophys. **319**, 699-719.

Ambruster, C.W., Sciortino, S., and Golub, L. 1987, *Rapid, low-level X-ray variability in active late-type dwarfs*, Astrophys. J. Suppl. Ser. **65**, 273-305.

Anastasiadis, A. and Vlahos, L. 1991, *Particle Acceleration inside a gas of shock waves*, Astron. Astrophys. **245**, 271-278.

Anastasiadis, A. and Vlahos, L. 1993, *Particle acceleration by multiple shocks at the hot spots of extragalactic radio sources* Astron. Astrophys. **275**, 427-432.

Anastasiadis, A. and Vlahos, L. 1994, *Particle acceleration in an evolving active region by an ensemble of shock waves*, Astrophys. J. **428**, 819-826.

Anastasiadis, A., Vlahos, L., and Georgoulis, K. 1997, *Electron acceleration by random DC electric fields*, Astrophys. J. **489**, 367-374.

Anastasiadis, A., Gontikakis, C., Vilmer, N., and Vlahos, L. 2004, *Electron acceleration and radiation in evolving complex active regions*, Astron. Astrophys. **422**, 323-330.

Andersen, J.V., Jensen, H.J., and Mouritsen, O.G. 1991, *Crossover in the power spectrum of a driven diffusive lattice-gas model*, Phys. Rev. B. **44**, 439-442.

Andrade, R.F.S., Schellnhuber, H.J., and Claussen, M. 1998, *Analysis of rainfall records: possible relation to self-organized criticality*, Physica A **254**, 3/4, 557-568.

Angelopoulos, V., Coroniti, F.V., Kennel, C.F., Kivelson, M.G., Walker, R.J., Russell, C.T., McPherron, R.L., Sanchez, E., Meng, C.I., Baumjohann, W., Reeves, G.D., Belian, R.D., Sato, N., Friis-Christensen, E., Sutcliffe, P.R., Yumoto,K., Harris,T. 1996, *Multipoint analysis of a bursty bulk flow event on April 11, 1985*, J. Geophys. Res. **101/A3**, 4967-4990.

Angelopoulos, V., Mukai, T., and Kokubun, S. 1999, *Evidence for intermittency in Earth's plasma sheet and implications for self-organized criticality*, Phys. Plasmas **6/11**, 4161-4168.

Ansari, M.H. and Smolin, L. 2008, *Self-organized criticality in quantum gravity*, Classical and Quantum Gravity **25/9**, pp. 095016.

Antipov, S.V., Nezlin, M.V., Snezhkin, E.N., and Trubnikov, A.S. 1985, *The Rossby autosoliton and a laboratory model of the Jupiter Red Spot*, Zhurnal Eksperimental'noi i Teoreticheskoi Fiziki **89**, 1905-1920 (in Russian).

Antonova, E.E. 2004, *Magnetostatic equilibrium and current systems in the Earth's magnetosphere*, Adv. Space Res. **33/5**, 752-760.

Argyle, E. and Gower, J.F.R. 1972, *The pulse-height distribution for NP 0532*, Astrophys. J. **175**, L89-L91.

Armstrong, J.W., Rickett, B.J., and Spangler, S.R. 1995, *Electron density power spectrum in the local interstellar medium*, Astrophys. J. **443**, 209-221.

Arzner, K. and Güdel, M. 2004, *Are coronae of magnetically active stars heated by flares? III. Analytical distribution of superposed flares*, Astrophys. J. **602**, 363-376.

Arzner, K., Güdel, M., Briggs, K., Telleschi, A., and Audard, M. 2007, *Statistics of superimposed flares in the Taurus molecular cloud*, Astron. Astrophys. **468**, 477-484.

Aschwanden, M.J., Benz, A.O., Dennis,B.R., and Kundu,M.R. 1994, *Pulsed acceleration in solar flares*, Astrophys. J. Suppl. Ser. **90**, 631-638.

Aschwanden, M.J., Benz, A.O., Dennis, B.R., and Schwartz, R.A. 1995, *Solar electron beams detected in hard X-rays and radio waves*, Astrophys. J. **455**, 347-365.

Aschwanden, M.J., Kliem, B., Schwarz, U., Kurths, J., Dennis, B. R. and Schwartz, R. A. 1998a, *Wavelet Analysis of Solar Flare Hard X-Rays*, Astrophys. J. **505**, 941-956.

Aschwanden, M.J., Dennis, B.R., and Benz, A.O. 1998b, *Logistic avalanche processes, elementary time structures, and frequency distributions of flares*, Astrophys. J. **497**, 972-993.

Aschwanden, M.J., Nightingale, R., Tarbell, T., and Wolfson, C.J. 2000a, *Time variability of the quiet Sun observed with TRACE: I. Instrumental effects, event detection, and discrimination of EUV nanoflares*, Astrophys. J. **535**, 1027-1046.

Aschwanden, M.J., Tarbell, T., Nightingale, R., Schrijver, C.J., Title, A., Kankelborg, C.C., Martens, P.C.H., and Warren, H.P. 2000b, *Time variability of the quiet Sun observed with TRACE: II. Physical parameters, temperature evolution, and energetics of EUV nanoflares*, Astrophys. J. **535**, 1047-1065.

Aschwanden, M.J. and Alexander, D. 2001, *Flare plasma cooling from 30 MK down to 1 MK modeled from Yohkoh, GOES, and TRACE observations during the Bastille-Day event (2000 July 14)*, Solar Phys. **204**, 91-121.

Aschwanden, M.J. and Parnell, C.E. 2002, *Nanoflare statistics from first principles: fractal geometry and temperature synthesis*, Astrophys. J. **572**, 1048-1071.

Aschwanden, M.J. and Charbonneau P. 2002, *Effects of temperature bias on nanoflare statistics*, Astrophys. J. **566**, L59-L62.

Aschwanden, M.J. 2004, *Physics of the Solar Corona. An Introduction*, PRAXIS, Chichester, UK, and Springer, Berlin, 842p.

Aschwanden, M.J. 2007, *From solar nanoflare to stellar giant flares – Scaling laws and non-implications for coronal heating*, Adv. Space Res. **39**, 1867-1875.

Aschwanden, M.J., Winebarger, A., Tsiklauri, D., and Peter, H. 2007, *The coronal heating paradox*, Astrophys. J. **659**, 1673-1681.

Aschwanden, M.J. and Aschwanden P.D. 2008a, *Solar flare geometries: I. The area fractal dimension*, Astrophys. J. **574**, 530-543.

Aschwanden, M.J. and Aschwanden P.D. 2008b, *Solar flare geometries: II. The volume fractal dimension*, Astrophys. J. **574**, 544-553.

Aschwanden, M.J., Stern, R.A., and Güdel, M. 2008c, *Scaling laws of solar and stellar flares*, Astrophys. J. **672**, 659-673.

Aschwanden, M.J. 2009, *Image processing techniques and feature recognition in solar physics*, Solar Phys. **262**, 235-275.

Aschwanden, M.J. and McTiernan, J.M. 2010, *Reconciliation of waiting time statistics of solar flares observed in hard X-rays*, Astrophysical J. **717**, 683-692.

Audard, M., Güdel, M., and Guinan, E.F. 1999, *Implications from extreme-ultraviolet observations for coronal heating of active stars*, Astrophys. J. **513**, L53-L56.

Audard, M., Güdel, M., Drake, J.J., and Kashyap, V.L. 2000, *Extreme-ultraviolet flare activity in late-type stars*, Astrophys. J. **541**, 396-409.

Audard, M., Güdel, M., and Skinner, S.L. 2003, *Separating the X-ray emissions of UV Ceti A and B with Chandra*, Astrophys. J. **589**, 983-987.

Babcock, K.L. and Westervelt, R.M. 1990, *Avalanches and self-organization in cellular magnetic-domain patterns*, Phys. Rev. Lett. **64/18**, 2168-2171.

Bai, T. 1993, *Variability of the occurrence frequency of solar flares as a function of peak hard X-ray rate*, Astrophys. J. **404**, 805-809.

Baiesi, M., Paczuski, M., and Stella, A.L. 2006, *Intensity thresholds and the statistics of the temporal occurrence of solar flares*, Phys. Rev. Lett. **96/5**, 051103.

Bak, P., Tang, C., and Wiesenfeld, K. 1987, *Self-organized criticality: An explanation of 1/f noise*, Physical Review Lett. **59/27**, 381-384.

Bak, P., Tang, C., and Wiesenfeld, K. 1988, *Self-organized criticality*, Physical Rev. A **38/1**, 364-374.

Bak, P., and Tang, C. 1989, *Earthquakes as a self-organized critical phenomena*, J. Geophys. Res. **94/11**, 15,635-15,637.

Bak, P. and Chen, K. 1989, *The physics of fractals*, Physica D **38**, 5-12.

Bak, P., Chen, K. and Creutz, M. 1989, *Self-organized criticality in the "Game of Life"*, Nature **342**, 780-781.

Bak, P., Chen, K., and Tang, C. 1990, *A forest-fire model and some thoughts on turbulence*, Phys. Lett A **147/5-6**, 297-300.

Bak, P. and Chen, K. 1991, *Self-organized criticality*, Scientific American, **264**, January 1991 issue, p.46-53.

Bak, P. and Sneppen, K. 1993, *Punctuated equilibrium and criticality in a simple model of evolution*, Phys. Rev. Lett. **71/24**, 4083-4086.

Bak, P., Chen, K., Scheinkman, J.A., and Woodford, M.A. 1993, *Aggregate fluctuations from independent sectoral shocks: Self-organized criticality in a model of production and inventory dynamics*, Ricerche Economichi **47**, 3-30.

Bak, P. and Chen, K. 1995, *Fractal in the Earth Sciences*, (ed. C. Barton and P. La Pointe), Plenum: New York, pp. 227-236.

Bak, P. and Paczuski, M. 1995, *Complexity, contingency, and criticality*, Proc. National Academy of Science, USA **92**, 6689-6696.

Bak, P. 1996, *How nature works*, Copernicus, Springer-Verlag, New York.

Bak, P., Paczuski, M., anbd Shubik, M. 1997, *Price variations in a stock market with many agents*, Physica A **246**, 430-453.

Bak, P., Christensen, K., Danon, L., and Scanlon, T. 2002, *Unified scaling law for earthquakes*, Phys. Rev. Lett. **88/17**, 178,501:1-4.

Baker, D.N., Klimas, A.J., McPherron, R.L., and Buechner, J. 1990, *The evolution from weak to strong geomagnetic activity – An interpretation in terms of deterministic chaos*, Geophys. Res. Lett. **17**, 41-44.

Balbus, S.A. and Hawley, J.F. 1991, *A powerful local shear instability in weakly magnetized disks. I. Linear analysis*, Astrophys. J. **376**, 214-233.

Balke,A.C., Schrijver, C.J., Zwaan,C., and Tarbell,T.D. 1993, *Percolation theory and the geometry of photospheric magnetic flux concentrations*, Solar Phys. **143**, 215-227.

Barabasi, A.L., Buldyrev, S.V., Stanley, H.E., and Suki, B. 1996, *Avalanches in the Lung: a statistical mechanical model*, Phys. Rev. Lett. **76**, 2192-2195.

Bargatze, L.F., Baker, D.N., Hones, E.W.,Jr., and McPherron, R.L. 1985, *Magnetospheric impulse response for many levels of geomagnetic activity*, J. Geophys. Res. **90**, 6387-6394.

Barnsley, M. 1988, *Fractals everywhere*, Academic Press, Inc., New York.

Bartolozzi, M., Leinweber, D.B., and Thomas, A.W. 2005, *Self-organized criticality and stock market dynamics: an empirical study*, Physica A **350**, 451-465.

Baryshev, Y. and Teerikorpi, P. 2002, *Discovery of cosmic fractals*, World Scientific Publishing, Co. Pte. Ltd, Singapore.

Bassler, I.E. and Paszuski, M. 1998, *Simple model of superconducting vortex avalanches*, Phys. Rev. Lett. **81**, 3761-3764.

Bastian, T.S. and Vlahos, L. 1997, *Energy release in the solar corona*, Lecture Notes in Physics **483**, (ed. G.Trottet, Springer: Berlin), p. 68-92.

Baumann, I. and Solanki, S.K. 2005, *On the size distribution of sunspot groups in the Greenwich sunspot record 1874-1976*, Astron. Astrophys. **443**, 1061-1066.

Baumjohann, W. and Treuman, R.A. 1996, *Basic Space Plasma Physics*, Imperial College Press: London.

Belanger, E., Vincent, A., and Charbonneau, P. 2007, *Predicting Solar Flares by Data Assimilation in Avalanche Models. I. Model Design and Validation*, Solar Phys. **245**, 141-165.

Belovsky, M.N., and Ochelkov, Yu. P. 1979, *Some features of solar-flare electromagnetic and corpuscular radiation production*, Izvestiya AN SSR, Phys. Ser. **43**, 749-752.

Beltrami, E. 1987, *Mathematics for dynamic modeling*, Academic Press, Inc.: Boston.

Benz, A.O. 1993, *Plasma astrophysics, kinetic processes in solar and stellar coronae*, Kluwer Academic Publishers, Dordrecht, The Netherlands.

Benz, A.O., Krucker, S., Acton, L.W., and Bastian, T.S. 1997, *Fine structure of the X-ray and radio emissions of the quiet solar corona*, Astron. Astrophys. **320**, 993-1000.

Benz, A.O. and Krucker, S. 2002, *Energy distribution of microevents in the quiet solar corona*, Astrophys. J. **568**, 413-421.

Berger, M.A. 1991, *Generation of coronal magnetic fields by random surface motions. I. Mean square twist and current density*, Astron. Astrophys. **252**, 369-376.

Berger, M.A. 1993, *Energy-crossing number relations for braided magnetic fields*, Phys. Rev. Lett. **70/6**, 705-708.

Berger, M.A. and Asgari-Targhi, M. 2009, *Self-organized braiding and the structure of coronal loops*, Astrophys. J. **705**, 347-355.

Berghmans, D., Clette, F., and Moses, D. 1998, *Quiet Sun EUV transient brightenings and turbulence. A panoramic view by EIT on board SOHO*, Astron. Astrophys. **336**, 1039-1055.

Bernardes, A.T., and Moreira, J.G. 1995, *Self-organized criticality in a model for fracture on fibrous materials*, J. Physique I **5/9**, 1135-1141.

Berrilli, F., Del Moro, D., Russo, S., Consolini, G., and Straus, Th. 2005, *Spatial clustering of photospheric structures*, Astrophys. J. **632**, 677-683.

Bevington, P. R. and Robinson, D. K. 1969 (2007 2nd ed.), *Data reduction and error analysis for the physical sciences*, McGraw-Hill, 2nd edition.

Biesecker, D.A., Ryan, J.M., and Fishman, G.J. 1993, *A search for small solar flares with BATSE*, Lecture Notes in Physics **432**, 225-230.

Biesecker, D.A. 1994, *On the occurrence of solar flares observed with the Burst and Transient Source Experiment (BATSE)*, PhD Thesis, University of New Hampshire.

Biesecker, D.A., Ryan, J.M., and Fishman, G.J. 1994, *Observations of small small solar flares with BATSE*, in *High-Energy Solar Phenomena – A New Era of Spacecraft Measurements*, (eds. J.M. Ryan and W.T.Vestrand), American Inst. Physics: New York, p.183-186.

Biesecker, D.A. and Thompson, B.J. 2000, *Sympathetic flaring with BATSE, GOES, and EIT data*, J. Atmos. Solar-Terr. Phys. **62/16**, 1449-1455.

Birkeland, K.W. and Landry, C.C. 2002, *Power-laws of snow avalanches*, Geophys. Res. Lett. **29/11**, 49.

Bocchialini, K., and Baudin, F. 1995, *Wavelet analysis of chromospheric solar oscillations*, Astron. Astrophys. **299**, 893-896.

Boffetta, G., Carbone, V., Giuliani, P., Veltri, P., and Vulpiani, A. 1999, *Power laws in solar flares: self-organized criticality or turbulence*, Phys. Rev. Lett. **83**/2, 4662-4665.

Bornmann, P.L. 1990, *Limits to derived flare properties using estimates for the background fluxes: examples from GOES*, Astrophys. J. **356**, 733-742.

Borovsky, J.E., Nemzek, R.J., and Belian, R.D. 1993, *The occurrence rate of magnetospheric-substorm onsets: Random and periodic substorms*, J. Geophys. Res. **98**, 3807-3813.

Borovsky, J.E. and Steinberg, J.T. 2006, *The "calm before the storm" in CIR/magnetosphere interactions: Occurrence statistics, solar wind statistis, and magnetospheric preconditioning*, J. Geophys. Res. **111.A7**, CiteID A07S10.

Bortkiewicz, L.J. 1898, *The Law of Small Numbers*, in German *Das Gesetz der kleinen Zahlen*.

Böttcher, S. and Paczuski, M. 1996, *Exact results for spatiotemporal correlations in a self-organized critical model of punctuated equilibrium*, Phys. Rev. Lett. **76/3**, 348-351.

Böttcher, S. and Paczuski, M. 1997, *Aging in a model of self-organized criticality*, Phys. Rev. Lett. **79/5**, 889-892.

Bovelet, B. and Wiehr, E. 2001, *A new algorithm for pattern recognition and its application to granulation and limb faculae*, Solar Phys. **201**, 13-26.

Boyd, P.T., Mindlin, G.B., Gilmore, R., and Solari, H.G. 1994, *Topological analysis of chaotic orbits: revisiting Hyperion*, Astrophys. J. **431**, 425-431.

Brault, J.W. and White, O.R. 1971, *The analysis and restoration of astronomical data via the Fast Fourier Transform*, Astron. Astrophys. **13**, 169-189.

Bretz, M., Cunningham, J.B., Kurczynsky, P.L., and Nori,E. 1992, *Imaging of avalanches in granular materials*, Phys.Rev.Lett. **69**, 2431-2434.

Bristow, W. 2008, *Statistics of velocity fluctuations observed by SuperDARN under steady interplanetary magnetic field conditions*, J. Geophys. Res. **113**, CiteID:A11202.

Bromund, K.R., McTiernan, J.M., and Kane, S.R. 1995, *Statistical studies of ISEE3/ICE observations of impulsive hard X-ray solar flares*, Astrophys. J., **455**, 733-745.

Brown, J.C. 1971, *The deduction of energy spectra of non-thermal electrons in flares from the observed dynamic spectra of Hard X-Ray bursts*, Solar Phys. **18**, 489-502.

Brown, S.R., Scholz, C.H., and Rundle, J.B. 1991, *A simplified spring-block model of earthquakes*, Geophys. Res. Lett. **18**, 215-218.

Bruch, A. 1992, *Flickering in cataclysmic variables: its properties and origins*, Astron. Astrophys. **266**, 237-265.

Bruch, A. 1995, *Flickering in cataclysmic variables: inventory and perspectives*, Lecture Notes in Physics **454**, 288-295.

Bruno, R., D'Amicis, R., Bavassano, B., Carbone, V., and Sorriso-Valvo, L. 2007, *Scaling laws and coherent structures in the solar wind*, Planet. Space Sci. **55**, 2233-2238.

Buchler, J.R., Kollath, Z., Serre, T., and Mattei, J. 1996, *Nonlinear analysis of the light curve of the variable star R Scuti*, Astrophys. J. **462**, 489-501.

Buchlin, E., Galtier, S., and Velli, M. 2005, *Influence of the definition of dissipative events on their statistics*, Astron. Astrophys. **436**, 355-362.

Buchlin, E., Vial, J.C., and Lemaire, P. 2006, *A statistical study of SUMER spectral images: events, turbulence, and intermittency*, Astron. Astrophys. **451**, 1091-1099.

Bumba, V. and Klvana, M. 1993, *Questions concerning the existence of sympathetic flaring*, Solar Phys. **199**, 45-52.

Burlaga, L.F., and Lazarus, A.J. 2000, *Lognormal distributions and spectra of solar wind plasma fluctuations: Wind 1995-1998.*, J. Geophys. Res. **105/A2**, 2357-2364.

Burridge R. and Knopoff L. 1967, *Model and theoretical seismicity*, Seis. Soc. Am. Bull. **57**, 341-347.

Butler, C.J., Rodono, M., Foing, B.H., and Haisch, B.M. 1986, *Coordinated EXOSAT and spectroscopic observations of flare stars and coronal heating*, Nature **321**, 679-682.

Cadavid, A.C., Lawrence, J.K., Ruzmaikin, A., and Kayleng-Knight, A. 1994, *Multifractal models of small-scale magnetic fields*, Astrophys. J. **429**, 391-399.

Cadavid, A.C., Lawrence, J. K., and Ruzmaikin, A. 2008, *Principal Components and Independent Component Analysis of Solar and Space Data*, Solar Phys. **248**, 247-261.

Cairns, I.H. and Robinson, P.A. 1999, *Strong Evidence for Stochastic Growth of Langmuir-like Waves in Earth's Foreshock*, Physical Review Letters **82**, 3066-3069.

Cairns, I.H. 2004, *Properties and interpretations of giant micropulses and giant pulses from pulsars*, Astrophys. J. **610**, 948-955.

Cairns, I.H., Johnston, S., and Das, P. 2004, *Intrinsic variability and field statistics for pulsars B1641-45 and B0950-08*, MNRAS **353**, 270-286.

Caldarelli, G., Di Tolla, F.D., and Petri, G.A. 1996, *Self-organization and annealed disorder in a fracturing process*, Phys. Rev. Lett. **77**, 2503-2506.

Camazine, S., Deneubourg, J.L., Frank, N.R., Sneyd, J., Theraulaz, G., and Gonabeau, E. 2001, *Self-organization in biological systems*, Princeton University Press, Princeton.

Cannizzo, J.K., Goodings, D.A., and Mattei, J.A. 1990, *A search for chaotic behavior in the light curves of three long-term variables*, Astrophys. J. **357**, 235-242.

Carbone, V., Roberto, B., Sorriso-Valvo, L., and Lepreti F. 2004, *Intermittency of magnetic turbulence in slow solar wind*, Planet. Space Sci. **52**, 953-956.

Carbone, V., Sorriso-Valvo, L., Harabaglia, P., and Guerra, I. 2005, *Unified scaling law for waiting times between seismic events*, Europhysics Lett. **71/6**, 1036-1042.

Carlson, J.M., and Langer, J.S. 1989a, *Mechanical model of an earthquake fault*, Physical Review A – General Physics **40**, 6470-6484.

Carlson, J.M., and Langer, J.S. 1989b, *Properties of earthquakes generated by fault dynamics*, Phys. Rev. Lett. **62**, 2632-2635.

Carlson, J.M., Chayes, J.T., Grannan, E.R., and Swindle, G.M. 1990, *Self-organized criticality in sandpiles: Nature of the critical phenomenon*, Phys. Rev. A **42**/4, 2467-2470.

Carlson, J.M. 1991a, *The intervals between characteristic earthquakes and correlations with smaller events: An analysis based on a mechanical model of a fault*, J. Geophys. Res. **96/B3**, 4255-4267.

Carlson, J.M. 1991b, *Two-dimensional model of a fault*, Physical Rev. A **44/10**, 6626-6232.

Carlson, J.M., Langer, J.S., Shaw, B.E., and Tang, C. 1991, *Intrinsic properties of a Burridge-Knopoff model of an earthquake fault*, Physical Rev. A **44/2**, 884-897.

Carlson, J.M., Langer, J.S., and Shaw, B.E. 1994, *Dynamics of earthquake faults*, Rev. Mod. Phys. **66**, 657-670.

Carreras, B.A., Newman, D., Lynch, V.E., and Diamond, P.H., 1996, *A model realization of self-organized criticality for plasma confinement*, Phys. Plasmas **3/8**, 2903-2911.

Cassak, P.A., Mullan, D.J., and Shay, M.A. 2008, *From solar and stellar flares to coronal heating: Theory and observations of how magnetic reconnection regulates coronal conditions*, Astrophys. J. **676**, L69-L72.

Castleman, K.R. 1996, *Digital Image Processing*, Prentice Hall, Upper Saddle River, New Jersey.

Chae, J., Schuehle, U. and Lemaire, P. 1998, *SUMER measurements of nonthermal motions: constraints on coronal heating mechanisms*, Astrophys. J. **505**, 957-973.

Challet, D. and Zhang, Y.C. 1997, *Emergence of cooperation and organization in an evolutionary game*, Physica A **246/3-4**, 407-418.

Chamel, N. and Haensel, P. 2008, *Physics of neutron star crusts*, Living Reviews in Relativity **11/10**, 1-182.

Chan, Y.T. 1995, *Wavelet basics*, Kluwer Academic Publishers, Dordrecht, The Netherlands.

Chang, H.K., Chen, K., Fenimore, E.E., and Ho, C. 1996, *Spectral studies of magnetic photon splitting in the March 5 event and SGR 1806-20*, AIP Conf. Proc. **384**,, 921-925.

Chang, S. 1988, *Planetary environment and the conditions of life*, Philosophical Transactions of the Royal Society of London, Series A, Math. and Phys. Sci. **325**, 601-610.

Chang, T.S. 1992, *Low-dimensional behavior and symmetry breaking of stochastic systems near criticality – Can these effects be observed in space and in the laboratory*, IEEE Trans. Plasma Sci. **20/6**, 691-694.

Chang, T.S. 1998a, *Sporadic, Localized reconnections and mnultiscale intermittent turbulence in the magnetotail*, in *Geospace Mass and Energy Flow* (eds. Horwitz, J.L., Gallagher, D.L., and Peterson, W.K.), AGU Geophysical Monograph **104**, p.193.

Chang, T.S. 1998b, *Multiscale intermittent turbulence in the magnetotail*, in *Proc. 4th Intern. Conf. on Substorms*, (eds. Kamide, Y. et al.), Kluwer Academic Publishers, Dordrecht, and Terra Scientific Company, Tokyo, p.431.

Chang, T.S. 1999a, *Self-organized criticality, multi-fractal spectra, and intermittent merging of coherent structures in the magnetotail*, Astrophys. Space Sci. **264**, 303-316.

Chang, T.S. 1999b, *Self-organized criticality, multi-fractal spectra, sporadic localized reconnections and intermittent turbulence in the magnetotail*, Phys. Plasmas **6/11**, 4137-4145.

Chang, T.S., Tam, S.W.Y., Wu, C.C., and Consolini, G. 2003, *Complexity, forced and/or self-organized criticality and topological phase transitions in space plasmas*, Space Sci. Rev. **107**, 425-445.

Chapman, S.C., Watkins, N.W., Dendy, R.O., Helander, P., and Rowlands, G. 1998, *A simple avalanche model as an analogue for magnetospheric activity*, Geophys. Res. Lett. **25/13**, 2397-2400.

Chapman, S.C., Dendy, R.O., and Rowlands, G. 1999, *A sandpile model with dual scaling for laboratory, space and astrophysical plasmas*, Phys. Plasmas **6/11**, 4169-4177.

Chapman, S.C. and Watkins, N. 2001, *Avalanching and self-organised criticality, a paradigm for geomagnetic activity?*, Space Sci. Rev. **95**, 293-307.

Chapman, S.C., Watkins, N., and Rowlands, G. 2001, *Signatures of dual scaling regimes in a simple avalanche model for magnetospheric activity*, J. Atmos. Solar-Terr. Phys. **63**, 1361-1370.

Chapman, S.C. and Nicol, R.M. 2009, *Generalized similarity in finite range solar wind magnetohydrodynamic turbulence*, Phys. Rev. Lett. **103/24**, CiteID 241101.

Charbonneau, P., McIntosh, S.W., Liu, H.L., and Bogdan, T.J. 2001, *Avalanche models for solar flares*, Solar Phys. **203**, 321-353.

Charbonneau, P. 2001, *Multiperiodicity, Chaos, and Intermittency in a Reduced Model of the Solar Cycle*, Solar Phys. **199**, 385-404.

Charbonneau, P., Blais-Laurier, G., and St-Jean, C. 2004, *Intermittency and phase persistence in a Babcock-Leighton model of the solar cycle*, Astrophys. J. **616**, L183-186.

Charbonneau, P., Joseph, R., and Pirot, D. 2007, *Deterministically-driven avalanche models of solar flares*, Solar Phys. (subm.).

Chatterjee, T.N. 1999, *On the existence of a low-dimensional chaotic attractor in the short-term solar UV time series*, Solar Phys. **186**, 421-429.

Che, X. and Suhl, H., 1990, *Magnetic domain pattern as self-organizing critical systems*, Phys. Rev. Lett. **64/14**, 1670-1673.

Chen, K., Bak, P. and Obukhov, S.P. 1991, *Self-organized criticality in a crack-propagation model of earthquakes*, Phys. Rev. A **43**, 625-630.

Chepurnov, A. and Lazarian, A. 2010, *Extending the big power law in the sky with turbulence spectra from Wisconsin Hα mapper data*, Astrophys. J. **710**, 853-858.

Chessa, A., Stanley, H.E., Bespignani, A., and Zapperi, S. 1999, *Universality in sandpiles*, Phys. Rev. E **59/1**, R12-R15.

Chialvo, D.R. and Bak, P. 1999, *Learning from mistakes*, Neuroscience **90**, 1137-1148.

Chian, A.C., Rempel, E.L., and Borotto, F.A. 2002, *Nonlinear dynamics of Alfven waves in the solar atmosphere*, in *Magnetic Coupling of the Solar Atmosphere*, Proc. Euroconference and IAU Colloquium 188, European Space Agency (ESA) Special Publication **SP-505**, (ed. Huguette Sawaya-Lacoste), ESTEC Noordwijk, Netherlands, p.301-304.

Chou, Y.P. 1999, *What affects the power-law distribution of the X-ray solar flares? A theoretical study based on a model of uniform normal field*, Astrophys. J. **527**, 958-966.

Chou, Y.P. 2001, *The Effect of Helicity Dissipation on the Critical State of an Avalanche Model for Solar Flares*, Solar Phys. **199**, 345-369.

Christe, S., Hannah, I.G., Krucker, S., McTiernan, J., and Lin, R.P. 2008, *RHESSI microflare statistics. I. Flare-finding and frequency distributions*, Astrophys. J. **677**, 1385-1394.

Christensen, K. and Olami, Z. 1992a, *Variation of the Gutenberg-Richter b values and nontrivial temporal correlations in a spring-block model for earthquakes*, J. Geophys. Res. **97/B6**, 8729-8735.

Christensen, K. and Olami, Z. 1992b, *Scaling, phase transitions, and nonuniversality in a self-organized critical cellular-automaton model*, Physical Rev. A **46/4**, 1829-1838.

Christensen, K., Olami, Z., and Bak, P. 1992, *Deterministic 1/f noise in nonconservative models of self-organized criticality*, Phys. Rev. Lett. **68/16**, 2417-2420.

Christensen, K. and Olami, Z. 1993, *Sandpile models with and without an underlying spatial structure*, Phys. Rev. E **48/5**, 3361-3372.

Christensen, K., Flyvbjerg, H., and Olami, Z. 1993, *Self-organized critical forest-fire model: mean-field theory and simulation results in 1 to 6 dimensions*, Phys. Rev. Lett. **71/17**, 2737-2740.

Chumak, O.V. and Zhang, H.Q. 2003, *Size-flux relation in solar active regions*, Chin. J. Astron. Astrophys. **3/2**, 175-182.

Ciliberto, S. and Laroche, C. 1994, *Experimental evidence of self-organized criticality in the stick-slip dynamics of two rough elastic surfaces*, J. Physique I **4/2**, 223-235.

Ciprini, S., Fiorucci, M., Tosti, G., and Marchili, N. 2003, *The optical variability of the blazar GV 0109+224. Hints of self-organized criticality*, in *High energy blazar astronomy*, ASP Conf. Proc. **229**, (eds. L.O. Takalo and E. Valtaoja), ASP: San Francisco, p.265.

Clar, S., Drossel, B., and Schwabl, F. 1994, *Scaling laws and simulation results for the self-organized critical forest-fire model*, Phys. Rev. E **50/2**, 1009-1018.

Clar, S., Drossel, B., and Schwabl, F. 1996, *Review article: Forest fires and other examples of self-organized criticality*, Journal of Physics: Condensed Matter **8/37**, 6803-6824.

Clar, S., Drossel, B., Schenk, K., and Schwabl, F. 1999, *Self-organized criticality in forest-fire models*, Physica A **266**, 153-159.

Cliver, E., Reames, D., Kahler, S., and Cane, H. 1991, *Size distribution of solar energetic particle events*, Internat. Cosmic Ray Conf. 22nd, Dublin, LEAC A92-36806 15-93, NASA:Greenbelt, p. 2:1-4.

Cognard, I., Shrauner, J.A., Taylor, J.H., and Thorsett, S.E. 1996, *Giant radio pulses from a millisecond pulsar*, Astrophys. J. **457**, L81-L84.

Colless, M., Dalton,G., Maddox,S., Sutherland, W., Norberg, P., Cole,S., Bland-Hawthorn,J., Bridges, T., et al. 2001, MNRAS **328/4**, 1039-1063.

Collura, A, Pasquini, L., Schmitt, J.H.M.M. 1988, *Time variability in the X-ray emission of dM stars observed by EXOSAT*, Astron. Astrophys. **205**, 197-206.

Conlon, P.A., Gallagher, P.T., McAteer, R.T.J., Ireland, J., Young, C.A., Kestener, P., Hewett, R.J., and Maguire, K. 2008, *Multifractal properties of evolving active regions*, Solar Phys. **248**, 297-309.

Consolini, G., Marcucci, M.F., and Candidi, M. 1996, *Multifractal structure of auroral electrojet data*, Phys. Rev. Lett. **76**, 4082-4085.

Consolini, G. 1997, *Sandpile cellular automata and magnetospheric dynamics*, in (Proc, Cosmic Physics in the year 2000), (eds. S.Aiello, N.Iucci, G.Sironi, A.Treves, and U.Villante), SIF: Bologna, Italy, Vol. **58**, 123-126.

Consolini, G. and Lui, A.T.Y. 1999, *Sign-singularity analysis of current disruption*, Geophys. Res. Lett. **26/12**, 1673-1676.

Consolini, G., and De Michelis, P. 2001, *A revised forest-fire cellular automaton for the nonlinear dynamics of the Earth's magnetotail*, J. Atmos. Solar-Terr. Phys. **63/13**, 1371-1377.

Consolini, G. and Chang, T.S. 2001, *Magnetic field topology and criticality in geotail dynamics: Relevance to substorm phenomena*, Space Sci. Rev. **95**, 309-321.

Consolini, G. 2002, *Self-organized criticality: A new paradigm for the magnetotail dynamics*, Fractals **10**, 275-283.

Consolini, G., and De Michelis, P. 2002, *Fractal time statistics of AE-index burst waiting times: evidence of metastability*, Nonlinear Proc. Geophys. **9**, 419-423.

Cont, R. and Bouchard, J.P. 2000, *Herd behavior and aggregate fluctuations in speculative markets*, Macroeconom. Dyn. **4**, 170-196.

Corral, A., Perez, C.J., Diaz-Guilera, A. and Arenas A. 1995, *Self-organized criticality and synchronization in a lattice model of integrate-and-fire oscillators*, Phys. Rev. Lett. **74**, 118-121.

Cote, P.J. and Meisel, L.V. 1991, *Self-organized criticality and the Barkhausen effect*, Phys. Rev. Lett. **67**, 1334-1337.

Cottrell, A.H. 1996, *Strain hardening in Andrade creep*, Phil. Mag. A. **74**, 375-379.

Cowie, P.A., Banneste, C., and Sornette, D. 1993, *Statistical physics model for the spatiotemporal evolution of faults*, J. Geophys. Res. **98/B12**, 21809-21821.

Cox, D. and Isham, V. 1980, *Point Processes*, London: Chapman and Hall.

Craig, I.J.D. and Wheatland, M.S. 2002, *Interpretation of Statistical Flare Data using Magnetic Reconnection Models*, Solar Phys. **211**, 275-287.

Crosby, N.B., Aschwanden, M.J., and Dennis, B.R. 1993, *Frequency distributions and correlations of solar X-ray flare parameters*, Solar Phys. **143**, 275-299.

Crosby, N.B. 1996, *Contribution à l'Etude des Phénomènes Eruptifs du Soleil en Rayons Z à partir des Observations de l'Expérience WATCH sur le Satellite Granat*, PhD Thesis, University Paris VII, Meudon, Paris, 348 p.

Crosby, N.B., Vilmer, N., Lund, N., and Sunyaev, R. 1998, *Deka-keV X-ray observations of solar bursts with WATCH/Granat: frequency distributions of burst parameters*, Astrophys. J. **334**, 299-313.

Crosby, N.B., Georgoulis, M., and Vilmer, N. 1999, *A comparison between the WATCH flare data statistical properties and predictions of the statistical flare model*, in *Plasma dynamics in the solar transition region and corona*, Proc. 8th SoHO Workshop (eds. J.C.Vial and B.Kaldeich-Schuermann), European Space Agency (ESA) **SP-446**, ESTEC Noordwijk, Netherlands, p.247-250.

Crosby, N.B., Meredith, N.P., Coates, A.J., and Iles, R.H.A. 2005, *Modelling the outer radiation belt as a complex system in a self-organised critical state*, Nonlinear Processes in Geophysics **12**, 993-1001.

Cross, C.A. 1966, *The size distribution of lunar craters*, MNRAS **134**, 245-252.

Crow, E.L. and Shimizu, K. 1998, *Lognormal Distributions: Theory and Applications*, New York: Marcel Dekker Inc.

Da Rocha, D. and Nottale, L. 2003, *Gravitational structure in scale relativity*, Chaos, Solitons and Fractals **16/4**, 565-595.

Das, T.K., Tarafdar, G., and Sen, A.K. 1997, *Validity of power law for the distribution of intensity of radio bursts*, Solar Phys. **176**, 181-184.

Das, T.K., De, B.K., and Bhattacharyya, J. 2004, *Different distribution functions of solar x-ray flares*, Bull. Astr. Soc. India **32**, 15-23.

da Silva, L., Papa, A.R.R. and de Souza A.M.C. 1998, *Criticality in a simple model for brain functioning*, Phys. Lett. A **242**, 343-348.

Datlowe, D.W., Elcan, M.J., and Hudson, H.S. 1974, *OSO-7 observations of solar X-rays in the energy range 10-100 keV*, Solar Phys. **39**, 155-174.

Daubechies, I. 1992, *Ten lectures on wavelets*, Soc. Industr. Appl. Math., Philadelphia, Vol. **61**, 357p.

Dauphin, C. 2007, *Particle acceleration in solar flares: linking magnetic energy release with the acceleration process*, Astron. Astrophys. **471**, 993-998.

Davidsen, J. and Goltz, C. 2004, *Are seismic waiting time distributions universal?*, Geophys. Res. Lett. **31/21**, CiteID L21612.

Deb, S. and Singh, H.P. 2009, *Light curve analysis of variable stars using Fourier decomposition and principal component analysis*, Astron. Astrophys. **507**, 1729-1737.

de Boer, J., Derrida, B., Glyvbjerg, H., Jackson, A.D., and Wettig, T. 1994, *Simple model of self-organized biological evolution*, Phys. Rev. Lett. **73/6**, 906-909.

de Boer, J., Jackson, A.D., and Wettig, T. 1995, Phys. Rev. E **51/2**, 1059-1074.

Dendy, R.O. and Helander, P. 1997, *Sandpiles, silos and tokamak phenomenology: a brief review*, Plasma Phys. Control. Fusion **39**, 1947-1961.

Dendy, R.O., Helander, P., and Tagger, M. 1998, *On the role of self-organised criticality in accretion systems*, Astron. Astrophys. **337**, 962-965.

Dennis, B.R. 1985, *Solar hard X-ray bursts*, Solar Phys. **100**, 465-490.

Dennis, B.R. and Zarro, D.M. 1993, *The Neupert effect: what can it tell us about the impulsive and gradual phases of solar flares*, Solar Phys. **146**, 177-190.

de Sousa Vieira, M., Vasconcelos, G.L., and Nagel, S.R. 1993, *Dynamics of spring-block models: Tuning to criticality*, Phys. Rev. E **47/4**, R2221-R2224.

Dhar, D. and Ramaswamy, R. 1989, *Exactly solved model of self-organized critical phenomena*, Phys. Rev. Lett. **63/16**, 1659-1662.

Dhar, D. and Majumdar, S.N. 1990, *Abelian sandpile model of the Bethe lattice*, J. Physics A **23/19**, 4333-4350.

Dhar, D. 1999, *The Abelian sandpile and related models*, Physica A **263**, 4-25.

Dickman, R., Vespignani, A., and Zapperi, S. 1998, *Self-organized criticality as an absorbing-state phase transition*, Phys. Rev. Lett. **57**/5, 5095-5105.

Diodati, P., Marchesoni, E., and Piazza, S. 1991, *Acoustic emission from volcanic rocks: an example of self-organized criticality*, Phys. Rev. Lett. **67**, 2239-2243.

Diodati, P., Bak, P., and Marchesoni,E. 2000, *Acoustic emission at the Stromboli volcano: scaling laws and seismic activity*, Earth Planet. Sci. Lett. **182**, 253-258.

Dmitruk, P. and Gomez, D.O. 1997, *Turbulent coronal heating and the distribution of nanoflares*, Astrophys. J. **484**, L83-L85.

Dmitruk, P., Gomez, D.O., and DeLuca, E.E. 1998, *Magnetohydrodynamic turbulence of coronal active regions and the distribution of nanoflares*, Astrophys. J. **505**, 974-983.

Dmitruk, P. and Gomez, D.O. 1999, *Scaling law for the heating of solar coronal loops*, Astrophys. J. **527**, L63-L66.

Dmitruk, P., Matthaeus, W.H., Seenu, N., and Brown, M.R. 2003, *Test particle acceleration in three-dimensional magnetohydrodynamic turbulence*, Astrophys. J. **597**, L81-L84.

Dmitruk, P. and Matthaeus, W.H. 2007, *Low-frequency 1/f fluctuations in hydrodynamic and magnetohydrodynamic turbulence*, Phys. Rev. E **76/3**, id. 036305.

Dominguez Cerdena, I., and Sanchez Almeida J. 2006, *The distribution of quiet Sun magnetic field strengths from 0 to 1800 G*, Astrophys. J. **636**, 496-509.

Drake, J.F. 1971, *Characteristics of soft solar X-ray bursts*, Solar Phys. **16**, 152-185.

Drossel, B. and Schwabl, F. 1992a, *Self-organized critical forest-fire model*, Phys. Rev. Lett. **69**, 1629-1632.

Drossel, B. and Schwabl, F. 1992b, *Self-organized criticality in a forest-fire model*, Physica A **191**, 47-50.

Drossel, B., Clar, S., and Schwabl, F. 1993, *Exact results for the one-dimensional self-organized critical forest-fire model*, Phys. Rev. Lett. **71**/23, 3739-3742.

Drossel, B. and Schwabl, F. 1995, *Self-organized critical limit of autocatalytic surface reactions*, Appl. Phys. **60**, 597-600.

Dulk, G.A. 1985, *Radio emission from the sun and stars*, Annual Reviews Astron. Astrophys. **23**, 169-224.

Duncan, R.C. and Thompson, C. 1992, *Formation of very strongly magnetized neutron stars: implications for gamma-ray bursts*, Astrophys. J. **392**, L9-L13.

Eastwood, J.P., Wheatland, M.S., Hudson, H.S., Krucker, S., Bale, S.D., Maksimovic, M., Goetz, K. 2010, *On the brightness and waiting-time distributions of a type III radio storm observed by STEREO/WAVES*, Astrophys. J. **708**, L95-L99.

Eggen, O.J., Lynden-Bell, D., and Sandage, A.R. 1962, *Evidence from the motions of old stars that the galaxy collapsed*, Astrophys. J. **136**, 748-766.

Einaudi, G., Velli, M., Politano, H., and Pouquet, A. 1996a, *Energy release in a turbulent corona*, Astrophys. J. **457**, L113-L116.

Einaudi, G., Califano, F., and Chiuderi, C. 1996b, *Induced deposition of magnetic energy in the solar corona*, Astrophys. J. **472**, 853-863.

Einaudi, G. and Velli M. 1999, *The distribution of flares, statistics of magnetohydrodynamic turbulence and coronal heating*, Phys. Plasmas **6/11**, 4146-4153.

Feder, J. 1988, *Fractals*, Plenum Press: New York, 283 p.

Feder, H.J.S. and Feder, J. 1991, *Self-organized criticality in a stick-slip process*, Phys. Rev. Lett. **66/20**, 2669-2672.

Feigenbaum, J.A. and Freund P.G.O. 1996, *Discrete scale invariance in stock markets before crashes*, Int. J. Mod. Phys. B **10/27**, 3737-3745.

Feigenbaum, J.A. 2003, *Financial physics*, Rep. Prog. Phys. **66**, 1611-1649.

Feldman, U., Doschek, G.A., and Klimchuk, J.A. 1997, *The occurrence rate of soft X-ray flares as a function of solar activity*, Astrophys. J. **474**, 511-517.

Fermi, E. 1949, *On the origin of the cosmic radiation*, Phys. Rev. Lett. **75**, 1169-1174.

Field, S., Witt, J., Nori, F., and Ling, X. 1995, *Superconducting vortex avalanches*, Phys. Rev. Lett. **74**, 1206-1209.

Fiig, T. and Jensen, H.J. 1993, *Diffusive description of lattice gas models*, J. Stat. Phys. **71/3-4**, 653-682.

Fisher, D.S., Dahmen, K., Ramanbatham, S. and Ben-Zion, Y. 1997, *Statistics of earthquakes in simple models of heterogeneous faults*, Phys. Rev. Lett. **78**, 4885-4888.

Fitzenreiter, R.J., Fainberg, J., and Bundy, R.B. 1976, *Directivity of low frequency solar type III radio bursts*, Solar Phys. **46**, 465-473.

Flyvbjerg, H., Sneppen, K., and Bak, P. 1993, *Mean field theory for a simple model of evolution*, Phys. Rev. Lett. **71/24**, 4087-4090.

Focke, W.B. 1998, *X-ray timing properties of Cygnus X-1 and Cygnus X-2*, PhD Thesis, University of Maryland, 166p.

Focke, W.B., Wai, L.L., and Swank, J.H. 2005, *Time domain studies of X-ray shot noise in Cygnus X-1* Astrophys. J. **633**, 1085-1094.

Fogedby, H.C., Jensen, M.H., Zhang, Y.C., Bohr, T., Jensen, H.J., and Rugh, H.H. 1991, *Temporal fluctuations of AN ideal brownian gas* Modern Phys. Lett. B **5/27**, 1837-1842.

Fragos, T., Rantsiou, E., and Vlahos, L. 2004, *On the distribution of magnetic energy storage in solar active regions*, Astron. Astrophys. **420**, 719-728.

Freeman, M.P., Watkins, N.W., and Riley, D.J. 2000a, *Power law distributions of burst duration and interburst interval in the solar wind: Turbulence of dissipative self-organized criticality?* Phys. Rev. E **62/6**, 8794-8797.

Freeman, M.P., Watkins, N.W., and Riley, D.J. 2000b, *Evidence for a solar wind origin of the power law burst lifetime distribution of the AE indices*, Geophys. Res. Lett. **27**, 1087-1090.

Freeman, M.P. and Morley, S.K. 2004, *A minimal substorm model that explains the observed statistical distribution of times between substorms*, Geophys. Res. Lett. **31**, CiteID L12807.

French, R.G. and Nicholson, P.D. 2000, *Saturn's rings. II. Partice sizes inferred from stellar occultation data*, Icarus **145**, 502-523.

Frette, V., Christensen, K., Malthe-Sorenssen, A., Feder, J., Jessang, T., and Meakin, P. 1996, *Avalanche dynamics in a pile of rice*, Nature **379**, 49-52.

Frick, P., Balkunas, S.L., Galyagin, D., Sokoloff, D., and Soon, W. 1997, *Wavelet analysis of stellar chromospheric activity variations*, Astrophys. J. **483**, 462-434.

Fritzova-Svestkova, L., Chase, R.C., and Svestka, Z. 1976, *On the occurrence of sympathetic flares*, Solar Phys. **48**, 275-286.

Fröhlich, C., Andersen, B.N., Appourchaux, T., Berthmieu, G., et al. 1997, *First results from VIRGO, the experiment for helioseismology and solar irradiance monitoring on SOHO*, Solar Phys. **170**, 1-25.

Frontera F., and Fuligni, F. 1979, *Shot-noise character of hard X-ray emission in a solar flare*, Astrophys. J. **323**, 590-594.

Fuyii, Y. 1969, Frequency distribution of the magnitude of landslides caused by heavy rainfall, Seismol. Soc. Japan J. **22**, 244-247.

Gabriel, S.B. and Feynman, J. 1996, *Power-law distribution for solar energetic proton events*, Solar Phys. **165**, 337-346.

Gabriel, S.B. and Patrick, G.J. 2003, *Solar and energetic particle events: phenomenology and prediction*, Space Science Rev. **107**, 55-62.

Gaite, J. 2007, *Scale invariance of dark matter clustering*, in *Frontiers of fundamental physics: 8th Internat. Symp. FFF8*, AIP Conf. Proc. **905**, p. 23-26.

Gallagher, P.T., Phillips, K.J.H., Harra-Murnion, L.K., and Keenan, F.P. 1998, *Properties of the quiet Sun EUV network*, Astron. Astrophys. **335**, 733-745.

Galsgaard, K. 1996, *Investigations of numerical avalanches in a 3-D vector field*, Astron. Astrophys. **315**, 312-318.

Galsgaard, K. and Nordlund, A., 1996,. *Heating and activity of the solar corona. I. Boundary shearing of an initially homogeneous magnetic field.* J. Geophys. Res. **101/A6**, 13445-13460.

Galsgaard, K. and Nordlund, A., 1997, *Heating and activity of the solar corona. 2. Kink instability in a flux tube*, J. Geophys. Res. **102**, 219-230.

Galtier, S., and Pouquet, A. 1998, *Solar flare statistics with a one-dimensional MHD model*, Solar Phys. **179**, 141-165.

Galtier, S., 1999, *A one-dimensional magnetohydrodynamic model of solar flares: Emergence of a population of weak events, and a possible road toward nanoflares*, Astrophys. J. **521**, 483-489.

Galtier, S., 2001, *Statistical study of short quiescent times between solar flares in a 1-D MHD model*, Solar Phys. **201**, 133-136.

Gardner, M. 1970, *Some mathematical curiosities embedded in the solar system*, Scientific American **222/4**, 108-112.

Gaskell, C.M. 2004, *Lognormal X-ray flux variations in an extreme narrow-line Seyfert 1 galaxy*, Astrophys. J. **612**, L21-L24.

Geisel T. 1985, *Chaos and noise*, in *Chaos in Astrophysics* (eds. Buchler, J.R. et al.), Reidel Publishing Company: Dordrecht, p.165-183.

Georgoulis, M.K., Kluivin,R. and Vlahos,L. 1995, *Extended instability criteria in isotropic and anisotropic energy avalanches*, Physica A **218**, 191-213.

Georgoulis, M.K. and Vlahos, L. 1996, *Coronal heating by nanoflares and the variability of the occurrence frequency distribution in solar flares*, Astrophys. J. **469**, L135-L138.

Georgoulis, M.K. and Vlahos, L. 1998, *Variability of the occurrence frequency of solar flares and the statistical flare*, Astron. Astrophys. **336**, 721-734.

Georgoulis, M.K., Vilmer,N., and Crosby,N.B. 2001, *A Comparison Between Statistical Properties of Solar X-Ray Flares and Avalanche Predictions in Cellular Automata Statistical Flare Models*, Astron. Astrophys. **367**, 326-338.

Georgoulis, M.K., Rust, D.M., Bernasconi, P.N., and Schmieder, B. 2002, *Statistics, morphology, and energetics of Ellerman bombs*, Astrophys. J. **575**, 506-528.

Gergely, T., and Erickson, W.C. 1975, *Decameter storm radiation. I.* Solar Phys. **42**, 467-486.

Gerontidou, M., Vassilaki, A., Mavromichalaki, H., and Kurt, V. 2002, *Frequency distributions of solar proton events*, J. Atmos. Solar-Terr. Physics **64/5-6**, 489-496.

Gil, L. and Sornette, D. 1996, *Landay-Ginzburg theory of self-organized criticality*, Phys. Rev. Lett. **76/21**, 3991-3994.

Giuliani, P. and Carbone, V. 1998, *A note on shell models for MHD turbulence*, Europhys. Lett. **43**, 527-532.

Gogus, E., Woods, P.M., Kouveliotou, C., van Paradijs, J., Briggs, M.S., Duncan, R.C., and Thompson, C. 1999, *Statistical properties of SGR 1900+14 bursts*, Astrophys. J. **526**, L93-L96.

Gogus, E., Woods, P.M., Kouveliotou, C., and van Paradijs, J. 2000, *Statistical properties of SGR 1806-20 bursts*, Astrophys. J. **532**, L121-L124.

Goldberger, A.L., Rigney, D.R., and West, B.J. 1990, *Chaos and fractals in human physiology*, Scientific American (February 1990 issue), p.43-49.

Goldberger, A.L., Amaral, L.A.N., Hausdorff, J.M., Ivanov, P.C., Peng, C.K., and Stanley, H.E. 2002, *Fractal dynamics in physiology: alterations with disease and aging*, in *Self-organized complexity in the physical, biological, and social sciences*, Arthur M. Sackler Colloquia, (eds. Turcotte, D., Rundle, J., and Frauenfelder, H.), The National Academy of Sciences: Washington DC, p.2466-2472.

Golitsyn, G.S. 1997, *The spectrum of cosmic rays from the point of view of similarity theory*, Astronomy Lett. **23**, 127-132.

Gomez, D.O., Martens, P.C.H., and Golub, L. 1993a, *Normal Incidence X-ray Telescope power spectra of X-ray emission from solar active regions: I. Observations*, Astrophys. J **405**, 767-772.

Gomez, D.O. and Golub, L. 1993b, *Normal Incidence X-ray Telescope power spectra of X-ray emission from solar active regions: II. Theory*, Astrophys. J **405**, 773-781.

Gonzalez, R.C. and Woods, R.E. 2008 (3rd Edition), *Digital Image Processing*, Pearson Prentice Hall, Upper Saddle River, New Jersey.

Gould, S.J. and Eldredge, N. 1977, *Punctuated equilibria: The tempo and mode of evolution reconsidered*, Paleobiology **3**, 115-151.

Gould, S.J. and Eldredge, N. 1993, *Punctuated equilibrium comes of age*, Nature **366**, 223-227.

Grassberger, P. and de La Torre, A. 1979, *Reggeon field theory (Schlögl's first model) on a lattice: Monte Carlo calculations of critical behaviour*, Annals of Physics **122**, 373-396.

Grassberger, P. and Procaccia, I. 1983a, *Characterization of strnage attractors*, Phys. Rev. Lett. **50**, 346-349.

Grassberger, P. and Procaccia, I. 1983b, *Measuring the strangeness of strange attractors*, Physica **9D**, 189-208.

Grassberger, P. 1985, *Information aspects of strange attractors*, in *Chaos in Astrophysics* (eds. Buchler, J.R. et al.), Reidel Publishing Company: Dordrecht, p.193-222.

Grassberger, P. 1993, *On a self-organized critical forest-fire model*, Journal of Physics A: Mathematical and General **26/9**, 2081-2089.

Grassberger, P. 1994, *Efficient large-scale simulations of a uniformly driven system*, Phys. Rev. E **49/3**, 2436-2444.

Grasso, J.R. and Bachelery, P. 1995, *Hierarchical organization as a diagnostic approach to volcano mechanics: validation on Piton de la Fournaise*, Geophys. Res. Lett. **22**, 2897.

Greco, A., Matthaeus, W.H., Servidio, S., and Dmitruk, P. 2009a, *Waiting-time distributions of magnetic discontinuities: Clustering or Poisson process?*, Phys. Rev. E **80**, CiteID 046401.

Greco, A., Matthaeus, W.H., Servidio, S., Chuychai, P., and Dmitruk, P. 2009b, *Statistical analysis of discontinuities in solar wind ACE data and comparison with intermittent MHD turbulence*, Astrophys. J. **69**, L111-L114.

Greenhough, J., Chapman, S.C., Dendy, R.O., Nakariakov, V.M. and Rowlands, G. 2003, *Statistical characterisation of full-disk EUV/XUV solar irradiance and correlation with solar activity*, Astron. Astrophys. **409**, L17-L20.

Grieger, B. 1992, *Quaternary climatic fluctuations as a consequence of self-organized criticality*, Physica A **191**, 51-56.

Grigolini, P., Leddon, D., and Scafetta, N. 2002, *Diffusion entropy and waiting time statistics of hard X-ray solar flares*, Phys. Rev. E **65**, 046203.

Grinstein, G., Lee, D.H., and Sachdev, S. 1990, *Conservation laws, anisotropy, and "self-organized criticality" in noisy nonequilibrium systems*, Phys. Rev. Lett. **64/16**, 1927-1930.

Guckenheimer J. 1985, *Clues to strange attractors*, in *Chaos in Astrophysics* (eds. Buchler, J.R. et al.), Reidel Publishing Company: Dordrecht, p.185-191.

Gudiksen, B.V., and Nordlund, A. 2005a, *An ab initio approach to the solar coronal heating problem*, Astrophys. J. **618**, 1020-1030.

Gudiksen, B.V., and Nordlund, A. 2005b, *An ab initio approach to solar coronal loops*, Astrophys. J. **618**, 1031-1038.

Güdel, M., Audard, M., Skinner, S.L., and Horvath, M.I. 2002, *X-ray evidence for flare density variations and continual chromospheric evaporation in Proxima Centauri*, Astrophys. J. **580**, L73-L76.

Güdel, M., Audard, M., Kashyap, V.L., and Guinan, E.F. 2003, *Are coronae of magnetically active stars heated by flares? II. Extreme Ultraviolet and X-ray flare statistics and the differential emission measure distribution*, Astrophys. J. **582**, 423-442.

Güdel, M. 2004, *X-ray astronomy of stellar coronae*, Astron. and Astrophys. Rev. **12**, 71-237.

Gutenberg, B. and Richer, C.F. 1954, *Seismicity of the Earth and Associated Phenomena*, Princeton University Press, Princeton, NJ, p.310 (2nd ed.).

Hagenaar, H.J., Schrijver, C.J., and Title,A.M. 1997, *The distribution of cell sizes of the solar chromospheric networks*, Astrophys. J. **481**, 988-995.

Hagenaar,H.J. 2001, *Ephemeral regions on a sequence of full-disk Michelson Doppler Imager (MDI) Magnetograms*, Astrophys. J. **555**, 448-461.

Hagenaar, H.J., Schrijver, C.J., and Title, A.M. 2003, *The Properties of Small Magnetic Regions on the Solar Surface and the Implications for the Solar Dynamo(s)*, Astrophys. J. **584**, 1107-1119.

Hagenaar, H.J. and Shine, R.A. 2005, *Moving magnetic features around sunspots*, Astrophys. J. **635**, 659-669.

Hall, M., Christensen, K., di Collobiano, S.A., and Jensen, H.J. 2002, *Time-dependent extinction rate and species abundance in a tangled-nature model of biological evolution*, Phys. Rev. E **66**, 011904:1-9.

Hamon, D., Nicodemi,M., and Jensen,H.J., 2002, *Continuously driven OFC: A simple model of solar flare statistics*, Astron. Astrophys. **387**, 326-334.

Harding, A.K., Shinbrot, T., and Cordes, J.M. 1990, *A chaotic attractor in timing noise from the Vela pulsar?*, Astrophys. J. **353**, 588-596.

Harvard Centroid-Moment Tensor Data Base 1997, Harvard University: Cambridge.

Harvey, K.L. and Zwaan,C. 1993, *Properties and emergence patterns of bipolar active regions*, Solar Phys. **148**, 85-118.

Hathaway, D.H., Beck, J.G., Bogart, R.S., Bachmann, K.T., Khatri, G., Petitto, J.M., Han, S., and Raymond, J. 2000, *The photospheric convection spectrum* Solar Phys. **193**, 299-312.

Held,G.A., Solina, D.H., Solina, H., Keane, D.T., Haag, W.J., Horn, P.M. and Grinstein, G. 1990, *Experimental study of critical-mass fluctuations in an evolving sandpile*, Phys. Rev. Lett. **65**, 1120-1123.

Henley, C.L. 1993, *Statics of a "self-organized" percolation model*, Phys. Rev. Lett. **71**, 2741-2744.

Henley, R.W. and Berger, B.R. 2000, *Self-ordering and complexity in epizonal mineral deposits*, Ann. Rev. Earth Planet. Sci. **28**, 669-719.

Hergarten, S. and Neugebauer, H.J. 1998, *Self-organized criticality in a sandslide model*, Geophys. Res. Lett. **25/6**, 801-804.

Hergarten, S. 2002, *Self-organized criticality in Earth systems*, Springer: New York, 272p.

Herz, A.V.M., and Hopfield J.J. 1995, *Earthquake cycles and neural reverberations: collective oscillations in systems with pulse-coupled threshold elements*, Phys. Res. Lett. **75**, 1222-1225.

Hewett, R.J., Gallagher, P.T., McAteer, R.T.J., Young, C.A., Ireland, J., Conlon, P.A., and Maguire, K. 2008, *Multiscale analysis of active region evolution*, Solar Phys. **248**, 311-322.

Heyerdahl, E., Swanson, F., Berry, D., and Agee, K. 1994, *Fire History Database of the Western United States*, Electronic data at H.J. Andrews LTER database (www.fsl.orst.edu/lter/ datafr.htm).

Heyvaerts, J. and Priest, E.R. 1992, *A self-consistent turbulent model for solar coronal heating*, Astrophys. J. **390**, 297-308.

Higuchi, T. 1988, *Approach to an irregular time series on the basis of the fractal theory*, Physica D: Nonlinear phenomena **31**, 277-283.

Hirzberger, J., Vazquez, M., Bonet, J.A., Hanslmeier, A., and Sobotka, M. 1997, *Time series of solar granulation images. I. Differences between small and large granules in quiet regions*, Astrophys. J. **480**, 406-419.

Hiscott, R.N., Colella, A., Pezard, P., Lovell, M.A. and Malinverno, A. 1992, in *Proc. Ocean Drill. Program, Sci. Results* **126**, 75-96.

Hnat, B., Chapman, S.C., Kiyani, K., Rowlands, G., Watkins, N.W. 2007, *On the fractal nature of the magnetic field energy density in the solar wind*, Geophys. Res. Lett. **34/15**, CiteID L15108.

Hopfield, J.J. 1994, *Neurons, dynamics and computation*, Phys. Today **47**, 40-47.

Horbury, T.S. and Balogh, A. 1997, *Structure function measurements of the intermittent MHD turbulent cascade*, Nonlinear Processes in Geophysics **4/3**, 185-199.

Horton, W. and Doxas, I. 1996, *A low-dimensional energy-conserving state space model for substorm dynamics*, J. Geophys. Res. **101/A2**, 27,223-27,238.

Hoshino, M., Nishida, A., Yamamoto, T., and Kokubrun, S. 1994, *Turbulent magnetic field in the distant magnetotail: bottom-up process of plasmoid formation?*, Geophys. Res. Lett. **21/25**, 2935-2938.

Hovius, N., Stark, C.P., Allen, P.A. 1997, *Sediment flux from a mountain belt derived by landslide mapping*, Geology **25**, 231-234.

Hovius, N., Stark, C.P., Chu, H.T., and Lin, J.C. 2000, *Supply and removal of sediment in a landslide-dominated mountain belt: central range, Taiwan*, J. Geol. **108**, 73-89.

Howes, G.G., Dorland, W., Cowley, S.C., Hammett, G.W., Quataert, E., Schekochihin, A.A., and Tatsuno, T. 2008, *Kinetic simulations of magnetized turbulence in astrophysical plasmas*, Phys. Rev. Lett. **100/6**, 065004.

Hoyng, P. 1976, *An error analysis of power spectra*, Astron. Astrophys. **47**, 449-452.

Huang, J., Narkounskaia, G., and Turcotte, D.L. 1992, *A cellular-automata, slider-block model for earthquakes. II. Demonstration of self-organized criticality for a 2-D system*, Geophys. J. Internat. **111**, 259-269.

Huang, J. and Turcotte, D.L. 1990, *Evidence for chaotic fault interactions in the seismicity of the San Andreas fault and Nankai through*, Nature **348**, 234-236.

Huang, Y., Saleur, H., Sammis, C. and Sornette, D. 1998, *Precursors, aftershocks, criticality and self-organized criticality*, Europhys. Lett. **41**, 43-48.

Huberman, B.A. and Adamic, L. 1999, *Growth dynamics of the World-Wide Web*, Nature **401**, 131.

Hudson, H.S., Peterson, L.E., and Schwartz, D.A. 1969, *The hard X-ray spectrum observed from the third orbiting solar observatory*, Astrophys. J. **157**, 389-415.

Hudson, H.S., Canfield, R.C., and Kane, S.R. 1978, *Indirect estimation of energy disposition by non-thermal electrons in solar flares*, Solar Phys. **60**, 137-142.

Hudson, H.S. 1991, *Solar flares, microflares, nanoflares, and coronal heating*, Solar Phys. **133**, 357-369.

Hudson, H.S. 2007, *The unpredictability of the most energetic solar events*, Astrophys. J. **663**, L45-L48.

Hufnagel, B.R. and Bregman, J.N. 1992, *Optical rand radio variability in blazars*, Astrophys. J. **386**, 473-484.

Hughes, D.W., Paczuski, M., Dendy, R.O., Helander, P., and McClements, K.G. 2003, *Solar flares as cascades of reconnecting magnetic loops*, Phys. Rev. Lett. **90/13**, id. 131101.

Hundhausen, A.J. 1993, *Sizes and locations of coronal mass ejections – SMM observations from 1980 and 1984-1989*, J. Geophys. Res. **98/A8**, 13,177-13,200.

Hurst, H.E. 1951, *Long-term storage capacity of reservoirs*, Trans. Am. Soc. Civil Eng. **116**, 770-799.

Inverarity, G.W., Priest, E.R., and Heyvarts, J. 1995, *Turbulent coronal heating. I. Sheared arcade*, Astron. Astrophys. **293**, 913-917.

Inverarity, G.W. and Priest, E.R. 1995, *Turbulent coronal heating. II. Twisted flux tube*, Astron. Astrophys. **296**, 395-404.

Isliker, H. 1992a, *A scaling test for correlation dimensions*, Phys. Lett. A **169**, 313-322.

Isliker, H. 1992b, *Structural properties of the dynamics in flare fragmentation*, Solar Phys. **141**, 325-334.

Isliker, H. and Kurths, J. 1993, *A test for stationarity: finding parts in time series apt for correlation dimension estimates*, Internat. J. Bifurc. Chaos **3/6**, 1573-1579.

Isliker, H. and Benz, A.O. 1994a, *Nonlinear properties of the dynamics of bursts and flares in solar and stellar coronae*, Astron. Astrophys. **285**, 663-674.

Isliker, H. and Benz, A.O. 1994b, *On deterministic chaos, stationarity, periodicity and intermittency in coronal bursts and flares* Space Science Rev. **68**, 185-192.

Isliker, H. 1994, *Dynamical properties of bursts and flares: An inquiry on deterministic chaos in the solar and stellar coronae*, PhD Thesis, ETH Zurich, No. 10495, 234 p.

Isliker, H. 1996, *Are solar flares random processes ?* Astron. Astrophys. **310**, 672-680.

Isliker, H., Anastasiadis, A., Vassiliadis, D., and Vlahos, L. 1998a, *Solar flare cellular automata interpreted as discretized MHD equations*, Astron. Astrophys. **335**, 1085-1092.

Isliker, H., Vlahos, L., Benz, A.O., and Raoult, A. 1998b, *A stochastic model for solar type III bursts*, Astron. Astrophys. **336**, 371-380.

Isliker, H., Anastasiadis, A., and Vlahos, L. 2000, *MHD consistent cellular automata (CA) models: I. Basic Features*, Astron. Astrophys. **363**, 1134-1144.

Isliker, H., Anastasiadis, A., and Vlahos, L. 2001, *MHD consistent cellular automata (CA) models: II. Applications to solar flares*, Astron. Astrophys. **377**, 1068-1080.

Isliker, H. and Benz, A.O. 2001, *On the reliability of peak-flux distributions, with an application to solar flares*, Astron. Astrophys. **375**, 1040-1048.

Ito, K. and Matsuzaki,M. 1990, *Earthquakes as self-organized critical phenomena*, J. Geophys. Res. **95**, 6854-6860.

Ivezic, Z., Tabachnik, S., Rafikov, R., Lupton, R.H., Quinn, T., Hammergren, M., Eyer, L., Chu, J., Armstrong, J.C., Fan, X., Finlator, K., Geballe, T.R., Gunn, J.E., Hennessy, G.S., Knapp, G.R., et al. (SDSS Collaboration) 2001, *Solar system objects observed in the Sloan Digital Sky Survey Commissioning Data*, Astronomical J. **122**, 2749-2784.

Jackson, E.A. 1989, *Perspectives of nonlinear dynamics*, Cambridge University Press: Cambridge.

Jaeger, H.M., Liu, C.H., and Nagel, S.R. 1989, *Relaxation at the angle of repose*, Phys. Rev. Lett. **62/1**, 40-43.

Jähne, B. 2005: (6th edition), *Digital Image Processing*, Springer, Berlin, 607 p.

Jain, A.K. 1989, *Fundamentals of Digital Image Processing*, Prentice Hall, Englewood Cliffs, 569 p.

Janssen, K., Voegler, A., and Kneer, F. 2003, *On the fractal dimension of small-scale magnetic structures in the Sun*, Astron. Astrophys. **409**, 1127-1134.

Jaynes, E.T. 2003, *Probability Theory: The Logic of Science*, Cambridge University Press: Cambridge, 758p.

Jedicke, R. and Metcalfe, T.S. 1998, *The orbital and absolute magnitude distributions of main belt asteroids*, Icarus **131/2**, 245-260.

Jensen, H.J. 1990, *1/f noise from the linear diffusion equation*, Physica Scripta **43**, 593.

Jensen, H.J. 1998, *Self-Organized Criticality. Emergent complex behavior in physical and biological systems*, Cambridge University Press, Cambridge UK, 153 p.

Jess, D.B., Andic, A., Mathioudakis, M., Bloomfield, D.S., and Keenan, F.P. 2007, *High-frequency oscillations in a solar active region observed with the RAPID DUAL IMAGER*, Astron. Astrophys. **473**, 943-950.

Jogi, P. and Sornette, D. 1998, *Self-organized critical random directed polymers*, Phys. Rev. E **57**, 6936-6943.

Johansen, A. 1994, *Spatio-temporal self-organization in a model of disease spreading*, Physica D **78**, 186-193.

Johnston, A.C. and Kanter, L.R. 1990, *Earthquakes in stable continental crust*, Scientific American (March 1990 issue), p.68-75.

Juergens, H., Peitgen, H.O., and Saupe, D. 1990, *The language of fractals*, Scientific American (Aug 1990 issue), p.60-67.

Kaiser, G. 1994, *A friendly guide to wavelets*, Birkhäuser: Boston, 325 p.

Kakinuma, T., Yamashita, T., and Enome, S. 1969, *A statistical study of solar radio bursts a microwave frequencies*, Proc. Res. Inst. Atmos. Nagoya Univ. Japan, Vol. **16**, 127-141.

Kaladze, T.D. and Shukla, P.K. 1987, *Self-organization of electromagnetic waves into vortices in a magnetized electron-positron plasma*, Astrophys. Space Sci. **137**, p.293-296.

Kalapotharakos, C., Voglis, N., and Contopoulos, G. 2004, *Chaos and secular evolution of triaxial N-body galactic models due to an imposed central mass*, Astron. Astrophys. **428**, 905-923.

Karlicky, M., Sobotka, M., and Jiricka, K. 1996, *Narrowband dm-spikes in the 2 GHz frequency range and MHD cascading waves in reconnection outflows*, Solar Phys. **168**, 375-383.

Karlicky ,M., Jiricka, K., and Sobotka, M. 2000, *Power-law spectra of 1-2 GHz narrowband dm-spikes*, Solar Phys. **195**, 165-174.

Kashyap, V.L. and Drake, J.J. 1999, *On X-ray variability in active binary stars*, Astrophys. J. **524**, 988-999.

Kashyap, V.L., Drake, J.J., Güdel, M., and Audard, M. 2002, *Flare heating in stellar coronae*, Astrophys. J. **580**, 1118-1132.

Kasischke, E.S. and French, N.H.F. 1995, Remote Sens. Environ. **51**, 263-275.

Kato, T., Ishioka, R., and Uemura, M. 2002, *Photometric study of KR Aurigae during the high state in 2001*, Publ. Astron. Soc. Japan **54**, 1033-1039.

Katsukawa, Y. and Tsuneta, S. 2001, *Small fluctuation of coronal X-ray intensity and a signature of nanoflares*, Astrophys. J. **557**, 343-350.

Katz, J.I. 1986, *A model of propagating brittle failure in heterogeneous media*, J. Geophys. Res. **91/B10**, 10412-10420.

Kawasaki, K. and Okuzono, T. 1996, *Self-organized critical behavior of two-dimensional foams*, Fractals **4**, 339-348.

Kennel, C.F. 1995, *Convection and Substorms*, Oxford University Press; New York, 408 p.

Kishimoto, Y., Tajima, T., Horton, W., LeBrun, M.J., and Kim, J.Y. 1996, *Theory of self-organized critical transport in tokamak plasmas*, Phys. Plasmas **3**, 1289-1307.

Klimas, A.J., Baker, D.N., Roberts, D.A., Fairfield, D.H., and Buechner, J. 1992, *A nonlinear dynamical analogue model of geomagnetic activity*, J. Geophys. Res. **97/A8**, 12,353-12,266.

Klimas, A.J., Valdivia, J.A., Vassiliadis, D., Baker, D.N., Hesse, M., and Takalo, J. 2000, *Self-organized criticality in the substorm phenomenon and its relation to localized reconnection in the magnetosphere plasma sheet*, J. Geophys. Res. **105/A8**, 18,765-18,780.

Klimas, A.J., Uritsky, V.M., Vassiliadis, D., and Baker, D.N. 2004, *Reconnection and scale-free avalanching in a driven current-sheet model*, J. Geophys. Res. **109/A2**, CiteID A02218.

Kolmogorov, A.N. 1941, The local structure of turbulence in incompressible viscous fluid for very large Reynolds' number, Dokl. Acad. Nauk SSSR **30**, 301-305.

Kopnin, S.I., Kosarev, I.N., Popel, S.I., and Hyu, M.Y. 2004, *Localized structures of nanosize charged dust in Earth's middle atmosphere*, Planet. Space Sci. **52/13**, 1187-1194.

Kouveliotou, C., Dieters, S., Strohmayer, T., van Paradijs, J., Fishman, G.J., Meegan, C.A., Hurley, K., Kommers, J., Smith, I., Frail, D., Muakami, T. 1998, *An X-ray pulsar with a superstrong magnetic field in the soft γ-ray repeater SGR 1806-20*, Nature **393**, 235-237.

Kouveliotou,C., Strohmayer, T., Hurley, K., van Paradijs, J., Finger, M.H., Dieters, S., Woods, P., Thomson, C., and Duncan, R.C. 1999, *Discovery of a magnetar associated with the soft gamma ray repeater SGR 1900+14*, Astrophys. J. **510**, L115-L118.

Kovacs, P., Carbone, V., and Voros, Z. 2001, *Wavelet-based filtering of intermittent events from geomagnetic time series*, Planetary and Space Science **49/12**, 1219-1231.

Kozelov, B.V., Uritsky, V.M., and Klimas, A.J. 2004, *Power law probability distributions of multiscale auroral dynamics from ground-based TV observations*, Geophys. Res. Lett. **31/20**, CiteID L20804.

Krasnoselskikh, V.V., Podladchikova, O., Lefebvre, B., and Vilmer, N. 2002, *Quiet Sun coronal heating: A statistical model*, Astron. Astrophys. **382**, 699-712.

Kremliovsky, M.N. 1994, *Can we understand time scales of solar activity?*, Solar Phys. **151**, 351-370.

Krucker, S. and Benz, A.O. 1998, *Energy distribution of heating processes in the quiet solar corona*, Astrophys. J. **501**, L213-L216.

Kucera, T.A., Dennis, B.R., Schwartz, R.A., and Shaw, D. 1997, *Evidence for a cutoff in the frequency distribution of solar flares from small active regions*, Astrophys. J. **475**, 388-347.

Kuijpers, J. 1995, *Flares in accretion disks*, Lecture Notes in Physics **444**, 135-158.

Kundu, M.R. 1965, *Solar radio astronomy*, Interscience Publication: New York, 660 p.

Kurths, J. and Herzel, H. 1986, *Can a solar pulsation event be characterized by a low-dimensional chaotic attractor*, Solar Phys. **107**, 39-45.

Kurths, J. 1987, *Estimating parameters of attractors in some astrophysical time series*, in *Nonlinear Oscillations*, Proc. 11th Internat. Conf. (ed. M. Farkas), p.664-667.

Kurths, J. and Herzel, H. 1987, *An attractor in a solar time series*, Physica **25D**, 165-172.

Kurths, J. and Karlicky, M. 1989, *The route to chaos during a pulsation event*, Solar Phys. **119**, 399-411.

Kurths, J., Benz, A.O., and Aschwanden, M.J. 1990, *The attractor dimension of solar decimetric radio pulsations*, in *The Dynamic Sun*, Proc. 6th European Meeting on Solar Physics, (ed. L.Dezso), Publications of Debrecen Heliophysical Observatory of the Hungarian Academy of Sciences, Vol. **7**, p.196-199.

Kurths, J., Benz, A.O., and Aschwanden, M.J. 1991, *The attractor dimension of solar decimetric radio pulsations*, Astron. Astrophys. **248**, 270-276.

Kurths, J. and Brandenburg, A. 1991, *Lyapunov exponents for hydrodynamic convection*, Phys. Rev. A. **44/6**, 3427-3429.

Kurths, J. and Schwarz, U. 1994, *Chaos theory and radio emission*, Space Science Rev. **68**, 171-184.

Kurths, J. and Schwarz, U. 1995, *On nonlinear signal processing*, in *Cluster Workshop on Data Analysis Tools* and *Physical Measurements and Mission-Oriented Theory* (eds. Glassmeier, K.H., Motschmann, U., and Schmidt, R.), European Space Agency (ESA) **SP-371**, ESTEC Noordwijk, Netherlands, p.15-22.

Kurths, J., Schwarz, U. and Witt, A. 1995, *Nonlinear data analysis and statistical techniques*, Lecture Notes in Physics (Springer: Berlin) **444**, 159-172.

Lawrence, A., Watson, M.G., Pounds, K.A., and Elvis, M. 1987, *Low-frequency divergent X-ray variability in the Seyfert galaxy NGC4501*, Nature **325**, 694-696.

Lawrence, J.K. 1991, *Diffusion of magnetic flux elements on a fractal geometry*, Solar Phys. **135**, 249-259.

Lawrence, J.K. and Schrijver, C.J. 1993, *Anomalous diffusion of magnetic elements across the solar surface*, Astrophys. J. **411**, 402-405.

Lawrence, J.K., Ruzmaikin, A., and Cadavid, A.C. 1993, *Multifractal measure of the solar magnetic field*, Astrophys. J. **417**, 805-811.

Lawrence, J.K., Cadavid, A.C., and Ruzmaikin, A. 1995, *Turbulent and chaotic dynamics underlying solar magnetic variability*, Astrophys. J. **455**, 366-375.

Lawrence, J.K., Cadavid, A., and Ruzmaikin, A. 1996, *On the multifractal distribution of solar fields*, Astrophys. J. **465**, 425-435.

Lawrence, J.K., Cadavid, A., and Ruzmaikin, A. 2005, *Principal component analysis of the solar magnetic field I: The axisymmetric field at the photosphere*, Solar Phys. **225**, 1-19.

Lee, T.T., Petrosian, V., and McTiernan, J.M. 1993, *The distribution of flare parameters and implications for coronal heating*, Astrophys. J. **412**, 401-409.

Lee, T.T., Petrosian, V., and McTiernan, J.M. 1995, *The Neupert effect and the chromospheric evaporation model for solar flares*, Astrophys. J. **418**, 915-924.

Lee, J.W., Chae, J.C., Yun, H.S., and Zirin, H. 1997, *Power spectra of solar network and non-network fields*, Solar Phys. **171**, 269-282.

Leighly, K. M. and O'Brien, P.T. 1997, *Evidence for nonlinear X-ray variability from the broad-line radio galaxy 3C 390.3*, Astrophys. J. **481**, L15-L18.

Lepreti, F., Fanello, P.C., Zaccaro, F., and Carbone, V. 2000, *Persistence of solar activity on small scales: Hurst analysis of time series coming from Hα flares*, Solar Phys. **197**, 149-156.

Lepreti, F., Carbone, V., and Veltri, P. 2001, *Solar flare waiting time distribution: varying-rate Poisson or Levy function?*, Astrophys. J. **555**, L133-L136.

Lepreti, F., Carbone, V., Giuliani, P., Sorriso-Valvo, L., and Veltri, P. 2004, *Statistical properties of dissipation bursts within turbulence: solar flares and geomagnetic activity*, Planet. Space Science **52**, 957-962.

Letellier, C., Aguirre, L.A., Maquet, J., and Gilmore, R. 2006, *Evidence for low dimensional chaos in sunspot cycles*, Astron. Astrophys. **449**, 379-388.

Levy, J.S. 1983, *War in the Modern Great Power System 1495-1975*, (Lexington, KY: University of Kentucky Press), p. 215.

Li, T.P. and Muraki, Y. 2002, *Power spectra of X-ray binaries*, Astrophys. J. **578**, 374-384.

Lin, R.P., Schwartz, R.A., Kane, S.R., Pelling, R.M., and Hurley, K.C. 1984, *Solar hard X-ray microflares*, Astrophys. J. **283**, 421-425.

Lin, R.P., Feffer,P.T., and Schwartz,R.A. 2001, *Solar hard X-ray bursts and electron acceleration down to 8 keV*, Astrophys. J. **557**, L125-L128.

Lise, S. and Jensen, H.J. 1996, *Transitions in nonconserving models of self-organized criticality*, Phys. Rev. Lett. **76**/13, 2326-2329.

Litvinenko, Y.E. 1994, *An explanation for the flare frequency energy dependence*, Solar Phys. **151**, 195-198.

Litvinenko, Y.E. 1996, *A new model for the distribution of flare energies*, Solar Phys. **167**, 321-331.

Litvinenko, Y.E. 1998a, *Analytical results in cellular automaton model of solar flare occurrence*, Astron. Astrophys. **339**, L57-L60.

Litvinenko, Y.E. 1998b, *Dimensional analysis of the flare distribution problem*, Solar Phys. **180**, 393-396.

Litvinenko, Y.E. and Wheatland, M.S. 2001, *Modeling the rate of occurrence of solar flares*, Astrophys. J. **550**, L109-L112.

Liu, H., Charbonneau, P., Pouquet, A., Bogdan, T., and McIntosh, S.W. 2002, *Continuum analysis of an avalanche model for solar flares*, Phys. Rev. E **66**, 056111.

Liu, W.W., Charbonneau, P., Thibault, K., and Morales, L. 2006, *Energy avalanches in the central plasma sheet*, Geophys. Res. Lett. **33/19**, CiteID L19106.

Longcope, D.W. and Sudan, R.N. 1994, *Evolution and statistics of current sheets in coronal magnetic loops*, Astrophys. J. **437**, 491-504.

Longcope, D.W. and Noonan, E.J. 2000, *Self-organized criticality from separator reconnection in solar flares*, Astrophys. J. **542**, 1088-1099.

Lorenz, E.N. 1963, *Deterministic nonperiodic flow*, J. Atmos. Sci. **20/2**, 130-148.

Loreto, V., Pietronero, L., Vespignani, A., and Zapperi, S. 1995, *Renormalization group approach to the critical behavior of the forest-fire model*, Phys. Rev. Lett. **75**/3, 465-468.

Lu, E.T. and Hamilton, R.J. 1991, *Avalanches and the distribution of solar flares*, Astrophys. J. **380**, L89-L92.

Lu, E.T., Hamilton, R.J., McTiernan, J.M., and Bromund, K.R. 1993, *Solar flares and avalanches in driven dissipative systems*, Astrophys. J. **412**, 841-852.

Lu, E.T. 1995a, *The statistical physics of solar active regions and the fundamental nature of solar flares*, Astrophys. J. **446**, L109-L112.

Lu, E.T. 1995b, *Constraints on energy storage and release models for astrophysical transients and solar flares*, Astrophys. J. **447**, 416-418.

Lu, E.T. 1995c, *Avalanches in continuum driven dissipative systems*, Phys. Rev. Lett. **74/13**, 2511-2514.

Lui, A.T.Y. 1996, *Current disruption in the Earth's magnetosphere: Observations and models*, J. Geophys. Res. **101**, 13067-13088.

Lui, A.T.Y. 1998, *Plasma sheet behavior associated with auroral breakups*, in *Proc. 4th Intern. Conf. on Substorms* (ed. Kamide Y.), Kluwer Academic Publishers, Dordrecht, and Terra Scientific Publishing Company, Tokyo, p. 183.

Lui, A.T.Y., Lopez, R.E., Krimigis, S.M., McEntire, R.W., Zanetti, L.J., and Potemra, T.A. 1988, *A case study of magnetotail current sheet disruption and diversion*, Geophys. Res. Lett. **15**, 721-724.

Lui, A.T.Y., Chapman, S.C., Liou,K., Newell, P.T., Meng, C.I., Brittnacher, M., and Parks, G.K. 2000, *Is the dynamic magnetosphere an avalanching system?*, Geophys. Res. Lett. **27/7**, 911-914.

Lundgren, S.C., Cordes,J.M., Ulmer, M., Matz, S.M., Lomatch,S., Foster, R.S., and Hankins,T. 1995, *Giant pulses from the Crab pulsar: A joint radio and gamma-ray study*, Astrophys. J. **453**, 433-445.

Macek, W.M. and Szczepaniak, A. 2008, *Generalized two-scale weighted Cantor set model for solar wind turbulence*, Geophys. Res. Lett. **35/2**, CiteID L02108.

Macek, W.M. and Wawrzaszek, A. 2009, *Evolution of asymmetric multifractal scaling of solar wind turbulence in the outer heliosphere*, J. Geophys. Res. (Space Physics) **114/A4**, CiteID A03108.

MacKinnon, A.L., MacPherson, K.P., and Vlahos, L. 1996, *Cellular automaton models of solar flare occurrence*, Astron. Astrophys. **310**, L9-L12.

MacKinnon, A.L. and MacPherson, K.P. 1997, *Nonlocal communication in self-organizing models of solar flare occurrence*, Astron. Astrophys. **326**, 1228-1234.

Macpherson, K.P. and MacKinnon, A.L. 1999, *Extended cellular automaton models of solar flare occurrence*, Astron. Astrophys. **350**, 1040-1050.

Majumdar, S.N., and Dhar, D. 1991, *Height correlations in the Abelian sandpile model*, J. Physics A **24/7**, L357-L362.

Majumdar, S.N., and Dhar, D. 1992, *Equivalence between the Abelian sandpile model and the* $q \mapsto 0$ *limit of the Potts model*, Physica A **185**, 129-145.

Makishima, K. 1988, in *Physics of Neutron Stars and Black Holes*, (ed. Y. Tanaka), Tokyo: Universal Academy Press, p.175.

Malamud, B.D., Morein, G., and Turcotte D.L. 1998, *Forest fires: An example of self-organized critical behavior*, Science **281**, 1840-1842.

Malamud, B.D., Guzzetti, F., Turcotte, D.L., and Reichenbach, P. 2001, *Power-law correlations of Italian landslide areas*, American Geophysical Union, Fall Meeting 2001, abstract #NG52A-10.

Malkov, M.A., Diamond, P.H., and Völk, H.J. 2000, *Critical self-organization of astrophysical shocks*, Astrophys. J. **533**, L171-L174.

Mallat, S.G. 1989, *A theory for multiresolution signal decomposition: the wavelet representation*, IEEE Trans. Pattern Anal. Machine Intelligence (ITPAM), **11**, 674-693.

Mallat, S. 2008 (3rd Edition), *A Wavelet Tour of Signal Processing*, The Sparse Way, Academic Press: New York, 663 p.

Mandelbrot, B.B. 1963, *The variation of certain speculative prices*, Journal of Business of the University of Chicago **36/4**, 394-419.

Mandelbrot, B.B. 1977, *Fractals: form, chance, and dimension*, Translation of *Les objects fractals*, W.H. Freeman, San Francisco.

Mandelbrot, B.B. 1983, *The fractal geometry of nature*, W.H. Freeman, San Francisco.

Mandelbrot, B.B. 1985, *Self-affine fractals and fractal dimension*, Physica Scripta **32**, 257-260.

Manna, S.S. 1991a, *Critical exponents and the sand pile models in two dimensions*, Physica A **179**/2, 249-268.

Manna, S.S. 1991b, *Two-state model of self-organized criticality*, J. Physics A, **24**/7, L363-L369.

Mantegna, R.N. and Stanley, H.E. 1997, *The physics of complex systems*, (eds. F. Mallamace and H.E.Stanley), IOS:Amsterdam, pp. 473-489.

Marchesoni, F. and Patriarca, M. 1994, *Self-organized criticality in dislocation networks*, Phys. Rev. Lett. **72**, 4101-4104.

Martens, P.C.H. and Gomez, D.O. 1992, *Spatial power spectra from Yohkoh soft X-ray images*, Publ. Astron. Soc. Japan **44**, L187-L191.

Maslov, S., Paczuski, M., and Bak, P. 1994, *Avalanches and 1/f noise in evolution and growth models*, Phys. Rev. Lett. **73**, 2162.

Matthaeus, W.H. and Goldstein, M.L. 1986, *Low-frequency 1/f noise in the interplanetary magnetic field*, Phys. Rev. Lett. **57**, 495-498.

Matthaeus, W.H. 2000, *MHD: Magnetic Reconnection and Turbulence*, in *Encyclopedia of Astronomy and Astrophysics*, (ed. Murdin, P.), Nature Publishing Group, Institute of Physics Publishing: Bristol, UK, and Grove's Dictionaries, Inc.: New York.

Matthaeus, W.H., Servidio, S., and Dmitruk, P. 2008, *Comment on "Kinetic simulations of magnetized turbulence in astrophysical plasmas*, Phys. Rev. Lett. **101/14**, id. 149501.

May R.M. 1974. *Model Ecosystems*, Princeton: Princeton, New Jersey.

McAteer, R.T.J., Young, C.A., Ireland, J., and Gallagher, P.T. 2007, *The bursty nature of solar flare X-ray emission*, Astrophys. J. **662**, 691-700.

McHardy, I. and Czerny, B. 1987, *Fractal X-ray time variability and spectral invariance of the Seyfert galaxy NGC5506*, Nature **325**, 696-698.

McIntosh, S.W. and Gurman, J.B. 2005, *Nine years of EUV bright points*, Solar Phys. **228**, 285-299.

McIntosh, S.W. and Charbonneau, P. 2001, *Geometrical effects in avalanche models for solar flares: Implications for coronal heating*, Astrophys. J. **563**, L165-L169.

McIntosh, S.W., Charbonneau, P., Bogdan, T.J., Liu, H.L., and Norman, J.P. 2002, *Geometrical properties of avalanches in self-organized critical model of solar flares*, Phys. Rev. E **65/4**, id. 046125.

Medvedev, M.V., Diamond, P.H., and Carreras, B.A. 1996, *On the statistical mechanics of self-organized profiles*, Phys. Plasmas **3**, 3745-3753.

Melatos, A., Peralta, C., and Wyithe, J.S.B. 2008, *Avalanche Dynamics of radio pulsar glitches*, Astrophys. J. **672**, 1103-1118.

Melendez, J.L., Sawant, H.S., Fernandes, F.C.R., and Benz, A.O. 1999, *Statistical analysis of high-frequency decimetric type III bursts*, Solar Phys. **187**, 77-88.

Melnick, J. and Selman, F.J. 2000, *Self-organized criticality and the IMF of starbursts*, in *Cosmic evolution and galaxy formation, structure, interactions, and feedback*, ASP Conf. Ser. **215**, 159-165.

Mendoza, B., Melendez-Venancio, R., Miroshnichenko, L.I., and Perez-Enriquez, R. 1997, *Frequency distributions of solar proton events*, Proc. 25th Int. Cosmic Ray Conf. **1**, 81.

Mercier, C. and Trottet, G. 1997, *Coronal radio bursts: A signature of nanoflares ?*, Astrophys. J. **484**, 920-926.

Meszarosova, H., Karlicky, M., Veronig, A., Zlobec, P. and Messerotti, M. 1999, *Power-law and exponential distributions of narrowband dm-spikes observed during the June 15, 1991 flare*, in *Magnetic fields and solar processes*, European Space Agency (ESA), **SP-448**, ESTEC Noordwijk, Netherlands, p.1025-1032.

Meszarosova, H., Karlicky, M., Veronig, A., Zlobec, P. and Messerotti, M. 2000, *Linear and nonlinear statistical analysis of narrow-band dm-spikes observed during the June 15, 1991 flare*, Astron. Astrophys. **360**, 1126-1138.

Meunier, N. 1999, *Fractal analysis of Michelson Doppler Imager magnetograms: a contribution to the study of the formation of solar active regions*. Astrophys. J. **515**, 801-811.

Meunier, N. 2003, *Statistical properties of magnetic structures: Their dependence on scale and solar activity*, Astron. Astrophys. **405**, 1107-1120.

Meunier, N. 2004, *Complexity of magnetic structures: flares and cycle phase dependence*, Astron. Astrophys. **420**, 333-342.

Meyer, Y. and Ryan, R.D. 1993, *Wavelets: Algorithms and applications*, Soc. Industr. App. Math. (SIAM), Philadelphia, 133p.

Middleton, A.A. and Tang, C. 1995, *Self-organized criticality in nonconserved systems*, Phys. Rev. Lett. **74/5**, 742-745.

Milano, L.J., Gomez, D.O., and Martens, P.C.H. 1997, *Solar coronal heating: AC versus DC*, Astrophys. J. **490**, 442-451.

Milano, L.J., Dmitruk, P., Mandrini, C.H., Gomez, D.O., and Demoulin, P. 1999, *Quasi-separatrix layers in a reduced magnetohydrodynamic model of a coronal loop*, Astrophys. J. **521**, 889-897.

Miller, J.A., Cargill, P.J., Emslie, A.G., Holman, G.D., Dennis, B.R., LaRosa, T.N., Winglee, R.M., Benka, S.G., and Tsuneta, S. 1997, *Critical issues for understanding particle acceleration in impulsive solar flares*, J. Geophys. Res. **102/A7**, 14631-14659.

Milovanov, A.V., and Zelenyi, L.M. 1999, *Fracton excitations as a driving mechanism for the self-organized dynamical structuring in the solar wind*, Astrophys. Space Sci. **264**, 317-345.

Milovanov, A.V., Zelenyi, L.M., Zimbardo, G., and Veltri, P. 2001, *Self-organized branching of magnetotail current systems near the percolation threshold*, J. Geophys. Res. **106/A4**, 6291-6308.

Milshtein, E., Biham, O., and Solomon, S. 1998, *Universality classes in isotropic, Abelian, and non-Abelian sandpile models*, Phys. Rev. E **58**, 303-310.

Mineshige, S., Ouchi,N.B., and Nishimori, H. 1994a, *On the generation of 1/f fluctuations in X-rays from black-hole objects*, Publ. Astron. Soc. Japan **46**, 97-105.

Mineshige, S., Takeuchi, M., and Nishimori, H. 1994b, *Is a black hole accretion disk in a self-organized critical state ?*, Astrophys. J. **435**, L125-L128.

Mineshige, S. 1999, *Self-organized criticality in accretion disks*, in *Disk instabilities in close binary systems*, (eds. S. Mineshige and J.C. Wheeler), Frontiers Science Series **26**, Universal Academy Press, Inc., p.295.

Mineshige, S. and Negoro, H. 1999, *Accretion disks in the context of self-organized criticality: How to produce 1/f fluctuations?*, in *High energy processes in accreting black holes*, ASP Conf. Ser. **161**, 113-128.

Mininni, P.D., Gomez, D.O., and Mindlin, G.B. 2002, *Biorthogonal decomposition techniques unveil the nature of the irregularities observed in the solar cycle*, Phys. Rev. Lett. **89**, p. 061101.

Miroshnichenko, L.I. 1995, *On the threshold effect of proton acceleration in solar flares*, Solar Phys. **156**, 119-129.

Miroshnichenko, L.I., Mendoza, B., and Perez-Enriquez R. 2001, *Size distributions of the >10 MeV solar proton events*, Solar Phys. **202**, 151-171.

Mitra-Kraev, U. and Benz, A.O. 2001, *A nanoflare heating model for the quiet solar corona*, Astron. Astrophys. **373**, 318-328.

Miyamoto, S., Kimura, K., Kitamoto, S.,, Dotani, T., and Ebisawa, K. 1991, *X-ray variability of GX 339-4 in its very high state*. Astrophys. J. **383**, 784-807.

Moffat, J.W. 1997, *Stochastic gravity and self-organized critical cosmology*, in *Very High Energy Phenomena in the Universe*, Morion Workshop, (eds. Y. Giraud-Heraud and J.T. Than Van), p.353.

Moon, Y.J., Choe, G.S., Yun, H.S., and Park, Y.D. 2001, *Flaring time interval distribution and spatial correlation of majore X-ray solar flares*, J. Geophys. Res. **106/A12**, 29951-29962.

Moon, Y.J., Choe, G.S., Park, Y.D., Wang, H., Gallagher, P.T., Chae, J.C., Yun, H.S., and Goode, P.R. 2002, *Statistical evidence for sympathetic flares*, Astrophys. J. **574**, 434-439.

Moon, Y.J., Choe, G.S., Wang, H., and Park, Y.D. 2003, *Sympathetic coronal mass ejections*, Astrophys. J. **588**, 1176-1182.

Morales, L. and Charbonneau, P. 2008a, *Self-organized critical model of energy release in an idealized coronal loop*, Astrophys. J. **682**, 654-666.

Morales, L. and Charbonneau, P. 2008b, *Scaling laws and frequency distributions of avalanche areas in a SOC model of solar flares*, Geophys. Res. Lett. **35**, 4108.

Morales, L. and Charbonneau, P. 2009, *Geometrical properties of avalanches in a pseudo 3-D coronal loop*, Astrophys. J. **698**, 1893-1902.

Morley, P.D. and Garcia-Pelayo, R. 1993, *Scaling law for pulsar glitches*, Europhys. Lett. **23/3**, 185-189.

Mossner, W.K., Drossel, B., and Schwabl, F. 1992, *Computer simulations of the forest-fire model*, Physica A **190/3-4**, 205-217.

Mousseau, N. 1996, *Synchronization by disorder in coupled systems*, Phys. Rev. Lett. **77**, 968-971.

Nagatani, T. 1995a, *Self-organized criticality in 1-D traffic flow model with inflow and outflow*, J. Physics A: Math. Gen. **28**, L119-L124.

Nagatani, T. 1995b, *Self-organized criticality and scaling in lifetime of traffic jams*, J. Phys. Soc. Japan **64/1**, 31.

Nagatani, T. 1995c, *Self-organized criticality in 2-D traffic flow model with jam-avoiding drive*, J. Phys. Soc. Japan **64/4**, 1421.

Nagatani, T. 1995d, *Self-organized criticality in asymmetric exclusion model with noise for freeway traffic*, Physica A **218**, 145-154.

Nagatani, T. 1995e, *Bunching of cars in asymmetric exclusion models for freeway traffic*, Phys. Rev. E **51/2**, 922-928.

Nagatani, T. 1995f, *Creation and annihilation of traffic jams in a stochastic asymmetric exclusion model with open boundaries: a computer simulation*, J. Physics A **28/24**, 7079-7088.

Nagel, K. and Raschke, E. 1992, *Self-organized criticality in cloud formation?*, Physica A **182/4**, 519-531.

Nagel, K. and Schreckenberg M. 1992, *A cellular automaton model for freeway traffic*, J. de Physique I **2/12**, 2221-2229.

Nagel, K. and Herrmann, H.J. 1993, *Deterministic models for traffic jams*, Physica A, **199/2**, 254-269.

Nagel, K. and Paczuski, M. 1995, *Emergent traffic jams*, Phys. Rev. E. **51/4**, 2909-2918.

Nakanishi, H. 1990, *Cellular-automaton model of earthquakes with deterministic dynamics*, Physical Review A **41/12**, 7086-7089.

Nakanishi, H. 1991, *Statistical properties of the cellular-automaton model for earthquakes*, Physical Review A **43/12**, 6613-6621.

Nathan, A. and Barbosa, V.C. 2006, "V-like formation in flocks of artificial birds", Artificial Life **14** (2008), 179–188.

Negoro, H. 1992, Master Thesis, Osaka University.

Negoro, H., Kitamoto, S., Takeuchi, M., and Mineshige, S. 1995, *Statistics of X-ray fluctuations from Cygnus X-1: Reservoirs in the disk ?* Astrophys. J. **452**, L49-L52.

Negoro, H., Kitamoto, S., and Mineshige, S. 2001, *Temporal and spectral variations of the superposed shot as causes of power spectral densities and hard X-ray time lags of Cygnus X-1*, Astrophys. J. **554**, 528-533.

Negoro, H. and Mineshige, S. 2002, *Log-normal distributions in Cygnuys X-1: Possible physical link with gamma-ray bursts and blazars* PASJ **54**, L69-L72.

Newman, M.E.J. and Sneppen, K. 1996, *Avalanches, scaling, and coherent noise*, Phys. Rev. E **54/6**, 6226-6231.

Newman, M.E.J., Watts, D.J., and Strogatz, S.H. 2002, *Random graph models of social networks*, in *Self-organized complexity in the physical, biological, and social sciences*, Arthur M. Sackler Colloquia, (eds. Turcotte, D., Rundle, J., and Frauenfelder, H.), The National Academy of Sciences: Washington DC, p.2566-2572.

Ning, Z., Wu, H., Xu, F., and Meng, X. 2007, *Frequency distributions of microwave pulses for the 18 March 2007 solar flare*, Solar Phys. **242**, 101-109.

Nishizuka, N., Asai, A., Takasaki, H., Kurokawa, H., and Shibata, K. 2009, *The power-law distribution of flare kernels and fractal current sheets in a solar flare*, Astrophys. J. **694**,. L74-L78.

Nita, G.M., Gary, D.E., Lanzerotti, L.J., and Thomson, D.J. 2002, *The peak flux distribution of solar radio bursts*, Astrophys. J. **570**, 423-438.

Nita, G.M., Gary, D.E., and Lee, J.W. 2004, *Statistical study of two years of solar flare radio spectra obtained with the Owens Valley solar array*, Astrophys. J. **605**, 528-545.

Noever, D.A. 1993, *Himalayan sandpiles*, Phys. Rev. E. **47**, 724-725.

Norman, J.P., Charbonneau, P., McIntosh, S.W., and Liu, H.L. 2001, *Waiting-time distributions in lattice models of solar flares*, Astrophys. J. **557**, 891-896.

Norris, J.P. and Matilsky, T.A. 1989, *Is Hercules X-1 a strange attractor?* Astrophys. J. **346**, 912-918.

Norris, J.P. 1995, *Gamma-ray bursts: the time domain*, Astrophysics and Space Science **231**, 95-102.

Norris, J.P., Nemiroff, R.J., Bonnell, J.T., Scargle, J.D., Kouveliotou,C., Paciesas, W.S., Meegan, C.A., and Fishman, G.J. 1996, *Attributes of pulses in long bright gamma-ray bursts*, Astrophys. J. **459**, 393-412.

Nozakura, T. and Ikeuchi, S. 1988, *Spiral patterns on a differentially rotating galactic disk – Self-organized structures in galaxies*, Astrophys. J. **333**, 68-77.

Nurujjaman, Md. and Sekar-Iyenbgar, A.N. 2007, *Realization of SOC behavior in a DC glow discharge plasma*, Phys. Lett. A **360**, 717-721.

O'Brien, K.P. and Weissman, M.B. 1994, *Statistical characterization of Barkhausen noise*, Phys. Rev. E **50/5**, 3446-3452.

Okuzono, T., and Kawasaki, K. 1995, *Intermittent flow behavior of random foams: A computer experiment on foam rheology*, Phys. Rev. E **51/2**, 1246-1253.

Olami, Z., Feder, H.J.S., and Christensen, K. 1992, *Self-organized criticality in a continuous cellular automaton modeling earthquakes*, Phys. Rev. Lett. **68/8**, 1244-1247.

Olmedo, O., Zhang, J., Wechsler, H., Poland, A., and Borne, K. 2008, *Automatic detection and tracking of coronal mass ejections in coronagraph time series*, Solar Phys. **248**, 485-499.

Olson, C.J., Reichhardt, C., and Nori, F. 1997, *Superconducting vortex avalanches, voltage bursts, and vortex plastic flow: effect of the microscopic pinning landscape on the macroscopic properties*, Phys. Rev. B **56/10**, 6175-6194.

Omori, F., 1895, J. Coll. Sci. Imper. Univ. Tokyo **7**, 111.

Osorio, I., Frei, M.G., Sornette, D., and Milton, J. 2009a, *Pharmaco-resistant seizures: self-triggering, capacity, scale-free properties and predictability?*, European J. Neuroscience **30**, 1554-1558.

Osorio, I., Frei, M.G., Sornette, D., Milton, J., and Lai, Y.C. 2009b, *Epileptic seizures, quakes in the brain*, (archiv.org/abs/0712.3929), preprint.

Osten, R.A. and Brown, A. 1999, *Extreme Ultraviolet Explorer photometry of RS Canum Venaticorum systems: four flaring megaseconds*, Astrophys. J. **515**, 746-761.

O'Toole, D.V., Robinson, P.A. and Myerscough, 1999, *Self-organized criticality and emergent oscillations in models of termite architecture*, J. Theor. Biol. **198**, 305-307.

Otsuka, M. 1972, *A simulation of earthquake occurrence*, Phys. Earth and Planet interiors **6**/4, 311-315.

Otter, R. 1949, Annals of Math. Stat. **20**, 206.

Paczuski, M. and Bak, P. 1993, *Theory of the one-dimensional forest-fire model*, Phys. Rev. E **48/5**, 3214-3216.

Paczuski, M., Maslov, S., and Bak, P. 1994, *Field theory for a model of self-organized criticality*, Europhysics Letters (EPL) **27/2**, 97-102.

Paczuski, M. and Böttcher, S. 1996, *Universality in sandpiles, interface depinning, and earthquake models*, Phys. Rev. Lett. **77**, 111-114.

Paczuski, M., Maslov, S., and Bak, P., 1996, *Avalanche dynamics in evolution. Growth and depinning models*, Phys. Rev. E **53**, 414.

Paczuski, M., Bassler, K.E., and Corral, A. 2000, *Self-organized networks of competing Boolean agents*, Phys. Rev. Lett. **48/14**, 3185-3188.

Paczuski, M., Böttcher, S., and Baiesi, M. 2005, *Inter-occurrence times in the Bak-Tang-Wiesenfeld sandpile model: a comparison with the turbulent statistics of solar flares*, Phys. Rev. Lett. **95/18**, id. 181102.

Paniveni, U., Krishan, V., Sing, J., and Srikanth, R. 2005, *On the fractal structure of solar supergranulation*, Solar Phys. **231**, 1-10.

Pankine, A.A. and Ingersoll, A.P. 2004, *Interannual variability of Mars global dust storms: an example of self-organized criticality?* Ikarus **170**, 514-518.

Parker, E.N. 1963, *Interplanetary Dynamical Processes*, Wiley: New York.

Parker, E.N. 1972, *Topological dissipation and the small-scale fields in turbulent gases*, Astrophys. J. **174**, 499-510.

Parker, E.N. 1979, *Cosmical Magnetic Fields*, Oxford: Oxford University Press, Chap. 15.

Parker, E.N. 1983, *Magnetic neutral sheets in evolving fields. II. Formation of the solar corona*, Astrophys. J. **264**, 642-647.

Parker, E.N. 1988, *Nanoflares and the solar X-ray corona*, Astrophys. J. **330**, 474-479.

Parnell, C.E. and Jupp, P.E. 2000, *Statistical analysis of the energy distribution of nanoflares in the quiet Sun* Astrophys. J. **529**, 554-569.

Parnell, C.E., DeForest, C.E., Hagenaar, H.J., Johnston, B.A., Lamb, D.A., and Welsch, B.T. 2009, *A power-law distribution of solar magnetic fields over more than five decades in flux*, Astrophys. J. **698**, 75-82.

Pavlidou, V., Kuijpers, J., Vlahos, L., and Isliker, H. 2001, *A cellular automaton model for the magnetic activity in accretion disks*, Astron. Astrophys. **372**, 326-337.

Pearce, G. and Harrison, R.A. 1990, *Sympathetic flaring*, Astron. Astrophys. **228**, 513-516.

Pearce, G., Rowe, A.K., and Yeung, J. 1993, *A statistical analysis of hard X-ray solar flares*, Astrophys. Space Science **208**, 99-111.

Peitgen, H.O. and Richter, P. 1986, *The beauty of fractals*, Springer, Berlin.

Peitgen, H.O. and Saupe, D. 1988, *The science of fractal images*, Springer, Berlin.

Pelletier, J.D., Malamud, B.D., Blodgett, T. and Turcotte, D.L. 1997, *Scale-invariance of soil moisture variability and its implications for the frequency-size distribution of sandslides*, Engin. Geol. **49**, 255-268.

Perez-Enriquez, R. and Miroshnichenko, L.I. 1999, *Frequency distributions of solar famma ray events related and not related with SPEs 1989-1995*, Solar Phys. **188**, 169-185.

Petri, A., Paparo, G., Vespignani, A., Alippi, A., and Constantini, M. 1994, *Experimental evidence for critical dynamics in microfracturing processes*, Phys. Rev. Lett. **73**, 3423-3426.

Petrosyan, A., Balogh, A., Goldstein, M.L., Leorat, J., Marsch, E., Petrovay, K., Roberts, B., von Steiger, R., and Vial, J.C. 2010, *Turbulence in the solar atmosphere and solar wind*, Space Science Rev. (subm.).

Pietronero, L., Vispignani, A., and Zapperi, S. 1994, *Renormalization scheme for self-organized criticality in sandpile models*, Phys. Rev. Lett. **72/11**, 1690-1693.

Pizzochero, P.M., Viverit, L., and Broglia, R.A. 1997, *Vortex-nucleus interaction and pinning forces in neutron stars*, Phys. Rev. Lett. **79/18**, 3347-3350.

Pla, O. and Nori, F. 1991, *Self-organized critical behavior in pinned flux lattices*, Phys. Rev. Lett. **67**, 919-922.

Podesta, J.J., Roberts, D.A., and Goldstein, M.L. 2006a, *Power spectrum of small-scale turbulent velocity fluctuations in the solar wind*, J. Geophys. Res. **111/A10**, CiteID A10109.

Podesta, J.J., Roberts, D.A., and Goldstein, M.L. 2006b, *Self-similar scaling of magnetic energy in the inertial range of solar wind turbulence*, J. Geophys. Res. **111/A9**, CiteID A09105.

Podesta, J.J., Roberts, D.A., and Goldstein, M.L. 2007, *Spectral exponents of kinetic and magnetic energy spectra in solar wind turbulence*, Astrophys. J. **664**, 543-548.

Podesta, J.J. 2007, *Self-similar scaling of kinetic energy density in the inertial range of solar wind turbulence*, J. Geophys. Res. (Space Physics) **112/11**, CiteID A11104.

Podladchikova, O.V., Krasnoselskikh, V., and Lefebvre, B. 1999, *Quiet Sun coronal heating: sand pile reconnection model*, in "Magnetic Fields and Solar Processes", Proc. of the 9th European Meeting (ed. Wilson, A.), European Space Agency Special Publication Vol. **448**, ESA, ESTEC Noordwijk, The Netherlands, p.553-559.

Podladchikova, O. 2002, *Statistical model of quiet Sun coronal heating*, PhD Thesis, Université de Orléans, France.

Podladchikova, O.V. and Berghmans,D. 2005, *Automated detection of EIT waves and dimmings*, Solar Phys. **228**, 265-284.

Podlazov, A.V., and Osokin, A.R. 2002, *Self-organized criticality model of solar plasma eruption processes*, Astrophys. Space Sci. **282**, 221-226.

Poliakov, A.N.B. and Herrmann, H.J. 1994, *Self-organized criticality of plastic shear bands in rocks*, Geophys. Res. Lett. **21**, 2143-2146.

Polygiannakis, J.M. and Moussas, X. 1994, *On experimental evidence of chaotic dynamics over short time scales in solar wind and cometary data using nonlinear prediction techniques*, Solar Phys. **151**, 341-350.

Polygiannakis, J.M., Preka-Papadema, P., and Moussas, X. 2003, *On signal-noise decomposition of time-series using the continuous wavelet transform: application to sunspot index*, Monthly Notices Royal Astron. Soc. **343**, 725-734.

Press, W.H. 1978, *Flicker noises in astronomy and elsewhere*, Comments on Modern Physics, Part C, **7/4**, 103-119.

Procaccia, I. 1985, *The static and dynamic invariants that characterize chaos and the relations between them in theory and experiments*, Phys. Scripta **T9**, 40-46.

Prozorov, R. and Giller, D. 1999, *Self-organization of vortices in type-II superconductors during magnetic relaxation*, Phys. Rev. B **59**, 14687-16691.

Qiu, J. and Wang, H. 2006, *On the temporal and spatial properties of elementary bursts*, Solar Phys. **236**, 293-311.

Quilligan, F., McBreen, B., Hanlong, L., McBreen, S., Hurley, K., and Watson, D. 2002, *Temporal properties of gamma-ray bursts as signatures of jets from the central engine*, Astron. Astrophys. **385**, 377-398.

Raup, M.D. 1986, *Biological extinction in Earth history*, Science **231**, 1528.

Raup, M.D. 1991, *Extinction: bad genes or bad luck?* New York: Norton, p. 210.

Ray, T.S. and Jan, N. 1994, *Anomalous approach to the self-organized critical state in a model for "life at the edge of chaos"*, Phys. Rev. Lett. **72/25**, 4045-4048.

Reed, W.J. and Hughes, B.D. 2002, *From gene families and genera to incomes and internet file sizes: Why power laws are so common in nature*, Phys. Rev. Lett. E **66**, 067103.

Rhodes, C.J. and Anderson, R.M. 1996, *Power laws governing epidemics in isolated population* , Nature **381**, 600-602.

Rhodes, C.J., Jensen, H.J. and Anderson, R.M. 1997, *On the critical behaviour of simple epidemics*, Proc. R. Soc. B **264**, 1639-1646.

Rhodes, T.L., Moyer, R.A., Groebner, R., Doyle, E.J., Lehmer, R., Peebles, W.A. and Rettig, C.L. 1999, *Experimental evidence for self-organized criticality in tokamak plasma turbulence*, Phys. Lett. A. **253/3-4**, 181-186.

Rice, J.R., 1995, *Mathematical Statistics and Data Analysis* (Second Ed.), Duxbury Press.

Richardson, L.F. 1941, *Frequency occurrence of wars and other fatal quarrels*, Nature **148**, 598.

Richardson, L.F. 1960, *Statistics of Deadly Quarrels*, Pittsburgh, PA: Boxwood, p. 373.

Rieutord, M., Meunier, N., Roudier, T., Rondi, S., Beigbeder, F., and Pares, L. 2008, *Solar supergranulation revealed by granule tracking*, Astron. Astrophys. **479**, L17-L20.

Rieutord, M., Roudier, T., Rincon, F., Malherbe, J.-M., Berger, T., and Frank, Z. 2010, *On the power spectrum of solar surface flows*, Astron. Astrophys. **512**, id. A4.

Rinaldo, A., Maritan, A., Colaiori, E., Flammini, A., Rigon, R., Ignacio, I., Rodriguez-Iturbe, I. and Banavan, J.R. 1996, *Thermodynamics of fractal river networks*, Phys. Rev. Lett. **76**, 3364-3367.

Robbrecht, E. and Berghmans, D. 2004, *Automated recognition of coronal mass ejections (CMEs) in near-real-time data*, Astron. Astrophys. **425**, 1097-1106.

Robbrecht, E., Berghmans, D., and Van der Linden, R.A.M. 2009, *Automated LASCO CME catalog for solar cycle 23: Are CMEs scale invariant ?*, Astrophys. J. **691**, 1222-1234.

Roberts, D.C. and Turcotte, D.L. 1998, *Fractality and self-organized criticality of wars*, Fractals **6**, 351-357.

Robinson, R.D., Carpenter, K.G., and Percival, J.W. 1999, *A search for microflaring activity on dMe flare stars. II. Observations of YZ Canis Minoris*, Astrophys. J. **516**, 916-923.

Robinson, P.A. 1993, *Stochastic-growth theory of Langmuir growth-rate fluctuations in type III solar radio sources*, Solar Phys. **146**, 357-363.

Rosenshein, E.B. 2003, *Applicability of complexity theory to Martian fluvial systems: A preliminary analysis*, 34th Annual Lunar and Planetary Science Conference, League City, Texas, abstract no. 1660.

Rosner, R., and Vaiana, G.S. 1978, *Cosmic flare transients: constraints upon models for energy storage and release derived from the event frequency distribution*, Astrophys. J. **222**, 1104-1108.

Rothman, D.H., Grotziger, J.P., and Flemings, P. 1994, *Scaling in turbidite deposition*, J. Sed. Petrol. A **64**, 59-67.

Roudier, T. and Muller, R. 1987, *Structure of solar granulation*, Solar Phys. **107**, 11-26.

Rouppe Van Der Voort, L.H.M., Loefdahl, M.G., Kiselman, D., and Scharmer, G.B. 2004, *Penumbral structure at 0.1 arcsec resolution. I. General appearance and power spectra*, Astron. Astrophys. **414**, 717-726.

Roux, S. and Guyon, E. 1989, *Temporal development of invasion percolation*, J. Physics A **22/17**, 3693-3705.

Rundle, J.B. and Klein, W. 1989, *Nonclassical nucleation and growth of cohesive tensile cracks*, Phys. Rev. Lett. **63/2**, 171-174.

Rundle, J.B., Tiampo, K.F., Klein, W., and Sa Martins, J.S. 2002, *Self-organization in leaky threshold systems: The influence of near-mean field dynamics and its implications for earthquakes, neurobiology, and forecasting*, in *Self-organized complexity in the physical, biological, and social sciences*, Arthur M. Sackler Colloquia, (eds. Turcotte, D., Rundle, J., and Frauenfelder, H.), The National Academy of Sciences: Washington DC, p.2514-2521.

Ryabov, V.B., Stepanov, A.V., Usik, P.V., Vavriv, D.M., Vinogradov, V.V., and Yurovsky, Y.F. 1997, *From chaotic to 1/f processes in solar microwave bursts*, Astron. Astrophys. **324**, 750-762.

Sahraoui, F., Goldstein, M.L., Robert, P., Khotyzintsev, Y.V. 2009, *Evidence of a cascade and dissipation of solar-wind turbulence at the electron gyroscale*, Phys. Rev. Lett. **102**, 231102:1-4.

Saichev, A. and Sornette,D. 2006, *"Universal" distribution of inter-earthquake times explained*, Phys. Rev. Lett. **97/7**, id. 078501.

Saichev, A., Malevergne, Y., and Sornette,D. 2009, *Theory of Zipf's Law and Beyond*, Lecture Notes in Economics and Mathematical Systems, Vol. **632**, Springer: Berlin.

Sammis, C., King, G. and Biegel, R. 1987, *The kinematics of gouge deformation*, Pure Appl. Geophys. **125**, 777-812.

Sammis, I. 1999, *Global versus local flare energy distributions*, Solar Phys. **189**, 173-179.

Sammis, I., Tang, F., and Zirin, H. 2000, *The dependence of large flare occurrence on the magnetic structure of sunspots*, Astrophys. J. **540**, 583-587.

Sanz, J.L., Herranz, D., and Martinez-Gonzalez, E. 2001, *Optimal detection of sources on a homogeneous and isotropic background*, Astrophys. J. **552**, 484-492.

Scargle, J. 1981, *Studies in Astronomical Time Series Analysis. I. Modeling random processes in the time domain*, Astrophys. J. Suppl. Ser. **45**, 1-71.

Scargle, J. 1982, *Studies in astronomical time series analysis. II. Statistical aspects of spectral analysis of unevenly spaced data*, Astrophys. J. **263**, 835-853.

Scargle, J. 1989, *Studies in astronomical time series analysis. III. Fourier transforms, autocorrelation functions, and cross-correlation functions of unevenly spaced data*, Astrophys. J. **343**, 874-887.

Scargle, J. 1990, *Studies in astronomical time series analysis. IV. Modeling chaotic and random processes with linear filters*, Astrophys. J. **359**, 469-482.

Scargle, J. 1993, *Wavelet methods in astronomical time series analysis*, Internat. Conf. on *Applications of time series analysis in astronomy and meteorology*, (ed. O. Lessi), Padova, Italy.

Scargle, J., Steiman-Cameron, T., Young, K., Donoho, D.L., Crutchfiled, J.P. Imamura, J. 1993, *The quasi-periodic oscillations and very low frequency noise of Scorpius X-1 as transient chaos: a dripping handrail?* Astrophys J. **411**, L91-L94.

Scargle, J. 1998, *Studies in astronomical time series analysis. V. Bayesian blocks, a new method to analyze structure in photon counting data*, Astrophys. J. **504**, 405-418.

Schatten, K.H. 2007, *Percolation and the solar dynamo*, Astrophys. J. Suppl. Ser. **169**, 137-153.

Scheinkman, J.A. and Woodford, M. 1994, *Self-organized criticality and economic fluctuations*, Am. Econ. Rev. **84**, 417-421.

Schrijver, C.J. 2009, *Driving major solar flares and eruptions: A review*, Adv. Space Res. **43**, 739-755.

Schroeder, M. 1991, *Fractals, Chaos, Power Laws: Minutes from an Infinite Paradise*, New York: Freeman.

Schulman, L.S. and Seiden, P.E. 1986, *Propagating stochastic star formation and galactic structure*, Ann. Israel Phys. Soc. **5**, 251-296.

Schuster, H.G. 1988, *Deterministic Chaos: An Introduction*, Weinheim (Germany): VCH Verlagsgesellschaft, 270 p.

Schwartz, R.A., Dennis, B.R., Fishman, G.J., Meegan, C.A., Wilson, R.B., and Paciesas, W.S. 1992, *BATSE flare observations in solar cycle 22*, in *The Compton Observatory Science Workshop*, (eds. C.R.Shrader, N.Gehrels, and B.R.Dennis), NASA CP 3137 (NASA: Washington DC), p.457-460.

Schwarz, U., Benz, A.O., Kurths, J. and Witt, A. 1993, *Analysis of solar spike events by means of symbolic dynamics methods*, Astron. Astrophys. **277**, 215-224.

Schwarz, U., Kurths, J., Kliem, B., Krueger, A., and Urpo, S. 1998, *Multiresolution analysis of solar mm-wave bursts*, Astron. Astrophys. Suppl. Ser. **127**, 309-318.

Searle, L. and Zinn, R. 1978, *Compositions of halo clusters and the formation of the galactic halo*, Astrophys. J. **225**, 357-379.

Seiden, P.E. and Gerola, H. 1982, *Propagating star formation and the structure and evolution of galaxies*, Fund. Cosmic Phys. **7**, 241-311.

Seiden, P.E. and Wentzel, D.G. 1996, *Solar active regions as a percolation phenomenon II*, Astrophys. J. **460**, 522-529.

Sen, A.K. 2007, *Multifractality as a measure of complexity in solar flare activity*, Solar Phys. **241**, 67-76.

Sepkoski, J.J. Jr. 1993, *Ten years in the library: new data confirm paleontological patterns*, Paleobiology **19**, 43-51.

Serre, T. and Nesme-Ribes, E. 2000, *Nonlinear analysis of solar cycles*, Astron. Astrophys. **360**, 319-330.

Shakura, N.I. and Sunyaev, R.A. 1973, *Black holes in binary systems. Observational appearance*, Astron. Astrophys. **24**, 337-355.

Shaw, B.E., Carlson, J.M., and Langer, J.S. 1992, *Patterns of seismic activity preceding large earthquakes*, J. Geophys. Res. **97**, 479-488.

Shaw, B.E. 1995, *Frictional weakening and slip complexity in earthquake faults*, J. Geophys. Res. **100/B9**, 18239-18252.

Shibata, K. and Yokoyama T. 1999, *Origin of the universal correlation between the flare temperature and the emission measure for solar and stellar flares*, Astrophys. J. **526**, L49-L52.

Shibata, K. and Yokoyama T. 2002, *A Hertzsprung-Russell-like diagram for solar/stellar flares and corona: emission measure versus temperature diagram*, Astrophys. J. **577**, 422-432.

Shimizu, T. 1995, *Energetics and occurrence rate of active-region transient brightenings and implications for the heating of the active-region corona*, Publ. Astron. Soc. Japan **47**, 251-263.

Shimojo, M. and Shibata, K. 1999, *Occurrence rate of microflares in an X-ray bright point within an active region*, Astrophys. J. **516**, 934-938.

Simnett, G.M. 1974, *A correlation between time-overlapping solar flares and the release of energetic particles*, Solar Phys. **34**, 377-391.

Sitnov, M.I., Sharma, A.S., Papadopoulos, K., Vassiliadis, D., Valdivia, J.A., Klimas, A.J., and Baker, D.N. 2000, J. Geophys. Res. **105/A6**, 12,955-12,974.

Sivia, D.S. and Skilling, J. 2006, (2nd ed.), *Data Analysis – A Bayesian Tutorial*, Oxford University Press, 264p.

Sivron, R. 1998, *Self-organized criticality in compact plasmas*, Astrophys. J. **503**, L57-L61.

Smalley, R.F. Jr., Turcotte, D.L., and Solla, S.A. 1985, *A renormalization group approach to the stick-slip behavior of faults*, J. Geophys. Res. **90**, 1894-1900.

Smart, D.F. and Shea, M.A. 1997, *Comment on the use of solar proton spectra in solar proton dose calculations*, in Proc. *Solar-Terrestrial Prediction Workshop V*, Hiraiso Solar-Terrestrial Research Center, Japan, p.449.

Sneppen, K. and Jensen, M.H. 1993, *Colored activity in self-organized critical interface dynamics*, Phys. Rev., Lett. **71**, 101-104.

Sneppen, K., Bak, P., Flyvbjerg, H., and Jensen, M.H. 1995, *Evolution as a self-organized critical phenomenon*, Proc. National Academy of Science, USA **92**, 5209-5213.

Socolar, J.E.S., Grinstein, G., and Jyaprakash, C. 1993, *On self-organized criticality in nonconserving systems*, Phys. Rev. E **47/4**, 2366-2376.

Sole, R.V. and Manrubia S.C. 1996, *Extinction and self-organized criticality in a model of large-scale evolution*, Phys. Rev. E **54/1**, R42-R45.

Sole, R.V., Manrubia, S.C., Benton, M., Kauffman, S. and Bak, P. 1999, *Self-organized criticality in evolutionary ecology*, Trends Ecol. Evol. **14**, 156-160.

Somfai, E., Czirok, A., and Vicsek,T. 1994a, *Self-affine roughening in a model experiment on erosion in Geomorphology*, J. Physics A **205**, 355-366.

Somfai, E., Czirok, A., and Vicsek,T. 1994b, *Power-law distribution of landslides in an experiment on the erosion of a granular pile*, J. Physics A **27**, L757-L763.

Sornette, A. and Sornette D. 1989, *Self-organized criticality and earthquakes*, Europhysics Letters **9**, 197-202.

Sornette, D., Sornette A., and Davy,P. 1990, *Structuration of the lithosphere in plate tectonics as a self-organized critical phenomenon*, J. Geophy. Res. **95**, 17,353-17,361.

Sornette, D., Johansen, A., and Bouchard, J.P. 1996, *Stock market crashes, precursors and replicas*, J. Physique I **6**, 167-175.

Sornette, D. and Johansen, A. 1997, *Large financial crashes*, Physica A **245**, 411-422.

Sornette, D. 2004, *Critical phenomena in natural sciences: chaos, fractals, self-organization and disorder: concepts and tools*, Springer, Heidelberg, 528 p.

Sotolongo-Costa, O., Antoranz, J.C., Posadas, A., Vidal, F., Vasquez, A. 2000, *Lévy flights and earthquakes*, J. Geophys. Res. **27/13**, 1965-1968.

Spiegel, E.A. 2009, *Chaos and Intermittency in the Solar Cycle*, Space Sci. Rev. **144**, 25-51.

Stanley, H.E., Amaral, L.A.N., Buldyrev, S.V., Gopikrishnan, P., Plerou, V., and Salinger, M.A. 2002, *Self-organized complexity in economics and finance*, in *Self-organized complexity in the physical, biological, and social sciences*, Arthur M. Sackler Colloquia, (eds. Turcotte, D., Rundle, J., and Frauenfelder, H.), The National Academy of Sciences: Washington DC, p.2361-2365.

Starck, J.-L., Murtagh, F., and Bijaoui, A., 1998, *Image Processing and Data Analysis*, Cambridge University Press, Cambridge. Electronic version available at *http://www.multi resolution.com/cupbook.html.*

Starck, J.-L. and Murtagh, F., 2002 (2nd edition 2006): *Astronomical image and data analysis*, Springer, Berlin; Electronic version available at *http://www.multiresolution.com/cupbook .html.*

Stassinopoulos, D. and Bak, P. 1995, *Democratic reinforcement. A principle for brain function*, Phys. Rev. E **51**, 5033-5039.

Stauffer, D. and Aharony, A. 1994, *Introduction to Percolation Theory*, 2nd ed., (London: Taylor and Francis).

Stein, R.S. and Yeats, R.S. 1989, *Hidden earthquakes*, Scientific American (June 1989 issue), p.48-57.

Stelzer, B., Flaccomio, E., Briggs, K., Micela, G., Scelsi, L, Audard, M., Pillitteri, I., and Güdel, M. 2007, *A statistical analysis of X-ray variability in pre-main sequence objects of the Taurus molecular cloud*, Astron. Astrophys. **468**, 463-475.

Stern, B.E. and Svennson, R. 1996, *Evidence for "chain reaction" in the time profiles of gamma-ray bursts*, Astrophys. J. **469**, L109-L113.

Sturrock, P.A. and Uchida, Y. 1981, *Coronal heating by stochastic magnetic pumping*, Astrophys. J. **246**, 331-336.

Su, Y., Gan, W.Q., and Li, Y.P. 2006, *A statistical study of RHESSI flares*, Solar Phys. **238**, 61-72.

Sundkvist, D., Krasnoselskikh, V., Shukla, P.K., Vaivads, A., Andre, M., Buchert, S., Reme, H. 2005, *In situ multi-satellite detection of coherent vortices as a manifestation of Alfvénic turbulence*, Nature **436**, Issue 7052, 825-828.

Sutherland, P.G., Weisskopf, M.C., and Kahn, S.M. 1978, *Short-term time variability of Cygnus X-1. II* Astrophys. J. **219**, 1029-1037.

Tadic, B. and Dhar, D. 1997, *Emergent spatial structures in critical sandpiles*, Phys. Rev. Lett. **79**/8, 1519-1522.

Tainaka, K., Fukawa, S., and Mineshige, S. 1993, *Spatial pattern formation of an interstellar medium*, Publ. Astron. Soc. Japan **45**, 57-64.

Takalo, J. 1993, *Correlation dimension of AE data*, Ph. Lic. Thesis, *Laboratory report 3*, Dept. Physics, University of Jyväskylä.

Takalo, J., Timonen, J., Koskinen, H. 1993, *Correlation dimension and affinity of AE data and biocolored noise*, Geophys. Res. Lett. **20/15**, 1527-1530.

Takalo, J., Timonem, J., Klimas, A., Valdivia, J., and Vassiliadis, D. 1999a, *Nonlinear energy dissipation in a cellular automaton magnetotail field model* Geophys. Res. Lett. **26/13**, 1813-1816.

Takalo, J., Timonem, J., Klimas, A., Valdivia, J., and Vassiliadis, D. 1999b, *A coupled-map model for the magnetotail current sheet*, Geophys. Res. Lett. **26/19**, 2913-2916.

Takeuchi, M., Mineshige, S., and Negoro, H. 1995, *X-ray fluctuations from black-hole objects and self organization of accretion disks*, Publ. Astron. Soc. Japan **47**, 617-627.

Takeuchi, M. and Mineshige, S. 1996, *X-ray fluctuations from black hole object: disk in a self-organized criticality*, in *Internat. Workshop on Basic Physics of Accretion Disks*, p.159-162.

Takeuchi, M. and Mineshige, S. 1997, *X-ray fluctuations from advection-dominated accretion disks with a critical behavior*, Astrophys. J. **486**, 160-168.

Tam, S.W.Y., Chang, T., Chapman, S.C., and Watkins, N.W. 2000, *Analytical determination of power-law index for the Chapman et al. sandpile (FSOC) analog for magnetospheric activity*, Geophys. Res. Lett. **27/9**, 1367.

Tanaka, Y. 1989, *Black holes in X-ray binaries: X-ray properties of galactic black hole candidates*, in ESA 23rd ESLAB Syposium on Two Topics in X-Ray Astronomy. Vol. **1**: X-Ray Binaries, p. 3-13.

Tang, C., Wiesenfeld, K., Bak, P., Coppersmith, S., and Littlewood, P., 1987, *Phase organization*, Phys. Rev. Lett. **58/12**, 1161-1164.

Tang, C. and Bak, P. 1988, *Critical exponents and scaling relations for self-organized critical phenomena*, Phys. Rev. Lett. **60**, 2347-2350.

Tang, F., Howard, R., and Adkins, J.M. 1984, *A statistical study of active regions 1967-1981*, Solar Phys. **91**, 75-86.

Tashiro, M., Makishima, K., Ohashi, T., Sakao, T., and Sansom, A.E. 1991, *Nonperiodic intensity variations in the X-ray pulsar GX 301-2*, MNRAS **252**, 156-162.

Terradas, J., Oliver, R., and Ballester, J.L. 2004, *Application of statistical techniques to the analysis of solar coronal oscillations*, ApJ **614**, 435-447.

Terrell, N.J. Jr. 1972, *Shot-noise character of Cygnus X-1 pulsations*, Astrophys. J. **174**, L35-L38.

Thompson, C. and Duncan, R.C. 1996, *The soft gamma repeaters as very strongly magnetized neutron stars. II. Quiescent neutrino, X-ray, and Alfven wave emission* Astrophys. J. **473**, 322-342.

Tranquille, C., Hurley, K., and Hudson, H. S. 2009, *The Ulysses Catalog of Solar Hard X-Ray Flares*, Solar Phys. **258**, 141-166.

Tsuneta, S. 1995, *Particle acceleration and magnetic reconnection in solar flares*, Publ. Astron. Soc. Japan **47**, 691-697.

Turcotte, D.L. 1997, *Fractals and Chaos in Geology and Geophysics*, Cambridge University Press: Cambridge, New York (2nd edition).

Turcotte, D.L. 1999, *Self-organized criticality*, Rep. Prog. Phys. **62**, 1377-1429.

Ueno, S., Mineshige, S., Negoro, H., Shibata, K., and Hudson, H.S. 1997, *Statistics of fluctuations in the solar soft X-ray emission*, Astrophys. J. **484**, 920-926.

Uritsky, V.M. and Pudovkin, M.I. 1998, *Low frequency 1/f-like fluctuations of the AE index as a possible manifestation of self-organized criticality in the magnetosphere*, Annal. Geophys. **16/12**, 1580-1588.

Uritsky, V.M., Klimas, A.J., Vassiliadis, D., Chua, D., and Parks, G. 2002, *Scale-free statistics of spatiotemporal auroral emission as depicted by Polar UVI images: dynamic magnetosphere is an avalanching system*, J. Geophys. Res. **1078/A12**, SMP 7-1, CiteID 1426.

Uritsky, V.M., Klimas, A.J., and Vassiliadis, D. 2003, *Evaluation of spreading critical exponents from the spatiotemporal evolution of emission regions in the nighttime aurora*, Geophys. Res. Lett. **30/15**, SSC 7-1, CiteID 1813.

Uritsky, V.M., Klimas, A.J., and Vassiliadis, D. 2006, *Critical finite-size scaling of energy and lifetime probability distributions of auroral emissions*, Geophys. Res. Lett. **33/8**, CiteID L08102.

Uritsky, V.M., Paczuski, M., Davila, J.M., and Jones, S.I. 2007, *Coexistence of self-organized criticality and intermittent turbulence in the solar corona*, Phys. Rev. Lett. **99/2**, 25001-25004.

Uttley, P. and McHardy, I.M. 2001, *The flux-dependent amplitude of broadband noise variability in X-ray binaries and active galaxies*, MNRAS **323**, L26-L30.

Uttley, P., McHardy, I.M.M., and Vaughan, S. 2005, *Non-linear X-ray variability in X-ray binaries and active galaxies*, MNRAS **359**, 345-362.

Uzdensky, D.A. 2007, *The fast collisionless reconnection condition and the self-organization of solar coronal heating*, Astrophys. J. **671**, 2139-3153.

Vainio, R., Laitinen,T., and Fichtner, H. 2003, *A simple analytical expression for the power spectrum of cascading Alfven waves in the solar wind*, Astron. Astrophys. **407**, 713-723.

Van der Ziel, A. 1950, *On the noise spectra of semi-conductor noise and of flicker effect*, Physica **16/4**, 359-372.

Vandewalle, N. and Ausloos M. 1995, *Self-organized criticality in phylogenetic-like tree growths*, J. Physique I **5**, 1011-1025.

Van Houten, C.J., van Houten-Groeneveld, I., Herget, P., and Gehrels, T. 1970, *The Palomar-Leiden survey of faing minor planets*, Astron. Astrophys. Suppl. Ser. **2/5**, 339-448.

VanHollebeke, M.A.I., Ma Sung L.S., and McDonald F.B. 1975, *The variation of solar proton energy spectra and size distribution with heliolongitude*, Solar Phys. **41**, 189-223.

Vassiliadis, D., Anastasiadis, A., Georgoulis, M., and Vlahos L. 1998, *Derivation of solar flare cellular automata models from a subset of the magnetohydrodynamic equations*, Astrophys. J. **509**, L53-L56.

Vaughan, B.A. and Nowak, M.A. 1997, *X-ray variability coherence: How to compute it, what it means, and how it constrains models of GX 339-4 and Cygnus X-1*, Astrophys. J. **474**, L42-L46.

Vecchio, A., Primavera, L., Carbone, V., and Sorriso-Valvo, L. 2005a, *Periodic behavior and stochastic fluctuations of solar activity: proper orthogonal decomposition analysis*, Solar Phys. **229**, 359-372.

Vecchio, A., Carbone, V., Lepreti, F., Primavera, L., Sorriso-Valvo, L., Veltri, P., Alfonsi, G., and Straus, T. 2005b, *Proper orthogonal decomposition of solar photospheric motions*, Phys. Rev. Lett. **95/6**, id. 061102.

Veltri, P. 1999, *MHD turbulence in the solar wind: self-similarity, intermittency and coherent structures*, Plasma Phys. Controlled Fusion **41**, A787-A795.

Veltri, P., Malara, F., and Primavera, L. 1999, *Nonlinear Alfven Wave Interaction with Large-Scale Heliospheric Current Sheet*, Lect. Notes Physics (Springer Verlag: Berlin), **536**, 222-250.

Verdes, P.F., Parodi, M.A., Granitto, P.M., Navone, H.D., Piacentini, R.D. and Ceccatto, H.A. 2000, *Predictions of the maximum amplitude for solar cycle 23 and its subsequent behavior using nonlinear methods*, Solar Phys. **191**, 419-425.

Veronig, A., Messerotti, M., and Hanslmeier, A. 2000, *Determination of fractal dimensions of solar radio bursts*, Astron. Astrophys. **357**, 337-350.

Veronig, A., Temmer, M., Hanslmeier, A., Otruba, W., and Messerotti, M. 2002a, *Temporal aspects and frequency distributions of solar X-ray flares*, Astron. Astrophys. **382**, 1070-1080.

Veronig, A., Vrsnak, B., Dennis, B.R., Temmer, M., Hanslmeier, A., and Magdalenic, J. 2002b, *The Neupert effect in solar flares and implications for coronal heating*, in *Magnetic coupling of the solar atmosphere*, (eds. Huguette Sawaya-Lacoste), European Space Agency (ESA) **SP-505**, ESTEC Noordwijk, The Netherlands, p.599-602.

Veronig, A., Temmer, M., and Hanslmeier, A. 2002d, *Frequency distributions of solar flares*, Hvar Observatory Bulletin **26/1**, 7-12.

Veronig, A., Vrsnak, B., Dennis, B.R., Temmer, M., Hanslmeier, A., and Magdalenic, J. 2002c, *Investigation of the Neupert effect in solar flares*, Astron. Astrophys. **392**, 699-712.

Vespignani, A., Zapperi, S., and Pietronero, L. 1995, *Renormalization approach to the self-organized critical behavior of sandpile models*, Phys. Rev. E **51/3**, 1711-1724.

Vespignani, A. and Zapperi, S. 1997, *Order parameter and scaling fields in self-organized criticality*, Phys. Rev. Lett. **78/25**, 4793-4796.

Vespignani, A. and Zapperi, S. 1998, *How self-organized criticality works: A unified mean-field picture*, Phys. Rev. E. **57/6**, 6345-6362.

Vieira, L.E.A., Gonzalez, W.D., Echer, E., Guarnieri, F.L., Prestes, A., Gonzalez, A.L.C., Santos, J.C., Dal Lago, A. and Schuch, N.J. 2003, *Multi-Scale Analysis of the Geomagnetic Symmetric Index (sym)*, Solar Phys. **217**, 383-394.

Vigoroux, A. and Delache, Ph. 1993, *Fourier versus wavelet analysis of solar diameter variability*, Astron. Astrophys. **278**, 607-616.

Vilmer, N. and Trottet, G. 1997, *Solar flare radio and hard X-ray observations and the avalanche model*, Lecture Notes in Physics **483** (ed. G. Trottet, Springer: Berlin), p.28-52.

Vlahos, L., Georgoulis, M., Kliuiving, R., and Paschos, P. 1995, *The statistical flare*, Astron. Astrophys. **299**, 897-911.

Vlahos, L. 2002, *Statistical properties of the evolution of solar magnetic fields*, in *SOLMAG 2002. Proc. Magnetic Coupling of the Solar Atmosphere*, Euroconf. and IAU Coll 188, (ed. H. Sawaya-Lacoste), European Space Agency (ESA) **SP-505**, ESTEC Noordwijk, Netherlands, p. 105-112.

Vlahos, L., Fragos, T., Isliker, H., and Georgoulis, M. 2002, *Statistical properties of the energy release in emerging and evolving active regions*, Astrophys. J. **575**, L87-L90.

Vlahos, L. and Georgoulis, M.K. 2004, *On the self-similarity of unstable magnetic discontinuities in solar active regions*, Astrophys. J. **603**, L61-L64.

Voges, W., Atmanspacher, H., and Scheingraber, H. 1987, *Deterministic chaos in accretion systems: Analysis of the X-ray variability of Hercules X-1*, Astrophys. J. **320**, 794-802.

Walsh, R.M., Bell, G.E., and Hood, A.W. 1997, *Discrete random heating events in coronal loops*, Solar Phys. **171**, 81-91.

Wang, H.N., Cui, Yl.M., and He,H. 2009, *A logistic model for magnetic energy storage in solar active regions*, Research in Astron. Astrophys. **9/6**, 687-693.

Wang, J.X., Wang, H., Tang, F., Lee, J.W., and Zirin, H. 1995, *Flux distribution of solar intranetwork magnetic fields*, Solar Phys. **160**, 277-288.

Warzawski, L. and Melatos, A. 2008, *A cellular automaton model of pulsar glitches*, MNRAS **390**, 175-191.

Watkins, N.W., Chapman, S.C., Dendy, R.O., and Rowlands, G. 1999, *Robustness of collective behavior in strongly driven avalanche models: magnetospheric implications*, Geophys. Res. Lett. **26/16**, 2617-2620.

Watkins, N.W., Oughton, S., and Freeman, M.P. 2001a, *What can we infer about the underlying physics from burst distributions observed in an RMHD simulation*, Planet. Space Science **49**, 1233-1237.

Watkins, N.W., Freeman, M.P., Chapman, S.C., and Dendy, R.O. 2001b, *Testing the SOC hypothesis for the magnetosphere*, J. Atmos. Solar-Terr. Phys. **63**, 1435-1445.

Watkins, N.W. 2002, *Scaling in the space climatology of the auroral indices: Is SOC the only possible description?*, Nonlin. Processes Geophys. **9**, 389-397.

Watkins, N.W., Chapman, S.C., Rosenberg, S.J., Uritsky, V.M., Davila, J.M., and Jones, S.I. 2009, *Comment on Coexistence of Self-Organized Criticality and Intermittent Turbulence in the Solar Corona*, Phys. Rev. Lett. **103**, 039501-1.

Watari, S. 1995, *Fractal dimensions of solar activity*, Solar Phys. **158**, 365-377.

Watari, S. 1996a, *Fractal dimensions of the time variation of solar radio emission*, Solar Phys. **163**, 371-388.

Watari, S. 1996b, *Chaotic behavior of the north-south asymmetry of sunspots?* Solar Phys. **163**, 259-266.

Wentzel, D.G. and Seiden, P.E. 1992, *Solar active regions as a percolation phenomenon*, Astrophys. J. **390**, 280-289.

Westbrook, C.D., Ball, R.C., Field, P.R., and Heymsfield, A.J. 2004, *Theory of growth by differential sedimentation, with application to snowflake formation*, Phys. Rev. E **70/2**, id. 021403.

Wheatland, M.S. and Sturrock, P.A. 1996, *Avalanche models of solar flares and the distribution of active regions*, Astrophys. J. **471**, 1044-1048.

Wheatland, M.S., Sturrock, P.A., and McTiernan, J.M. 1998, *The waiting-time distribution of solar flare hard X-ray bursts*, Astrophys. J. **509**, 448-455.

Wheatland, M.S. and Eddey, S.D. 1998, *Models for flare statistics and the waiting-time distribution of solar flare hard X-ray bursts*, in Proc. Nobeyama Symposium, "Solar Physics with Radio Observations", (eds. Bastian, T., Gopalswamy, N., and Shibasaki, K.), NRO Report **479**, p.357-360.

Wheatland, M.S. and Glukhov, S. 1998, *Flare frequency distributions based on a master equation*, Astrophys. J. **494**, 858-863.

Wheatland, M.S. and Uchida, Y. 1999, *Frequency-energy distributions of flares and active region transient brightenings*, Solar Phys. **189**, 163-172.

Wheatland, M.S. 2000a, *The origin of the solar flare waiting-time distribution*, Astrophys. J. **536**, L109-L112.

Wheatland, M.S. 2000b, *Do solar flares exhibit an interval-size relationship?* Solar Phys. **191**, 381-389.

Wheatland, M.S. 2000c, *Flare frequency-size distributions for individual active regions*, Astrophys. J. **532**, 1209-1214.

Wheatland, M.S. 2001, *Rates of flaring in individual active regions*, Solar Phys. **203**, 87-106.

Wheatland, M.S. and Litvinenko, Y.E. 2001, *Energy balance in the flaring solar corona*, Astrophys. J. **557**, 332-336.

Wheatland, M.S. 2002, *Distribution of flare energies based on independent reconnecting structures*, Solar Phys. **208**, 33-42.

Wheatland, M.S. and Litvinenko, Y.E. 2002, *Understanding solar flare waiting-time distributions*, Solar Phys. **211**, 255-274.

Wheatland, M.S. 2003, *The coronal mass ejection waiting-time distribution*, Solar Phys. **214**, 361-373.

Wheatland, M.S. and Craig, I.J.D. 2003, *Toward a reconnection model for solar flare statistics*, Astrophys. J. **595**, 458-464.

Wheatland, M.S. 2004, *A Bayesian approach to solar flare prediction*, Astrophys. J. **609**, 1134-1139.

Wheatland, M.S. 2006, *A rate-independent test for solar flare sympathy*, Solar Phys. **236**, 313-324.

Wheatland, M.S. and Craig, I.J.D. 2006, *Including flare sympathy in a model for solar flare statistics*, Solar Phys. **238**, 73-86.

Wheatland, M.S. 2008, *The energetics of a flaring solar active region and observed flare statistics*, Astrophys. J. **679**, 1621-1628.

Wheatland, M.S. 2009, *Monte Carlo simulation of solar active-region energy*, Solar Phys. **255**, 211-227.

Wiesenfeld, K., Theiler, J., and McNamara, B. 1990, *Self-organized criticality in a deterministic automaton*, Phys. Rev. Lett. **65/8**, 949-952.

Wiita, P.J. and Xiong, Y. 1998, *Self-organized criticality in accretion discs*, in *Theory of black hole accretion disks*, eds. M.A. Abramowicz, G. Gjornsson, and J.E. Pringle), Cambridge University Press, p.274-283.

Willinger W., Govindan, R., Jamin, S., Paxson, V., and Shenker, S. 2002, *Scaling phenomena in the Internet: critically examining criticality*, in *Self-organized complexity in the physical, biological, and social sciences*, Arthur M. Sackler Colloquia, (eds. Turcotte, D., Rundle, J., and Frauenfelder, H.), The National Academy of Sciences: Washington DC, p.2573-2580.

Willis, J.C. and Yule, G.U. 1922, *Some statistics of evolution and geographical distribution in plants and animals, and their significance*, Nature **109**, 177-179.

Willson, R.C. and Mordvinov, A.V. 1999, *Time-frequency analysis of total solar irradiance variations*, Geophys. Res. Lett. **26/24**, 3613-3616.

Wolfram, S. 2002, *A new kind of science*, Wolfram Media, ISBN 1-57955-008-8.

Woods, J.W. 2006, *Multi-Dimensional Signal, Image, and Video Processing and Coding*, Elsevier and Academic Press, New York.

Wu, S. and Zhang, J. 1992, *Self-organized rock textures and multiring structures in the Duolun crater*, LPI contributions **790**, 81.

Xapsos, M.A., Stauffer, C., Barth, J.L., and Burke, E.A. 2006, *Solar particle events and self-organized criticality: Are deterministic predictions of events possible?* IEEE Transactions on nuclear science **53/4**, 1839-1843.

Xiong, Y., Wiita, P.J., and Bao, G. 2000, *Models for accretion-disk fluctuations through self-organized criticality including relativistic effects*, from Publ. Astron. Soc. Japan **52**, 1097-1107.

Xu, R.X., Tao, D.J., and Yang, Y. 2006, *The superflares of soft gamma-ray repeaters: giant quakes in solid quark stars?*, Mon. Not. Royal Astron. Soc. **373**, L85-L89.

Yashiro, S., Akiyama, S., Gopalswamy, N. and Howard, R.A. 2006, *Different power-law indices in the frequency distributions of flares with and without coronal mass ejections*, Astrophys. J. **650**, L143-L146.

Yeh, W.J., and Kao, Y.H. (1984), *Measurements of flux-flow and 1/f noise in superconductors*, Phys. Rev. Lett. **53/16**, 1590-1593.

Yeh,, W.J., Ding, M., and Chen, P. 2005, *Waiting time distribution of coronal mass ejections*, Chinese J. Astron. Astrophys. bf 5, 193-197.

Yonehara, A., Mineshige, S., and Welsh, W.F. 1997, *Cellular-automaton model for flickering of cataclysmic variables*, Astrophys. J. **486**, 388-396.

Yoshida, F., Nakamura, T., Watanab, J., Kinoshita, D., and Yamamoto, N., 2003, *Size and spatial distributions of sub-km main-belt asteroids*, Publ. Astron. Soc. Japan **55**, 701-715.

Yoshida, F. and Nakamura T. 2007, *Subary Main Belt Asteroid Survey (SMBAS) – Size and color distributions of small main-belt asteroids*, Planet. Space Science **55**, 113-1125.

Young, C.A., and Gallagher, P.T. 2008, *Multiscale edge detection in the corona*, Solar Phys. **248**, 457-469.

Young, K. and Scargle, J.D. 1996, *The dripping handrail model: Transient chaos in accretion disks*, Astrophys. J. **468**, 617-632.

Young, M.D.T. and Kenny, B.G. 1996, *Are giant pulses evidence of self-organized criticality?*, in *Pulsars: Problems and progress*, ASP Conf. Ser. **105**, (eds. S. Johnston, M.A. Walker, and M.Bailes), ASP: San Francisco, p.179.

Young, C.A., Meredith, D.C., and Ryan,J.M. 1995, *A compact representation of gamma-ray burst time series*, Astrophys. J. Suppl. Ser. **231**, 119-122.

Yurovsky, Y. and Magun, A. 1996, *The signs of non-periodic acceleration of electrons in solar active regions*, Solar Phys. **166**, 433-436.

Yurovsky, Y. 1997, *On mechanisms for modulating the radio emission of solar flares*, Astron. Rep. **41/6**, 845-856.

Yurovsky, Y. and Magun, A. 1998, *On the nature of modulation of radio emission during solar flares*, Solar Phys. **180**, 409-426.

Zanette, D.H. 2007, *Multiplicative processes and city sizes*, in The Dynamics of Complex Urban Systems. An Interdisciplinary Approach, (eds. S. Albeverio, D. Andrey, P. Giordano, and A. Vancheri, eds. (Springer, Berlin, 2007).

Zapperi,S., Lauritsen, B.K., and Stanley, H.E. 1995, *Self-organizing branching processes: Mean-field theory for avalanches*, Phys. Rev. Lett. **75/22**, 4071-4074.

Zebker, H.A., Maroufm, E.A., and Tyler, G.L. 1985, *Saturn's rings particle size distributions for a thin layer model*, Ikarus **64**, 531-548.

Zhang, Y.C. 1989, *Scaling theory of self-organized criticality*, Phys. Rev. Lett. **63/5**, 470-473.

Zhou, Y., Matthaeus, W.H., and Dmitruk, P. 2004, *Magnetohydrodynamic turbulence and time scales in astrophysical and space plasmas*, Rev. Mod. Phys. **76**, 1015-1035.

Zieve, R.J., Rosenbaum, T.F., Jaeger, H.M., Seidler, G.T., Crabtree, G.W., and Welp, U. 1996, *Vortex avalanches at one thousandth the superconducting transition temperature*, Phys. Rev. B. **53/17**, 11849-11854.

Zipf, G.K. 1949, *Human Behavior and the Principle of Least Effort*, Cambridge MA, Addison-Wesley.

Zirker, J.B. and Cleveland, F.M. 1993a, *Nanoflare mechanisms: twisting and braiding*, Solar Phys. **144**, 341-347.

Zirker, J.B. and Cleveland, F.M. 1993b, *Avalanche models of active region heating and flaring*, Solar Phys. **145**, 119-128.

Index